Lecture Notes in Physics

The Editorial Policy for Monographs

The series Lecture Notes in Physics reports new developments in physical research and teaching - quickly, informally, and at a high level. The type of material considered for publication includes monographs presenting original research or new angles in a classical field. The timeliness of a manuscript is more important than its form, which may be preliminary or tentative. Manuscripts should be reasonably self-contained. They will often present not only results of the author(s) but also related work by other people and will provide sufficient motivation, examples, and applications.

Acceptance

The manuscripts or a detailed description thereof should be submitted either to one of the series editors or to the managing editor. The proposal is then carefully refereed. A final decision concerning publication can often only be made on the basis of the complete manuscript, but otherwise the editors will try to make a preliminary decision as definite as they can on the basis of the available information.

Contractual Aspects

Authors receive jointly 30 complimentary copies of their book. No royalty is paid on Lecture Notes in Physics volumes. But authors are entitled to purchase directly from Springer other books from Springer (excluding Hager and Landolt-Börnstein) at a $33\frac{1}{3}$% discount off the list price. Resale of such copies or of free copies is not permitted. Commitment to publish is made by a letter of interest rather than by signing a formal contract. Springer secures the copyright for each volume.

Manuscript Submission

Manuscripts should be no less than 100 and preferably no more than 400 pages in length. Final manuscripts should be in English. They should include a table of contents and an informative introduction accessible also to readers not particularly familiar with the topic treated. Authors are free to use the material in other publications. However, if extensive use is made elsewhere, the publisher should be informed. As a special service, we offer free of charge LaTeX macro packages to format the text according to Springer's quality requirements. We strongly recommend authors to make use of this offer, as the result will be a book of considerably improved technical quality. The books are hardbound, and quality paper appropriate to the needs of the author(s) is used. Publication time is about ten weeks. More than twenty years of experience guarantee authors the best possible service.

LNP Homepage (springerlink.com)

On the LNP homepage you will find:
–The LNP online archive. It contains the full texts (PDF) of all volumes published since 2000. Abstracts, table of contents and prefaces are accessible free of charge to everyone. Information about the availability of printed volumes can be obtained.
–The subscription information. The online archive is free of charge to all subscribers of the printed volumes.
–The editorial contacts, with respect to both scientific and technical matters.
–The author's / editor's instructions.

A.N. Gorban I.V. Karlin

Invariant Manifolds
for Physical
and Chemical Kinetics

 Springer

Authors

Alexander N. Gorban
Multidisciplinary Research Institute 213
Department of Mathematics
University of Leicester
University Road
LE1 7RH Leicester
United Kingdom
ag153@leicester.ac.uk
and
Institute of Computational Modeling
Russian Academy of Sciences
660036 Krasnoyarsk
Russia

Ilya V. Karlin
Institute of Energy Technology
Clausiusstrasse 33
ETH Center CLT A 6
8092 Zürich
Switzerland
karlin@lav.mavt.ethz.ch
and
Institute of Computational Modeling
Russian Academy of Sciences
660036 Krasnoyarsk
Russia

A. N. Gorban, I. V. Karlin, *Invariant Manifolds for Physical and Chemical Kinetics*,
Lect. Notes Phys. **660** (Springer, Berlin Heidelberg 2005), DOI 10.1007/b98103

Library of Congress Control Number: 2004113306

ISSN 0075-8450
ISBN 3-540-22684-2 Springer Berlin Heidelberg New York

Springer is a part of Springer Science+Business Media

springeronline.com

© Springer-Verlag Berlin Heidelberg 2005
Printed in Germany

The use of general descriptive names, registered names, trademarks, etc. in this publication does not imply, even in the absence of a specific statement, that such names are exempt from the relevant protective laws and regulations and therefore free for general use.

Typesetting: by the authors and TechBooks using a Springer LaTeX macro package
Cover design: *design & production*, Heidelberg

Printed on acid-free paper
55/3141/jl - 5 4 3 2 1 0

To our parents

Preface

This book is about model reduction in kinetics. Is this physics or mathematics? There are at least four reasonable answers to this question:

- It is physics, it is not mathematics;
- It is mathematics, it is not physics;
- It is both physics *and* mathematics;
- It is neither physics, nor mathematics, it is something else (but what could that be?).

Of course, it is physics. Model reduction in kinetics requires physical concepts and structures; it is impossible to make an expedient reduction of a kinetic model without thermodynamics, for example. The entropy, the Legendre transformation generated by the entropy, and the Riemann structure defined by the second differential of the entropy provide the elementary geometrical basis for the first approximation. The physical sense of the models gives many hints for their further processing. So, it is not mathematics; we care about the physical sense more than about rigorous proofs. We should deal with equations even in the absence of theorems about existence and uniqueness of solutions. Mathematics assimilates the physical notions with a considerable delay in time, but any such an assimilation leads to further insights.[1]

But, without doubt, it is mathematics. The story about invariant manifolds for differential equations began inside mathematics. The first significant steps were taken by two great mathematicians, A.M. Lyapunov and H. Poincaré, at the end of the XIXth century. Then N.M. Krylov and N.N. Bogolyubov, A.N. Kolmogorov, V.I. Arnold and J. Moser, J.E. Marsden, M.I. Vishik, R. Temam, and many other mathematicians developed this field of science, and many elegant theorems and useful methods were created. This is not only pure mathematics, the wide field of applications was developed too, from hydrodynamics to process engineering and control theory and methods. This is *pure and applied dynamics*. The language of model reduction, the basic notions that we use, the theorems and methods, all this either came from

---

[1] The closest example: after mathematicians discovered how the entropy functional may be important for the theory of the Boltzmann equation, then they proved the existence theorem (P.L. Lions and R. DiPerna, this work was awarded the Fields medal in 1994).

pure and applied dynamics directly, or bears the visible imprint of its ideas and methods. Maybe the book presents a specific chapter on this subject?

But, of course, the problems came from physics, from engineering. Maybe it is both physics *and* mathematics? Or perhaps it is something different, but what can it be? It is not so easy to answer the question, what is the subject of our book, even for the authors. But we can say what we want it to be. We want it to be a special "meeting point" of pure and applied dynamics, of physics, and of engineering sciences. This meeting point has a sufficient number of specific problems, methods and results to deserve a special name. We propose the name *Model Engineering*. As long as it is engineering, it is synthetic subject: if it is possible to prove something exactly, this is great, and we should follow this possibility, but if the physical sense gives us a seminal hint, well, we should use it even if the rigorous foundations are far from complete. *The result is the model that works.* In this enormous field of intellectual activity our book tends to be in the theoretical corner; we focus our study on constructive *methods*, and the examples that fill up more than three-quarters of the book are used for motivation, demonstration and development of the methods.

Which scientific disciplines should meet at the meeting point we build in our book? The last century demonstrated the emergence of two disciplines, of the theory of dynamical systems in mathematics, and of statistical physics. Nonequilibrium statistical physics, in short, is a science about slow-fast motion decomposition. Dynamic theory is about general features of long-time typical behaviour. Our book is about what dynamic theory has to say about nonequilibrium systems. The very brief answer is – it makes the theory of nonequilibrium systems the theory of slow invariant manifolds. But the reverse impact of physics on methods is also significant. Applied mathematics and computational physics create a "second (computational) reality". This is a beautiful intellectual building, but in each element of this building, at each step of the work, we should take into account the basic physics; the violation of a physical law at one place can destroy an important part of the whole construction.

The presented methods to construct slow invariant manifolds certainly reflect the authors' preference and their own work. Much effort was spent to coordinate the developed methods with the basic physics at each step.

The book can be used for various purposes:

- As a collection of tools for model reduction in kinetics;
- As a source of mathematical problems;
- As a guide to physical concepts useful for model reduction;
- As a collection of successful examples of model reduction;
- As a source of recent literature on model reduction, invariant manifolds and related topics.

We wrote the book for our colleagues and for our students in order to avoid in the future the usual excessive explanation: to explain the basic notions and physical sense, to answer the common questions about invariant manifolds and model reduction, about our point of view, about the balance between physics, mathematics (dynamics) and engineering in our work. Now we can simply hand over this book and suggest reading approaches. There are many possible approaches for different purposes. Some of them are presented in the introduction.

As useful background for reading the book, three graduate courses should be mentioned: differential equations and dynamical systems, kinetics and thermodynamics, and elementary functional analysis.

Once upon a time Lev Landau gave the following advice: If the Contents of a book is interesting to you, close the book and try to write it. If it is too difficult a task, then look through the first chapter and try to write it. If it is still too hard, go ahead and try to write a section, a subsection, a paragraph, a formula. We completely agree with this advice with just one addition: please send us your results, because your book will contain another point of view, and will be highly interesting.

Acknowledgements

First of all, we are grateful to our collaborators, S. Ansumali (Zürich), V.I. Bykov (Krasnoyarsk), C. Frouzakis (Zürich), M. Deville (Lausanne), G. Dukek (Ulm), P.A. Gorban (Krasnoyarsk–Omsk–Zürich), P. Ilg (Zürich–Berlin), R.G. Khlebopros (Krasnoyarsk), T.F. Nonnenmacher (Ulm), V.A. Okhonin (Krasnoyarsk–Toronto), H.C. Öttinger (Zürich), A.A. Rossiev (Krasnoyarsk), S. Succi (Rome), L.L. Tatarinova (Krasnoyarsk–Zürich), G.S. Yablonskii (Novosibirsk–Saint-Louis), D.C. Wunsch (Lubbock–Missouri-Rolla), A.Yu. Zinovyev (Krasnoyarsk–Bures-sur-Yvette), and V.B. Zmievskii (Krasnoyarsk–Lausanne–Montreal), for years of collaboration, stimulating discussions and support. We appreciate the comments from T. Kaper (Boston), H. Lam (Princeton), and A. Santos (Badajoz) about various parts of the book. We thank M. Grmela (Montreal) for detailed and encouraging discussions of the geometrical foundations of nonequilibrium thermodynamics. M. Shubin (Moscow-Boston) explained to us some important chapters on the pseudodifferential operators theory. Finally, it is our pleasure to thank Misha Gromov (Bures-sur-Yvette) for encouragement and the spirit of Geometry.

Zürich,   *Alexander Gorban*
November 2004   *Iliya Karlin*

Contents

1 Introduction

1.1 Ideas and References

In this book, we present a collection of constructive methods to study slow (stable) positively invariant manifolds of dynamic systems. The main objects of our study are dissipative dynamic systems (finite or infinite) which arise in various problems of kinetics. Some of the results and methods presented herein may have a more general applicability, and can be useful not only for dissipative systems but also, for example, for conservative systems.

Nonequilibrium statistical physics is a collection of ideas and methods for the extraction of slow invariant manifolds. Reduction of description for dissipative systems assumes (explicitly or implicitly) the following picture: There exists a manifold of slow motions in the phase space of the system. From the initial conditions the system goes quickly in a small neighborhood of the manifold, and after that moves slowly along this manifold (see, for example, [1]). The manifold of slow motion (slow manifold, for short) must be positively invariant: if a motion starts on the manifold at t_0, then it stays on the manifold at $t > t_0$. The frequently used wording "invariant manifold" is not really precise: for dissipative systems, the possibility of extending the solutions (in a meaningful way) backwards in time is limited. So, in nonequilibrium statistical physics we study *positively invariant* (or inward invariant) slow manifolds. The necessary invariance condition can be written explicitly as the differential equation for the manifold immersed into the phase space. This picture is directly applicable to *dissipative* systems.

Time separation for *conservative* systems and the way from the reversible mechanics (for example, from the Liouville equation) to dissipative systems (for example, to the Boltzmann equation) requires some additional ideas and steps. For any conservative system, a restriction of its dynamics onto any invariant manifold is conservative again. We should represent a dynamics of a large conservative system as a result of dynamics in its small subsystems, and it is necessary to take into account that a *macroscopically* small interval of time can be considered as an infinitely large interval for a small subsystem, i.e. microscopically. It allows us to represent the relaxation of such large systems as an ensemble of *indivisible events* (for example, collisions). The Bogolyubov–Born–Green–Kirkwood–Yvon (BBGKY) hierarchy

Alexander N. Gorban and Iliya V. Karlin: *Invariant Manifolds for Physical and Chemical Kinetics*, Lect. Notes Phys. **660**, 1–19 (2005)
www.springerlink.com   © Springer-Verlag Berlin Heidelberg 2005

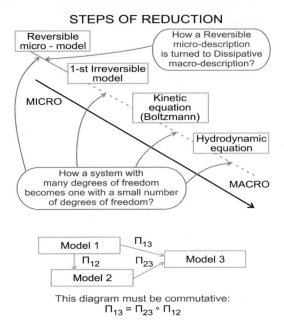

Fig. 1.1. The stairs of reduction, step by step

and Bogolyubov's method of derivation of the Boltzmann equation give us the unexcelled realization of this approach [2].

The "stairs of reduction" (Fig. 1.1) lead from the reversible microdynamics to irreversible macrokinetics. The most mysterious is the first step: the emergence of irreversibility. We discuss this problem in Chap. 12, but the main focus of our attention in the book is the model reduction for dissipative systems.

For dissipative systems, we always keep in mind the following picture (Fig. 1.2). The vector field $J(x)$ generates the motion on the phase space U: $dx/dt = J(x)$. An ansatz manifold Ω is given, it is the current approximation to the invariant manifold. This manifold Ω is described as the image of the map $F : W \to U$. The choice of the space of macroscopic variables W is the important step of the model reduction: all corrections of the current ansatz manifold are described as images of various F for given W.

The projected vector field $PJ(x)$ belongs to the tangent space T_x, and the equation $dx/dt = PJ(x)$ describes the motion along the ansatz manifold Ω (if the initial state belongs to Ω). The induced dynamics on the space W is generated by the vector field

$$\frac{dy}{dt} = (D_y F)^{-1} PJ(F(y)) .$$

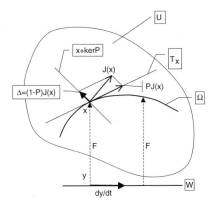

Fig. 1.2. The main geometrical structures of model reduction: U is the phase space, $J(x)$ is the vector field of the system under consideration: $dx/dt = J(x)$, Ω is an ansatz manifold, W is the space of macroscopic variables (coordinates on the manifold), the map $F : W \to U$ maps any point $y \in W$ into the corresponding point $x = F(y)$ on the manifold Ω, T_x is the tangent space to the manifold Ω at the point x, $PJ(x)$ is the projection of the vector $J(x)$ onto tangent space T_x, the vector field dy/dt describes the induced dynamics on the space of parameters, $\Delta = (1 - P)J(x)$ is the defect of invariance, the affine subspace $x + \ker P$ is the plain of fast motions, and $\Delta \in \ker P$

Here the inverse linear operator $(D_y F)^{-1}$ is defined on the tangent space $T_{F(y)}$, because the map F is assumed to be immersion, that is the differential $(D_y F)$ is the isomorphism onto the tangent space $T_{F(y)}$.

The main focus of our analysis is the *invariance equation*[1]:

$$\Delta = (1 - P)J = 0 \, ,$$

the *defect of invariance* Δ should vanish. It is a differential equation for an unknown map $F : W \to U$. Solutions of this equation are invariant in the sense that the vector field $J(x)$ is tangent to the manifold $\Omega = F(W)$ for

---

[1] A.M. Lyapunov studied analytical solutions of similar equations near a fixed point [3]. He found these solutions in a form of the Taylor series expansion and proved the convergency of those power series near the non-resonant fixed point (the Lyapunov auxiliary theorem). In 1960s the invariance equations approach was developed, first of all, in the context of the Kolmogorov–Arnold–Moser theory for invariant tori computation [4–6], as a special analytical perturbation theory [7,8]. Recently, the main task is to develop constructive non-perturbative methods, because the series of perturbations theory diverge and, moreover, the high–order terms loose the physical sense for most interesting applications. The seminal Kolmogorov's idea was to use Newton's method for solution of the invariance equation (instead of the Taylor series expansion) [4]. In this book we discuss the methods for invariant manifold construction that exploit the thermodynamic properties of the kinetic equations.

each point $x \in \Omega$. But this condition says nothing about the slowness of the manifold Ω.

How to choose the projector P? Another form of this question is: how to define the plain of fast motions $x + \ker P$? The choice of the projector P is ambiguous, from the formal point of view, but the second law of thermodynamics gives a good hint [9]: the entropy should grow in the fast motion, and the point x should be the point of entropy maximum on the plane of fast motion $x + \ker P$. That is, the subspace $\ker P$ should belong to the kernel of the entropy differential:

$$\ker P_x \subset \ker D_x S .$$

Of course, this rule is valid for closed systems with entropy, but it can be also extended onto open systems: the projection of the "thermodynamic part" of $J(x)$ onto T_x should have the positive entropy production. If this thermodynamic requirement is valid for any ansatz manifold not tangent to the entropy levels and for any thermodynamic vector field, then the thermodynamic projector is unique [10]. Let us describe this projector P for given point x, subspace $T_x = \text{im}P$, differential $D_x S$ of the entropy S at the point x and the second differential of the entropy at the point x, the bilinear functional $(D_x^2 S)_x$. We need the positively definite bilinear form $\langle z|p\rangle_x = -(D_x^2 S)_x(z, p)$ (the entropic scalar product). There exists a unique vector g such that $\langle g|p\rangle_x = D_x S(p)$. It is the Riesz representation of the linear functional $D_x S$ with respect to entropic scalar product. If $g \neq 0$ then the thermodynamic projector is

$$P(J) = P^\perp(J) + \frac{g^\|}{\langle g^\||g^\|\rangle_x} \langle g^\perp|J\rangle_x ,$$

where $P^\perp$ is the orthogonal projector onto T_x with respect the entropic scalar product, and the vector g is splitted onto tangent and orthogonal components:

$$g = g^\| + g^\perp; \ g^\| = P^\perp g; \ g^\perp = (1 - P^\perp)g .$$

This projector is defined if $g^\| \neq 0$.

If $g = 0$ (the equilibrium point) then $P(J) = P^\perp(J)$.

For given T_x, the *thermodynamic projector* (5.25) depends on the point x through the x-dependence of the scalar product $\langle|\rangle_x$, and also through the differential of S in x.

A dissipative system may have many closed positively invariant sets. For example, for every set of initial conditions K, union of all the trajectories $\{x(t), t \geq 0\}$ with initial conditions $x(0) \in K$ is positively invariant. Thus, the selection of the slow (stable) positively invariant manifolds becomes an important problem[2].

---

[2] Nevertheless, there exists a different point of view: "Non–uniqueness, when it arises, is irrelevant for modeling" [13], because the differences between the possible manifolds are of the same order as the differences we set out to ignore in

One of the difficulties in the problem of reducing the description is due to the fact that there exists no commonly accepted formal definition of a slow (and stable) positively invariant manifold. This difficulty is resolved in Chap. 4 of our book in the following way: First, we consider manifolds immersed into a phase space and study their motion along trajectories. Second, we subtract from this motion the motion of immersed manifolds along themselves, and obtain a new equation for dynamics of manifolds in the phase space: the manifold Ω moves by the vector field Δ. It is *the film extension of the dynamics*:

$$\frac{\mathrm{d}F_t(y)}{\mathrm{d}t} = \Delta \,,$$

where the defect of invariance, $\Delta = (1 - P)J$, depends on the point $x = F(y)$ and on the tangent space to the manifold $\Omega = F(W)$ at this point. Invariant manifolds are fixed points for this extended dynamics, and *slow* invariant manifolds are *Lyapunov stable fixed points*.

The main body of this book is about how to actually compute the slow invariant manifold. We present three approaches to constructing slow (stable) positively invariant manifolds.

– *Iteration method* for solution of the invariance equation (Newton method subject to incomplete linearization);
– *Relaxation methods* based on the film extension of the original dynamic system;
– *The method of natural projector* that projects not the vector fields, but rather finite segments of trajectories.

The Newton method (with incomplete linearization) is the iteration method for solving the invariance equation. On each iteration we linearize the invariance equation and solve obtained linear equation. In the defect of invariance $\Delta = (1 - P)J(x)$ both the vector field $J(x) = J(F(y))$ ($y \in W$) and the projector P depend on the unknown map F (P depends on the point $x \in W$ and on the tangent space $T_x = \mathrm{im}D_yF$). On each iteration we use for $J(F(y))$ the first-order (linear in F) approximation, and for P only the zero-order (constant) one. The iteration method with this *incomplete linearization* leads to the slowest invariant manifold [11]. The Newton method (with incomplete linearization) is convenient for obtaining the explicit formulas – even one iteration can give a good approximation.

Relaxation methods are directed more towards the numerical implementation. Nevertheless, several first steps also can give appropriate analytical approximations, competitive with other methods. These methods are based on the stepwise solution of the differential equation $\mathrm{d}F(y)/\mathrm{d}t = \Delta$ (the *film extension of the dynamics*).

---

establishing the low-dimensional model. We do not share this viewpoint because it may be relevant only if there exists a small parameter, and, moreover, only asymptotically when this small parameter tends to zero.

Finally, the natural projector method constructs not the manifold itself but a projection of slow dynamics onto some set of variables. This method is the successor of two important methods: the Ehrenfests' coarse-graining [15] and the Hilbert method for solution of the Boltzmann equation [16]. It can by applied to reversible and irreversible systems, and allows us to make the first step of reduction (see Fig. 1.1) as well as the following steps.

The Newton method subject to incomplete linearization was developed for the construction of slow (stable) positively invariant manifolds in the following problems:

– Derivation of the post–Navier–Stokes hydrodynamics from the Boltzmann equation [11, 12, 14, 17].
– Description of the dynamics of polymers solutions [12, 106].
– Correction of the moment equations [12, 21].
– Reduced description for chemical kinetics [12, 22, 23, 105].

Relaxation methods based on the film extension of the original dynamic system were applied to the Fokker–Planck equation [12, 24]. Applications of these methods in the theory of the Boltzmann equation can benefit from the estimations, obtained in the papers [26, 27].

The method of natural projector was originally applied to derivation of the dissipative equations of macroscopic dynamics from the conservative equations of the microscopic dynamics [12, 29–35]. Using this method, new equations were obtained for the post–Navier–Stokes hydrodynamics, equations of plasma hydrodynamics and others [30, 34]. This short-memory approximation was applied to the Wigner formulation of quantum mechanics [36–38]. The dissipative dynamics of a single quantum particle in a confining external potential is shown to take the form of a damped oscillator whose effective frequency and damping coefficients depend on the shape of the quantum-mechanical potential [35]. Further examples of the coarse-graining quantum fields dynamics can be found in [39]. The natural projector method can also be applied effectively to dissipative systems: instead of the Chapman–Enskog method in theory of the Boltzmann equation, for example.

The most natural initial approximation for the methods under consideration is a quasiequilibrium manifold. It is the manifold of conditional maxima of the entropy. The majority of works on nonequilibrium thermodynamics deal with corrections to quasi-equilibrium approximations, or with applications of these approximations (with or without corrections). The construction of the quasi-equilibrium allows for the following generalization: almost every manifold can be represented as a set of minimizers of the entropy under linear constraints. However, in contrast to the standard quasiequilibrium, these linear constraints will depend on the point on the manifold. We describe the quasiequilibrium manifold and the quasiequilibrium projector on the tangent space of this manifold. This projector is orthogonal with respect to the entropic scalar product (the bilinear form defined by the negative second differential of the entropy). We construct the thermodynamical projector, which transforms the arbitrary vector field equipped with the given Lyapunov function (the entropy) into a vector field with the same Lyapunov function for an arbitrary anzatz manifold which is not tangent to the level of the Lyapunov function. The uniqueness of this construction is demonstrated.

Here, a comment on the status of most of the statements in this book is in order. Just like the absolute majority of claims concerning such things as general solutions of the Navier–Stokes or the Boltzmann equation, they have the status of being plausible. They can become theorems only if one restricts essentially the set of the objects under consideration. Among such restrictions we should mention cases of the exact reduction, for example, exact derivation of hydrodynamics from kinetics [40, 42]. In these (still infinite-dimensional) examples one can compare different methods, for example, the Newton method with the methods of series summation in the perturbation theory [42, 43].

Also, it is necessary to stress here, that even if in the limit all the methods lead to the same results, they can give rather different approximations "on the way".

The rigorous foundation of the constructive methods of invariant manifolds should, in particular, include theorems about *persistence of invariant manifolds under perturbations*. For instance, the compact normally hyperbolic invariant manifolds persist under small perturbations for finite-dimensional dynamical systems [46, 47]. The most well-known result of this type is the Kolmogorov–Arnold–Moser theory about persistence of almost all invariant tori of completely integrable system under small perturbations [4–6].

Such theorems exist for some classes of infinite dimensional dissipative systems too [48]. Unfortunately, it is not proven until now that many important systems (the Boltzmann equation, the three-dimensional Navier–Stokes equations, the Grad equations, etc.) belong to these classes. So, it is necessary to act with these systems without a rigorous basis.

The new quantum field theory formulation of the problem of persistence of invariant tori in perturbed completely integrable systems was obtained [68], and a new proof of the KAM theorem for analytic Hamiltonians based on the renormalization group method was given.

Two approaches to the construction of the invariant manifolds are widely used: the *Taylor series expansion* for the solution of the invariance equation [3, 50–52] and the method of *renormalization group* [53, 54, 56–59]. The advantages and disadvantages of the Taylor series expansion are well-known: constructivity versus the absence of physical meaning for the high-order terms (often), and divergence in the most interesting cases (often).

In the paper [56], a geometrical formulation of the renormalization group method for global analysis was given. It was shown that the renormalization group equation can be interpreted as an envelope equation. Recently [57] the renormalization group method was formulated in terms of invariant manifolds. This method was applied to derive kinetic and transport equations from the respective microscopic equations [58]. The derived equations include the Boltzmann equation in classical mechanics (see also the paper [55], where it was shown for the first time that kinetic equations such as the Boltzmann equation can be understood naturally as renormalization group equations), the Fokker–Planck equation, a rate equation in a quantum field theoretical model.

From the point of view of the authors of the paper [55], the relation of renormalization group theory and reductive perturbation theory has simultaneously been recognized: renormalization group equations are actually the slow-motion equations which are usually obtained by reductive perturbation methods.

The renormalization group approach was applied to the stochastic Navier–Stokes equation in order to model fully developed fluid turbulence [60–62]. For the evaluation of the relevant degrees of freedom the renormalization group technique was revised for discrete systems in the recent paper [59].

The kinetic theory approach to subgrid modeling of fluid turbulence became more popular recently. [63–66]. A mean-field approach (filtering out subgrid scales) was applied to the Boltzmann equation in order to derive a subgrid turbulence model based on kinetic theory. It was demonstrated [66] that the only Smagorinsky type model which survives in the hydrodynamic limit on the viscosity time scale is the so-called tensor-diffusivity model [67].

The first systematic and successful method of constructing invariant manifolds for dissipative systems was the celebrated *Chapman-Enskog method* [70] for the Boltzmann kinetic equation. The Chapman–Enskog method results in a series development of the so-called normal solution (the notion introduced by Hilbert [16]) where the one-body distribution function depends on time and space only through its locally conserved moments. To the first approximation, the Chapman–Enskog method leads to hydrodynamic equations with transport coefficients expressed in terms of molecular scattering cross-sections. However, the higher order terms of the Chapman–Enskog expansion bring in the "ultra-violet catastrophe" (noticed first by Bobylev [72]) and negative viscosity. This drawback pertinent to the Taylor series expansion disappears as soon as the Newton method is used to construct the invariant manifold [11].

The Chapman–Enskog method was generalized many times [76] and gave rise to a host of subsequent works and methods, such as the famous method of the *quasi-steady state* in chemical kinetics, pioneered by Bodenstein and Semenov and explored in considerable detail by many authors (see, for example, [22, 77–81]), and the theory of *singularly perturbed* differential equations [77, 82–87].

There exists a set of methods to construct an ansatz for the invariant manifold based on the spectral decomposition of the Jacobian. The idea to use the spectral decomposition of Jacobian fields in the problem of separating the motions into fast and slow originates from analysis of stiff systems [88], and from methods of sensitivity analysis in control theory [89, 90]. One of the currently most popular methods based on the spectral decomposition of Jacobian fields is the construction of the so-called *intrinsic low-dimensional manifold* (ILDM) [93].

These methods were thoroughly analyzed in two papers [94, 95]. It was shown that the successive applications of the Computational Singular Perturbation (CSP) algorithm (developed in [90]) generate, order by order, the asymptotic expansion of a slow manifold, and the manifold identified by the ILDM technique (developed in [93]) agrees with the invariant manifold to some order. An explicit algorithm based on the CSP method is designed for the integration of stiff systems of PDEs by means of explicit schemes [91]. The CSP analysis of time scales and manifolds in a transient flame-vortex interaction was presented in [92].

The theory of *inertial manifold* is based on the special linear dominance in higher dimensions. Let an infinite-dimensional system have a form: $\dot{u} + Au = R(u)$, where A is self-adjoint, and has a discrete spectrum $\lambda_i \to \infty$ with sufficiently big gaps between λ_i, and let $R(u)$ be continuous. One can build the slow manifold as the graph over a root space of A [96]. The textbook [100] provides an exhaustive introduction to the main ideas and methods of this theory. Systems with linear dominance have limited utility in kinetics. Often there are no big spectral gaps between λ_i, and even the sequence $\lambda_i \to \infty$ might be bounded (for example, this is the case for the model Bhatnagar–Gross–Krook (BGK) equations, or for the Grad equations). Nevertheless, the concept of the inertial attracting manifold has wider field of applications than the theory, based on the linear dominance assumption.

The Newton method with incomplete linearization and the relaxation method allow us to find an approximate slow invariant manifolds without Jacobian field spectral decomposition. Moreover, a necessary slow invariant subspace of the Jacobian at the equilibrium point appears as a by-product of the Newton iterations (with incomplete linearization), or of the relaxation method.

It is of importance to search for minimal (or subminimal) sets of natural parameters that uniquely determine the long-time behaviour of a system. This problem was first discussed by Foias and Prodi [97] and by Ladyzhenskaya [98] for the two-dimensional Navier–Stokes equations. They have proved that the long-time behaviour of solutions is completely determined by the dynamics of sufficiently large number of Fourier modes. A general approach to the problem on the existence of a finite number of determining parameters has been discussed [99, 100].

The past decade has witnessed a rapid development of the so-called *set oriented* numerical methods [101]. The purpose of these methods is to compute attractors, invariant manifolds (often, computation of stable and unstable manifolds in hyperbolic systems [102–104]). Also, one of the central tasks of these methods is to gain statistical information, i. e. computations of physically observable invariant measures. The distinguished feature of the modern set-oriented methods of numerical dynamics is the use of ensembles of trajectories within a relatively short propagation time instead of a long time single trajectory.

In this book we systematically consider a discrete analog of the slow (stable) positively invariant manifolds for dissipative systems, *invariant grids*. These invariant grids were introduced in [22]. Here we shall describe the Newton method subject to incomplete linearization and the relaxation methods for the invariant grids [105].

It is worth mentioning that the problem of the grid correction is fully decomposed into the tasks of the grid's nodes correction. The edges between the nodes appear only in the calculation of the tangent spaces at the nodes. This fact determines the high computational efficiency of the invariant grids method.

Let the (approximate) slow invariant manifold for a dissipative system be found. *Why have we constructed it?* One important part of the answer to this question is: *We have constructed it to create models of open system dynamics in the neighborhood of this manifold.* Different approaches for this modeling are described.

We apply these methods to the problem of reduced description in polymer dynamics and derive the universal limit in dynamics of dilute polymeric solutions. It is represented by the *revised Oldroyd 8 constants* constitutive equation [106] for the polymeric stress tensor. Coefficients of this constitutive equation are expressed in terms of the microscopic parameters. This limit of dynamics of dilute polymeric solutions is universal, and any physically consistent equation should contain the obtained equation as a limit, or one should explain why it is not achieved. Such universal limit equations are well-known in various fields of physics. For example, the Navier–Stokes equation in fluid dynamics is an universal limit for dynamics of simple gas described by the Boltzmann equation, the Korteweg–De-Vries equation is universal in the description of the dispersive dissipative nonlinear waves, etc.

The phenomenon of *invariant manifold explosion* in driven open systems is demonstrated on the example of dumbbell models of dilute polymeric solutions [109]. This explosion gives us a possible mechanism of drag reduction in dilute polymeric solutions [110].

Suppose that for the kinetic system the approximate invariant manifold has been constructed and the slow motion equations have been derived. Suppose that we have solved the slow motion system and obtained $x_{sl}(t)$. We consider the following two questions:

– How well does this solution approximate the true solution $x(t)$ given the same initial conditions?
– How is it possible to use the solution $x_{sl}(t)$ for its refinement without solving the slow motion system (or its modifications) again?

These two questions are interconnected. The first question states the problem of the *accuracy estimation*. The second one states the problem of *post-processing* [348–351]. We propose various algorithms for post-processing and accuracy estimation, and give an example of application.

Our collection of methods and algorithms can be incorporated into recently developed technologies of computer-aided multiscale analysis which enable "level jumping" between microscopic and macroscopic (system) levels. It is possible both for the traditional technique based on transition from microscopic equations to macroscopic equations and for the "equation-free" approach [107]. This approach developed in recent work [108], when successful, can bypass the derivation of the macroscopic evolution equations when these equations conceptually exist but are not available in closed form. The mathematics-assisted development of a computational superstructure may enable alternative descriptions of the problem physics (e.g. Lattice Boltzmann (LB), kinetic Monte- Carlo (KMC) or Molecular Dynamics (MD) microscopic simulators, executed over relatively short time and space scales) to perform systems level tasks (integration over relatively large time and space scales, coarse bifurcation analysis, optimization, and control) directly. It is possible to use macroscopic invariant manifolds in this environment without explicit equations.

1.2 Content and Reading Approaches

The present book comprises sections of two kinds. The first includes the sections that contain basic notions, methods and algorithms. Another group of sections entitled "Examples" contain various case studies where the methods are applied to specific equations. Exposition in the "Examples" sections is not as consequent as in the basic sections. Most of the examples can be read more or less independently. Logical connections between chapters are presented in Fig. 1.3.

The main results and notions presented in the book are as follows. In this Chap. 1 we present the main ideas, references, abstracts of chapters, and the possible reading plans.

Chapter 2 is the second introduction, it introduces the main equations of kinetics: the Boltzmann equation, equations of chemical kinetics, and the Fokker–Planck equation. The main methods of reduction for these equations are also discussed: from the Chapman–Enskog and Hilbert methods to quasiequilibrium and quasi-steady state approximations.

In Chap. 3 we write down the *invariance equation* in the differential form. This equation gives the necessary conditions of invariance of a manifold immersed into the phase space of a dynamical system. In order to estimate the discrepancy of an ansatz manifold, the *defect of invariance* if defined. The introduction of this defect of invariance requires a *projector field*. These

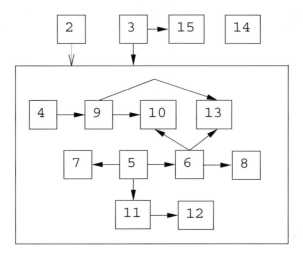

Fig. 1.3. Logical connections between chapters. All the chapters depend on Chap. 3. For understanding examples and problems it may be useful (but not always necessary) to read Chap. 2

notions, defect of invariance and projector field, as well as the invariance equation play the central role in the whole book.

Chapter 4 is devoted to the definition of *slowness* of a positively invariant manifold. The equation of motion of the manifold (the "film") immersed into the phase space of the dynamical system is discussed (equation for the film motion). A slow positively invariant manifold is defined as a stable fixed point for this motion. The projector field introduced in Chap. 3 is crucial for the definition of the stability.

The main thermodynamic structures, the entropy, the entropic scalar product, quasiequilibrium, and the thermodynamic projector, are introduced in Chap. 5. The quasiequilibrium manifold is the manifold of conditional entropy maxima for given values of macroscopic variables. These values parametrize this manifold. Most of the works on nonequilibrium thermodynamics deal with corrections to quasiequilibrium approximations, or with applications of these approximations (with or without corrections). This viewpoint is not the only possible, but it proves very efficient for the construction of a variety of useful models, approximations and equations, as well as methods to solve them.

The entropic scalar product is generated by the second differential of the entropy. It endows the space of states by the unique distinguished Riemannian structure. The thermodynamic projector is the operator which transforms the arbitrary vector field equipped with the given Lyapunov function into a vector field with the same Lyapunov function. Uniqueness of such projector is proved.

In Chap. 5 we start the series of examples for the Boltzmann equations. First, we analyze the defect of invariance for the Local Maxwellian manifold: the manifold of the locally equilibrium distributions. Second, we present the quasi-equilibrium closure hierarchies for the Boltzmann equation. In 1949, Harold Grad [201] extended the basic assumption behind the Hilbert and Chapman–Enskog methods (the space and time dependence of the normal solutions is mediated by the five hydrodynamic moments). A physical rationale behind the Grad moment method is an assumption of the decomposition of motion. (i) During the time of order τ, a set of distinguished moments M' (which include the hydrodynamic moments and a subset of higher-order moment) does not change significantly as compared to the rest of the moments M'' (the fast evolution). (ii) Towards the end of the fast evolution, the values of the moments M'' become unambiguously determined by the values of the distinguished moments M'. (iii) On the time of order $\theta \gg \tau$, dynamics of the distribution function is determined by the dynamics of the distinguished moments while the rest of the moments remains to be determined by the distinguished moments (the slow evolution period).

An important generalization of the Grad moment method is the concept of quasiequilibrium approximations. The quasiequilibrium distribution function for a set of distinguished moments M' maximizes the entropy density S for fixed M'. The quasiequilibrium manifold is the collection of the quasiequilibrium distribution functions for all admissible values of M. The quasiequilibrium approximation is the simplest and very useful (not only in the kinetic theory itself) implementation of the hypothesis about time separation.

The quasiequilibrium approximation does not exist if the highest order moment is an odd polynomial of velocity (therefore, there exists no quasi-equilibrium for thirteen Grad's moments). The Grad moment approximation is the first-order expansion of the quasiequilibrium around the local Maxwellian. An explicit method of constructing of approximations (the Triangle Entropy Method) is developed for strongly nonequilibrium problems of Boltzmann–type kinetics, i.e. when standard moment variables are insufficient. This method enables one to treat any complicated nonlinear functionals that fit the physics of a problem (such as, for example, rates of processes) as new independent variables.

The method is applied to the problem of derivation of hydrodynamics from the Boltzmann equation. New macroscopic variables are introduced (moments of the Boltzmann collision integral, or collision moments). They are treated as independent variables rather than as infinite moment series. This approach gives the complete account of the rates of scattering processes. Transport equations for scattering rates are obtained (the second hydrodynamic chain), similar to the usual moment chain (the first hydrodynamic chain). Using the triangle entropy method, three different types of macroscopic description are considered. The first type involves only moments of distribution functions, and the results coincide with those of the Grad method in the Maximum Entropy version. The second type of description involves only collision moments. Finally, the third type involves both the moments and the collision moments (the mixed description). The second and the mixed hydrodynamics are sensitive to the choice of the collision model. The second hydrodynamics is equivalent to the first hydrodynamics only for Maxwell molecules, and the mixed hydrodynamics exists for all types of collision models excluding Maxwell molecules. Various examples of the closure of the first, of the second, and of the mixed hydrodynamic chains are considered for the hard spheres model. It is shown, in particular, that the complete account of scattering processes leads to a renormalization of transport coefficients.

We apply the developed method to a classical problem: determination of molecular dimensions (as diameters of equivalent hard spheres) from experimental viscosity data. It is the third example in Chap. 5.

The first non-perturbative method for solution of the invariance equation is developed in Chap. 6. It is the *Newton method with incomplete linearization*. The incomplete linearization means that in the Newton–type iteration for the invariance equation we do not use the whole differential of the right-hand side of the invariance equation: the differential of the projector field is excluded. This modification of the Newton method leads to selection of the slowest invariant manifold. The series of examples for the Boltzmann equations is continued in this chapter. The non-perturbative correction to the Local Maxwellian manifold is constructed, and the equations of the high-order (the post–Navier–Stokes) hydrodynamics are obtained.

In Chap. 5 we use the second law of thermodynamics – existence of the entropy – in order to equip the problem of constructing slow invariant manifolds with a geometric structure. The requirement of the entropy growth (universally, for all the reduced models) significantly restricts the form of the thermodynamic projectors. In Chap. 7 we introduce a different but equally important argument – the *micro-reversibility* (T-invariance), and its macroscopic consequences, the *Onsager reciprocity relations*. The main idea in this chapter is to use the reciprocity relations for the fast motions. In order to appreciate this idea, we should mention that the decomposition of motions into fast and slow is not unique. Requirement of the Onsager reciprocity relations for any equilibrium point of fast motions implies the selection (filtration) of the fast motions. We term this the *Onsager filter*. Equilibrium points of fast motions are all the points on manifolds of slow motions. The formalism of the *quasi-chemical representation* is one of the most developed means of modelling, it makes it possible to "assemble" complex processes out of elementary processes. This formalism is very natural for representation of the reciprocity relations. And again, the Example to this chapter continues the "Boltzmann series". It is the quasi-chemical representation and the self-adjoint (i.e. Onsager) linearization of the Boltzmann collision operator in the slow, but not obligatory equilibrium states.

In Chap. 8 a new class of exactly solvable problems in nonequilibrium statistical physics is described. The systems that allow the exact solution of the reduction problem are presented. Up to now, the problem of the exact relationship between kinetics and hydrodynamics remains unsolved. All the methods used to establish this relationship are not rigorous, and involve approximations. In this chapter, we consider situations where hydrodynamics is the exact consequence of kinetics, and in that respect, a new class of exactly solvable models of statistical physics has been established. The Chapman–Enskog method is treated as the Taylor series expansion approach to solving the appropriate invariance equation. A detailed treatment of the classical Chapman–Enskog derivation of hydrodynamics is given in the framework of Grad's moment equations. Grad's systems are considered as the minimal kinetic models where the Chapman–Enskog method can be studied exactly, thereby providing the basis to compare various approximations in extending the hydrodynamic description beyond the Navier–Stokes approximation. Various techniques, such as the method of partial summation, the Padé approximants, and the invariance principle are compared both in linear and nonlinear situations.

In Chap. 9 the "large stepping" *relaxation method* for solution of the invariance equation is developed. The relaxation method is an alternative to the Newton iteration method described in Chap. 6: The initial approximation to the invariant manifold is moved with the film extension of the dynamics described in Chap. 4. The proposed step in time for the stepwise solution of the film extension equation is the maximal possible step that does not violate the thermodynamic conditions. In the examples, the idea of the large stepping is applied to the Fokker–Planck equation and to the initial layer problem for the Boltzmann equation. The obtained approximate solutions of the initial layer problem are compared to the exact solutions.

How can we represent invariant manifolds numerically? How can we use the numerical representation in all the methods for invariant manifold refinement? Chapter 10 is devoted to answering these questions. A grid-based version of the method of invariant manifold is developed. The most essential element of this chapter is the systematic consideration of a discrete analogue of the slow (stable) positively invariant manifolds for dissipative systems, *invariant grids*. The invariant grid is defined as a mapping of finite-dimensional grids into the phase space of a dynamic system. We define the differential operators on the grid as difference operators, hence, it is possible to define the tangent space at each point of the grid mapped into the phase space. If the tangent space is constructed, then the invariance equation can be written down. We describe the Newton method and the relaxation method for solution of this discrete analogue of the invariance equation. Examples for this chapter are taken from the chemical kinetics. One attractive feature of two-dimensional invariant grids is the possibility to use them as a screen, on which one can display different functions and dynamic of the system.

P. and T. Ehrenfest suggested in 1911 a model of dynamics with a coarse-graining of the original conservative system in order to introduce irreversibility [15]. The Ehrenfests considered a partition of the phase space into small cells, and they have suggested combining the motions of the phase space ensemble due to the Liouville equation with coarse-graining "shaking" steps – averaging of the density of the ensemble over the phase cells. This generalizes to the following: combination of the motion of the phase ensemble due to microscopic equations with returns to the quasiequilibrium manifold while preserving the values of the macroscopic variables. In Chap. 11 we develop the method of natural projector, a formalism of nonequilibrium thermodynamics based on this generalization.

The method of natural projector can be considered as a development of the ideas of the Hilbert method from the theory of the Boltzmann equation. The main new element in the method of natural projector with respect to the Hilbert method is the construction of the macroscopic equations from the microscopic equations, not just a "normal solution" to a microscopic equation. The obtained macroscopic equations contain one unknown parameter, the time between coarse-graining (shaking) steps (τ). This parameter can be obtained from the experimental data, or from independent microscopic or phenomenological consideration.

In the first example to this chapter the microscopic dynamics is given by the one-particle Liouville equation. The set of macroscopic variables is density, momentum density, and the density of average kinetic energy. The correspondent quasiequilibrium distribution is the local Maxwell distribution. For the hydrodynamic equations, the zeroth (quasiequilibrium) approximation is given by the Euler equations of compressible nonviscous fluid. The next order approximation gives the Navier–Stokes equations which have dissipative terms. Higher-order approximations to the hydrodynamic equations, when they are derived from the Boltzmann kinetic equation by the Chapman–Enskog expansion (so-called Burnett approximation), are prone to various difficulties, in particular, they exhibit instability of sound waves at sufficiently short wave length (see Chap. 8). Here we demonstrate how model hydrodynamic equations, including the post–Navier–Stokes approximations, can be derived on the basis of the coarse-graining idea, and find that the resulting equations are stable, contrary to the Burnett equation.

In the second example the fluctuation-dissipation formula is derived by the method of natural projector and is illustrated by the explicit computation for the exactly solvable McKean kinetic model [285]. It is demonstrated that the result is identical, on the one hand, to the sum of the Chapman–Enskog expansion, and, on the other hand, to the exact solution of the invariance equation.

In Chap. 12 the general geometrical framework of nonequilibrium thermodynamics is developed. It is the generalization of the method of natural projector (Chap. 11) to large steps in time. The notion of *macroscopically definable ensembles* is introduced. The thesis about macroscopically definable ensembles is suggested. This thesis should play the same role in the nonequilibrium thermodynamics, as the Church–Turing thesis in the theory of computability. The *primitive macroscopically definable ensembles* are described. These are ensembles with macroscopically prepared initial states.

The method for computing trajectories of primitive macroscopically definable nonequilibrium ensembles is elaborated. These trajectories are represented as sequences of deformed equilibrium ensembles and simple quadratic models between them. The primitive macroscopically definable ensembles form a manifold in the space of ensembles. We call this manifold the *film of nonequilibrium states*. The equation for the film and the equation for the ensemble motion on the film are written down. The notion of the invariant film of non-equilibrium states, and the method of its approximate construction transform the problem of nonequilibrium kinetics into a series of problems of equilibrium statistical physics. The developed methods allow us to solve the problem of macro-kinetics even when there are no autonomous equations of macro-kinetics.

The slow invariant manifold for a closed system has been found. What next? Chapter 13 gives the answer to this question. The theory of invariant manifolds is developed for weakly open systems. In the first example the method of invariant manifold for driven systems is developed for a derivation of a reduced description in kinetic equations of dilute polymeric solutions. The method applies to any models of polymers and is consistent with basic physical requirements: frame invariance and dissipativity of resulting constitutive equation. It is demonstrated that this reduced description becomes universal in the limit of small Deborah and Weissenberg numbers, and it is represented by the *revised Oldroyd 8 constants constitutive equation* for the polymeric stress tensor. This equation differs from the classical Oldroyd 8 constants constitutive equation by one additional term. Coefficients of this constitutive equation are expressed in terms of the microscopic parameters of the polymer model. A systematic procedure of corrections to the revised Oldroyd 8 constants equations is developed. Results are tested with simple flows.

In the second example in this chapter the derivation of macroscopic equations from the simplest dumbbell models is revisited. It is demonstrated that the onset of the macroscopic description is sensitive to the flows. For the FENE-P model it is shown that there is a possibility of "explosion" of the Gaussian manifold: with a small initial deviation, solution of the kinetic equation very quickly deviate from the manifold, and then slowly come back to the stationary point located on the Gaussian manifold. Nevertheless, the Gaussian manifold remains invariant. Some consequences of these observations are discussed. A new class of closures is introduced, the kinetic multipeak polyhedra. Distributions of this type are expected in kinetic models with multidimensional instability as universally, as the Gaussian distribution appears for stable systems. The number of possible relatively stable states of a nonequilibrium system grows as 2^m, and the number of macroscopic parameters is of the order mn, where n is the dimension of configuration space, and m is the number of independent unstable directions in this space. The elaborated class of closures and equations pretends to describe the effects of so-called "molecular individualism".

How can we prove that all the attractors of a infinite-dimensional system belong to a finite-dimensional manifold? How can we estimate the dimension of this manifold? There are two methods for such estimations, discussed in Chap. 14. First, if we find that *k-dimensional volumes are contracted* due to dynamics, then (after some additional technical steps concerning existence of the positively–invariant bounded set and uniformity of the k-volume contraction on this set) we can state that the *Hausdorff dimension* of the maximal attractor is less, then k. Second, if we find the representation of our system as a nonlinear kinetic system with *conservation of supports* of distributions, then (again, after some additional technical steps) we can state that the asymptotics is finite-dimensional. This conservation of support has a *quasi-biological interpretation, the inheritance* (if a gene is not presented in an isolated population without mutations, then it cannot appear in time). The finite-dimensional asymptotics demonstrates the effects of *"natural" selection.*

The post-processing (Chap. 15) is a very simple, but attractive idea. In the method of invariant manifold we improve the whole manifold on each iteration. If we need only one or several solutions, this whole manifold may be too big for our goals, and we can restrict our activity by refinement of a given solution: a curve instead of a multi-dimensional manifold. The classical Picard iteration for a solution of a differential equation gives the simplest post-processing. Various forms of post-processing are presented. In the example to this chapter the method which recognizes the onset and breakdown of the macroscopic description in microscopic simulations is presented. The method is based on the invariance of the macroscopic dynamics relative to the microscopic dynamics, and it is demonstrated for a model of dilute polymeric solutions where it decides switching between Direct Brownian Dynamics simulations and integration of constitutive equations.

The list of cited literature is by no means complete although we spent effort in order to reflect at least the main directions of studies related to computations of the invariant manifolds. We think that this list is more or less exhaustive in the second-order approximation.

There are many different roads of reading this book. Chapter 3 is necessary for reading all of the other chapters, as is shown in the flowchart (Fig. 1.3). Here we propose several possible roads. This is not the exhaustive list, and everybody can invent his own road.

The *short formal road*: Chap. 3, Sects.: 4.1, 5.1–5.3, 6.1, 7.1, 9.1, 10.1, 11.1, 13.1–13.4, 15.1. If you are ready to look at the formal ordinary differential equation $\frac{\mathrm{d}x}{\mathrm{d}t} = J(x)$, $x \in U$, and to imagine in this form all the kinetic equations, from the Boltzmann equation to the Fokker–Planck equation, then this formal road is the best way to start. After that, you can choose various examples and chapters. Before reading the examples sections, it may be useful to look through Chap. 2.

The *long formal road*: Chaps. 3, 4, Sects.: 5.1–5.3, 6.1, 7.1, 9.1, 10.1, 11.1, 13.1–13.4, 14.1, 14.2, 15.1.

The *short Boltzmann road*: Chap. 2 (including chemical kinetics), Chap. 3, Sects. 5.1–5.3, 5.5, 6.1–6.3, Chap. 8. This road gives the invariance equation, the Newton method with incomplete linearization for solution of this equation, the theory of Local Maxwellian manifold, and the application of this method to correction of these manifolds. Chapter 8 adds the exact solutions of the reduction problem and the test of the developed methods on these solutions.

The *long Boltzmann road*: Chap. 2 (including chemical kinetics), Chap. 3, Sects. 5.1–5.3, 5.5, 5.6, 5.7, Chaps. 6–8, Sects. 4.1, 9.1, 9.3. Exhaustive reading: everything concerning the Boltzmann equation.

The *nonequilibrium thermodynamic road*: Chap. 2, Chap. 3, Sects. 4.1, 5.1–5.4, 7.1, 9.1, Chaps. 11, 12, 14. This road can be naturally supplemented by some sections from the Boltzmann roads.

The *short Grad road*: Chaps. 2, 3, Sects. 5.1–5.6, 6.1, Chap. 8.

2 The Source of Examples

2.1 The Boltzmann Equation

2.1.1 The Equation

The *Boltzmann equation* is the first and the most celebrated nonlinear kinetic equation introduced by the great Austrian scientist Ludwig Boltzmann in 1872 [111]. This equation describes the dynamics of a moderately rarefied gas, taking into account two processes: the free flight of the particles, and their collisions. In its original version, the Boltzmann equation has been formulated for particles represented by hard spheres. The physical condition of rarefaction means that only pair collisions are taken into account, a mathematical specification of which is given by the *Grad–Boltzmann limit* [200]: If N is the number of particles, and σ is the diameter of the hard sphere, then the Boltzmann equation is expected to hold when N tends to infinity, σ tends to zero, $N\sigma^3$ (the volume occupied by the particles) tends to zero, while $N\sigma^2$ (the total collision cross section) remains constant. The microscopic state of the gas at time t is described by the one-body distribution function $P(\boldsymbol{x}, \boldsymbol{v}, t)$, where $\boldsymbol{x}$ is the position of the center of the particle, and $\boldsymbol{v}$ is the velocity of the particle. The distribution function is the probability density of finding the particle at time t within the infinitesimal phase space volume centered at the phase point $(\boldsymbol{x}, \boldsymbol{v})$. The collision mechanism of two hard spheres is presented by a relation between the velocities of the particles before [$\boldsymbol{v}$ and $\boldsymbol{w}$] and after [$\boldsymbol{v}'$ and $\boldsymbol{w}'$] their impact:

$$\boldsymbol{v}' = \boldsymbol{v} - \boldsymbol{n}(\boldsymbol{n}, \boldsymbol{v} - \boldsymbol{w}) \,,$$
$$\boldsymbol{w}' = \boldsymbol{w} + \boldsymbol{n}(\boldsymbol{n}, \boldsymbol{v} - \boldsymbol{w}) \,,$$

where $\boldsymbol{n}$ is the unit vector along $\boldsymbol{v} - \boldsymbol{v}'$. Transformation of the velocities conserves the total momentum of the pair of colliding particles ($\boldsymbol{v}' + \boldsymbol{w}' = \boldsymbol{v} + \boldsymbol{w}$), and the total kinetic energy ($\boldsymbol{v}'^2 + \boldsymbol{w}'^2 = \boldsymbol{v}^2 + \boldsymbol{w}^2$). The Boltzmann equation reads:

$$\frac{\partial P}{\partial t} + \left(\boldsymbol{v}, \frac{\partial P}{\partial \boldsymbol{x}}\right) = N\sigma^2 \int_{R^3} \int_{B^-} (P(\boldsymbol{x}, \boldsymbol{v}', t)P(\boldsymbol{x}, \boldsymbol{w}', t)$$
$$- P(\boldsymbol{x}, \boldsymbol{v}, t)P(\boldsymbol{x}, \boldsymbol{w}, t)) \mid (\boldsymbol{w} - \boldsymbol{v}, \boldsymbol{n}) \mid \mathrm{d}\boldsymbol{w} \, \mathrm{d}\boldsymbol{n} \,, \quad (2.1)$$

Alexander N. Gorban and Iliya V. Karlin: *Invariant Manifolds for Physical and Chemical Kinetics*, Lect. Notes Phys. **660**, 21–63 (2005)
www.springerlink.com

where integration in w is carried over the whole space R^3, while integration in n is over a hemisphere $B^- = \{n \in S^2 \mid (w-v, n) < 0\}$. This inequality $(w - v, n) < 0$ corresponds to the particles entering the collision. The nonlinear integral operator in the right hand side of (2.1) is nonlocal in the velocity variable, and local in space. The Boltzmann equation for arbitrary hard-core interaction is a generalization of the Boltzmann equation for hard spheres under the proviso that the true infinite-range interaction potential between the particles is cut off at some distance. This generalization amounts to a replacement,

$$\sigma^2 \mid (w - v, n) \mid \, \mathrm{d}n \rightarrow B(\theta, \mid w - v \mid) \, \mathrm{d}\theta \, \mathrm{d}\varepsilon \,, \tag{2.2}$$

where the function B is determined by the interaction potential, and the vector n is identified with two angles, θ and ε. In particular, for potentials proportional to the n-th inverse power of the distance, the function B reads

$$B(\theta, \mid v - w \mid) = \beta(\theta) \mid v - w \mid^{\frac{n-5}{n-1}} \,. \tag{2.3}$$

In the special case $n = 5$, function B is independent of the magnitude of the relative velocity (Maxwell molecules). Maxwell molecules occupy a distinct place in the theory of the Boltzmann equation: they provide exact results. Three most important findings for the Maxwell molecules should be mentioned: (a) The exact spectrum of the linearized Boltzmann collision integral, found by Truesdell and Muncaster [261], (b) Exact transport coefficients found by Maxwell even before the Boltzmann equation was formulated, (c) Exact solutions to the space-free version of the nonlinear Boltzmann equation. Galkin [71] found the general solution to the system of moment equations in a form of a series expansion, Bobylev, Krook and Wu [255, 256, 262] found an exact solution of a particular elegant closed form, and Bobylev demonstrated the complete integrability of this dynamic system [73]. The review of relaxation of spatially uniform dilute gases for several types of interaction models, of exact solutions and related topics was given in [75].

A broad review of the Boltzmann equation and analysis of analytical solutions to kinetic models is presented in the book of Cercignani [112]. A modern account of rigorous results on the Boltzmann equation is given in the book [113]. Proof of the existence theorem for the Boltzmann equation was given by DiPerna and Lions [119].

It is customary to write the Boltzmann equation using another normalization of the distribution function, $f(x, v, t) \, \mathrm{d}x \, \mathrm{d}v$, taken in such a way that the function f is compliant with the definition of the hydrodynamic fields: the mass density ρ, the momentum density ρu, and the energy density e:

$$\int f(x, v, t) m \, \mathrm{d}v = \rho(x, t) \,,$$

$$\int f(x, v, t) m v \, \mathrm{d}v = \rho u(x, t) \,, \tag{2.4}$$

$$\int f(\boldsymbol{x}, \boldsymbol{v}, t) m \frac{v^2}{2} \, d\boldsymbol{v} = e(\boldsymbol{x}, t) \ .$$

Here m is the particle mass.

The Boltzmann equation for the distribution function f reads,

$$\frac{\partial f}{\partial t} + \left(\boldsymbol{v}, \frac{\partial}{\partial \boldsymbol{x}} f \right) = Q(f, f) \ , \tag{2.5}$$

where the nonlinear integral operator at the right hand side is the Boltzmann collision integral,

$$Q = \int_{R^3} \int_{B^-} (f(\boldsymbol{v}') f(\boldsymbol{w}') - f(\boldsymbol{v}) f(\boldsymbol{w})) B(\theta, \boldsymbol{v}) \, d\boldsymbol{w} \, d\theta \, d\varepsilon \ . \tag{2.6}$$

Finally, we mention the following form of the Boltzmann collision integral (sometimes referred to as the *scattering* or the *quasi-chemical* representation),

$$Q = \int W(\boldsymbol{v}, \boldsymbol{w} \mid \boldsymbol{v}', \boldsymbol{w}') [(f(\boldsymbol{v}') f(\boldsymbol{w}') - f(\boldsymbol{v}) f(\boldsymbol{w}))] \, d\boldsymbol{w} \, d\boldsymbol{w}' \, d\boldsymbol{v}' \ , \tag{2.7}$$

where W is a generalized function which is called the probability density of the elementary event,

$$W = w(\boldsymbol{v}, \boldsymbol{w} \mid \boldsymbol{v}', \boldsymbol{w}') \delta(\boldsymbol{v} + \boldsymbol{w} - \boldsymbol{v}' - \boldsymbol{w}') \delta(v^2 + w^2 - v'^2 - w'^2) \ . \tag{2.8}$$

2.1.2 The Basic Properties of the Boltzmann Equation

The generalized function W has the following symmetries:

$$W(\boldsymbol{v}', \boldsymbol{w}' \mid \boldsymbol{v}, \boldsymbol{w}) \equiv W(\boldsymbol{w}', \boldsymbol{v}' \mid \boldsymbol{v}, \boldsymbol{w})$$
$$\equiv W(\boldsymbol{v}', \boldsymbol{w}' \mid \boldsymbol{w}, \boldsymbol{v}) \equiv W(\boldsymbol{v}, \boldsymbol{w} \mid \boldsymbol{v}', \boldsymbol{w}') \ . \tag{2.9}$$

The first two identities reflect the symmetry of the collision process with respect to labeling the particles, whereas the last identity is the celebrated *detailed balance* condition which is underpinned by the time-reversal symmetry of the microscopic (Newton's) equations of motion. The basic properties of the Boltzmann equation are:

1. *Additive invariants of the collision operator:*

$$\int Q(f, f)\{1, \boldsymbol{v}, v^2\} \, d\boldsymbol{v} = 0 \ , \tag{2.10}$$

for any function f, assuming the integrals exist. Equality (2.10) reflects the fact that the number of particles, the three components of particle's momentum, and the particle's energy are conserved in collisions. Conservation laws (2.10) imply that the local hydrodynamic fields (2.4) can change in time only due to redistribution over space.

2. The zero point of the integral ($Q = 0$) satisfies the equation (which is also called the *detailed balance*): For almost all velocities,

$$f(v', x, t)f(w', x, t) = f(v, x, t)f(w, x, t) .$$

3. Boltzmann's *local entropy production inequality*:

$$\sigma(x, t) = -k_B \int Q(f, f) \ln f \, dv \geq 0 , \qquad (2.11)$$

for any function f, assuming integrals exist. The dimensional *Boltzmann's constant* ($k_B \approx 1.3806503 \cdot 10^{-23}$ J/K) in this expression serves for a recalculation of the energy units into absolute temperature units. Moreover, equality holds if $\ln f$ is a linear combination of the additive invariants of collision.

Distribution functions f whose logarithm is a linear combination of additive collision invariants with coefficients dependent on x, are called *local Maxwell distribution functions* f_{LM},

$$f_{LM} = \frac{\rho}{m} \left(\frac{2\pi k_B T}{m} \right)^{-3/2} \exp \left(\frac{-m(v - u)^2}{2k_B T} \right) . \qquad (2.12)$$

Local Maxwellians are parametrized by values of five hydrodynamic variables, ρ, u and T. This parametrization is consistent with the definitions of the hydrodynamic fields (2.4), $\int f_{LM}\{m, mv, mv^2/2\} \, dv = (\rho, \rho u, e)$, provided the relation between the energy and the kinetic temperature T holds, $e = \frac{3\rho}{2m} k_B T$.

4. Boltzmann's H *theorem*: The function

$$S[f] = -k_B \int f \ln f \, dv , \qquad (2.13)$$

is called the *entropy density*[1]. The *local H theorem* for distribution functions independent of space states that the rate of the entropy density increase is equal to the nonnegative entropy production,

$$\frac{dS}{dt} = \sigma \geq 0 . \qquad (2.14)$$

Thus, if no space dependence is considered, the Boltzmann equation describes relaxation to the unique global Maxwellian (whose parameters are fixed by initial conditions), and the entropy density grows monotonically along the solutions. Mathematical specifications of this property has been

---

[1] From the physical point of view the value of the function f can be treated as dimensional quantity, but if one changes the scale and multiplies f by a positive number ν then $S[f]$ transforms into $\nu S[f] + \nu \ln \nu \int f \, dv$. For a closed system the corresponding transformation of the entropy is an inhomogeneous linear transformation with constant coefficients.

initialized by Carleman [259], and many estimations of the entropy growth were obtained over the past two decades. In the case of space-dependent distribution functions, the local entropy density obeys the *entropy balance equation*:

$$\frac{\partial S(\boldsymbol{x},t)}{\partial t} + \left(\frac{\partial}{\partial \boldsymbol{x}}, \boldsymbol{J}_s(\boldsymbol{x},t)\right) = \sigma(\boldsymbol{x},t) \geq 0 , \qquad (2.15)$$

where $\boldsymbol{J}_s$ is the entropy flux, $\boldsymbol{J}_s(\boldsymbol{x},t) = -k_{\mathrm{B}} \int \ln f(\boldsymbol{x},t) \boldsymbol{v} f(\boldsymbol{x},t) \, \mathrm{d}\boldsymbol{v}$. For suitable boundary conditions, such as specularly reflecting or at infinity, the entropy flux gives no contribution to the equation for the *total entropy*, $S_{tot} = \int S(\boldsymbol{x},t) \, \mathrm{d}\boldsymbol{x}$ and its rate of changes is then equal to the nonnegative total entropy production $\sigma_{tot} = \int \sigma(\boldsymbol{x},t) \, \mathrm{d}\boldsymbol{x}$ (the *global $\boldsymbol{H}$ theorem*). For more general boundary conditions which maintain the entropy influx, the global H theorem needs to be modified. A detailed discussion of this question is given by Cercignani [112]. The local Maxwellian is also specified as the maximizer of the Boltzmann entropy function (2.13), subject to fixed hydrodynamic constraints (2.4). For this reason, the local Maxwellian is also termed the local equilibrium distribution function.

2.1.3 Linearized Collision Integral

Linearization of the Boltzmann integral around the local equilibrium results in the linear integral operator,

$$Lh(\boldsymbol{v}) = \int W(\boldsymbol{v}, \boldsymbol{w} \mid \boldsymbol{v}', \boldsymbol{w}') f_{\mathrm{LM}}(\boldsymbol{v}) f_{\mathrm{LM}}(\boldsymbol{w})$$

$$\times \left[\frac{h(\boldsymbol{v}')}{f_{\mathrm{LM}}(\boldsymbol{v}')} + \frac{h(\boldsymbol{w}')}{f_{\mathrm{LM}}(\boldsymbol{w}')} - \frac{h(\boldsymbol{v})}{f_{\mathrm{LM}}(\boldsymbol{v})} - \frac{h(\boldsymbol{w})}{f_{\mathrm{LM}}(\boldsymbol{w})} \right] \mathrm{d}\boldsymbol{w}' \, \mathrm{d}\boldsymbol{v}' \, \mathrm{d}\boldsymbol{w} . \quad (2.16)$$

The *linearized collision integral* is symmetric with respect to the scalar product defined by the second derivative of the entropy functional,

$$\int f_{\mathrm{LM}}^{-1}(\boldsymbol{v}) g(\boldsymbol{v}) Lh(\boldsymbol{v}) \, \mathrm{d}\boldsymbol{v} = \int f_{\mathrm{LM}}^{-1}(\boldsymbol{v}) h(\boldsymbol{v}) Lg(\boldsymbol{v}) \, \mathrm{d}\boldsymbol{v} .$$

The operator L is nonpositive definite,

$$\int f_{\mathrm{LM}}^{-1}(\boldsymbol{v}) h(\boldsymbol{v}) Lh(\boldsymbol{v}) \, \mathrm{d}\boldsymbol{v} \leq 0 ,$$

where equality holds if the function $h f_{\mathrm{LM}}^{-1}$ is a linear combination of collision invariants which characterize the null-space of the operator L. The spectrum of the linearized collision integral is well studied in the case of the small angle cut-off.

2.2 Phenomenology and Quasi-Chemical Representation of the Boltzmann Equation

Boltzmann's original derivation of his collision integral was based on a phenomenological "bookkeeping" of the gain and loss of probability density in the collision process. This derivation postulates that the rate of gain G^+ equals

$$G^+ = \int W^+(\boldsymbol{v}, \boldsymbol{w} \mid \boldsymbol{v}', \boldsymbol{w}') f(\boldsymbol{v}') f(\boldsymbol{w}') \, \mathrm{d}\boldsymbol{v}' \, \mathrm{d}\boldsymbol{w}' \, \mathrm{d}\boldsymbol{w} \, ,$$

while the rate of loss L^- is

$$L^- = \int W^-(\boldsymbol{v}, \boldsymbol{w} \mid \boldsymbol{v}', \boldsymbol{w}') f(\boldsymbol{v}) f(\boldsymbol{w}) \, \mathrm{d}\boldsymbol{v}' \, \mathrm{d}\boldsymbol{w}' \, \mathrm{d}\boldsymbol{w} \, .$$

The form of the gain and of the loss, containing products of one-body distribution functions in place of the two-body distribution, constitutes the famous Stosszahlansatz. The Boltzmann collision integral follows now as ($Q = G^+ - L^-$), subject to the detailed balance for the rates of individual collisions,

$$W^+(\boldsymbol{v}, \boldsymbol{w} \mid \boldsymbol{v}', \boldsymbol{w}') = W^-(\boldsymbol{v}, \boldsymbol{w} \mid \boldsymbol{v}', \boldsymbol{w}') \, .$$

This representation $Q = G^+ - L^-$ for interactions different from hard spheres requires also the cut-off of functions β (2.3) at small angles. The gain$-$loss form of the collision integral makes it evident that the detailed balance for the rates of individual collisions is sufficient to prove the local H theorem. A weaker condition which is also sufficient to establish the H theorem was first derived by Stueckelberg [114] (so-called *semi-detailed balance*), and later generalized *to inequalities of concordance* [115]:

$$\int \mathrm{d}\boldsymbol{v}' \int \mathrm{d}\boldsymbol{w}' (W^+(\boldsymbol{v}, \boldsymbol{w} \mid \boldsymbol{v}', \boldsymbol{w}') - W^-(\boldsymbol{v}, \boldsymbol{w} \mid \boldsymbol{v}', \boldsymbol{w}')) \geq 0 \, ,$$

$$\int \mathrm{d}\boldsymbol{v} \int \mathrm{d}\boldsymbol{w} (W^+(\boldsymbol{v}, \boldsymbol{w} \mid \boldsymbol{v}', \boldsymbol{w}') - W^-(\boldsymbol{v}, \boldsymbol{w} \mid \boldsymbol{v}', \boldsymbol{w}')) \leq 0 \, .$$

The semi-detailed balance follows from these expressions if the inequality signs are replaced by equalities.

The pattern of Boltzmann's phenomenological approach is often used to construct nonlinear kinetic models. In particular, nonlinear *equations of chemical kinetics* are based on this idea: If n chemical species A_i participate in a complex chemical reaction,

$$\sum_i \alpha_{si} A_i \leftrightarrow \sum_i \beta_{si} A_i \, ,$$

where α_{si} and β_{si} are nonnegative integers (*stoichiometric coefficients*) then equations of chemical kinetics for the concentrations of species c_j are written

$$\frac{dc_i}{dt} = \sum_{s=1}^{n} (\beta_{si} - \alpha_{si}) \left[\varphi_s^{+} \exp\left(\sum_{j=1}^{n} \frac{\partial G}{\partial c_j} \alpha_{sj} \right) - \varphi_s^{-} \exp\left(\sum_{j=1}^{n} \frac{\partial G}{\partial c_j} \beta_{sj} \right) \right] .$$

Functions φ_s^{+} and φ_s^{-} are interpreted as constants of the forward and reverse reactions, respectively, while the function G is an analog of the Boltzmann's H-function.

Modern derivations of the Boltzmann equation, initialized by the seminal work of Bogoliubov [2], seek a replacement condition for the Stosszahlansatz which would be more closely related to many-particle dynamics. Different conditions has been formulated by Zubarev [195], Lewis [281] and others. The advantage of these formulations is the possibility to systematically find corrections not included in the Stosszahlansatz.

2.3 Kinetic Models

Mathematical complications caused by the nonlinear Boltzmann collision integral are traced back to the Stosszahlansatz. Several approaches were developed in order to simplify the Boltzmann equation. Such simplifications are termed kinetic models. Various kinetic models preserve only certain features of the Boltzmann equation, while sacrificing the rest of them. The best known kinetic model is the nonlinear Bhatnagar–Gross–Krook model (BGK) [116]. The BGK collision integral reads:

$$Q_{\text{BGK}} = -\frac{1}{\tau}(f - f_{\text{LM}}(f)) .$$

The time parameter $\tau > 0$ is interpreted as a characteristic relaxation time to the local Maxwellian. The BGK collision integral is a nonlinear operator: The parameters of the local Maxwellian (ρ, $\boldsymbol{u}$ and T, see (2.12)) are the values of the corresponding moments of the distribution function f. This nonlinearly is of "lower dimension" than in the Boltzmann collision integral because $f_{\text{LM}}(f)$ is a nonlinear function of only the moments of f whereas the Boltzmann collision integral is nonlinear in the distribution function f itself. This type of simplification introduced by the BGK approach is closely related to the family of the so-called mean-field approximations in statistical mechanics.

By its construction, the BGK collision integral preserves the following three properties of the Boltzmann equation: additive invariants of collision, uniqueness of the equilibrium, and the H theorem.

A class of kinetic models which generalized the BGK model to quasi-equilibrium approximations of a general form is described as follows: The quasiequilibrium f^* for the set of linear functionals $M(f)$ is a distribution function $f^*(M)(\boldsymbol{x}, \boldsymbol{v})$ which maximizes the entropy under fixed values of the functionals M. The quasiequilibrium (QE) models are characterized by the collision integral [117],

$$Q_{\mathrm{QE}}(f) = -\frac{1}{\tau}[f - f^*(M(f))] + Q(f^*(M(f)), f^*(M(f))) \,. \qquad (2.17)$$

The first term in (2.17) describes the relaxation *to* the quasiequiulibrium manifold $\{f^*(M)(\boldsymbol{x}, \boldsymbol{v})\}$ (parametrized by the values of the moments M), and the second term is the quasiequilibrium approximation for the Boltzmann collision integral, that is, the value of the Boltzmann collision integral *on* the quasiequilibrium distribution. If the set of moment M is ρ, $\boldsymbol{u}$ and T then the quasiequilibrium model (2.17) turns into the BGK model (2.17)

Same as in the case of the BGK collision integral, operator Q_{QE} is nonlinear in the moments M only. The QE models preserve the following properties of the Boltzmann collision operator: additive invariants, uniqueness of the equilibrium, and the H theorem, provided the relaxation time τ to the quasiequilibrium is sufficiently small [117].

A different nonlinear model was proposed by Lebowitz, Frisch and Helfand [118]:

$$Q_D = D\left(\frac{\partial}{\partial \boldsymbol{v}}\frac{\partial}{\partial \boldsymbol{v}}f + \frac{m}{k_{\mathrm{B}}T}\frac{\partial}{\partial \boldsymbol{v}}(\boldsymbol{v} - \boldsymbol{u}(f))f\right) \,.$$

The collision integral has the form of the self-consistent Fokker–Planck operator, describing diffusion (in the velocity space) in the self-consistent potential. Diffusion coefficient $D > 0$ may depend on the distribution function f. Operator Q_D preserves the same properties of the Boltzmann collision operator as the BGK model.

The kinetic BGK model has been used to obtain exact solutions of gasdynamic problems, especially for stationary problems. The linearized BGK collision model has been extended to model more precisely the linearized Boltzmann collision integral [112].

2.4 Methods of Reduced Description

One of the major issues raised by the Boltzmann equation is the problem of the reduced description. The equations of hydrodynamics constitute a closed set of equations for the hydrodynamic fields (local density, local momentum, and local temperature). From the standpoint of the Boltzmann equation, these quantities are low-order moments of the one-body distribution function, or, in other words, macroscopic variables. The problem of the reduced description consists in the following questions:

1. What are the conditions under which the macroscopic description is valid?
2. What macroscopic variables are relevant for this description?

3. How can we derive equations for the macroscopic variables from the kinetic equations?

The classical methods of reduced description for the Boltzmann equation are the Hilbert method, the Chapman–Enskog method, and the Grad moment method.

2.4.1 The Hilbert Method

In 1911, David Hilbert introduced the notion of normal solutions,

$$f_H(v,\, n(x,t),\, u(x,t),\, T(x,t))\,,$$

that is, solutions to the Boltzmann equation which depend on space and time only through five hydrodynamic fields [16]

$$\int f(x,v,t)\,\mathrm{d}v = n(x,t),\quad \int v f(x,v,t)\,\mathrm{d}v = n(x,t)u(x,t)\,,$$

$$\int \frac{mv^2}{2} f(x,v,t)\,\mathrm{d}v = \frac{3}{2}n(x,t)k_B T\,.$$

The normal solutions are found from a singularly perturbed Boltzmann equation,

$$D_t f = \frac{1}{\varepsilon}Q(f,f)\,, \tag{2.18}$$

where ε is a small parameter, and

$$D_t f \equiv \frac{\partial}{\partial t}f + \left(v, \frac{\partial}{\partial x}\right)f\,.$$

Physically, parameter ε corresponds to the Knudsen number, the ratio between the mean free path of the molecules between collisions, and the characteristic scale of variation of the hydrodynamic fields. In the Hilbert method, one seeks functions $n(x,t)$, $u(x,t)$, $T(x,t)$, such that the normal solution in the form of the Hilbert expansion,

$$f_H = \sum_{i=0}^{\infty} \varepsilon^i f_H^{(i)} \tag{2.19}$$

satisfies (2.18) order by order. Hilbert was able to demonstrate that this is formally possible. Substituting (2.19) into (2.18), and matching various order in ε, we obtain the sequence of integral equations

$$Q(f_H^{(0)}, f_H^{(0)}) = 0\,, \tag{2.20}$$

$$L f_H^{(1)} = D_t f_H^{(0)}\,, \tag{2.21}$$

$$L f_H^{(2)} = D_t f_H^{(1)} - 2Q(f_H^{(0)}, f_H^{(1)})\,, \tag{2.22}$$

and so on for higher orders. Here L is the linearized collision integral. From (2.20), it follows that $f_{\mathrm{H}}^{(0)}$ is the local Maxwellian with parameters not yet determined. The Fredholm alternative, as applied to (2.21) results in:

(a) Solvability condition,

$$\int D_t f_{\mathrm{H}}^{(0)} \{1, \boldsymbol{v}, v^2\} \, d\boldsymbol{v} = 0 \,,$$

which is the set of the compressible Euler equations of the non-viscous hydrodynamics. The solution of the Euler equation determines the parameters of the Maxwellian f_{H}^0.

(b) General solution $f_{\mathrm{H}}^{(1)} = f_{\mathrm{H}}^{(1)1} + f_{\mathrm{H}}^{(1)2}$, where $f_{\mathrm{H}}^{(1)1}$ is the special solution to the linear integral equation (2.21), and $f_{\mathrm{H}}^{(1)2}$ is a yet undetermined linear combination of the additive invariants of collision.

(c) Solvability condition to the next equation (2.22) determines coefficients of the function $f_{\mathrm{H}}^{(1)2}$ in terms of solutions to linear hyperbolic differential equations,

$$\int D_t (f_{\mathrm{H}}^{(1)1} + f_{\mathrm{H}}^{(1)2}) \{1, \boldsymbol{v}, v^2\} \, d\boldsymbol{v} = 0 \,.$$

Hilbert was able to demonstrate that this procedure of constructing the normal solution can be carried out to arbitrary order n, where the function $f_{\mathrm{H}}^{(n)}$ is determined from the solvability condition at the next, $(n+1)$-th order. In order to summarize, implementation of the Hilbert method requires solutions for the functions $n(\boldsymbol{x}, t)$, $\boldsymbol{u}(\boldsymbol{x}, t)$, and $T(\boldsymbol{x}, t)$ obtained from a sequence of partial differential equations.

2.4.2 The Chapman–Enskog Method

A completely different approach to the reduced description was invented in 1917 by David Enskog [120], and independently by Sidney Chapman [70]. The key idea was to seek an expansion of the time derivatives of the hydrodynamic variables rather than seeking the time-space dependence of these functions, as in the Hilbert method.

The Chapman–Enskog method starts also with the singularly perturbed Boltzmann equation, and with the expansion

$$f_{\mathrm{CE}} = \sum_{n=0}^{\infty} \varepsilon^n f_{\mathrm{CE}}^{(n)} \,.$$

However, the procedure of evaluation of the functions $f_{\mathrm{CE}}^{(n)}$ differs from the Hilbert method:

$$Q(f_{\mathrm{CE}}^{(0)}, f_{\mathrm{CE}}^{(0)}) = 0 \,, \tag{2.23}$$

$$L f_{\mathrm{CE}}^{(1)} = -Q(f_{\mathrm{CE}}^{(0)}, f_{\mathrm{CE}}^{(0)}) + \frac{\partial^{(0)}}{\partial t} f_{\mathrm{CE}}^{(0)} + \left(\boldsymbol{v}, \frac{\partial}{\partial \boldsymbol{x}}\right) f_{\mathrm{CE}}^{(0)} \,. \tag{2.24}$$

The operator $\partial^{(0)}/\partial t$ is defined from the expansion of the right hand side of the hydrodynamic equations,

$$\frac{\partial^{(0)}}{\partial t}\{\rho, \rho\boldsymbol{u}, e\} \equiv -\int \left\{m, m\boldsymbol{v}, \frac{mv^2}{2}\right\}\left(\boldsymbol{v}, \frac{\partial}{\partial \boldsymbol{x}}\right) f_{\mathrm{CE}}^{(0)}\, d\boldsymbol{v}\ . \qquad (2.25)$$

From (2.23), function $f_{\mathrm{CE}}^{(0)}$ is again the local Maxwellian, whereas (2.25) are the Euler equations, and $\partial^{(0)}/\partial t$ acts on various functions $g(\rho, \rho\boldsymbol{u}, e)$ according to the chain rule,

$$\frac{\partial^{(0)}}{\partial t}g = \frac{\partial g}{\partial \rho}\frac{\partial^{(0)}}{\partial t}\rho + \frac{\partial g}{\partial(\rho\boldsymbol{u})}\frac{\partial^{(0)}}{\partial t}(\rho\boldsymbol{u}) + \frac{\partial g}{\partial e}\frac{\partial^{(0)}}{\partial t}e\ ,$$

while the time derivatives $\frac{\partial^{(0)}}{\partial t}$ of the hydrodynamic fields are expressed using the right hand side of (2.25).

The result of the Chapman–Enskog definition of the time derivative $\frac{\partial^{(0)}}{\partial t}$, is that the Fredholm alternative is satisfied by the right hand side of (2.24). Finally, the solution to the homogeneous equation is set to zero by the requirement that the hydrodynamic variables as defined by the function $f^{(0)} + \varepsilon f^{(1)}$ coincide with the parameters of the local Maxwellian $f^{(0)}$:

$$\int \{1, \boldsymbol{v}, v^2\} f_{\mathrm{CE}}^{(1)}\, d\boldsymbol{v} = 0\ .$$

The first correction $f_{\mathrm{CE}}^{(1)}$ of the Chapman–Enskog method adds the terms

$$\frac{\partial^{(1)}}{\partial t}\{\rho, \rho\boldsymbol{u}, e\} = -\int \left\{m, m\boldsymbol{v}, \frac{mv^2}{2}\right\}\left(\boldsymbol{v}, \frac{\partial}{\partial \boldsymbol{x}}\right) f_{\mathrm{CE}}^{(1)}\, d\boldsymbol{v}$$

to the time derivatives of the hydrodynamic fields. These terms correspond to the dissipative hydrodynamics where viscous momentum transfer and heat transfer are in the Navier–Stokes and Fourier form. The Chapman–Enskog method was the first true success of the Boltzmann equation since it made it possible to derive macroscopic equations without a priori guessing (the generalization of the Boltzmann equation onto mixtures predicted existence of the thermodiffusion before it has been found experimentally), and to express transport coefficients in terms of microscopic particles interaction.[2]

However, higher-order corrections of the Chapman–Enskog method, resulting in hydrodynamic equations with higher derivatives (Burnett hydrodynamic equations) face severe difficulties both from the theoretical, as well as from the practical point of view. In particular, they result in unphysical instabilities of the equilibrium.

---

[2] For all of the reduction methods many properties of the gas, from the characteristics of the velocity distribution function to the transport coefficients, may be expressed in terms of functions of the collision integral (kinetic integrals). Although the evaluation of these functions is conceptually straightforward, technically it is frequently rather cumbersome. Now the methods for the analytical evaluation of kinetic integrals using computer algebra are developed [121].

2.4.3 The Grad Moment Method

In 1949, Harold Grad extended the basic assumption of the Hilbert and the Chapman–Enskog methods (the space and time dependence of normal solutions is mediated by the five hydrodynamic moments) [201]. A physical rationale behind the Grad moment method is an assumption of the decomposition of motions:

1. During the time of order τ, a set of distinguished moments M' (which include the hydrodynamic moments and a subset of higher-order moments) does not change significantly in comparison to the rest of the moments M'' (the fast dynamics).
2. Towards the end of the fast evolution, the values of the moments M'' become unambiguously determined by the values of the distinguished moments M'.
3. On the time of order $\theta \gg \tau$, dynamics of the distribution function is determined by the dynamics of the distinguished moments while the rest of the moments remain to be determined by the distinguished moments (the slow evolution period).

Implementation of this picture requires an ansatz for the distribution function in order to represent the set of states visited in the course of the slow evolution. In Grad's method, these representative sets are finite-order truncations of an expansion of the distribution functions in terms of Hermite velocity tensors:

$$f_G(M', v) = f_{LM}(\rho, u, e, v) \left[1 + \sum_{(\alpha)}^{N} a_{(\alpha)}(M') H_{(\alpha)}(v - u) \right] , \quad (2.26)$$

where $H_{(\alpha)}(v - u)$ are Hermite tensor polynomials, orthogonal with the weight f_{LM}, while coefficient $a_{(\alpha)}(M')$ are known functions of the distinguished moments M'. Other moments are assumed to be functions of M': $M'' = M''(f_G(M'))$.

Slow evolution of distinguished moments is found upon substitution of (2.26) into the Boltzmann equation and finding the moments of the resulting expression (*Grad's moment equations*). Following Grad, this very simple approximation can be improved by extending the list of distinguished moments. The best known is Grad's thirteen-moment approximation where the set of distinguished moments consists of the five hydrodynamic moments, the five components of the traceless stress tensor $\sigma_{ij} = \int m[(v_i - u_i)(v_j - u_j) - \delta_{ij}(v - u)^2/3] f \, dv$, and of the three components of the heat flux vector $q_i = \int (v_i - u_i) m(v - u)^2/2 f \, dv$.

The decomposition of motions hypothesis cannot be evaluated for its validity within the framework of Grad's approach. It is not surprising therefore that Grad's methods failed to work in situations where it was (unmotivatedly) supposed to, primarily, in phenomena with sharp time-space dependence such as the strong shock waves. On the other hand, Grad's method was quite successful for describing transition between parabolic and hyperbolic propagation, in particular, the second sound effect in massive solids at low temperatures, and, in general, situations slightly deviating from the classical Navier–Stokes–Fourier domain. Finally, the Grad method has been important background for the development of phenomenological nonequilibrium thermodynamics based on a hyperbolic first-order equation, the so-called EIT (extended irreversible thermodynamics [235, 236]).

2.4.4 Special Approximations

Special approximations to the solutions of the Boltzmann equation were found for several problems, which perform better than the results of "regular" procedures. The best known is the Tamm–Mott-Smith ansatz introduced independently by Mott-Smith and Tamm for the strong shock wave problem: The (stationary) distribution function is represented as

$$f_{\mathrm{TMS}}(a(x)) = (1 - a(x))f_+ + a(x)f_- , \qquad (2.27)$$

where $f_\pm$ are upstream and downstream Maxwell distribution functions, and $a(x)$ is an undetermined scalar function of the coordinate along the shock tube.

Equation for the function $a(x)$ is obtained upon substitution of (2.27) into the Boltzmann equation, and integration with some velocity-dependent function $\varphi(\boldsymbol{v})$. Two general problems arise with the special approximation thus constructed: which function $\varphi(\boldsymbol{v})$ should be taken, and how to find a correction to an ansatz like (2.27)?

2.4.5 The Method of Invariant Manifold

The general problem of reduced description for dissipative system was recognized as the problem of finding stable invariant manifolds in the space of distribution functions [9, 11, 12, 14]. The notion of invariant manifold generalizes the normal solution in the Hilbert and in the Chapman–Enskog method, and the finite-moment sets of distribution function in the Grad method: If Ω is a smooth manifold in the space of distribution functions, and if f_Ω is an element of Ω, then Ω is invariant with respect to the dynamic system,

$$\frac{\mathrm{d}f}{\mathrm{d}t} = J(f) , \qquad (2.28)$$
$$\text{if } J(f_\Omega) \in T_{f_\Omega}\Omega, \text{ for all } f_\Omega \in \Omega , \qquad (2.29)$$

where $T_{f_\Omega}\Omega$ is the tangent space of the manifold Ω at the point f_Ω. Application of the invariant manifold idea to dissipative systems is based on iterations, progressively improving the initial approximation, and it involves the following steps: construction of the thermodynamic projector and iterations for the invariance condition

Thermodynamic Projector

Given a manifold Ω (not obligatory invariant), the macroscopic dynamics on this manifold is defined by the *macroscopic vector field*, which is the result of a projection of vectors $J(f_\Omega)$ onto the tangent bundle $T\Omega$. The thermodynamic projector $P^*_{f_\Omega}$ takes advantage of dissipativity:

$$\ker P^*_{f_\Omega} \subseteq \ker D_f S \,|_{f_\Omega} \,, \tag{2.30}$$

where $D_f S \,|_{f_\Omega}$ is the differential of the entropy evaluated in f_Ω.

This condition of thermodynamicity means that the projector $P^*_{f_\Omega}$ determines a decomposition of motion near Ω: $f_\Omega + \ker P^*_{f_\Omega}$ is the plane of fast motion, and $\operatorname{im} P^*_{f_\Omega}$ is the tangent space to f_Ω, we assume that the motion along Ω is slow. Each state of the manifold Ω can be considered as the result of the fast relaxation. During the fast motion the entropy should grow. Hence, the state f_Ω is the maximum entropy state on the plain of fast motions $f_\Omega + \ker P^*_{f_\Omega}$.

The condition of thermodynamicity (2.30) does not define the projector completely; rather, it is the condition that should be satisfied by any projector used to define the macroscopic vector field, $J'_\Omega = P^*_{f_\Omega} J(f_\Omega)$. For, once the condition (2.30) is met, the macroscopic vector field preserves dissipativity of the original microscopic vector field $J(f)$:

$$D_f S \,|_{f_\Omega} \cdot P^*_{f_\Omega}(J(f_\Omega)) \geq 0 \text{ for all } f_\Omega \in \Omega \,. \tag{2.31}$$

Nevertheless, the thermodynamic projector is uniquely defined by the requirement dissipativity preservation (2.31) for *all* the dissipative vector field with the given entropy (see Chap. 5 and [10]).

The thermodynamic projector is the formalization of the assumption that Ω is the manifold of slow motion: If a fast relaxation takes place at least in a neighborhood of Ω, then the states visited in this process before arriving at f_Ω belong to $\ker P^*_{f_\Omega}$. In general, $P^*_{f_\Omega}$ depends in a non-trivial way on f_Ω.

Iterations for the Invariance Condition

The invariance condition for the manifold Ω reads,

$$P_\Omega(J(f_\Omega)) - J(f_\Omega) = 0 \,,$$

here P_Ω is arbitrary (not obligatory thermodynamic) projector onto the tangent bundle of Ω. The invariance condition is considered as an equation which is solved iteratively, starting with an initial approximation Ω_0. On the

$(n + 1)$−st iteration, the correction $f^{(n+1)} = f^{(n)} + \delta f^{(n+1)}$ is found from linear equations,

$$D_f J_n^* \delta f^{(n+1)} = P_n^* J(f^{(n)}) - J(f^{(n)}) ,$$
$$P_n^* \delta f^{(n+1)} = 0 , \tag{2.32}$$

where $D_f J_n^*$ is the linear self-adjoint operator with respect to the scalar product by the second differential of the entropy $D_f^2 S \mid_{f^{(n)}}$.

Together with the above-mentioned principle of thermodynamic projection, the *self-adjoint linearization* implements the assumption about the decomposition of motions around the n'th approximation. The self-adjoint linearization of the Boltzmann collision integral Q (2.7) around a distribution function f is given by the formula,

$$D_f Q^{\mathrm{SYM}} \delta f = \int W(\boldsymbol{v}, \boldsymbol{w}, \mid \boldsymbol{v}', \boldsymbol{w}') \frac{f(\boldsymbol{v})f(\boldsymbol{w}) + f(\boldsymbol{v}')f(\boldsymbol{w}')}{2}$$
$$\times \left[\frac{\delta f(\boldsymbol{v}')}{f(\boldsymbol{v}')} + \frac{\delta f(\boldsymbol{w}')}{f(\boldsymbol{w}')} - \frac{\delta f(\boldsymbol{v})}{f(\boldsymbol{v})} - \frac{\delta f(\boldsymbol{w})}{f(\boldsymbol{w})} \right] \mathrm{d}\boldsymbol{w}' \, \mathrm{d}\boldsymbol{v}' \, \mathrm{d}\boldsymbol{w} .$$
$$\tag{2.33}$$

If $f = f_{\mathrm{LM}}$, the self-adjoint operator (2.33) becomes the linearized collision integral.

The method of invariant manifold is the iterative process:

$$(f^{(n)}, P_n^*) \to (f^{(n+1)}, P_n^*) \to (f^{(n+1)}, P_{n+1}^*)$$

On the each first step of the iteration, the linear equation (2.32) is solved with the projector known from the previous iteration. On the each second step, the projector is updated, following the thermodynamic construction. The method of invariant manifold can be further simplified if smallness parameters are known.

2.4.6 Quasiequilibrium Approximations

Important generalization of the Grad moment method is the concept of the *quasiequilibrium approximations* already mentioned above (we discuss this approximation in detail in Chap. 5). The quasiequilibrium distribution function for a set of distinguished moments $M = m(f)$ maximizes the entropy density S for fixed M. The quasiequilibrium manifold $\Omega^*(M)$ is the collection of the quasiequilibrium distribution functions for all admissible values of M. The quasiequilibrium approximation is the simplest and extremely useful (not only in the kinetic theory itself) implementation of the hypothesis about a decomposition of motions: If M are considered as slow variables, then states which could be visited in the course of rapid motion in the vicinity of $\Omega^*(M)$ belong to the planes

$$\Gamma_M = \{f \mid m(f - f^*(M)) = 0\} \ .$$

In that respect, the thermodynamic construction in the method of invariant manifold is a generalization of the quasiequilibrium approximation where the given manifold is equipped with a quasiequilibrium structure by choosing appropriately the macroscopic variables of the slow motion. In contrast to the quasiequilibrium, the macroscopic variables thus constructed are not obligatory moments. A textbook example of the quasiequilibrium approximation is the generalized Gaussian function for $M = \{\rho, \rho\boldsymbol{u}, P\}$, where $P_{ij} = \int v_i v_j f \, d\boldsymbol{v}$ is the pressure tensor.

The thermodynamic projector P^* for a quasiequilibrium approximation was first introduced by B. Robertson [126] (in a different context of conservative dynamics and for a special case of the Gibbs–Shannon entropy). It acts on a function Ψ as follows

$$P_M^* \Psi = \sum_i \frac{\partial f^*}{\partial M_i} \int m_i \Psi \, d\boldsymbol{v} \ ,$$

where $M = \int m_i f \, d\boldsymbol{v}$. The quasiequilibrium approximation does not exist if the highest order moment is an odd-order polynomial of velocity (therefore, there exists no quasiequilibrium for thirteen Grad's moments), and a regularization is then required. Otherwise, the Grad moment approximation is the first-order expansion of the quasiequilibrium around the local Maxwellian.

2.5 Discrete Velocity Models

If the number of microscopic velocities is reduced drastically to only a finite set, the resulting discrete velocity models, continuous in time and in space, can still mimic gas-dynamic flows. This idea was introduced in Broadwell's paper in 1963 to mimic the strong shock wave [122].

Further important development of this idea was due to Cabannes and Gatignol in the seventies who introduced a systematic class of discrete velocity models [129]. The structure of the collision operators in the discrete velocity models mimics the polynomial character of the Boltzmann collision integral. Discrete velocity models are implemented numerically by using the natural operator splitting in which each update due to free flight is followed by the collision update, the idea which dates back to Grad. One of the most important recent results is the proof of convergence of the discrete velocity models with pair collisions to the Boltzmann collision integral [124].

2.6 Direct Simulation

Besides the analytical approach, direct numerical simulation of Boltzmann-type nonlinear kinetic equations have been developed since the middle of

1960's, beginning with the seminal works of Bird [127, 128]. The basis of the approach is a representation of the Boltzmann gas by a set of particles whose dynamics is modeled as a sequence of free propagation and collisions. The modeling of collisions uses a random choice of pairs of particles inside the cells of the space, and changing the velocities of these pairs in such a way as to comply with the conservation laws, and in accordance with the kernel of the Boltzmann collision integral. At present, there exists a variety of models based on this scheme known as the Direct Simulation Monte-Carlo method (DSMC) [127, 128]. The DSMC, in particular, provides data to test various analytical theories.

2.7 Lattice Gas and Lattice Boltzmann Models

Since the mid 1980's, the kinetic-theory based approach to simulate complex macroscopic phenomena such as hydrodynamics has been developed. The main idea of the approach is the construction of a minimal kinetic system in such a way that their long-time and large-scale limit matches the desired macroscopic equations. For this purpose, the fully discrete (in time, space, and velocity) nonlinear kinetic equations are considered on sufficiently isotropic lattices, where the links represent the discrete velocities of fictitious particles. In the earlier version of the lattice methods, the particle–based picture has been exploited. These models obey the exclusion rule (one or zero particle per lattice link) (the lattice gas model [130]). Most of the present versions use the distribution function picture, where populations of the links are non-integer (the lattice Boltzmann model [131–135]). Discrete-time dynamics consists of a propagation step where populations are transmitted to adjacent links and collision step where populations of the links at each node of the lattice are equilibrated according a certain simple rule. Many of present versions use the BGK-type equilibration, where the local equilibrium is constructed in such a way as to match desired macroscopic equations. The lattice Boltzmann method is a useful approach for computational fluid dynamics, effectively compliant with parallel architectures. The proof of the H theorem for the Lattice gas models is based on the semi-detailed (or Stueckelberg's) balance principle. The proof of the H theorem in the framework of the lattice Boltzmann method has only very recently been achieved [136–141] (see below).

2.7.1 Discrete Velocity Models for Hydrodynamics

We start with a generic discrete velocity kinetic model. Let $f_i(\boldsymbol{x}, t)$ be the population of D-dimensional discrete velocities $\boldsymbol{c}_i$, $i = 1, \ldots, n_{\mathrm{d}}$, at position $\boldsymbol{x}$ and time t. The hydrodynamic fields are the first few moments of the populations, namely

$$\sum_{i=1}^{n_{\mathrm{d}}} \{1, \, \boldsymbol{c}_i, \, c_i^2\} f_i = \{\rho, \, \rho\boldsymbol{u}, \, \rho DT + \rho u^2\} \, , \tag{2.34}$$

where ρ is the mass density of the fluid, $\rho\boldsymbol{u}$ is the D-dimensional momentum density vector, and $e = \rho DT + \rho u^2$ is the energy density. Below, the index $\alpha = 1, \ldots, D$, denotes the spatial components. In the case of athermal hydrodynamics, the set of independent hydrodynamic fields contains only the mass and momentum densities. It is convenient to introduce n_{d}-dimensional population vectors $\boldsymbol{f}$, and the standard scalar product, $\langle \boldsymbol{f}|\boldsymbol{g}\rangle = \sum_{i=1}^{n_{\mathrm{d}}} x_i y_i$. We will describe here the construction of the discrete velocity models for the incompressible hydrodynamics (the most important field of applications), and will present the results for a weakly compressible case below. So, let the locally conserved fields be density and momentum density,

$$\langle \boldsymbol{1}|\boldsymbol{f}\rangle = \rho, \ \langle \boldsymbol{c}_\alpha|\boldsymbol{f}\rangle = \rho u_\alpha \, . \tag{2.35}$$

Here $\boldsymbol{1} = \{1\}_{i=1}^{n_{\mathrm{d}}}$, $\boldsymbol{v}_\alpha = \{c_{i\alpha}\}_{i=1}^{n_{\mathrm{d}}}$, $\alpha = 1, \ldots, D$. In this case, the construction of the kinetic simulation scheme begins with finding a convex function of populations H (entropy function), which satisfies the following condition: If $\boldsymbol{f}^{\mathrm{eq}}(\rho, \boldsymbol{u})$ (local equilibrium) minimizes H subject to the hydrodynamic constraints (2.35), then $\boldsymbol{f}^{\mathrm{eq}}$ also satisfies certain restrictions on the higher-order moments. For example, the equilibrium stress tensor must respect the Galilean invariance,

$$\sum_{i=1}^{n_{\mathrm{d}}} c_{i\alpha} c_{i\beta} f_i^{\mathrm{eq}}(\rho, \boldsymbol{u}) = \rho c_{\mathrm{s}}^2 \delta_{\alpha\beta} + \rho u_\alpha u_\beta \, . \tag{2.36}$$

Here c_{s} is the speed of sound. The corresponding entropy functions for the athermal and thermal cases are given below (see Table 2.1 and Table 2.2). For the time being, assume the convex function H is fixed.

The next step is to write down the set of kinetic equations,

$$\partial_t f_i + c_{i\alpha}\partial_\alpha f_i = \Delta_i \, . \tag{2.37}$$

Table 2.1. Reconstruction of macroscopic dynamics with the increase of the order of the Hermite polynomial

Order of Polynomial	Independent Variables	Discrete Velocities (1D)	Weights	Target Equation
2	ρ	± 1	$\frac{1}{2}$	Diffusion
3	$\rho, \rho\boldsymbol{u}$	$0, \pm\sqrt{3T_0}$	$\frac{2}{3}, \frac{1}{6}$	Athermal Navier–Stokes, $O(u^2)$
4	$\rho, \rho\boldsymbol{u}, e$	$\pm a, \pm b$	$\frac{T_0}{4a^2}, \frac{T_0}{4b^2}$	Thermal Navier–Stokes, $O(\theta^2)$
				Athermal Navier–Stokes, $O(u^3)$

Table 2.2. Reconstruction of higher-order moments, in comparison to the continuous case. Symbol Δ denotes the difference from the continuous case

	$\Delta P^{\mathrm{eq}}_{\alpha\beta}$	$\Delta Q^{\mathrm{eq}}_{\alpha\beta\gamma}$	$\Delta R^{\mathrm{eq}}_{\alpha\beta}$
Athermal case	$O(u^4)$	$O(u^3)$	
Thermal case*	$O(u^8)$	$O(u\theta^2)$, $O(u^3\theta)$, and $O(u^5)$	$O(\theta^2)$, $O(u^2\theta^2)$, and $O(u^4)$

*$\theta = (T_0 - T)/T_0$ is the deviation of the temperature from the reference value.

For a generic case of n_c locally conserved fields $M_i = \langle \boldsymbol{m}_i | \boldsymbol{f} \rangle$, $i = 1, \ldots, n_c$, $n_c < n_d$, the n_d-dimensional vector function $\boldsymbol{\Delta}$ (collision integral), must satisfy the conditions:

$$\langle \boldsymbol{m}_i | \boldsymbol{\Delta} \rangle = 0 \ \text{(local conservation laws)},$$

$$\sigma = \langle \boldsymbol{\nabla} H | \boldsymbol{\Delta} \rangle \leq 0 \ \text{(entropy production inequality)}.$$

Here $\boldsymbol{\nabla} H$ is the row-vector of partial derivatives $\partial H/\partial f_i$. Moreover, the local equilibrium vector $\boldsymbol{f}^{\mathrm{eq}}$ must be the only zero point of $\boldsymbol{\Delta}$, that is, $\boldsymbol{\Delta}(\boldsymbol{f}^{\mathrm{eq}}) = \boldsymbol{0}$, and, finally, $\boldsymbol{f}^{\mathrm{eq}}$ must be the only zero point of the local entropy production, $\sigma(\boldsymbol{f}^{\mathrm{eq}}) = 0$. Collision integral which satisfies all these requirements is called admissible. Let us discuss several possibilities of constructing admissible collision integrals.

BGK Model

Suppose the entropy function H known. If, in addition, the local equilibrium is also known as an explicit function of locally conserved variables (or some reliable approximation of this function is known), the simplest option is to use the Bhatnagar-Gross-Krook (BGK) model. In the case of athermal hydrodynamics, for example, we write

$$\boldsymbol{\Delta} = -\frac{1}{\tau}(\boldsymbol{f} - \boldsymbol{f}^{\mathrm{eq}}(\rho(\boldsymbol{f}), \boldsymbol{u}(\boldsymbol{f}))). \tag{2.38}$$

The BGK collision operator is sufficient for many applications. However, it becomes advantageous only if the local equilibrium is known in a closed form. In other cases only the entropy function is known but not its minimizer. For those cases one should construct collision integrals based solely on the knowledge of the entropy function. We here present two particular realizations of the collision integral based on the knowledge of the entropy only.

Quasi-Chemical Model

Let $\boldsymbol{m}_1, \ldots, \boldsymbol{m}_{n_c}$ be the n_d-dimensional vectors of locally conserved fields, $M_i = \langle \boldsymbol{m}_i | \boldsymbol{f} \rangle$, $i = 1, \ldots, n_c$, and let $\boldsymbol{g}_s$, $s = 1, \ldots, n_d - n_c$, be a basis of the subspace orthogonal (in the standard scalar product) to vectors of conservation laws. For each vector $\boldsymbol{g}_s$, we define a decomposition $\boldsymbol{g}_s = \boldsymbol{g}_s^+ - \boldsymbol{g}_s^-$, where all components of vectors $\boldsymbol{g}_s^\pm$ are nonnegative, and if $g_{si}^\pm \neq 0$, then $g_{si}^\mp = 0$. Let us consider the collision integral of the form:

$$\boldsymbol{\Delta} = \sum_{s=1}^{n_\mathrm{d}-n_\mathrm{c}} \gamma_s \boldsymbol{g}_s \left\{ \exp\left[\langle \boldsymbol{\nabla} H | \boldsymbol{g}_s^- \rangle\right] - \exp\left[\langle \boldsymbol{\nabla} H | \boldsymbol{g}_s^+ \rangle\right] \right\} . \tag{2.39}$$

Here $\gamma_s > 0$. By the construction, the collision integral (2.39) is admissible. If the entropy function is Boltzmann–like, and the components of vectors $\boldsymbol{g}_s$ are integers, the collision integral assumes the familiar Boltzmann–like (or mass action law) form.

Single Relaxation Time Gradient Model

The BGK collision model (2.38) has the important property: linearization of the operator (2.38) at the local equilibrium point has a very simple spectrum $\{0, -1/\tau\}$, where 0 is the n_c-times degenerated eigenvalue corresponding to the conservation laws, while the eigenvalue $-1/\tau$ corresponds to all the rest of the (kinetic) eigenvectors. Nonlinear collision operators which have this property of their linearizations at equilibrium are called single relaxation time models (SRTM). They play an important role in modelling because they allow for the simplest identification of transport coefficients.

The SRTM, based on the given entropy function H, is constructed as follows (single relaxation time gradient model, SRTGM). For the system with n_c local conservation laws, let $\boldsymbol{e}_s$, $s = 1, \dots, n_\mathrm{d} - n_\mathrm{c}$, be an orthonormal basis in the kinetic subspace, $\langle \boldsymbol{m}_i | \boldsymbol{e}_s \rangle = 0$, and $\langle \boldsymbol{e}_s | \boldsymbol{e}_p \rangle = \delta_{sp}$. Then the single relaxation time gradient model is

$$\boldsymbol{\Delta} = -\frac{1}{\tau} \sum_{s,p=1}^{n_\mathrm{d}-n_\mathrm{c}} \boldsymbol{e}_s K_{sp}(\boldsymbol{f}) \langle \boldsymbol{e}_p | \boldsymbol{\nabla} H \rangle , \tag{2.40}$$

where K_{sp} are elements of a positive definite $(n_\mathrm{d} - n_\mathrm{c}) \times (n_\mathrm{d} - n_\mathrm{c})$ matrix $\mathbf{K}$,

$$\mathbf{K}(\boldsymbol{f}) = \boldsymbol{C}^{-1}(\boldsymbol{f}) , \tag{2.41}$$
$$C_{sp}(\boldsymbol{f}) = \langle \boldsymbol{e}_s | \boldsymbol{\nabla}\boldsymbol{\nabla} H(\boldsymbol{f}) | \boldsymbol{e}_p \rangle .$$

Here $\boldsymbol{\nabla}\boldsymbol{\nabla} H(\boldsymbol{f})$ is the $n_\mathrm{d} \times n_\mathrm{d}$ matrix of second derivatives, $\partial^2 H / \partial f_i \partial f_j$. Linearization of the collision integral at equilibrium results in the form,

$$\boldsymbol{L} = -\frac{1}{\tau} \sum_{s=1}^{n_\mathrm{d}-n_\mathrm{c}} \boldsymbol{e}_s \boldsymbol{e}_s , \tag{2.42}$$

and is obviously single relaxation time. Use of the SRTGM instead of the BGK model results in the equivalent hydrodynamics even when the local equilibrium is not known in a closed form.

H-Functions of Minimal Kinetic Models

The Boltzmann H-function written in terms of the one-particle distribution function $F(\boldsymbol{x}, \boldsymbol{c})$ is $H = \int F \ln F \, d\boldsymbol{c}$, where $\boldsymbol{c}$ is the continuous velocity. Close to the local equilibrium, this integral can be approximated by using the Gauss–Hermite quadrature. This gives the entropy functions of the discrete-velocity models,

$$H_{\{w_i, c_i\}} = \sum_{i=1}^{n_\mathrm{d}} f_i \ln \left(\frac{f_i}{w_i} \right) . \tag{2.43}$$

Here w_i is the weight associated with the i-th discrete velocity $\boldsymbol{c}_i$, while the particles mass and Boltzmann's constant k_B are set equal to one. The discrete-velocity distribution functions (populations) $f_i(\boldsymbol{x})$ are related to the values of the continuous distribution function at the nodes of the quadrature by the formula,

$$f_i(\boldsymbol{x}) = w_i (2\pi T_0)^{(D/2)} \exp(c_i^2/(2T_0)) F(\boldsymbol{x}, \boldsymbol{c}_i) .$$

The discrete-velocity entropy functions (2.43) for various $\{w_i, \boldsymbol{c}_i\}$ is the single input for all the constructions of the minimal kinetic models. The set of discrete velocities corresponds to zeroes of the Hermite polynomials.

As the order of the Hermite polynomials used in the quadrature is increased (this corresponds to increasing the number of discrete velocities), the discrete H-functions $H_{\{w_i, c_i\}}$ (2.43) become a better approximation. Thus, with the increase of the order of the Hermite polynomials, a better approximation to the hydrodynamics is obtained as demonstrated in Table 2.1, where $a = \sqrt{3 - \sqrt{6}}(T_0)^{1/2}$, and $b = \sqrt{3 + \sqrt{6}}(T_0)^{1/2}$ are the absolute values of the roots of the fourth-order Hermite polynomial. In higher dimensions, the discrete velocities are products of the discrete velocities in one dimension, and the weights are constructed by multiplying the weights associated with each component direction.

Athermal Hydrodynamics

If the discrete velocities are formed using the roots of the third-order Hermite polynomials (see Table 2.1), the Navier–Stokes equation is reproduced up to the order $O(u^2)$, and which is sufficient for many hydrodynamic applications.

As the higher-order moments of the local equilibrium are not enforced by the construction, we need to check their behavior. Relevant higher-order moments of the equilibrium distribution, required to reproduce the hydrodynamics in the long-time large-scale limit are the equilibrium pressure tensor, $P_{\alpha\beta}^\mathrm{eq} = \sum_i f_i^\mathrm{eq} c_{i\alpha} c_{i\beta}$, the equilibrium third-order moments, $Q_{\alpha\beta\gamma}^\mathrm{eq} = \sum_i f_i^\mathrm{eq} c_{i\alpha} c_{i\beta} c_{i\gamma}$, and the equilibrium fourth order moment $R_{\alpha\beta}^\mathrm{eq} = \sum_i c_{i\alpha} c_{i\beta} c^2 f_i^\mathrm{eq}$. For the athermal hydrodynamics, only the equilibrium pressure tensor and the equilibrium third-order moments are required to be correctly reproduced in order to recover the Navier–Stokes equations. The deviation of these higher-order moments from the expression for the continuous case is reported in Table 2.2.

2.7.2 Entropic Lattice Boltzmann Method

If the set of discrete velocities forms the links of a Bravais lattice (with possibly several sub-lattices), then the discretization in time and space of the discrete velocity kinetic equations is particularly simple, and it leads to the entropic lattice Boltzmann scheme. This happens in the most important case of the athermal hydrodynamics. The equation of the entropic lattice Boltzmann scheme reads

$$f_i(\boldsymbol{x} + \boldsymbol{v}_i \delta t, t + \delta t) - f_i(\boldsymbol{x}, t) = \beta \alpha(\boldsymbol{f}(\boldsymbol{x}, t)) \Delta_i(\boldsymbol{f}(\boldsymbol{x}, t)) . \tag{2.44}$$

Here δt is the discretization time step, $\beta \in [0, 1]$ is a fixed parameter which matches the viscosity coefficient in the long-time large-scale dynamics of the kinetic scheme (2.44), while the function of the population vector α defines the maximal over-relaxation of the scheme, and is found from the entropy condition,

$$H(\boldsymbol{f}(\boldsymbol{x}, t) + \alpha \boldsymbol{\Delta}(\boldsymbol{f}(\boldsymbol{x}, t))) = H(\boldsymbol{f}(\boldsymbol{x}, t)) . \tag{2.45}$$

The nontrivial root of this equation is found for populations at each lattice site at each moment of discrete time. Equation (2.45) ensures the discrete-time H-theorem, and is required in order to stabilize the scheme if the relaxation parameter β is close to 1. The latter limit is of particular importance in the applications of the entropic lattice Boltzmann method because it corresponds to the vanishing viscosity, and hence to simulations of high Reynolds number flows. The geometrical sense of the over-relaxation is illustrated in Fig. 2.1.

2.7.3 Entropic Lattice BGK Method (ELBGK)

An important further simplifications happens in the case of athermal hydrodynamics when the entropy function is constructed using third-order Hermite polynomials (see Table 2.1). In this case the local equilibrium populations vector can be found in a closed form [141]. This enables the simplest entropic scheme – the entropic lattice BGK model – for simulation of athermal hydrodynamics. We present this model in the dimensionless lattice units.

Let D be the spatial dimension. For $D = 1$, the three discrete velocities are

$$\boldsymbol{c} = \{-1, 0, 1\} . \tag{2.46}$$

For $D > 1$, the discrete velocities are tensor products of the discrete velocities of the one-dimensional velocities (2.46). Thus, we have the 9-velocity model for $D = 2$ and the 27-velocity model for $D = 3$. The H function is Boltzmann-like,

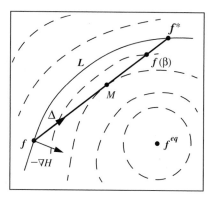

Fig. 2.1.   Entropic stabilization of the lattice Boltzmann scheme with over-relaxation. Curves represent entropy levels, surrounding the local equilibrium $\boldsymbol{f}^{\mathrm{eq}}$. The solid curve L is the entropy level with the value $H(\boldsymbol{f}) = H(\boldsymbol{f}^*)$, where $\boldsymbol{f}$ is the initial, and $\boldsymbol{f}^*$ is the maximally over-relaxed population $\boldsymbol{f} + \alpha \boldsymbol{\Delta}$. The vector $\boldsymbol{\Delta}$ represents the collision integral, the sharp angle between $\boldsymbol{\Delta}$ and the vector $-\boldsymbol{\nabla} H$ reflects the entropy production inequality, while $\boldsymbol{M}$ is the point of minimum of H on the segment between $\boldsymbol{f}$ and $\boldsymbol{f}^*$. The point $\boldsymbol{f}^* \, \boldsymbol{M}$ is the solution to (2.45). The result of the collision update is represented by the point $\boldsymbol{f}(\beta)$. The choice of β shown corresponds to the over-relaxation: $H(\boldsymbol{f}(\beta)) > H(\boldsymbol{M})$ but $H(\boldsymbol{f}(\beta)) < H(\boldsymbol{f})$. The particular case of the BGK collision (not shown) would be represented by a vector $\boldsymbol{\Delta}_{\mathrm{BGK}}$, pointing from $\boldsymbol{f}$ towards $\boldsymbol{f}^{\mathrm{eq}}$, in which case $\boldsymbol{M} = \boldsymbol{f}^{\mathrm{eq}}$.

$$H = \sum_{i=1}^{3^D} f_i \ln \left(\frac{f_i}{w_i} \right) . \tag{2.47}$$

The weights w_i are associated with the each of the ith discrete velocity c_i. For $D = 1$, the three-dimensional vector of the weights corresponding to the velocities (2.46) is

$$\boldsymbol{w} = \left\{ \frac{1}{6}, \frac{2}{3}, \frac{1}{6} \right\} . \tag{2.48}$$

For $D > 1$, the weights are constructed by multiplying the weights associated with each component direction.

The local equilibrium minimizes the H-function (2.43) subject to the fixed density and momentum,

$$\sum_{i=1}^{3^D} f_i = \rho, \quad \sum_{i=1}^{3^D} f_i c_{i\alpha} = \rho u_\alpha, \quad \alpha = 1, \dots, D . \tag{2.49}$$

The explicit solution to this minimization problem reads,

$$f_i^{\mathrm{eq}} = \rho w_i \prod_{\alpha=1}^{D} \left(2 - \sqrt{1 + 3 u_\alpha^2} \right) \left(\frac{2 u_\alpha + \sqrt{1 + 3 u_\alpha^2}}{1 - u_\alpha} \right)^{c_{i\alpha}} . \tag{2.50}$$

Note that the exponent, $c_{i\alpha}$, in (2.50) takes the values ± 1, and 0 only. The speed of sound, c_s, in this model is equal to $1/\sqrt{3}$. The factorization of the local equilibrium (2.50) over spatial components is quite remarkable, and resembles the familiar property of the local Maxwellians.

The entropic lattice BGK model for the local equilibrium (2.50) reads,

$$f_i(\boldsymbol{x} + \boldsymbol{c}_i\delta t, t + \delta t) - f_i(\boldsymbol{x}, t) = -\beta\alpha(f_i(\boldsymbol{x}, t) - f_i^{\text{eq}}(\rho(\boldsymbol{f}(\boldsymbol{x}, t)), \boldsymbol{u}(\boldsymbol{f}(\boldsymbol{x}, t)))) \;. \tag{2.51}$$

The parameter β is related to the relaxation time τ of the BGK model (2.38) by the formula,

$$\beta = \frac{\delta t}{2\tau + \delta t} \;. \tag{2.52}$$

Note that β depends on the discretization interval δt nonlinearly. The value of the over-relaxation parameter α is computed on each lattice site at every time from the entropy estimate,

$$H(\boldsymbol{f} - \alpha(\boldsymbol{f} - \boldsymbol{f}^{\text{eq}}(\boldsymbol{f}))) = H(\boldsymbol{f}) \;. \tag{2.53}$$

In the hydrodynamic limit, the model (2.51) reconstructs the Navier-Stokes equations with the viscosity

$$\mu = \rho c_s^2\tau = \rho c_s^2\delta t \left(\frac{1}{2\beta} - \frac{1}{2}\right) \;. \tag{2.54}$$

The zero-viscosity limit corresponds to $\beta \to 1$. It is the maximal over-relaxation (see Fig. 2.1).

Thermal Hydrodynamics

The minimal entropic kinetic model for the thermal case requires zeroes of fourth-order Hermite polynomials (see Table 2.1). This is an off-lattice model (discrete velocities at zeroes of the fourth-order Hermit polynomials do not form links of any lattice). Therefore, a discretization in space should use other methods familiar from the discretization of hyperbolic equations. However, the theory of the entropy estimate for the discretization in the time presented above is fully applicable in this case too. We here present the local equilibrium of the thermal model.

In order to evaluate Lagrange multipliers in the formal solution to the minimization problem,

$$f_i^{\text{eq}} = w_i \exp\left(A + B_\alpha c_{i\alpha} + C\, c_i^2\right),$$

we note that they can be computed exactly for $\boldsymbol{u} = 0$ and any temperature T within the positivity interval, $a^2 < T < b^2$:

$$B_\alpha = 0, \quad C_0 = \frac{1}{(b^2 - a^2)} \log\left(\frac{w_a (T - a^2)}{w_b (b^2 - T)}\right),$$

$$A_0 = \log\left(\frac{\rho (b^2 - T)^D}{(2w_a)^D (b^2 - a^2)^D}\right) - D a^2 C_0. \tag{2.55}$$

With this, the equilibrium at the zero value of the average velocity and the arbitrary temperature reads

$$f_i^{\text{eq}} = \frac{\rho w_i}{2^D (b^2 - a^2)^D} \prod_{\alpha=1}^{D} \left(\frac{b^2 - T}{w_a}\right)^{\left(\frac{b^2 - c_{i\alpha}^2}{b^2 - a^2}\right)} \left(\frac{T - a^2}{w_b}\right)^{\left(\frac{c_{i\alpha}^2 - a^2}{b^2 - a^2}\right)}. \tag{2.56}$$

The factorization over spatial components is again clearly visible. Once the exact solution for the zero velocity is found, the extension to $\mathbf{u} \neq 0$ is obtained perturbatively. The first few terms of the expansion of the Lagrange multipliers are:

$$A = A_0 - \frac{T}{(T - a^2)(b^2 - T)} u^2 + O(u^4),$$

$$B_\alpha = \frac{u_\alpha}{T} + \frac{(T - T_0)^2}{2DT^4}\left(D u_\beta u_\theta u_\gamma \delta_{\alpha\beta\gamma\theta} - 3u^2 u_\alpha\right) + O(u^5),$$

$$C = C_0 + \frac{a^2(b^2 - T) - T(b^2 - 3T)}{2DT^2(T - a^2)(b^2 - T)} u^2 + O(u^4).$$

For the numerical implementation, the equilibrium distribution function can be calculated analytically up to any order of accuracy required. The accuracy of the relevant higher-order moments in this case is shown in the Table 2.2. Once the errors in these terms are small, the minimal kinetic models reconstruct the full thermal hydrodynamic equations.

While in the athermal case the closeness of the resulting macroscopic equations to the Navier–Stokes equations is controlled solely by the deviations from zero of the average velocity (low Mach number flows), in the thermal regime deviations are also due to variations of the temperature away from the reference value. This means that not only the actual velocity should be much less than the heat velocity, but also that the temperature deviation from T_0 should be small, $|T - T_0|/T_0 \ll 1$. However, by increasing the reference temperature, one gets a wider range of validity of the present model. Another important remark is about the use of the thermal model for the Navier–Stokes equation. If the temperature is fixed at the reference value $T = T_0$, the pressure tensor becomes exact to any purposes of simulation, while the third moment $Q_{\alpha\beta\gamma}^{\text{eq}}$ becomes exact to the order $O(u^5)$.

In the construction of the discrete velocity model, the focus is on achieving a good approximation of the Boltzmann H-function. Thus, one can expect that the correct thermodynamics will be also preserved (within the accuracy of the discretization) even in the discrete case. Indeed, the local equilibrium entropy, $S = -k_{\text{B}} H_{\{w_i, \mathbf{c}\}}(f^{\text{eq}})$, for the thermal model satisfies the usual expression for the entropy of the ideal monatomic gas to the overall order of approximation of the method,

$$S = \rho\, k_{\mathrm{B}} \ln\left(T^{D/2}/\rho\right) + O(u^4, \theta^2)\,. \tag{2.57}$$

2.7.4 Boundary Conditions

The boundary (a solid wall) ∂R is specified at any point $\boldsymbol{x} \in \partial R$ by the inward unit normal $\boldsymbol{n}$, the wall temperature T_{w} and the wall velocity $\boldsymbol{u}_{\mathrm{w}}$. The simplest boundary condition for the minimal kinetic models is obtained upon evaluation of the diffusive wall boundary condition for the Boltzmann equation [112] with the help of the Gauss-Hermite quadrature [142]. The explicit expression for the diffusive wall boundary condition in the discrete velocity models is

$$f_i = \frac{\sum_{\boldsymbol{\xi}_{i'} \cdot \boldsymbol{n}\, <0} |(\boldsymbol{\xi}_{i'} \cdot \boldsymbol{n})| f_{i'}}{\sum_{\boldsymbol{\xi}_{i'} \cdot \boldsymbol{n}\, <0} |(\boldsymbol{\xi}_{i'} \cdot \boldsymbol{n})| f_{i'}^{\mathrm{eq}}(\rho_{\mathrm{w}}, \boldsymbol{u}_{\mathrm{w}})}\, f_i^{\mathrm{eq}}(\rho_{\mathrm{w}}, \boldsymbol{u}_{\mathrm{w}}), \qquad (\boldsymbol{\xi}_i \cdot \boldsymbol{n} > 0)\,, \tag{2.58}$$

Here $\boldsymbol{\xi}_i$ is the discrete velocity in the wall reference frame, $\boldsymbol{\xi}_i = \boldsymbol{c}_i - \boldsymbol{u}_{\mathrm{w}}$. Implementation of the diffusive wall boundary condition (2.58) in the context of the fully discrete entropic lattice Boltzmann method is given in the paper [143].

2.7.5 Numerical Illustrations of the ELBGK

The Kramers problem [112] is a limiting case of the plane Couette flow, where one of the plates is moved to infinity, while keeping a fixed shear rate. The analytical solution for the slip-velocity at the wall calculated for the linearized BGK collision model [112] are compared with the simulation of the entropic lattice BGK model in Fig. 2.2. This shows that the important feature of the original Boltzmann equation, the Knudsen number dependent slip at the wall is retained in the present model.

In another numerical experiment, the ELBGK method was tested in the setup of the two-dimensional Poiseuille flow. The time evolution of the computed profile as compared to the analytical result obtained from the incompressible Navier–Stokes equations is demonstrated in Fig. 2.3.

2.8 Other Kinetic Equations

2.8.1 The Enskog Equation for Hard Spheres

The Enskog equation for hard spheres is an extension of the Boltzmann equation to moderately dense gases. The Enskog equation explicitly takes into

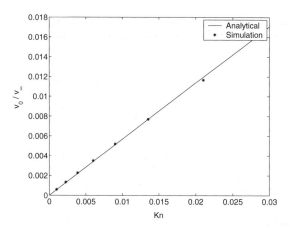

Fig. 2.2. Relative slip at the wall in the simulation of the Kramers problem for shear rate $a = 0.001$, box length $L = 32$, $v_\infty = a \times L = 0.032$ (See for details the paper [142])

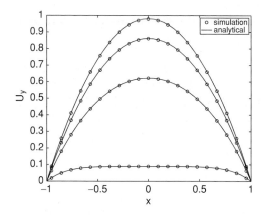

Fig. 2.3. Development of the velocity profile in the Poiseuille flow. Reduced velocity $U_y(x) = u_y/u_{y_{max}}$ is shown versus the reduced coordinate across the channel x. Solid line: Analytical solution. Different lines correspond to different instants of the reduced time $T = (\mu t)/(4R^2)$, increasing from bottom to top, R is the half-width of the channel. Symbol: simulation with the ELBGK algorithm. Parameters used are: viscosity $\mu = 5.0015 \times 10^{-5}$ ($\beta = 0.9997$), steady state maximal velocity $u_{y_{max}} = 1.10217 \times 10^{-2}$. Reynolds number $Re = 1157$. (See for details the paper [140])

account the nonlocality of collisions through a two-fold modification of the Boltzmann collision integral: First, the one-particle distribution functions are evaluated at the locations of the centers of spheres, separated by the nonzero distance at the impact. This makes the collision integral nonlocal in space. Second, the equilibrium pair distribution function at the contact of the spheres enhances the scattering probability.

Enskog's collision integral for hard spheres of radius r_0 is written in the following form [70]:

$$Q = \int_{R^3} \int_{B^-} [(v - w) \cdot n] \left[\chi(x, x + r_0 n) f(x, v') f(x + 2r_0 n, w') \right.$$
$$\left. - \chi(x, x - r_0 n) f(x, v) f(x - 2r_0 n, w) \right] dw \, dn , \qquad (2.59)$$

where $\chi(x, y)$ is the equilibrium pair-correlation function for given temperature and density, and integration in w is carried over the whole space R^3, while integration in n is over a hemisphere $B^- = \{n \in S^2 \mid (w - v, n) < 0\}$.

The proof of the H theorem for the Enskog equation has posed certain difficulties, and has led to a modification of the collision integral [145].

Methods of solution of the Enskog equation are immediate generalizations of those developed for the Boltzmann equation, but there is one additional difficulty. The Enskog collision integral is nonlocal in space. The Chapman–Enskog method, when applied to the Enskog equation, is supplemented with a gradient expansion around the homogeneous equilibrium state.

2.8.2 The Vlasov Equation

The Vlasov equation (or kinetic equation for a self-consistent force) is the nonlinear equation for the one-body distribution function, which takes into account a long-range interaction between particles:

$$\frac{\partial}{\partial t} f + \left(v, \frac{\partial}{\partial x} f \right) + \left(F, \frac{\partial}{\partial v} f \right) = 0 ,$$

where $F = \int \Phi(| x - x' |) \frac{x-x'}{|x-x'|} n(x') \, dx'$ is the self-consistent force. In this expression $\Phi(| x - x' |) \frac{x-x'}{|x-x'|}$ is the microscopic force between the two particles, and $n(x')$ is the density of particles, defined self-consistently, $n(x') = \int f(x', v) \, dv$.

The Vlasov equation is used for the description of collisionless plasmas in which case it is complemented by the set of Maxwell equations for the electromagnetic field [172]. It is also used for the description of gravitating gas.

The Vlasov equation is an infinite-dimensional Hamiltonian system [146]. Many special and approximate (wave-like) solutions to the Vlasov equation are known and they describe important physical effects [147]. One of the best known effects is the Landau damping [172]: The energy of a volume element dissipates with the rate

$$Q \approx - \mid E \mid^2 \frac{\omega(k)}{k^2} \left. \frac{df_0}{dv} \right|_{v=\frac{\omega}{k}} ,$$

where f_0 is the Maxwell distribution function, $\mid E \mid$ is the amplitude of the applied monochromatic electric field with the frequency $\omega(k)$, k is the wave vector. The Landau damping is thermodynamically reversible, and it is not accompanied with an entropy increase. Thermodynamically reversed to the Landau damping is the plasma echo effect.

2.8.3 The Fokker–Planck Equation

The Fokker–Planck equation (FPE) is a familiar model in various problems of nonequilibrium statistical physics [148–150]. We consider the FPE of the form

$$\frac{\partial W(\boldsymbol{x},t)}{\partial t} = \frac{\partial}{\partial \boldsymbol{x}}\left\{D\left[W\frac{\partial}{\partial \boldsymbol{x}}U + \frac{\partial}{\partial \boldsymbol{x}}W\right]\right\}. \qquad (2.60)$$

Here, $W(\boldsymbol{x},t)$ is the probability density over the configuration space x at time t, while $U(\boldsymbol{x})$ and $D(\boldsymbol{x})$ are the potential and the positively semi-definite $((\boldsymbol{y},D\boldsymbol{y}) \geq 0)$ diffusion matrix.

The FPE (2.60) is particularly important in studies of polymer solutions [151–153].

Let us recall the three properties of the FPE (2.60):

1. Conservation of the total probability:

$$\int W(\boldsymbol{x},t)\,\mathrm{d}x \equiv 1.$$

2. The equilibrium distribution,

$$W_{\mathrm{eq}} \propto \exp(-U),$$

   is the unique stationary solution to the FPE (2.60) for the given total probability.
3. The entropy,

$$S[W] = -\int W(\boldsymbol{x},t)\ln\left[\frac{W(\boldsymbol{x},t)}{W_{\mathrm{eq}}(\boldsymbol{x})}\right]\mathrm{d}x, \qquad (2.61)$$

   is a monotonically growing function due to the FPE (2.60), and it attaines the global maximum at equilibrium.

These properties become more elicit when the FPE (2.60) is rewritten as follows:

$$\partial_t W(\boldsymbol{x},t) = \hat{M}_W \frac{\delta S[W]}{\delta W(\boldsymbol{x},t)}, \qquad (2.62)$$

where

$$\hat{M}_W = -\frac{\partial}{\partial \boldsymbol{x}}\left[W(\boldsymbol{x},t)D(\boldsymbol{x})\frac{\partial}{\partial \boldsymbol{x}}\right]$$

is a positive semi-definite symmetric operator. The form (2.62) is the dissipative part of a structure termed GENERIC (the dissipative vector field is a metric transform of the entropy gradient) [154, 155].

Entropy does not depend on kinetic constants. It is the same for different details of kinetics, and depends only on the equilibrium data. Let us call this property *"universality"*. It is known that for the Boltzmann equation there exists only one universal Lyapunov functional: the entropy (we do not distinguish functionals which are related to each other by monotonic transformations). For the FPE there exists a whole family of universal Lyapunov functionals. Let $h(a)$ be a convex function of one variable $a \geq 0$, $h''(a) > 0$,

$$S_h[W] = -\int W_{\text{eq}}(\boldsymbol{x}) h\left[\frac{W(\boldsymbol{x},t)}{W_{\text{eq}}(\boldsymbol{x})}\right] \, \mathrm{d}x \ . \tag{2.63}$$

The density of production of the generalized entropy S_h, σ_h, is non-negative:

$$\sigma_h(\boldsymbol{x}) = W_{\text{eq}}(\boldsymbol{x}) h''\left[\frac{W(\boldsymbol{x},t)}{W_{\text{eq}}(\boldsymbol{x})}\right]\left(\frac{\partial}{\partial \boldsymbol{x}}\frac{W(\boldsymbol{x},t)}{W_{\text{eq}}(\boldsymbol{x})}, D\frac{\partial}{\partial \boldsymbol{x}}\frac{W(\boldsymbol{x},t)}{W_{\text{eq}}(\boldsymbol{x})}\right) \geq 0 \ . \tag{2.64}$$

The most important variants for the choice of h are:

- $h(a) = a \ln a$, and S_h is the Boltzmann–Gibbs–Shannon entropy (in the Kullback form [156, 157]),
- $h(a) = a \ln a - \epsilon \ln a$, $\epsilon > 0$, and S_h^{ϵ} is the maximal family of *additive* entropies [158–160] (these entropies are additive for the composition of independent subsystems).
- $h(a) = \frac{1-a^q}{1-q}$, and S_h^q is the family of Tsallis entropies [161, 162]. These entropies are not additive, but become additive after a nonlinear monotonous transformation. This property can serve as a definition of the Tsallis entropies in the class of generalized entropies (2.63) [160].

2.9 Equations of Chemical Kinetics and Their Reduction

2.9.1 Dissipative Reaction Kinetics

We begin with an outline of reaction kinetics (for details see, for example, the book [81]). Let us consider a closed system with n chemical species $A_1, \ldots, A_n$, participating in a complex reaction. The mechanism of complex reaction is represented by the following stoichiometric equations:

$$\alpha_{s1} A_1 + \ldots + \alpha_{sn} A_n \rightleftharpoons \beta_{s1} A_1 + \ldots + \beta_{sn} A_n \ , \tag{2.65}$$

where the index $s = 1, \ldots, r$ enumerates the reaction steps, and the integers, α_{si} and β_{si}, are the stoichiometric coefficients. For each reaction step s, we introduce n-dimensional vectors $\boldsymbol{\alpha}_s$ and $\boldsymbol{\beta}_s$ with components α_{si} and β_{si}. The *stoichiometric vector*, $\boldsymbol{\gamma}_s$, has integer components $\gamma_{si} = \beta_{si} - \alpha_{si}$.

For every A_i, an *extensive variable* N_i, "the number of particles of the i-th specie", is introduced. The concentration of A_i is then $c_i = N_i/V$, where V is the volume of the system.

Given the reaction mechanism (2.65), the kinetic equations read:

$$\dot{\boldsymbol{N}} = V\boldsymbol{J}(\boldsymbol{c}), \ \ \boldsymbol{J}(\boldsymbol{c}) = \sum_{s=1}^{r} \boldsymbol{\gamma}_s W_s(\boldsymbol{c}) \ , \tag{2.66}$$

where dot denotes the time derivative, and W_s is the reaction rate function of the sth reaction step. In particular, the *mass action law* suggests a polynomial form for the reaction rates:

$$W_s(\boldsymbol{c}) = W_s^+(\boldsymbol{c}) - W_s^-(\boldsymbol{c}) = k_s^+(T) \prod_{i=1}^{n} c_i^{\alpha_{si}} - k_s^-(T) \prod_{i=1}^{n} c_i^{\beta_{si}} \ , \tag{2.67}$$

where $k_s^+(T)$ and $k_s^-(T)$ are the constants of the forward and reverse reactions, respectively, of the sth reaction step, and T is the temperature. The (generalized) Arrhenius equation is the most popular expression for $k_s^\pm(T)$:

$$k_s^\pm(T) = a_s^\pm T^{b_s^\pm} \exp(S_s^\pm/k_{\mathrm{B}}) \exp(-H_s^\pm/k_{\mathrm{B}}T) \ , \tag{2.68}$$

where $a_s^\pm$, $b_s^\pm$ are constants, $H_s^\pm$ are activation enthalpies, and $S_s^\pm$ are activation entropies.

If the stoichiometric vectors $\{\boldsymbol{\gamma}_s\}$ are linearly dependent then the rate constants are not independent, but related through the *principle of detailed balance* gives the following connection between these constants: There exists a positive vector, $\boldsymbol{c}^{\mathrm{eq}}(T)$, such that

$$W_s^+(\boldsymbol{c}^{\mathrm{eq}}) = W_s^-(\boldsymbol{c}^{\mathrm{eq}}) \ \text{for all } s = 1,\ldots,r \ . \tag{2.69}$$

The necessary and sufficient conditions for the existence of such a $\boldsymbol{c}^{\mathrm{eq}}$ can be formulated as the system of polynomial equalities for $\{k_s^\pm\}$, (see, for example, [81]).

The reaction kinetics equations (2.66) do not form a closed system, because the dynamics of the volume V is not yet defined. Four classical conditions for closure of this system are well studied: U, $V = $ const (isolated system, U is the internal energy); H, $P = $ const (thermal isolated isobaric system, P is the pressure, $H = U + PV$ is the enthalpy), V, $T = $ const (isochoric isothermal conditions); P, $T = $ const (isobaric isothermal conditions). For V, $T = $ const no additional equations and data are needed. Equation (2.66) can be divided by the constant volume to obtain

$$\dot{\boldsymbol{c}} = \sum_{s=1}^{r} \boldsymbol{\gamma}_s W_s(\boldsymbol{c}) \ . \tag{2.70}$$

For non-isothermal and non-isochoric conditions addition formulae are needed to derive T and V. For all four classical conditions, the thermodynamic Lyapunov functions $G_\bullet$ for kinetic equations are known:

$$
\begin{aligned}
&U, V = \text{const}, \; G_{U,V} = -S/k_B \; ; \\
&V, T = \text{const}, \; G_{V,T} = F/k_B T = U/k_B T - S/k_B \; ; \\
&H, P = \text{const}, \; G_{H,P} = -S/k_B \; ; \\
&P, T = \text{const}, \; G_{P,T} = G/k_B T = H/k_B T - S/k_B \; ,
\end{aligned}
\tag{2.71}
$$

where $F = U - TS$ is the free energy (Helmholtz free energy), and $G = H - TS$ is the free enthalpy (Gibbs free energy). All the thermodynamic Lyapunov functions are normalized to the dimensionless scale (if the number of particles is expressed in moles, it is necessary to change k_B to R). All these functions decrease with time. For the classical conditions, the corresponding thermodynamic Lyapunov functions can be written in the form: $G_\bullet(\text{const}, \boldsymbol{N})$. The derivatives $\partial G_\bullet(\text{const}, \boldsymbol{N})/\partial N_i$ are the same functions of $\boldsymbol{c}$ and T for all classical conditions:

$$
\mu_i(\boldsymbol{c}, T) = \frac{\partial G_\bullet(\text{const}, \boldsymbol{N})}{\partial N_i} = \frac{\mu_i^{\text{chem}}(\boldsymbol{c}, T)}{k_B T} \; ,
\tag{2.72}
$$

where $\mu_i^{\text{chem}}(\boldsymbol{c}, T)$ is the chemical potential of species A_i.

Usual $G_\bullet(\text{const}, \boldsymbol{N})$ are strictly convex functions of $\boldsymbol{N}$, and the matrix $\partial \mu_i/\partial c_j$ is positively definite. The dissipation inequality

$$
\frac{1}{V} \frac{dG_\bullet}{dt} = (\boldsymbol{\mu}, \boldsymbol{J}) \le 0
\tag{2.73}
$$

holds. This inequality poses a restriction on possible kinetic laws and on possible values of the kinetic constants.

One of the most important generalizations of the mass action law (2.67) is the Marcelin–De Donder kinetic function. This generalization [243, 244] is based on ideas from the thermodynamic theory of affinity [245]. Within this approach, the functions W_s are constructed as follows [244]: For a given $\boldsymbol{\mu}(\boldsymbol{c}, T)$ (2.72), and for a given reaction mechanism (2.65), we define the gain $(+)$ and the loss $(-)$ rates of the sth reaction step as,

$$
W_s^+ = \varphi_s^+ \exp(\boldsymbol{\mu}, \boldsymbol{\alpha}_s), \quad W_s^- = \varphi_s^- \exp(\boldsymbol{\mu}, \boldsymbol{\beta}_s) \; ,
\tag{2.74}
$$

where $\varphi_s^\pm > 0$ are kinetic factors, $(\,,\,)$ is the standard inner product (the sum of coordinates products).

The Marcelin–De Donder kinetic function reads: $W_s = W_s^+ - W_s^-$, and the right hand side of the kinetic equation (2.66) becomes,

$$
\boldsymbol{J} = \sum_{s=1}^{r} \boldsymbol{\gamma}_s \{ \varphi_s^+ \exp(\boldsymbol{\mu}, \boldsymbol{\alpha}_s) - \varphi_s^- \exp(\boldsymbol{\mu}, \boldsymbol{\beta}_s) \} \; .
\tag{2.75}
$$

For the Marcelin–De Donder reaction rate (2.74), the dissipation inequality (2.73) is particularly elegant:

$$\dot{G} = \sum_{s=1}^{r}[(\boldsymbol{\mu},\boldsymbol{\beta}_s) - (\boldsymbol{\mu},\boldsymbol{\alpha}_s)]\left\{\varphi_s^+ e^{(\boldsymbol{\mu},\boldsymbol{\alpha}_s)} - \varphi_s^- e^{(\boldsymbol{\mu},\boldsymbol{\beta}_s)}\right\} \leq 0 . \tag{2.76}$$

The kinetic factors $\varphi_s^{\pm}$ should satisfy certain conditions in order to satisfy the dissipation inequality (2.76). A well known sufficient condition is the detailed balance:

$$\varphi_s^+ = \varphi_s^- . \tag{2.77}$$

Other sufficient conditions are discussed in detail elsewhere [81, 115, 163].

For ideal systems, the function $G_\bullet$ is constructed from the thermodynamic data of individual species. It is convenient to start from the isochoric isothermal conditions. The Helmholtz free energy for an ideal system is

$$F = k_{\mathrm{B}}T\sum_i N_i[\ln c_i - 1 + \mu_{0i}] + \mathrm{const}_{T,V} , \tag{2.78}$$

where the internal energy is assumed to be a linear function of N in a given interval of c, T:

$$U = \sum_i N_i u_i(T) = \sum_i N_i(u_{0i} + C_{Vi}T) ,$$

where $u_i(T)$ is the internal energy of species A_i per particle. It is well known that $S = -(\partial F/\partial T)_{V,N=\mathrm{const}}$, $U = F + TS = F - T(\partial F/\partial T)_{V,N=\mathrm{const}}$, hence, $u_i(T) = -k_{\mathrm{B}}T^2 d\mu_{0i}/dT$ and

$$\mu_{0i} = \delta_i + u_{0i}/k_{\mathrm{B}}T - (C_{Vi}/k_{\mathrm{B}})\ln T , \tag{2.79}$$

where $\delta_i = \mathrm{const}$, C_{Vi} is the heat capacity at constant volume (per particle) of species A_i.

In concordance with the form of ideal free energy (2.78) the expression for $\boldsymbol{\mu}$ is:

$$\mu_i = \ln c_i + \delta_i + u_{0i}/k_{\mathrm{B}}T - (C_{Vi}/k_{\mathrm{B}})\ln T . \tag{2.80}$$

For the function $\boldsymbol{\mu}$ of the form (2.80), the Marcelin–De Donder equation obtains the more familiar mass action law form (2.67). Taking into account the principle of detailed balance (2.77) we get the ideal rate functions:

$$W_s(\boldsymbol{c}) = W_s^+(\boldsymbol{c}) - W_s^-(\boldsymbol{c}) ,$$

$$W_s^+(\boldsymbol{c}) = \varphi_s(\boldsymbol{c},T)T^{-\sum_i \alpha_{si}C_{Vi}/k_{\mathrm{B}}}e^{\sum_i \alpha_{si}(\delta_i+u_{0i}/k_{\mathrm{B}}T)}\prod_{i=1}^{n} c_i^{\alpha_{si}} ,$$

$$W_s^-(\boldsymbol{c}) = \varphi_s(\boldsymbol{c},T)T^{-\sum_i \beta_{si}C_{Vi}/k_{\mathrm{B}}}e^{\sum_i \beta_{si}(\delta_i+u_{0i}/k_{\mathrm{B}}T)}\prod_{i=1}^{n} c_i^{\beta_{si}} . \tag{2.81}$$

where $\varphi_s(c, T)$ is an arbitrary positive function (from the thermodynamic point of view).

Let us discuss further the vector field $J(c)$ in the concentration space (2.70). Conservation laws (balances) impose linear constraints on admissible vectors dc/dt:

$$(b_i, c) = B_i = \text{const}, \quad \left(b_i, \frac{dc}{dt}\right) = 0, \; i = 1, \dots, l, \tag{2.82}$$

where b_i are fixed and linearly independent vectors. Let us denote as B the set of vectors which satisfy the conservation laws (2.82) for given B_i:

$$B = \{c | (b_1, c) = B_1, \dots, (b_l, c) = B_l\} .$$

The natural phase space X of the system (2.70) is the intersection of the cone of n-dimensional vectors with nonnegative components, with the set B, and $\dim X = d = n - l$. In the sequel, we term a vector $c \in X$ the state of the system. In addition, we assume that each of the conservation laws is supported by each elementary reaction step, that is

$$(\gamma_s, b_i) = 0 , \tag{2.83}$$

for each pair of vectors γ_s and b_i.

Reaction kinetic equations describe variations of the states in time. The phase space X is positive-invariant for system (2.70): If $c(0) \in X$, then $c(t) \in X$ for all times $t > 0$.

In the sequel, we assume that the kinetic equations (2.70) describe evolution towards the unique equilibrium state, c^{eq}, in the interior of the phase space X. Furthermore, we assume that there exists a strictly convex function $G(c)$ which decreases monotonically in time due to (2.70), ∇G is the vector of partial derivatives $\partial G/\partial c_i$, and the convexity means that the $n \times n$ matrix

$$H_c = \|\partial^2 G(c)/\partial c_i \partial c_j\| , \tag{2.84}$$

is positive definite for all $c \in X$. In addition, we assume that the matrix (2.84) is invertible if c is taken in the interior of the phase space.

Function G is the Lyapunov function for the system (2.66), and c^{eq} is the point of global minimum of G in the phase space X. Otherwise stated, the manifold of equilibrium states $c^{\text{eq}}(B_1, \dots, B_l)$ is the solution to the variational problem,

$$G \to \min \text{ for } (b_i, c) = B_i, \; i = 1, \dots, l . \tag{2.85}$$

For each fixed value of the conserved quantities B_i, the solution is unique. In many cases, however, it is convenient to consider the whole equilibrium manifold, keeping the conserved quantities as parameters.

For example, for perfect systems in a constant volume system at constant temperature, the Lyapunov function G reads:

$$G = \sum_{i=1}^{n} c_i [\ln(c_i/c_i^{\text{eq}}) - 1] \, . \qquad (2.86)$$

It is important to stress that c^{eq} in (2.86) is an *arbitrary* equilibrium of the system, under arbitrary values of the balances. In order to compute $G(c)$, it is unnecessary to calculate the specific equilibrium c^{eq} which corresponds to the initial state c. Let us compare the Lyapunov function G (2.86) with the classical formula for the free energy (2.78). This comparison gives a possible choice for c^{eq}:

$$\ln c_i^{\text{eq}} = -\delta_i - u_{0i}/k_{\text{B}}T + (C_{Vi}/k_{\text{B}}) \ln T \, . \qquad (2.87)$$

2.9.2 The Problem of Reduced Description in Chemical Kinetics

Reduction of a description of a chemical system means the following:

1. Reduce the number of species. This, in turn, can be achieved in two ways:
   - eliminate inessential species, or
   - lump some of the species into integrated components.
2. Reduce the number of reactions. This can also be done in several ways:
   - eliminate inessential reactions, those which do not significantly influence the reaction progress;
   - assume that some of the reactions "have already been completed", and that the equilibrium has been reached along their paths (this leads to dimensional reduction because the rate constants of the "completed" reactions are not used thereafter, what one needs are equilibrium constants only).
3. Decompose the motions into fast and slow, into independent (almost-independent) and slaved etc. As a result of such a decomposition, the system admits a study "in parts". At the end, the results are combined into a joint picture. There are several approaches which fall into this category. The famous method of the *quasi-steady state* (QSS), pioneered by Bodenstein and Semenov, follows the Chapman–Enskog method. The *partial equilibrium approximations* are predecessors of Grad's method and quasiequilibrium approximations in physical kinetics. These two family of methods have different physical backgrounds and mathematical forms.

2.9.3 Partial Equilibrium Approximations

Quasiequilibrium with respect to reactions is constructed as follows: From the list of reactions (2.65), one selects those which are assumed to equilibrate first. Let these reactions be indexed with the integers $s_1, \ldots, s_k$. The quasi-equilibrium manifold is defined by the system of equations,

$$W_{s_i}^+ = W_{s_i}^-, \quad i = 1, \ldots, k \, . \qquad (2.88)$$

This system looks particularly elegant when written in terms of conjugated (dual) variables, $\boldsymbol{\mu} = \boldsymbol{\nabla} G$:

$$(\boldsymbol{\gamma}_{s_i}, \boldsymbol{\mu}) = 0, \ i = 1, \ldots, k \ . \tag{2.89}$$

In terms of the conjugated variables, the quasiequilibrium manifold forms a linear subspace. This subspace, $L^\perp$, is the orthogonal completement to the linear envelope of vectors, $L = \mathrm{lin}\{\boldsymbol{\gamma}_{s_1}, \ldots, \boldsymbol{\gamma}_{s_k}\}$.

Quasiequilibrium with respect to species is constructed practically in the same way but without selecting the subset of reactions. For a given set of species, $A_{i_1}, \ldots, A_{i_k}$, one assumes that their concentrations evolve fast to equilibrium and remain there. Formally, this means that in the k-dimensional subspace of the space of concentrations with coordinates $c_{i_1}, \ldots, c_{i_k}$, one constructs the subspace L which is defined by the balance equations, $(\boldsymbol{b}_i, \boldsymbol{c}) = 0$. In terms of the conjugated variables, the quasiequilibrium manifold, $L^\perp$, is defined by the equations,

$$\boldsymbol{\mu} \in L^\perp, \ (\boldsymbol{\mu} = (\mu_1, \ldots, \mu_n)) \ . \tag{2.90}$$

The same quasiequilibrium manifold can also be defined with the help of fictitious reactions: Let $\boldsymbol{g}_1, \ldots, \boldsymbol{g}_q$ be a basis in L. Then (2.90) may be rewritten as follows:

$$(\boldsymbol{g}_i, \boldsymbol{\mu}) = 0, \ i = 1, \ldots, q \ . \tag{2.91}$$

Illustration: Quasiequilibrium with respect to reactions in hydrogen oxidation: Let us assume equilibrium with respect to the dissociation reactions, $H_2 \rightleftharpoons 2H$, and, $O_2 \rightleftharpoons 2O$, in some subdomain of reaction conditions. This gives:

$$k_1^+ c_{H_2} = k_1^- c_H^2, \ k_2^+ c_{O_2} = k_2^- c_O^2 \ .$$

Quasiequilibrium with respect to species: For the same reactions, let us assume equilibrium over H, O, OH, and H_2O_2, in a subdomain of reaction conditions. The subspace L is defined by the balance constraints:

$$c_H + c_{OH} + 2c_{H_2O_2} = 0, \ c_O + c_{OH} + 2c_{H_2O_2} = 0 \ .$$

The subspace L is twodimensional. Its basis, $\{\boldsymbol{g}_1, \boldsymbol{g}_2\}$, in the coordinates c_H, c_O, c_{OH}, and $c_{H_2O_2}$ reads:

$$\boldsymbol{g}_1 = (1, 1, -1, 0), \quad \boldsymbol{g}_2 = (2, 2, 0, -1) \ .$$

Correspondingly (2.91) becomes:

$$\mu_H + \mu_O = \mu_{OH}, \ 2\mu_H + 2\mu_O = \mu_{H_2O_2} \ .$$

General construction of the quasiequilibrium manifold: In the space of concentrations, one defines a subspace L which satisfies the balance constraints:

$$(\boldsymbol{b}_i, L) \equiv 0 \ .$$

The orthogonal complement of L in the space with coordinates $\boldsymbol{\mu} = \boldsymbol{\nabla} G$ defines then the quasiequilibrium manifold $\boldsymbol{\Omega}_L$. For the actual computations, one requires the inversion from $\boldsymbol{\mu}$ to $\boldsymbol{c}$. The duality structure $\boldsymbol{\mu} \leftrightarrow \boldsymbol{c}$ is well studied by many authors [163, 164].

Quasiequilibrium projector. It is not sufficient to just derive the manifold, it is also required to define a *projector* which transforms the vector field defined on the space of concentrations into a vector field on the manifold. The quasiequilibrium manifold consists of points which minimize G in affine spaces of the form $\boldsymbol{c} + L$. These affine planes are hypothetical planes of fast motions (G is decreasing in the course of the fast motions). Therefore, the quasiequilibrium projector maps the whole space of concentrations on $\boldsymbol{\Omega}_L$ parallel to L. The vector field is also projected onto the tangent space of $\boldsymbol{\Omega}_L$ parallel to L.

Thus, the quasiequilibrium approximation assumes the decomposition of motions into fast – parallel to L, and slow – along the quasiequilibrium manifold. In order to construct the quasiequilibrium approximation, the knowledge of reaction rate constants of "fast" reactions is not required (stoichiometric vectors of all these fast reaction are in L, $\boldsymbol{\gamma}_{\text{fast}} \in L$, thus, the knowledge of L suffices), one only needs some confidence in that they all are sufficiently fast [165]. The quasiequilibrium manifold itself is constructed based on the knowledge of L and G. The dynamics on the quasiequilibrium manifold is defined as the quasiequilibrium projection of the "slow component" of the kinetic equations (2.66).

2.9.4 Model Equations

The assumption behind quasiequilibrium is the hypothesis of the decomposition of motions into fast and slow. The quasiequilibrium approximation itself describes slow motions. However, sometimes it becomes necessary to restore the state of the whole system, and take into account the fast motions as well. With this, it is desirable to keep intact one of the important advantages of the quasiequilibrium approximation – its independence from the rate constants of the fast reactions. For this purpose, the detailed fast kinetics is replaced by a model equation (*single relaxation time approximation*).

Quasiequilibrium models (QEM) are constructed as follows: For each concentration vector $\boldsymbol{c}$, consider the affine manifold, $\boldsymbol{c} + L$. It intersects the quasiequilibrium manifold $\boldsymbol{\Omega}_L$ at a single point. This point delivers the minimum to G on $\boldsymbol{c} + L$. Let us denote this point as $\boldsymbol{c}_L^*(\boldsymbol{c})$. The equation of the quasiequilibrium model reads:

$$\dot{\boldsymbol{c}} = -\frac{1}{\tau}[\boldsymbol{c} - \boldsymbol{c}_L^*(\boldsymbol{c})] + \underbrace{\sum \boldsymbol{\gamma}_s W_s(\boldsymbol{c}_L^*(\boldsymbol{c}))}_{\text{slow}} , \qquad (2.92)$$

where $\tau > 0$ is the relaxation time of the fast subsystem. Rates of slow reactions are computed at the points $\boldsymbol{c}_L^*(\boldsymbol{c})$ (the second term in the right hand side of (2.92), whereas the rapid motion is taken into account by a simple relaxational term (the first term in the right hand side of (2.92). The most famous model kinetic equation is the BGK equation in the theory of the Boltzmann equation [116]. The general theory of the quasiequilibrium models, including proofs of their thermodynamic consistency, was constructed in the paper [117].

Single relaxation time gradient models (SRTGM) were introduced in the context of the lattice Boltzmann method for hydrodynamics [140,166]. These models are aimed at improving the obvious drawback of the quasiequilibrium model (2.92): In order to construct the QEM, one needs to compute the function,

$$c_L^*(c) = \arg \min_{x \in c+L, \ x>0} G(x) . \tag{2.93}$$

This is a convex programming problem, which does not always have a closed-form solution.

Let $g_1, \ldots, g_k$ be some orthonormal basis of L. We denote as $D(c)$ the $k \times k$ matrix with elements $(g_i, H_c g_j)$, where H_c is the matrix of second derivatives of G (2.84). Let $C(c)$ be the inverse of $D(c)$. The single relaxation time gradient model has the form:

$$\dot{c} = -\frac{1}{\tau} \sum_{i,j=1}^k g_i C(c)_{ij} (g_j, \nabla G) + \sum_{\text{slow}} \gamma_s W_s(c) . \tag{2.94}$$

The first term drives the system to the minimum of G on $c + L$, does not require solving problem (2.93), and its spectrum at quasiequilibrium is the same as in the quasiequilibrium model (2.92). Note that the slow component is evaluated at the "current" state c.

The first term of equation (2.94) has a simple form

$$\dot{c} = -\frac{1}{\tau} \text{grad} G + \sum_{\text{slow}} \gamma_s W_s(c) , \tag{2.95}$$

if one calculates the gradient $\text{grad} G \in L$ on the plane of fast motions $c + L$ with the entropic scalar product[3] $\langle x, y \rangle = (x, H_c y)$.

The models (2.92) and (2.94) lift the quasiequilibrium approximation to a kinetic equation by approximating the fast dynamics with a single "reaction rate constant" – the relaxation time τ.

---

[3] Let us remind that $\text{grad} G$ is the Riesz representation of the differential of G in the phase space X: $G(c+\Delta c) = G(c) + \langle \text{grad} G(c), \Delta c \rangle + o(\Delta c)$. It belongs to the tangent space of X and depends on the scalar product. From the thermodynamic point of view, there is only one distinguished scalar product in the concentration space, the entropic scalar product. The usual definition of $\text{grad} G$ as the vector of partial derivatives (∇G) corresponds to the standard scalar product $(\bullet, \bullet)$ and to the choice X being the whole concentration space. In equation (2.95), $X = c + L$ and we use the entropic scalar product.

2.9.5 Quasi-Steady State Approximation

The quasi-steady state approximation (QSS) is a tool used in a large number of works. Let us split the species in two groups: The basic and the intermediate (radicals etc). Concentration vectors are denoted accordingly, c^s (slow, basic species), and c^f (fast, intermediate species). The concentration vector c is the direct sum, $c = c^s \oplus c^f$. The fast subsystem is (2.66) for the concentrations c^f at fixed values of c^s. If it happens that the so-defined fast subsystem relaxes to a stationary state, $c^f \rightarrow c^f_{qss}(c^s)$, then the assumption that $c^f = c^f_{qss}(c)$ is precisely the QSS assumption. The slow subsystem is the part of system (2.66) for c^s, in the right hand side of which the component c^f is replaced with $c^f_{qss}(c)$. Thus, $J = J_s \oplus J_f$, where

$$\dot{c}^f = J_f(c^s \oplus c^f), \; c^s = \text{const}; \quad c^f \rightarrow c^f_{qss}(c^s) \; ; \tag{2.96}$$

$$\dot{c}^s = J_s(c^s \oplus c^f_{qss}(c^s)) \; . \tag{2.97}$$

Bifurcations of the system (2.96) under variation of c^s correspond to kinetic critical phenomena. Studies of more complicated dynamic phenomena in the fast subsystem (2.96) require various techniques of averaging, stability analysis of the averaged quantities etc.

Various versions of the QSS method are possible, and are actually used widely, for example, the hierarchical QSS method. There, one defines not a single fast subsystem but a hierarchy of them, $c^{f_1}, \ldots, c^{f_k}$. Each subsystem c^{f_i} is regarded as a slow system for all the foregoing subsystems, and it is regarded as a fast subsystem for the following members of the hierarchy. Instead of one system of equations (2.96), a hierarchy of systems of lower-dimensional equations is considered, each of these subsystem being easier to study analytically.

The theory of singularly perturbed systems of ordinary differential equations provides the mathematical background and refinements of the QSS approximation. In spite of a broad literature on this subject, it remains, in general, unclear, what is the smallness parameter that separates the intermediate (fast) species from the basic (slow). Reaction rate constants cannot be such a parameter (unlike in the case of quasiequilibrium). Indeed, intermediate species participate in the *same* reactions, as the basic species (for example, $H_2 \rightleftharpoons 2H$, $H + O_2 \rightleftharpoons OH + O$). It is therefore incorrect to state that c^f evolves faster than c^s. In the sense of reaction rate constants, c^f is not faster.

For catalytic reactions, it is not difficult to figure out what is the smallness parameter that separates the intermediate species from the basic, and which allows to upgrade the QSS assumption to a singular perturbation theory rigorously [81]. This smallness parameter is the ratio of balances: Intermediate species include a catalyst, and their total amount is simply significantly smaler than the amount of all the c_i's. After renormalizing to the variables of one order of magnitude, the small parameter appears explicitly. The simplest example is provided by the catalytic reaction $A + Z \rightleftharpoons AZ \rightleftharpoons P + Z$ (here Z is a catalyst, A and P are an initial substrate and a product). The kinetic equations are (in obvious notations):

$$\dot{c}_A = -k_1^+ c_A c_Z + k_1^- c_{AZ} ,$$
$$\dot{c}_Z = -k_1^+ c_A c_Z + k_1^- c_{AZ} + k_2^+ c_{AZ} - k_2^- c_Z c_P ,$$
$$\dot{c}_{AZ} = k_1^+ c_A c_Z - k_1^- c_{AZ} - k_2^+ c_{AZ} + k_2^- c_Z c_P ,$$
$$\dot{c}_P = k_2^+ c_{AZ} - k_2^- c_Z c_P . \tag{2.98}$$

The constants and the reactions rates are the same for concentrations c_A, c_P, and for c_Z, c_{AZ}, and they cannot be a reason for the relative slowness of c_A, c_P in comparison with c_Z, c_{AZ}. However, there may be another source of slowness. There are two balances for this kinetics: $c_A + c_P + c_{AZ} = B_A$, $c_Z + c_{AZ} = B_Z$. Let us switch to the dimensionless variables:

$$\varsigma_A = c_A/B_A, \; \varsigma_P = c_P/B_A, \; \varsigma_Z = c_Z/B_Z, \; \varsigma_{AZ} = c_{AZ}/B_Z .$$

The kinetic system (2.98) is then rewritten as

$$\dot{\varsigma}_A = B_Z \left[-k_1^+ \varsigma_A \varsigma_Z + \frac{k_1^-}{B_A} \varsigma_{AZ} \right] ,$$

$$\dot{\varsigma}_Z = B_A \left[-k_1^+ \varsigma_A \varsigma_Z + \frac{k_1^-}{B_A} \varsigma_{AZ} + \frac{k_2^+}{B_A} \varsigma_{AZ} - k_2^- \varsigma_Z \varsigma_P \right] ,$$

$$\varsigma_A + \varsigma_P + \frac{B_Z}{B_A} \varsigma_{AZ} = 1, \; \varsigma_Z + \varsigma_{AZ} = 1; \; \varsigma_\bullet \geq 0 . \tag{2.99}$$

For $B_Z \ll B_A$ (the total amount of the catalyst is much smaller than the total amount of the substrate) the slowness of ς_A, ς_P is evident from these equations (2.99).

For usual radicals, the origin of the smallness parameter is quite similar. There are much less radicals than basic species (otherwise, the QSS assumption is inapplicable). In the case of radicals, however, the smallness parameter cannot be extracted directly from the balances B_i (2.82). Instead, one can come up with a thermodynamic estimate: Function G decreases in the course of reactions, whereupon we obtain the limiting estimate of concentrations of any species:

$$c_i \leq \max_{G(c) \leq G(c(0))} c_i , \tag{2.100}$$

where $c(0)$ is the initial composition. If the concentration c_R of the radical R is small both initially and at equilibrium, then it should also remain small along the path to equilibrium. For example, in the case of ideal G (2.86) under relevant conditions, for any $t > 0$, the following inequality is valid:

$$c_R [\ln(c_R(t)/c_R^{eq}) - 1] \leq G(c(0)) . \tag{2.101}$$

Inequality (2.101) provides the simplest (but rather crude) thermodynamic estimate of $c_R(t)$ in terms of $G(\boldsymbol{c}(0))$ and c_R^{eq} *uniformly for $t > 0$*. The complete theory of thermodynamic estimates of reaction kinetics has been developed in the book [115].

One can also do computations without a priori estimations, if one accepts the QSS assumption as long as the values $\boldsymbol{c}^f$ stay sufficiently small. It is the simplest way to operate with QSS: Just use it *as long as $\boldsymbol{c}^f$ are small!*

Let us assume that an a priori estimate has been found, $c_i(t) \leq c_{i\ max}$, for each c_i. These estimates may depend on the initial conditions, thermodynamic data etc. With these estimates, we are able to renormalize the variables in the kinetic equations (2.66) in such a way that the renormalized variables take their values from the unit interval $[0,1]$: $\tilde{c}_i = c_i/c_{i\ max}$. Then the system (2.66) can be written as follows:

$$\frac{d\tilde{c}_i}{dt} = \frac{1}{c_{i\ max}} J_i(\boldsymbol{c}) . \tag{2.102}$$

The system of dimensionless parameters, $\epsilon_i = c_{i\ max}/\max_i c_{i\ max}$ defines a hierarchy of relaxation times, and with its help one can establish various realizations of the QSS approximation. The simplest version is the standard QSS assumption: Parameters ϵ_i are separated in two groups, the smaller ones, and those of order 1. Accordingly, the concentration vector is split into $\boldsymbol{c}^s \oplus \boldsymbol{c}^f$. Various hierarchical QSS are possible, rendering the problem more tractable analytically.

There exists a variety of ways to introduce the smallness parameter into kinetic equations, and one can find applications to each of the realizations. However, two particular realizations remain basic for chemical kinetics:

– Fast reactions (under a given thermodynamic data);
– Small concentrations.

In the first case, one is led to the quasiequilibrium approximation, in the second, to the classical QSS assumption. Both of these approximations allow for hierarchical realizations, those which include not just two but many relaxation time scales. Such a *multi-scale approach* essentially simplifies analytical studies of the problem.

2.9.6 Thermodynamic Criteria
for the Selection of Important Reactions

One of the problems addressed by sensitivity analysis is the selection of the important and unimportant reactions. In the paper [167] a simple idea was suggested to compare the importance of different reactions according to their contribution to the entropy production (or, which is the same, according to their contribution to dG/dt). Based on this principle, Dimitrov [170] described domains of parameters in which the reaction of hydrogen oxidation,

$H_2 + O_2 + M$, proceeds due to different mechanisms. For each elementary reaction, he has derived the domain inside which the contribution of this reaction cannot be neglected. Due to its simplicity, this entropy production principle is especially well suited for the analysis of complex problems. In particular, recently, a version of the entropy production principle was used in the problem of selection of boundary conditions for Grad's moment equations [168, 169]. For ideal systems (2.86), as well, as for the Marcelin–De Donder kinetics (2.76) the contribution of the sth reaction to $\dot{G}$ has a particularly simple form:

$$\dot{G}_s = -W_s \ln \left(\frac{W_s^+}{W_s^-} \right), \quad \dot{G} = \sum_{s=1}^{r} \dot{G}_s . \tag{2.103}$$

2.9.7 Opening

One of the problems to focus on when studying closed systems is to extend the result for open or driven by flows systems. External flows are usually taken into account by additional terms in the kinetic equations (2.66):

$$\dot{N} = V J(c) + \Pi(c, t) . \tag{2.104}$$

It is important to stress here that the vector field $J(c)$ in equations (2.104) is the same as for the closed system, with thermodynamic restrictions, Lyapunov functions, etc. The thermodynamic structures are important for the analysis of open systems (2.104), if the external flow Π is small in some sense, for example, if it is a linear function of c, has small time derivatives, etc. There are some general results for such "weakly open" systems, for example, the Prigogine minimum entropy production theorem [171] and the estimations of possible steady states and limit sets for open systems, based on thermodynamic functions and stoihiometric equations [115].

There are general results for another limiting case: for very intensive flows the dynamics becomes very simple again [81]. Let the flow have a natural structure: $\Pi(c, t) = v_{in}(t) c_{in}(t) - v_{out}(t) c(t)$, where v_{in} and v_{out} are the rates of inflow and outflow, $c_{in}(t)$ is the concentration vector for inflow. If v_{out} is sufficiently large, $v_{out}(t) > v_0$ for some critical value v_0 and all $t > 0$, then for the open system (2.104) the Lyapunov norm exists: for any two solutions $c^1(t)$ and $c^2(t)$ the function $\|c^1(t) - c^2(t)\|$ monotonically decreases in time. Such a critical value v_0 exists for any norm, for example, for usual Euclidian norm $\| \bullet \|^2 = (\bullet, \bullet)$.

For an arbitrary form of Π, the system (2.104) can loose all signs of being a thermodynamic one. Nevertheless, thermodynamic structures can often help in the study of open systems.

The crucial questions are: What happens with slow/fast motion separation after opening? Which slow invariant manifolds for the closed system can be deformed to the slow invariant manifolds of the open system? Which slow invariant manifold for the closed system can be used as approximate slow invariant manifold for the open system? There exists a more or less useful technique to seek answers for specific systems under consideration. We shall return to this question in Chap. 13.

The way to study an open system as the result of opening a closed system may be fruitful. Out of this way we have a general dynamical system (2.104) and no hints what to do with it.

***

The basic introductory textbook on physical kinetics of the Landau and Lifshitz Course of Theoretical Physics [172] contains many further examples and their applications.

Modern development of kinetics follows the route of specific numerical methods, such as direct simulations. An opposite tendency is also clearly observed, and kinetic theory based schemes are increasingly often used for the development of numerical methods and models in mechanics of continuous media.

3 Invariance Equation in Differential Form

Definition of invariance in terms of motions and trajectories assumes, at least, existence and uniqueness theorems for solutions of the original dynamical system. This prerequisite causes difficulties when one studies equations relevant to physical and chemical kinetics, such as, for example, equations of hydrodynamics. Nevertheless, there exists a necessary *differential condition of invariance*: The vector field of the original dynamic system touches the manifold at every point. Let us write down this condition in order to set up the notation.

Let E be a linear space, U (the phase space) be a domain in E, and let a vector field $J : U \to E$ be defined in U. This vector field defines the original dynamical system,

$$\frac{\mathrm{d}x}{\mathrm{d}t} = J(x), \ x \in U \ . \tag{3.1}$$

In the sequel, we consider submanifolds in U which are parameterized by a given set of parameters. Let a linear space of parameters L be defined, and let W be a domain in L. We consider differentiable maps, $F : W \to U$, such that, for every $y \in W$, the differential of F, $D_y F : L \to E$, is an isomorphism of L on a subspace of E. That is, F are the manifolds, immersed in the phase space of the dynamical system (3.1), and parametrized by the parameter set W.

Remark: One never discusses the choice of norms and topologies is such a general setting. It is assumed that the corresponding choice is made appropriately in each specific case.

We denote T_y the tangent space at point y, $T_y = (D_y F)(L)$. *The differential condition of invariance* has the following form: For every $y \in W$,

$$J(F(y)) \in T_y \ . \tag{3.2}$$

Let us rewrite the differential condition of invariance (3.2) in the form of a differential equation. In order to achieve this, one needs to define a projector $P_y : E \to T_y$ for every $y \in W$. Once a projector P_y is defined, then condition (3.2) takes the form:

$$\Delta_y = (1 - P_y)J(F(y)) = 0 \ . \tag{3.3}$$

Alexander N. Gorban and Iliya V. Karlin: *Invariant Manifolds for Physical and Chemical Kinetics*, Lect. Notes Phys. **660**, 65–67 (2005)
www.springerlink.com

Obviously, by $P_y^2 = P_y$ we have, $P_y \Delta_y = 0$. We refer to the function Δ_y as *the defect of invariance* at point y. The defect of invariance will be encountered often in what follows.

Equation (3.3) is the first-order differential equation for the function $F(y)$. Projectors P_y should be tailored to the specific physical features of the problem at hand, and separate chapter below will be devoted to their construction. There we shall demonstrate how to construct a projector, $P(x,T): E \to T$, given a point $x \in U$ and a specified subspace T. We then set $P_y = P(F(y), T_y)$ in equation (3.3)[1].

There are two possible meanings of the notion "approximate solution of the invariance equations" (3.3):

1. Approximation of the solution;
2. The map F with small defect of invariance (the right hand side approximation).

The approximation of the first kind requires theorems about existence of solutions for the initial system (3.1). In order to find this approximation one should estimate the deviations of exact solutions of (3.1) from the approximate invariant manifold. The second kind of approximations does not require the existence of solutions. Moreover, the manifold with sufficiently small defect of invariance can serve as a slow manifold by itself. *The defect of invariance should be small in comparison with the initial vector field J.*

So, we shall accept the concept of approximate invariant manifold (the manifold with small defect of invariance) instead of the approximation of the invariant manifold (see also [25, 349] and other works about approximate inertial manifolds). Sometime these approximate invariant manifolds provide approximations of the invariant manifolds, sometimes not, but it is additional and often difficult problem to make a distinction between these situations. In addition to the defect of invariance, Jacobians, the differentials of $J(x)$, play the key role in the analysis of motion separation into fast and slow. Some estimations of errors of this separation will be presented below in the subsection devoted to *post-processing*.

---

[1] One of the main routes to define the field of projectors $P(x,T)$ is to make use of a Riemannian structure. To this end, one defines a scalar product in E for every point $x \in U$, that is, a bilinear form $\langle p|q \rangle_x$ with a positive definite quadratic form, $\langle p|p \rangle_x > 0$, if $p \neq 0$. A good candidate for such a scalar product is the bilinear form defined by the negative second differential of the entropy at the point x, $-D^2 S(x)$. As we demonstrate later in this book, close to equilibrium this choice is essentially the only correct one. However, far from equilibrium, a refinement is required in order to guarantee the thermodymamicity condition, $\ker P_y \subset \ker(D_x S)_{x=F(y)}$, for the field of projectors, $P(x,T)$, defined for any x and T, if $T \not\subset \ker D_x S$. The thermodymamicity condition provides the preservation of the type of dynamics: if $dS/dt > 0$ for initial vector field (3.1) at point $x = F(y)$, then $dS/dt > 0$ at this point x for the projected vector field $P_y(J(F(y)))$, too.

Our discussion is focused on nonperturbative methods for computing invariant manifolds, but it should be mentioned that in many applications, the Taylor series expansion is in use, and sometimes works quite well. The main idea is the continuation of the slow manifold with respect to a small parameter: Let our system depend on the parameter ε, and let a manifold of steady states and fibers of motions towards these steady states exist for $\varepsilon = 0$, for example

$$\dot{x} = \varepsilon f(x, y); \quad \dot{y} = g(x, y) \,. \tag{3.4}$$

For $\varepsilon = 0$, the value of the (vector) variable x is a vector of conserved quantities. Let for every x the equation of fast motion, $\dot{y} = g(x, y)$, be globally stable: Its solution $y(t)$ tends to the unique (for given x) stable fixed point y_x. If the function $g(x, y)$ meets the conditions of the implicit function theorem, then the graph of the map $x \mapsto y_x$ forms a manifold $\Omega_0 = \{(x, y_x)\}$ of steady states. For small $\varepsilon > 0$ we can look for the slow manifold in a form of a series in powers of ε:

$$\Omega_\varepsilon = \{(x, y(x, \varepsilon))\}, \ y(x, \varepsilon) = y_x + \varepsilon y^1(x) + \varepsilon^2 y^2(x) + \dots \,.$$

The fibers of fast motions can be constructed in a form of a power series too (the zero term is the fast motion $\dot{y} = g(x, y)$ in the affine planes $x = \text{const}$). This analytic continuation with respect to the parameter ε for small $\varepsilon > 0$ is studied in the Fenichel's "Geometric singular perturbation theory" [352,353] (recent applications to chemical kinetics see in [95]). As it was mentioned above, the first successful application of such an approach for the construction of a slow invariant manifold in the form of Taylor series expansion in powers of small parameter ε was the Chapman-Enskog expansion [70].

It is wellknown in various applications that there are many different ways to introduce a small parameter into a system, there are many ways to include a given system in a one-parametric family. Different ways of specification of such a parameter result in different definitions of slowness of positively invariant manifold. Therefore it is desirable to study the notion of separation of motions without such an artificial specification. *The notion of slow positively invariant manifold should be intrinsic.* At least we should try to invent such a notion.

4 Film Extension of the Dynamics: Slowness as Stability

4.1 Equation for the Film Motion

One of the difficulties in the problem of reducing the description is caused by the fact that there exists no commonly accepted formal definition of slow (and stable) positively invariant manifolds. Classical definitions of stability and asymptotic stability of the invariant sets sound as follows: Let a dynamical system be defined in some metric space (so that we can measure distances between points), and let $x(t, x_0)$ be a motion of this system at time t with the initial condition $x(0) = x_0$ at time $t = 0$. The subset S of the phase space is called *invariant* if it is made of whole trajectories, that is, if $x_0 \in S$ then $x(t, x_0) \in S$ for all $t \in (-\infty, \infty)$.

Let us denote as $\rho(x, y)$ the distance between the points x and y. The distance from x to a closed set S is defined as usual: $\rho(x, S) = \inf\{\rho(x, y) | y \in S\}$. The closed invariant subset S is called *stable*, if for every $\epsilon > 0$ there exists $\delta > 0$ such that if $\rho(x_0, S) < \delta$, then for every $t > 0$ it holds $\rho(x(t, x_0), S) < \epsilon$. A closed invariant subset S is called *asymptotically stable* if it is stable and attractive, that is, there exists $\epsilon > 0$ such that if $\rho(x_0, S) < \epsilon$, then $\rho(x(t, x_0), S) \to 0$ as $t \to \infty$.

Formally, one can reiterate the definitions of stability and of the asymptotic stability for positively invariant subsets. Moreover, since in the definitions mentioned above it goes only about $t \geq 0$ or $t \to \infty$, it might seem that positively invariant subsets can be a natural object of study concerning stability issues. Such conclusion is misleading, however. The study of the classical stability of the positively invariant subsets reduces essentially to the notion of stability of invariant sets – maximal attractors.

Let Y be a closed positively invariant subset of the phase space. *The maximal attractor* for Y is the set M_Y,

$$M_Y = \bigcap_{t \geq 0} T_t(Y) , \qquad (4.1)$$

where T_t is the shift operator for the time t:

$$T_t(x_0) = x(t, x_0) .$$

Alexander N. Gorban and Iliya V. Karlin: *Invariant Manifolds for Physical and Chemical Kinetics*, Lect. Notes Phys. **660**, 69–78 (2005)
www.springerlink.com

The maximal attractor M_Y is invariant, and the stability of Y defined classically is equivalent to the stability of M_Y under any sensible assumption about uniform continuity (for example, it is so for a compact phase space).

For systems which relax to a stable equilibrium, the maximal attractor is simply one and the same for any bounded positively invariant subset, and it consists of a single stable point.

It is important to note that in the definition (4.1) one considers motions of a positively invariant subset to equilibrium *along itself*: $T_t Y \subset Y$ for $t \geq 0$. It is precisely this motion which is uninteresting from the perspective of the comparison of stability of positively invariant subsets. If one subtracts this *motion along itself* out of the vector field $J(x)$ (3.1), one obtains a less trivial picture.

We again assume submanifolds in U parameterized with a single parameter set $F : W \rightarrow U$. Note that there exists a wide class of transformations which do not alter the geometric picture of motion: For a smooth diffeomorphism $\varphi : W \rightarrow W$ (a smooth coordinate transform), maps F and $F \circ \varphi$ define the same geometric pattern in the phase space.

Let us consider motions of the manifold $F(W)$ along solutions of equation (3.1). Denote as F_t the time-dependent map, and write equation of motion for this map:

$$\frac{dF_t(y)}{dt} = J(F_t(y)) . \qquad (4.2)$$

Let us now subtract the component of the vector field responsible for the motion of the map $F_t(y)$ along itself from the right hand side of equation (4.2). In order to do this, we decompose the vector field $J(x)$ in each point $x = F_t(y)$ as

$$J(x) = J_\|(x) + J_\perp(x) , \qquad (4.3)$$

where $J_\|(x) \in T_{t,y}$ ($T_{t,y} = (D_y F_t(y)(L))$. If projectors are well defined, $P_{t,y} = P(F_t(y), T_{t,y})$, then decomposition (4.3) has the form:

$$J(x) = P_{t,y} J(x) + (1 - P_{t,y}) J(x) . \qquad (4.4)$$

Subtracting the component $J_\|$ from the right hand side of equation (4.2), we obtain,

$$\frac{dF_t(y)}{dt} = (1 - P_{t,y}) J(F_t(y)) . \qquad (4.5)$$

Note that the geometric pictures of motion corresponding to equations (4.2) and (4.5) are identical *locally* in y and t. Indeed, the infinitesimal shift of the manifold W along the vector field is easily computed:

$$(D_y F_t(y))^{-1} J_\|(F_t(y)) = (D_y F_t(y))^{-1} (P_{t,y} J(F_t(y))) . \qquad (4.6)$$

This defines a smooth change of the coordinate system (assuming all solutions exist). In other words, the component $J_\perp$ defines the motion of the manifold in U, while we can consider (locally) the component $J_\parallel$ as a component which locally defines motions in W (a coordinate transform).

The positive semi-trajectory of motion (for $t > 0$) of any submanifold in the phase space along the solutions of initial differential equation (3.1) (without subtraction of $J_\parallel(x)$) is the positively invariant manifold. The closure of such semi-trajectory is an invariant subset. The construction of the invariant manifold as a trajectory of an appropriate initial edge may be useful for producing invariant exponentially attracting set [173, 174]. Very recently, the notion of exponential stability of invariants manifold for ODEs was revised by splitting motions into tangent and transversal (orthogonal) components in [175].

We further refer to equation (4.5) as *the film extension* of the dynamical system (3.1). The phase space of the dynamical system (4.5) is the set of maps F (films). Fixed points of equation (4.5) are solutions to the invariance equation in the differential form (3.3). These include, in particular, all positively invariant manifolds. Stable or asymptotically stable fixed points of equation (4.5) are the slow manifolds we are interested in. It is the notion of stability associated with the film extension of the dynamics which is relevant to our study. In Chap. 9, we consider relaxation methods for constructing slow positively invariant manifolds on the basis of the film extension (4.5).

4.2 Stability of Analytical Solutions

When studying the Cauchy problem for equation (4.5), one should ask a question of how to choose the boundary conditions the function F must satisfy at the boundary of W. Without fixing the boundary conditions, the general solution of the Cauchy problem for the film extension equations (4.5) in the class of smooth functions on W is essentially ambiguous.

The boundary of W, ∂W, splits in two pieces: $\partial W = \partial W_+ \bigcup \partial W_-$. For a smooth boundary these parts can be defined as

$$\partial W_+ = \{y \in \partial W \,|\, (\nu(y), (DF(y))^{-1}(P_y J(F(y)))) < 0\} \,,$$
$$\partial W_- = \{y \in \partial W \,|\, (\nu(y), (DF(y))^{-1}(P_y J(F(y)))) \geq 0\} \,. \qquad (4.7)$$

where $\nu(y)$ denotes the unit outer normal vector at the boundary point y, and $(DF(y))^{-1}$ is the isomorphism of the tangent space T_y on the linear space of parameters L.

One can understand the boundary splitting (4.7) in such a way: The projected vector field $P_y J(F(y))$ defines dynamics on the manifold $F(W)$, this dynamics is the image of some dynamics on W. The corresponding vector field on W is $v(y) = (DF(y))^{-1}(P_y J(F(y)))$. The boundary part ∂W_+ consists of points y, where the velocity vector $v(y)$ points inside W, while for $y \in \partial W_-$ this vector $v(y)$ is directed outside of W (or is tangent to ∂W). The splitting $\partial W = \partial W_+ \bigcup \partial W_-$ depends on t with the vector field $v(y)$:

$$v_t(y) = (DF_t(y))^{-1}(P_y J(F_t(y))) \, ,$$

and the dynamics of $F_t(y)$ is determined by (4.5).

If we would like to derive a solution of the film extension (4.5) $F(y, t)$ for $(y, t) \in W \times [0, \tau]$ for some time $\tau > 0$, then it is necessary to fix some boundary conditions on ∂W_+ (for the "incoming from abroad" part of the function $F(y)$).

Nevertheless, there is a way to study equation (4.5) in W without introducing any boundary condition. It is in the spirit of the classical Cauchy-Kovalevskaya theorem [176–178] about analytical Cauchy problem solutions with analytical data, as well as in the spirit of the classical Lyapunov auxiliary theorem about analytical invariant manifolds in the neighborhood of a fixed point [3,52] and the Poincaré theorem [50] about analytical linearization of analytical non-resonant contractions (see [181]).

We note in passing that recently the interest to the classical analytical Cauchy problem is revived in the mathematical physics literature [179,180]. In particular, analogs of the Cauchy-Kovalevskaya theorem were obtained for the generalized Euler equations [179]. A technique to estimate the convergence radii of the series emerging therein was also developed.

Analytical solutions to equation (4.5) do not require boundary conditions on the boundary of W. The analyticity condition itself allows finding unique analytical solutions of the equation (4.5) with the analytical right hand side $(1 - P)J$ for analytical initial conditions F_0 in W (assuming that such solutions exist). Of course, the analytical continuation without additional regularity conditions is an ill-posed problem. However, it may be useful to switch from functions to germs[1]: we can solve chains of ordinary differential equations for Taylor coefficients instead of partial differential equations for functions (4.5), and it may be possible to prove the convergence of the Taylor series thus obtained. This is the way to prove the Lyapunov auxiliary theorem [3], and one of the known ways to prove the Cauchy-Kovalevskaya theorem.

Let us consider the system (3.1) with stable equilibrium point x^*, real analytical right hand side J, and real analytical projector field $P(x, T): E \to T$. We shall study real analytical sub-manifolds, which include the equilibrium point point x^* ($0 \in W, F(0) = x^*$). Let us expand F in a Taylor series in the neighborhood of zero:

$$F(y) = x^* + A_1(y) + A_2(y, y) + \ldots + A_k(y, y, \ldots, y) + \ldots \, , \qquad (4.8)$$

where $A_k(y, y, \ldots, y)$ is a symmetric k-linear operator ($k = 1, 2, \ldots$).

Let us expand also the right hand side of the film equation (4.5). Matching operators of the same order, we obtain a hierarchy of equations for $A_1, \ldots, A_k, \ldots$:

---

[1] The germ is the sequences of Taylor coefficients that represent an analytical function near a given point.

$$\frac{\mathrm{d}A_k}{\mathrm{d}t} = \Psi_k(A_1, \ldots, A_k) \, . \tag{4.9}$$

It is crucially important, that the dynamics of A_k does not depend on $A_{k+1}, \ldots$, and equations (4.9) can be studied in the following order: we first study the dynamics of A_1, then the dynamics of A_2 with the A_1 motion already given, then A_3 and so on.

Let the projector P_y in equation (4.5) be an analytical function of the derivative $D_y F(y)$ and of the deviation $x - x^*$. Let the corresponding Taylor series expansion at the point $(A_1^0(\bullet), x^*)$ have the form:

$$D_y F(y)(\bullet) = A_1(\bullet) + \sum_{k=2}^{\infty} k A_k(y, \ldots, \bullet) \, , \tag{4.10}$$

$$P_y = \sum_{k,m=0}^{\infty} P_{k,m} \underbrace{(D_y F(y)(\bullet) - A_1^0(\bullet), \ldots, D_y F(y)(\bullet) - A_1^0(\bullet)}_{k} \, ;$$

$$\underbrace{F(y) - x^*, \ldots, F(y) - x^*)}_{m} \, ,$$

where $A_1^0(\bullet)$, $A_1(\bullet)$, $A_k(y, \ldots, \bullet)$ are linear operators. $P_{k,m}$ is a $k+m$-linear operator $(k, m = 0, 1, 2, \ldots)$ with values in the space of linear operators $E \to E$. The operators $P_{k,m}$ depend on the operator $A_1^0(\bullet)$ as on a parameter. Let the point of expansion $A_1^0(\bullet)$ be the linear part of F: $A_1^0(\bullet) = A_1(\bullet)$.

Let us represent the analytical vector field $J(x)$ as a power series:

$$J(x) = \sum_{k=1}^{\infty} J_k(x - x^*, \ldots, x - x^*) \, , \tag{4.11}$$

where J_k is a symmetric k-linear operator $(k = 1, 2, \ldots)$.

Let us write, for example, the first two equations of the equation chain (4.9):

$$\frac{\mathrm{d}A_1(y)}{\mathrm{d}t} = (1 - P_{0,0}) J_1(A_1(y)) \, ,$$

$$\frac{\mathrm{d}A_2(y, y)}{\mathrm{d}t} = (1 - P_{0,0})[J_1(A_2(y, y)) + J_2(A_1(y), A_1(y))]$$

$$- [2P_{1,0}(A_2(y, \bullet)) + P_{0,1}(A_1(y))] J_1(A_1(y)) \, . \tag{4.12}$$

Here, operators $P_{0,0}$, $P_{1,0}(A_2(y, \bullet))$, $P_{0,1}(A_1(y))$ parametrically depend on the operator $A_1(\bullet)$; hence, the first equation is nonlinear, and the second is linear with respect to $A_2(y, y)$. The leading term in the right hand side has the same form for all equations of the sequence (4.9):

$$\frac{\mathrm{d}A_n(y, \ldots, y)}{\mathrm{d}t} \tag{4.13}$$

$$= (1 - P_{0,0}) J_1(A_n(y, \ldots, y)) - n P_{1,0}(A_n(\underbrace{y, \ldots, y}_{n-1}, \bullet)) J_1(A_1(y)) + \ldots \, .$$

There are two important conditions on P_y and $D_yF(y)$: $P_y^2 = P_y$, because P_y is a projector, and $\mathrm{im}P_y = \mathrm{im}D_yF(y)$, because P_y projects on the image of $D_yF(y)$. If we expand these conditions in the power series, then we get the conditions on the coefficients. For example, from the first condition we get:

$$P_{0,0}^2 = P_{0,0} \, ,$$
$$P_{0,0}[2P_{1,0}(A_2(y, \bullet)) + P_{0,1}(A_1(y))] + [2P_{1,0}(A_2(y, \bullet)) + P_{0,1}(A_1(y))]P_{0,0}$$
$$= 2P_{1,0}(A_2(y, \bullet)) + P_{0,1}(A_1(y)), \cdots . \tag{4.14}$$

After multiplication of the second equation in (4.14) with $P_{0,0}$ we get

$$P_{0,0}[2P_{1,0}(A_2(y, \bullet)) + P_{0,1}(A_1(y))]P_{0,0} = 0 . \tag{4.15}$$

Similar identities can be obtained for any oder of the expansion. These equalities allow us to simplify the stationary equation for the sequence (4.9). For example, for the first two equations of the sequence (4.12) we obtain the following stationary equations:

$$(1 - P_{0,0})J_1(A_1(y)) = 0 \, ,$$
$$(1 - P_{0,0})[J_1(A_2(y, y)) + J_2(A_1(y), A_1(y))]$$
$$-[2P_{1,0}(A_2(y, \bullet)) + P_{0,1}(A_1(y))]J_1(A_1(y)) = 0 . \tag{4.16}$$

The operator $P_{0,0}$ is the projector on the space $\mathrm{im}A_1$ (the image of A_1), hence, from the first equation in (4.16) it follows: $J_1(\mathrm{im}A_1) \subseteq \mathrm{im}A_1$. So, $\mathrm{im}A_1$ is a J_1-invariant subspace in E ($J_1 = D_xJ(x)|_{x^*}$) and $P_{0,0}(J_1(A_1(y)) \equiv J_1(A_1(y))$. It is equivalent to the first equation of (4.16). Let us multiply the second equation of (4.16) with $P_{0,0}$ from the left. As a result we obtain the condition:

$$P_{0,0}[2P_{1,0}(A_2(y, \bullet)) + P_{0,1}(A_1(y))]J_1(A_1(y)) = 0 \, ,$$

for solution of equations (4.16), because $P_{0,0}(1 - P_{0,0}) \equiv 0$. If $A_1(y)$ is a solution of the first equation of (4.16), then this condition becomes an identity, and we can write the second equation of (4.16) in the form

$$(1 - P_{0,0})[J_1(A_2(y, y)) + J_2(A_1(y), A_1(y)) - (2P_{1,0}(A_2(y, \bullet))$$
$$+P_{0,1}(A_1(y)))J_1(A_1(y))] = 0 . \tag{4.17}$$

It should be stressed, that the choice of the projector field P_y (4.10) has impact only on the $F(y)$ parametrization, whereas the invariant geometrical properties of the solutions of (4.5) do not depend on the projector field if some transversality and analyticity conditions hold. The conditions of thermodynamic structures preservation significantly reduce ambiguity of the projector choice. One of the most important condition is $\ker P_y \subset \ker D_xS$, where $x = F(y)$ and S is the entropy (see Chap. 5 about the entropy below). The thermodynamic projector is the unique operator which transforms the arbitrary vector field equipped with the given Lyapunov function into a vector field with the same Lyapunov function on the arbitrary submanifold which is not tangent to the level of the Lyapunov function. For the thermodynamic projectors P_y the entropy $S(F(y))$ is conserved on the solutions $F(y,t)$ of the equation (4.5) for any $y \in W$.

If the projectors P_y in equations (4.10)–(4.17) are thermodynamic, then $P_{0,0}$ is the orthogonal projector with respect to the entropic scalar product[2]. For orthogonal projectors the operator $P_{1,0}$ has a simple explicit form. Let $A : L \to E$ be an isomorphic injection (an isomorphism on the image), and $P : E \to E$ be the orthogonal projector on the image of A. The orthogonal projector on the image of the perturbed operator $A + \delta A$ is $P + \delta P$,

$$\delta P = (1 - P)\delta A A^{-1} P + (\delta A A^{-1} P)^{+}(1 - P) + o(\delta A),$$
$$P_{1,0}(\delta A(\bullet)) = (1 - P)\delta A(\bullet)A^{-1}P + (\delta A(\bullet)A^{-1}P)^{+}(1 - P) . \quad (4.18)$$

In (4.18), the operator A^{-1} is defined on $\mathrm{im}A$, $\mathrm{im}A = \mathrm{im}P$, and the operator $A^{-1}P$ acts on E.

Equation (4.18) for δP follows from the three conditions:

$$(P+\delta P)(A+\delta A) = A+\delta A, (P+\delta P)^2 = P+\delta P, (P+\delta P)^{+} = P+\delta P . \quad (4.19)$$

Every A_k is driven by $A_1, \ldots, A_{k-1}$. Stability of the germ of the positively invariant analytical manifold $F(W)$ at point 0 ($F(0) = x^*$) is defined as stability of the solution of the corresponding equations sequence (4.9). Moreover, the notion of the k-jet stability can be useful: let us call k-jet stable such a germ of a positively invariant manifold $F(M)$ at the point 0 ($F(0) = x^*$), if the corresponding solution of the equation sequence (4.9) is stable for $k = 1, \ldots, n$. The simple "triangle" structure of the equation sequence (4.9) with the form (4.13) of principal linear part makes the problem of jets stability very similar for all orders $n > 1$.

Let us demonstrate the stability conditions for the 1-jets in a n-dimensional space E. Let the Jacobian matrix $J_1 = D_x J(x)|_{x^*}$ be selfadjoint with a simple spectrum $\lambda_1, \ldots, \lambda_n$, and the projector $P_{0,0}$ be orthogonal (this is a typical "thermodynamic" situation). The eigenvectors of J_1 form a basis in E: $\{e_i\}_{i=1}^n$. Let a linear space of parameters L be the k-dimensional real space, $k < n$. We shall study the stability of operator A_1^0 which is a fixed point for the first equation of the sequence (4.9). The operator A_1^0 is a fixed point of this equation, if $\mathrm{im}A_1^0$ is a J_1-invariant subspace in E. We discuss full-rank operators, so, for some order of $\{e_i\}_{i=1}^n$ numbering, the matrix of A_1^0 should have a form: $a_{1ij}^0 = 0$, if $i > k$. Let us choose the basis in L: $l_j = (A_1^0)^{-1}e_j$, ($j = 1, \ldots, k$). For this basis $a_{1ij}^0 = \delta_{ij}$, ($i = 1, \ldots, n$, $j =$

---

[2] This scalar product is the bilinear form defined by the negative second differential of the entropy at the point x^*, $-D^2 S(x)$.

$1, \ldots, k$, where δ_{ij} is the Kronecker symbol). The corresponding projectors P and $1 - P$ have the matrices:

$$P = \mathrm{diag}(\underbrace{1, \ldots, 1}_{k}, \underbrace{0, \ldots, 0}_{n-k}), \ 1 - P = \mathrm{diag}(\underbrace{0, \ldots, 0}_{k}, \underbrace{1, \ldots, 1}_{n-k}), \qquad (4.20)$$

where $\mathrm{diag}(\alpha_1, \ldots, \alpha_n)$ is the $n \times n$ diagonal matrix with numbers $\alpha_1, \ldots, \alpha_n$ on the diagonal.

The equations of the linear approximation for the dynamics of the variation δA read:

$$\frac{\mathrm{d}\delta A}{\mathrm{d}t} = \mathrm{diag}(\underbrace{0, \ldots, 0}_{k}, \underbrace{1, \ldots, 1}_{n-k})[\mathrm{diag}(\lambda_1, \ldots, \lambda_n)\delta A - \delta A \mathrm{diag}(\underbrace{\lambda_1, \ldots, \lambda_k}_{k})] \ .$$

$$(4.21)$$

The time derivative of A is orthogonal to A: for any $y, z \in L$ the equality $(\dot{A}(y), A(x)) = 0$ holds, hence, for the stability analysis it is necessary and sufficient to study δA with $\mathrm{im}\delta A_1^0 \perp \mathrm{im}A$. The matrix for such a δA has the form:

$$\delta a_{ij} = 0, \text{ if } i \leq k \ .$$

For $i = k + 1, \ldots, n, \ j = 1, \ldots, k$ equation (4.21) gives:

$$\frac{\mathrm{d}\delta a_{ij}}{\mathrm{d}t} = (\lambda_i - \lambda_j)\delta a_{ij} \ . \qquad (4.22)$$

Therefore, the stability condition becomes:

$$\lambda_i - \lambda_j < 0 \text{ for all } i > k, \ j \leq k \ . \qquad (4.23)$$

This means that the relaxation *towards* $\mathrm{im}A$ (with the spectrum of relaxation times $|\lambda_i|^{-1}$ $(i = k+1, \ldots, n)$) is faster, than the relaxation *along* $\mathrm{im}A$ (with the spectrum of relaxation times $|\lambda_j|^{-1}$ $(j = 1, \ldots, k)$).

Let the condition (4.23) hold. For negative λ, it means that the relaxation time for the film (in the first approximation) is:

$$\tau = 1/(\min_{i>k} |\lambda_i| - \max_{j \leq k} |\lambda_j|) \ ,$$

thus it depends on the *spectral gap* in the spectrum of the operator $J_1 = D_x J(x)|_{x^*}$.

It is the gap between spectra of two restrictions of the operator J_1, $J_1^{\|}$ and $J_1^{\perp}$, respectively. The operator $J_1^{\|}$ is the restriction of J_1 on the J_1-invariant subspace $\mathrm{im}A_1^0$ (it is the tangent space to the slow invariant manifold at point x^*). The operator $J_1^{\perp}$ is the restriction of J_1 on the orthogonal complement to $\mathrm{im}A_1^0$. This subspace is also J_1-invariant, because J_1 is selfadjoint. The spectral gap between spectra of these two operators is the spectral gap between relaxation *towards* the slow manifold and relaxation *along* this manifold.

The stability condition (4.23) demonstrates that our formalization of the slowness of manifolds as the stability of fixed points for the film extension (4.5) of initial dynamics meets the intuitive expectations.

For the analysis of system (4.9) in the neighborhood of some manifold F_0 ($F_0(0) = x^*$), the following parametrization can be convenient. Let us consider

$$F_0(y) = A_1(y) + \ldots, \quad T_0 = A_1(L)$$

to be a tangent space to $F_0(W)$ at point x^*, $E = T_0 \oplus H$ is the direct sum decomposition.

We shall consider analytical sub-manifolds in the form

$$x = x^* + (y, \Phi(y)) , \tag{4.24}$$

where $y \in W_0 \subset T_0$, W_0 is a neighborhood of zero in T_0, $\Phi(y)$ is an analytical map of W_0 in H, and $\Phi(0) = 0$. Any analytical manifold close to F_0 can be represented in this form.

Let us define the projector P_y that corresponds to the decomposition (4.24), as the projector on T_y parallel to H. Furthermore, let us introduce the corresponding decomposition of the vector field $J = J_y \oplus J_z, J_y \in T_0, J_z \in H$. Then

$$P_y(J) = (J_y, (D_y\Phi(y))J_y) . \tag{4.25}$$

The corresponding equation of motion of the film (4.5) has the following form:

$$\frac{\mathrm{d}\Phi(y)}{\mathrm{d}t} = J_z(y, \Phi(y)) - (D_y\Phi(y))J_y(y, \Phi(y)) . \tag{4.26}$$

If J_y and J_z depend analytically on their arguments, then from (4.26) one can easily obtain a hierarchy of equations of the form (4.9) (of course, $J_y(x^*) = 0$, $J_z(x^*) = 0$).

Using these notions, it is convenient to formulate the *Lyapunov auxiliary theorem* [3]. Let $T_0 = R^m, H = R^p$, and in U an analytical vector field be defined $J(y, z) = J_y(y, z) \oplus J_z(y, z)$, $(y \in T_0, z \in H)$. Assume the following conditions are satisfied:

1. $J(0,0) = 0$;
2. $D_z J_y(y, z)\big|_{(0,0)} = 0$;
3. $0 \notin \mathrm{conv}\{k_1, \ldots, k_m\}$,
   where $k_1, \ldots, k_m$ are the eigenvalues of the operator $D_y J_y(y, z)\big|_{(0,0)}$, and $\mathrm{conv}\{k_1, \ldots, k_m\}$ is the convex hull of $\{k_1, \ldots, k_m\}$;
4. the numbers k_i and λ_j are not related by any equation of the form

$$\sum_{i=1}^{m} m_i k_i = \lambda_j , \tag{4.27}$$

where λ_j ($j = 1, \ldots, p$) are eigenvalues of $D_z J_z(y, z)\big|_{(0,0)}$, and $m_i \geq 0$ are integers, $\sum_{i=1}^{m} m_i > 0$.

Let us also consider an analytical manifold $(y, \Phi(y))$ in U in the neighborhood of zero ($\Phi(0) = 0$) and write for it the differential invariance equation with the projector (4.25):

$$(D_y\Phi(y))J_y(y, \Phi(y)) = J_z(y, \Phi(y)) \ . \tag{4.28}$$

Lyapunov auxiliary theorem. Given conditions 1-4, equation (4.24) has the unique analytical solution in the neighborhood of zero, satisfying the condition $\Phi(0) = 0$.

Recently, various new applications of this theorem were developed [52, 184–186].

In order to weaken the non-resonance condition in [49] the existence of invariant manifolds near fixed points tangent to invariant subspaces of the linearization was proved without assumption that the corresponding space for the linear map is a spectral subspace. (This proof was based on the graph transform method [46].)

Studying germs of invariant manifolds using Taylor series expansion in a neighborhood of a fixed point is definitely useful from the theoretical as well as from the practical perspective. But the well known difficulties pertinent to this approach, of convergence, of small denominators (connected with proximity to the resonances (4.27)) and others call for development of different methods. A hint can be found in the famous KAM theory: one should use iterative methods instead of the Taylor series expansion [4–6]. Below we present two such methods:

–   The Newton method subject to incomplete linearization;
–   The relaxation method which is the Galerkin-type approximation to Newton's method with projection on the defect of invariance (3.3), i.e. on the right hand side of equation (4.5).

5 Entropy, Quasiequilibrium, and Projectors Field

Projection operators P_y contribute both to the invariance equation (3.2), and to the film extension of the dynamics (4.5). Limiting results, exact solutions, etc. only weakly depend on the particular choice of projectors, or do not depend on it at all. However, validity of approximations obtained on each iteration step towards the limit strongly depends on the choice of the projector. Moreover, if we want each approximate solution to be consistent with such physically crucial conditions as the second law of thermodynamics (the entropy of the isolated systems increases), then the choice of the projector becomes practically unique.

In this chapter we consider the main ingredients for constructing the projector, based on the two additional structures:

(a) The moment parameterization,
(b) The entropy and the entropic scalar product.

5.1 Moment Parameterization

Same as in the previous chapters, let a regular map (projection) is defined, $\Pi : U \to W$. We consider only maps $F : W \to U$ which satisfy $\Pi \circ F = 1$. We seek slow invariant manifolds among such maps. (A remark is in order here: sometimes one has to consider F which are defined not on the whole W but only on some subset of it.) In this case, the unique projector consistent with the given structure is the superposition of the differentials (the chain rule):

$$P_y = (D_y F)_y \circ (D_x \Pi)_{F(y)} . \tag{5.1}$$

In the language of differential equations (5.1) has the following significance: First, equation (3.1) is projected,

$$\frac{dy}{dt} = (D_x \Pi)_{F(y)} J(F(y)) . \tag{5.2}$$

Second, the latter equation is lifted back to U with the help of F and its differential,

Alexander N. Gorban and Iliya V. Karlin: *Invariant Manifolds for Physical and Chemical Kinetics*, Lect. Notes Phys. **660**, 79–138 (2005)
www.springerlink.com

$$x(t) = F(y(t)) \, ; \tag{5.3}$$

$$\left. \frac{\mathrm{d}x}{\mathrm{d}t} \right|_{x=F(y)} = (D_y F)_y \left(\frac{\mathrm{d}y}{\mathrm{d}t} \right) = (D_y F)_y ((D_x \Pi)_{F(y)} J(F(y))) = P_y J(F(y)) \, .$$

The most standard example of the construction just described is as follows: x is the distribution density, $y = \Pi(x)$ is the set of selected moments of this density, $F : y \to x$ is a "closure assumption", a distribution density parameterized by the values of the moments y. Another standard example is relevant to problems of chemical kinetics: x is a detailed description of the reacting mixture (including all intermediates and radicals), y are concentrations of stable reactants and products of the reaction.

The moment parameterization and moment projectors (5.1) are often encountered in applications. However, they have certain shortcomings. In particular, it is by far not always the case that the moment projection *transforms a dissipative system into another dissipative system*. Of course, for invariant $F(y)$ *any* projector transforms the dissipative system into a dissipative system. However, for various approximations to invariant manifolds (closure assumptions) this is not readilyso[1]. The property of projectors to *preserve the type of the dynamics* will be imposed below as one of the requirements.

5.2 Entropy and Quasiequilibrium

The dissipation properties of the system (3.1) are described by specifying the *entropy* S, the distinguished *Lyapunov function* which monotonically increases along solutions of equation (3.1). In a certain sense, this Lyapunov function is more fundamental than the system (3.1) itself. That is, usually, the entropy is known much better than the right hand side of equation (3.1). For example, in chemical kinetics, the entropy is obtained from the *equilibrium* data. The same holds for other Lyapunov functions, which are defined by the entropy and by a specification of the reaction conditions (the free energy, $U - TS$, for the isothermal isochoric processes, the free enthalpy, $U - TH$, for the isothermal isobaric processes etc.). On physical grounds, all these entropic Lyapunov functions are proportional (up to additive constants) to the entropy of the minimal isolated system which includes the system under study [115]. In general, with some abuse of language, we term the Lyapunov functional S the entropy elsewhere below, although it may be a different functional for non-isolated systems.

Thus, we assume that a concave functional S is defined in U, such that it takes maximum in an inner point $x^* \in U$. This point is termed the equilibrium.

---

[1] See, e.g. a discussion of this problem for the Tamm–Mott-Smith approximation for the strong shock wave in [9].

For any dissipative system (3.1) under consideration in U, the derivative of S due to equation (3.1) must be nonnegative,

$$\left.\frac{\mathrm{d}S}{\mathrm{d}t}\right|_x = (D_x S)(J(x)) \geq 0 , \tag{5.4}$$

where $D_x S$ is the linear functional, the differential of the entropy, while the equality in (5.4) is attained only in the equilibrium $x = x^*$.

Most of the works on nonequilibrium thermodynamics deal with quasi-equilibrium approximations and corrections to them, or with applications of these approximations (with or without corrections). This viewpoint is not the only possible but it proves very efficient for the construction of a variety of useful models, approximations and equations, as well as methods to solve them. From time to time it is discussed in the literature, who was the first to introduce the quasiequilibrium approximations, and how to interpret them. At least a part of the discussion is due to a different role the quasiequilibrium plays in the entropy-conserving and the dissipative dynamics. The very first use of the entropy maximization dates back to the classical work of G. W. Gibbs [222], but it was first claimed for a principle of informational statistical thermodynamics by E. T. Jaynes [193]. Probably the first explicit and systematic use of quasiequilibria to derive dissipation from entropy-conserving systems was undertaken by D. N. Zubarev. Recent detailed exposition is given in [195]. For dissipative systems, the use of the quasiequilibrium to reduce description can be traced to the works of H. Grad on the Boltzmann equation [201]. A review of the informational statistical thermodynamics was presented in [227]. The connection between entropy maximization and (nonlinear) Onsager relations was also studied [164, 188]. The viewpoint of the present authors was influenced by the papers by L. I. Rozonoer and co-workers, in particular, [223–225]. A detailed exposition of the quasiequilibrium approximation for Markov chains is given in the book [115] (Chap. 3, *Quasiequilibrium and entropy maximum*, pp. 92–122), and for the BBGKY hierarchy in the paper [226]. The maximum entropy principle was applied to the description the universal dependence the three-particle distribution function F_3 on the two-particle distribution function F_2 in classical systems with binary interactions [229]. For a discussion the quasiequilibrium moment closure hierarchies for the Boltzmann equation [224] see the papers [230, 233, 234]. A very general discussion of the maximum entropy principle with applications to dissipative kinetics is given in the review [231]. Recently the quasiequilibrium approximation with some further correction was applied to description of rheology of polymer solutions [254, 266] and of ferrofluids [267, 268]. Quasiequilibrium approximations for quantum systems in the Wigner representation [36, 37] was discussed very recently [232]. We shall now introduce the quasiequilibrium approximation in the most general setting.

A *linear moment parameterization* is a linear operator, $\Pi : E \rightarrow L$, where $L = \mathrm{im}\Pi = E/\ker \Pi$, $\ker \Pi$ is a closed linear subspace of space E, and Π is the projection of E onto factor-space L. Let us denote $W = \Pi(U)$. *Quasiequilibrium* (or restricted equilibrium, or conditional equilibrium, or constrained equilibrium) is the embedding, $F^* : W \to U$, which puts into correspondence to each $y \in W$ the solution to the entropy maximization problem:

$$S(x) \to \max, \ \Pi(x) = y \ . \tag{5.5}$$

We assume that, for each $y \in \operatorname{int} W$, there exists the unique solution $F^*(y) \in \operatorname{int} U$ to the problem (5.5). This solution, $F^*(y)$, is called the quasi-equilibrium, corresponding to the value y of the macroscopic variables. The set of quasiequilibria $F^*(y)$, $y \in W$, forms a manifold in $\operatorname{int} U$, parameterized by the values of the macroscopic variables $y \in W$.

Let us specify some notations: E^T is the adjoint to the E space. Adjoint spaces and operators will be indicated by ^{T}, whereas notation * is earmarked for equilibria and quasiequilibria.

Furthermore, $[l, x]$ is the result of application of the functional $l \in E^T$ to the vector $x \in E$. We recall that, for an operator $A : E_1 \to E_2$, the adjoint operator, $A^T : E_1^T \to E_2^T$ is defined by the following relation: For any $l \in E_2^T$ and $x \in E_1$,

$$[l, Ax] = [A^T l, x] \ .$$

Next, $D_x S(x) \in E^T$ is the differential of the entropy functional $S(x)$, $D_x^2 S(x)$ is the second differential of the entropy functional $S(x)$. The corresponding quadratic functional $D_x^2 S(x)(z, z)$ on E is defined by the Taylor formula,

$$S(x + z) = S(x) + [D_x S(x), z] + \frac{1}{2} D_x^2 S(x)(z, z) + o(\|z\|^2) \ . \tag{5.6}$$

We keep the same notation for the corresponding symmetric bilinear form, $D_x^2 S(x)(z, p)$, and also for the linear operator, $D_x^2 S(x) : E \to E^T$, defined by the formula,

$$[D_x^2 S(x) z, p] = D_x^2 S(x)(z, p) \ .$$

In the latter formula, on the left hand side, there is the operator, on the right hand side there is the bilinear form. Operator $D_x^2 S(x)$ is symmetric on E, $D_x^2 S(x)^T = D_x^2 S(x)$.

Concavity of the entropy S means that, for any $z \in E$, the inequality holds,

$$D_x^2 S(x)(z, z) \le 0 \ ;$$

in the restriction onto the affine subspace parallel to $\ker \Pi$ we assume the strict concavity,

$$D_x^2 S(x)(z, z) < 0, \text{ if } z \in \ker \Pi, \text{ and if } z \ne 0 \ .$$

In the remainder of this section we are going to construct the important object, the projector onto the tangent space of the quasiequilibrium manifold.

Let us compute the derivative $D_y F^*(y)$. For this purpose, let us apply the method of Lagrange multipliers: There exists such a linear functional $\Lambda(y) \in (L)^T$, that

$$D_x S(x)\big|_{F^*(y)} = \Lambda(y) \cdot \Pi, \ \Pi(F^*(y)) = y \ , \tag{5.7}$$

or

$$D_x S(x)\big|_{F^*(y)} = \Pi^T \cdot \Lambda(y), \ \Pi(F^*(y)) = y \ . \tag{5.8}$$

From equation (5.8) we get,

$$\Pi(D_y F^*(y)) = 1_L \ , \tag{5.9}$$

where we have indicated the space in which the unit operator acts. Next, using the latter expression, we transform the differential of the equation (5.7),

$$D_y \Lambda = (\Pi (D_x^2 S)^{-1}_{F^*(y)} \Pi^T)^{-1} \ , \tag{5.10}$$

and, consequently,

$$D_y F^*(y) = (D_x^2 S)^{-1}_{F^*(y)} \Pi^T (\Pi (D_x^2 S)^{-1}_{F^*(y)} \Pi^T)^{-1} \ . \tag{5.11}$$

Notice that, elsewhere in equation (5.11), operator $(D_x^2 S)^{-1}$ acts on the linear functionals from L^T. These functionals are precisely those which become zero on $\ker \Pi$ or, that is the same, those which can be represented as linear functionals of macroscopic variables.

The tangent space to the quasiequilibrium manifold at the point $F^*(y)$ is the image of the operator $D_y F^*(y)$:

$$\mathrm{im}\,(D_y F^*(y)) = (D_x^2 S)^{-1}_{F^*(y)} L^T = (D_x^2 S)^{-1}_{F^*(y)} \mathrm{Ann}(\ker \Pi) \tag{5.12}$$

where $\mathrm{Ann}(\ker \Pi)$ is the set of linear functionals which become zero on $\ker \Pi$. Another way to write equation (5.12) is the following:

$$x \in \mathrm{im}\,(D_y F^*(y)) \Leftrightarrow (D_x^2 S)_{F^*(y)}(z,p) = 0, \ p \in \ker \Pi \ . \tag{5.13}$$

This means that $\mathrm{im}\,(D_y F^*(y))$ is the orthogonal complement of $\ker \Pi$ in E with respect to the scalar product,

$$\langle z|p\rangle_{F^*(y)} = -(D_x^2 S)_{F^*(y)}(z,p) \ . \tag{5.14}$$

The entropic scalar product (5.14) appears often in the constructions below. (Usually, it becomes the scalar product indeed after the conservation laws are excluded). Let us denote as $T_y = \mathrm{im}(D_y F^*(y))$ the tangent space to the quasiequilibrium manifold at the point $F^*(y)$. Important role in the construction of quasiequilibrium dynamics and its generalizations is played by the quasiequilibrium projector, an operator which projects E on T_y parallel to ker Π. This is the orthogonal projector with respect to the entropic scalar product, $P_y^* : E \to T_y$:

$$P_y^* = D_y F^*(y) \cdot \Pi = \left(D_x^2 S \big|_{F^*(y)} \right)^{-1} \Pi^T \left(\Pi \left(D_x^2 S \big|_{F^*(y)} \right)^{-1} \Pi^T \right)^{-1} \Pi \ . \tag{5.15}$$

It is straightforward to check the equality $P_y^{*2} = P_y^*$, and the self-adjointness of P_y^* with respect to the entropic scalar product (5.14). Thus, we have introduced the basic constructions: the quasiequilibrium manifold, the entropic scalar product, and the quasiequilibrium projector.

Quasiequilibrium entropy $S(y)$ is a functional on W. It is defined as the value of the entropy on the corresponding quasiequilibrium $x = F^*(y)$:

$$S(y) = S(F^*(y)) \tag{5.16}$$

Quasiequilibrium dynamics is a dynamics on W, defined by the equation (5.2) for the quasiequilibrium $F^*(y)$:

$$\frac{dy}{dt} = \Pi J(F^*(y)) \ . \tag{5.17}$$

Here Π is constant linear operator (in the general case (5.2), it may become nonlinear). The corresponding quasiequilibrium dynamics on the quasiequilibrium manifold $F^*(W)$ is defined using the projector (5.1):

$$\frac{dx}{dt} = P_y^* \big|_{x=F^*(y)} J(x) = (D_y F^*)_{x=F^*(y)} \Pi J(x), \ x \in F^*(W) \ . \tag{5.18}$$

The orthogonal projector P_y^* in the right hand side of equation (5.18) can be explicitly written using the second derivative of S and the operator Π (5.15). Let's remind that the only distinguished scalar product in E is the entropic scalar product (5.14):

$$\langle z, p \rangle_x = -(D_x^2 S)_x(z, p) \tag{5.19}$$

It depends on the point $x \in U$. This dependence $\langle | \rangle_x$ endows U with the structure of a Riemann space.

The most important property of the quasiequilibrium system (5.17), (5.18) is highlighted by the *conservation of the dynamics type* theorem: if for the original dynamic system (3.1) $\frac{dS}{dt} \geq 0$, then for the quasiequilibrium dynamics $\frac{dS}{dt} \geq 0$. If for the original dynamic system (3.1) $\frac{dS}{dt} = 0$ (conservative system), then for the quasiequilibrium dynamics $\frac{dS}{dt} = 0$ as well.

The construction of the quasiequilibrium allows for the following generalization: Almost every manifold can be represented as a set of minimizers of the entropy under linear constraints. However, in contrast to the standard quasiequilibrium, these linear constraints will depend, generally speaking, on the point on the manifold.

So, let the manifold $\Omega = F(W) \subset U$ be given. However, now macroscopic variables y are not functionals on R or U but just parameters identifying points on the manifold. The problem is how to extend the definitions of y onto a neighborhood of $F(W)$ in such a way that $F(W)$ will become a solution to the variational problem:

$$S(x) \to \max, \ \Pi(x) = y \ . \tag{5.20}$$

For each point $F(y)$, we identify $T_y \in E$, the tangent space to the manifold Ω in F_y, and the subspace $Y_y \subset E$, which depends smoothly on y, and which has the property, $Y_y \oplus T_y = E$. Let us define $\Pi(x)$ in the neighborhood of $F(W)$ in such a way, that

$$\Pi(x) = y, \ \text{if} \ x - F(y) \in Y_y \ . \tag{5.21}$$

The point $F(y)$ is the solution of the quasiequilibrium problem (5.20) if and only if

$$D_x S(x)\big|_{F(y)} \in \text{Ann} \ Y_y \ . \tag{5.22}$$

That is, if and only if $Y_y \subset \ker D_x S(x)\big|_{F(y)}$. It is always possible to construct subspaces Y_y with the properties just specified, at least locally, if the functional $D_x S\big|_{F(y)}$ is not identically equal to zero on T_y.

The construction just described allows to consider practically any manifold as a quasiequilibrium. This construction is required when one seeks the induced dynamics on a given manifold. Then the vector fields are projected on T_y parallel to Y_y, and this preserves the basic properties of the quasiequilibrium approximations.

5.3 Thermodynamic Projector without a Priori Parameterization

Quasiequilibrium manifolds is a place where the entropy and the moment parameterization "meet each other". The projector P_y for a quasiequilibrium manifold is nothing but the orthogonal with respect to the entropic scalar product $\langle | \rangle_x$ projector (5.15). The quasiequilibrium projector preserves the type of dynamics. Note that in order to preserve the type of dynamics we needed only one condition to be satisfied,

$$\ker P_y \subset \ker(D_x S)_{x=F(y)} \ . \tag{5.23}$$

Let us require that the field of projectors, $P(x, T)$, is defined for any x and T satisfying the following transversality condition holds

$$T \not\subset \ker D_x S .\tag{5.24}$$

It follows immediately from the condition (5.23) that in the equilibrium, $P(x^*, T)$ is the orthogonal projector onto T (ortogonality is with respect to the entropic scalar product $\langle | \rangle_{x^*}$).

The field of projectors was constructed in the neighborhood of the equilibrium following the requirement of the maximal smoothness of P as a function of $g_x = D_x S$ and x [22]. It turns out that to the first order in the deviations $x - x^*$ and $g_x - g_{x^*}$, the projector is defined uniquely. Let us first describe the construction of the projector, and next discuss its uniqueness [10].

Let the subspace $T \subset E$, the point x, and the differential of the entropy at this point, $g = D_x S$, be defined in such a way that the transversality condition (5.24) is satisfied. Let us define $T_0 = T \cap \ker g_x$. By the condition (5.24), $T_0 \neq T$. Let us denote, $e_g = e_g(T) \in T$ the vector in T, such that e_g is orthogonal to T_0, and is normalized by the condition $g(e_g) = 1$. The vector e_g is defined unambiguously. The projector $P_{S,x} = P(x, T)$ is defined as follows: For any $z \in E$,

$$P_{S,x}(z) = P_0(z) + e_g g_x(z) ,\tag{5.25}$$

where P_0 is the orthogonal projector on T_0 (orthogonality is with respect to the entropic scalar product $\langle | \rangle_x$). The *thermodynamic projector* (5.25) depends on the point x through the x-dependence of the scalar product $\langle | \rangle_x$, and also through the differential of S in x, the functional g_x. Further we shall often omit the index S in $P_{S,x}$.

Obviously, $P(z) = 0$ implies $g(z) = 0$, that is, the thermodynamicity requirement (5.23) is satisfied. Uniqueness of the thermodynamic projector (5.25) is supported by the requirement of the *maximal smoothness* (analyticity) [22] of the projector as a function of g_x and $\langle | \rangle_x$, and is done in two steps which we sketch here (detailed proof is given in the next section, following the paper [10]):

1. Considering the expansion of the entropy at the equilibrium up to the quadratic terms, one demonstrates that in the equilibrium the thermodynamic projector is the orthogonal projector with respect to the scalar product $\langle | \rangle_{x^*}$.
2. For a given g, one considers auxiliary dissipative dynamic systems (3.1), which satisfy the condition: For every $x' \in U$, it holds, $g_x(J(x')) = 0$, that is, g_x defines an additional linear conservation law for the auxiliary systems. For the auxiliary systems, the point x is the equilibrium. Eliminating the linear conservation law g_x, and using the result of the previous point, we end up with the formula (5.25).

Thus, the entropic structure defines unambiguously the field of projectors (5.25), for which the dynamics of *any* dissipative system (3.1) projected on *any* closure manifold remains dissipative.

5.4 Uniqueness of Thermodynamic Projector

In this section, the uniqueness theorem for thermodynamic projector will be proved.

5.4.1 Projection of Linear Vector Field

Let E be real Hilbert space with the scalar product $\langle\,|\,\rangle$, Q be a set of linear bounded operators in E with negatively definite quadratic form $\langle Ax \mid x\rangle \leq 0$ for every $A \in Q$, $T \subsetneq E$ be a nontrivial $(T \neq \{0\})$ closed subspace. For every projector $P : E \to T$ $(P^2 = P)$ and linear operator $A : E \to E$ we define the projected operator $P(A) : T \to T$ in such a way:

$$P(A)x = PAx \equiv PAPx \text{ for } x \in T \ . \tag{5.26}$$

The space T is the Hilbert space with the scalar product $\langle\,|\,\rangle$. Let Q_T be a set of linear bounded operators in T with negatively definite quadratic form $\langle Ax \mid x\rangle \leq 0$.

Proposition 1. *The inclusion $P(Q) \subseteq Q_T$ for a projector $P : E \to T$ holds if and only if P is the orthogonal projector with respect to the scalar product $\langle\,|\,\rangle$.*

Proof. If P is orthogonal (and, hence, selfadjoint) and $\langle Ax \mid x\rangle \leq 0$, then

$$\langle PAPx \mid x\rangle = \langle APx \mid Px\rangle \leq 0 \ .$$

If P is not orthogonal, then $Px \neq 0$ for some vector $x \in T^\perp$ in orthogonal complement of T. Let us consider the negatively definite selfadjoint operator

$$A_x = -\mid Px - ax\rangle\langle Px - ax \mid$$

$(A_x y = -(Px - ax)\langle Px - ax \mid y\rangle)$. The projection of A_x on T is:

$$P(A_x) = (a - 1)\mid Px\rangle\langle Px \mid \ .$$

This operator is not negatively definite for $a > 1$. $\square$

Immediately from this proof follows the Corollary 1.

Corollary 1. *Let $Q^{\mathrm{sym}} \subset Q$ be a subset of selfadjoint operators in E. The inclusion $P(Q^{\mathrm{sym}}) \subseteq Q_T$ for a projector $P : E \to T$ holds if and only if P is the orthogonal projector with respect to the scalar product $\langle\,|\,\rangle$.* $\square$

Corollary 2. *Let $Q_T^{\mathrm{sym}} \subset Q_T$ be a subset of selfadjoint operators in T. If $P(Q) \subseteq Q_T$ for a projector $P : E \to T$, then $P(Q^{\mathrm{sym}}) \subseteq Q_T^{\mathrm{sym}}.$* $\square$

It follows from the Proposition 1 and the obvious remark: If operators A and P are selfadjoint, then operator PAP is selfadjoint too.

The Proposition 1 means that a projector which transforms every linear vector field Ax with Lyapunov function $\langle x \mid x \rangle$ into projected vector field $PAPx$ with the same Lyapunov function is orthogonal with respect to the scalar product $\langle \mid \rangle$.

According to the Corollary 1, the conditions of the Proposition 1 can be made weaker: A projector which transforms every *selfadjoint* linear vector field Ax with Lyapunov function $\langle x \mid x \rangle$ into projected vector field $PAPx$ with the same Lyapunov function is orthogonal with respect to the scalar product $\langle \mid \rangle$. In physical applications it means, that we can deal with requirement of dissipation persistence for vector field with Onsager's reciprocity relations. The consequence of such a requirement will be the same, as for the class of all continuous linear vector field: The projector should be orthogonal.

The Corollary 2 is a statement about persistence of the reciprocity relations.

5.4.2 The Uniqueness Theorem

In this subsection we discuss finite-dimensional systems. There are technical details which make the theory of nonlinear infinite-dimensional case too cumbersome: the Hilbert spaces equipped with entropic scalar product $\langle \mid \rangle_x$ (5.14) for different x consist of different functions. Of course, there exists a common dense subspace, and geometrical sense remains the same, as for the finite-dimensional space, but we defer the discussion of all the details till a special mathematical publication.

Let E be n-dimensional real vector space, $U \subset E$ be a domain in E, and a m-dimensional space of parameters L be defined, $m < n$, and let W be a domain in L. We consider differentiable maps, $F : W \to U$, such that, for every $y \in W$, the differential of F, $D_y F : L \to E$, is an isomorphism of L on a subspace of E. That is, F are the manifolds, immersed in the phase space of the dynamic system (3.1), and parametrized by parameter set W.

Let the twice differentiable function S on U be given (the entropy). We assume that S is strictly concave in the second approximation: The quadratic form defined by second differential of the entropy $D_x^2 S(y, y)$ is strictly negative definite in E for every $x \in U$. We will use the entropic scalar product (13.2). Let S have the interior point of maximum in U: $x^{\mathrm{eq}} \in \mathrm{int}U$.

The function S is Lyapunov function for a vector field J in U, if

$$(D_x S)(J(x)) \geq 0 \text{ for every } x \in U \ .$$

First of all, we shall study vector fields with Lyapunov function S in the neighborhood of x^{eq}. Let $0 \in \mathrm{int}W$, $F : W \to U$ be an immersion, and $F(0) = x^{\mathrm{eq}}$. Let us define $T_y = \mathrm{im}D_y F(y)$ for each $y \in W$. This T_y is the tangent space to $F(W)$ in the point y. Assume that the mapping F is sufficiently smooth, and $F(W)$ is not tangent to entropy levels:

$$T_y \not\subset \ker D_x S|_{x=F(y)}$$

for every $y \neq 0$. The thermodynamic projector for a given F is a projector-valued function $y \mapsto P_y$, where $P_y : E \to T_y$ is a projector. The *thermodynamic conditions* reads: *For every smooth vector field $J(x)$ in U with Lyapunov function S the projected vector field $P_y(J(F(y)))$ on $F(W)$ has the same Lyapunov function $S(F(y))$.*

Proposition 1 and Corollaries 1, 2 make it possible to prove uniqueness of the thermodynamic projector for the weakened thermodynamic conditions too: *For every smooth vector field $J(x)$ in U with Lyapunov function S and selfadjoint Jacobian operator for every equilibrium point (zero of $J(x)$) the projected vector field $P_y(J(F(y)))$ on $F(W)$ has the same Lyapunov function $S(F(y))$.* We shall not discuss it separately.

Proposition 2. *Let the thermodynamic projector P_y be a smooth function of y. Then*

$$P_0 = P_0^{\perp} \text{ and } P_y = P_y^{\perp} + O(y) , \tag{5.27}$$

where $P_y^{\perp}$ is orthogonal projector onto T_y with respect to the entropic scalar product $\langle \,|\, \rangle_{F(y)}$.

Proof. A smooth vector field in the neighborhood of $F(0) = x^{\mathrm{eq}}$ can be presented as $A(x - x^{\mathrm{eq}}) + o(\|x - x^{\mathrm{eq}}\|)$, where A is a linear operator. If S is the Lyapunov function for this vector field, then the quadratic form $\langle Ax \,|\, x \rangle_{x^{\mathrm{eq}}}$ is negatively definite. $P_y = P_0 + O(y)$, because P_y is a continuous function. Hence, for P_0 we have the problem solved by the Proposition 1, and $P_0 = P_0^{\perp}$. $\square$

Theorem 1. *Let the thermodynamic projector P_y be a smooth function of y. Then*

$$P_y = P_{0y} + e_g D_x S|_{x=F(y)} , \tag{5.28}$$

where notations of formula (13.4) are used: T_{0y} is the kernel of linear functional $D_x S|_{x=F(y)}$ in T_y, $P_{0y} : T_{0y} \to E$ is the orthogonal projector with respect to the entropic scalar product $\langle \,|\, \rangle_{F(y)}$ (5.14). Vector $e_g \in T$ is proportional to the Riesz representation g_y of linear functional $D_x S|_{x=F(y)}$ in T_y with respect to the entropic scalar product:

$$\langle g_y \,|\, x \rangle_{F(y)} = (D_x S|_{x=F(y)})(x)$$

for every $x \in T_y$, $e_g = g_y / \langle g_y \,|\, g_y \rangle_{F(y)}$.

Proof. Let $y \neq 0$. Let us consider an auxiliary class of vector fields J on U with additional linear balance $(D_x S)_{x=F(y)})(J) = 0$. If such a vector field has Lyapunov function S, then $x = F(y)$ is its equilibrium point: $J(F(y)) = 0$.

The class of vector fields with this additional linear balance and Lyapunov function S is sufficiently rich and we can use the Propositions 1, 2 for dynamics on the auxiliary phase space

$$\{z \in U | (D_x S|_{x=F(y)})(z - F(y)) = 0\} \, .$$

Hence, the restriction of P_y on the hyperplane $\ker D_x S|_{x=F(y)}$ is P_{0y}. Formula (5.28) gives the unique continuation of this projector on the whole E. $\Box$

5.4.3 Orthogonality of the Thermodynamic Projector and Entropic Gradient Models

In Euclidean spaces with the given scalar product, we often identify the differential of a function $f(x)$ with its gradient: in the orthogonal coordinate system $(\operatorname{grad} f(x))_i = \partial f(x)/\partial x_i$. However, when dealing with a more general setting, one can run into problems while making sense out of such a definition. What to do, if there is no distinguished scalar product, no preselected orthogonality?

For a given scalar product $\langle | \rangle$ the gradient $\operatorname{grad}_x f(x)$ of a function $f(x)$ at a point x is such a vector g that $\langle g|y \rangle = D_x f(y)$ for any vector y, where $D_x f$ is the differential of function f at a point x. The differential of function f is the linear functional that provides the best linear approximation near the given point.

In order to transform a vector into a linear functional one needs a *pairing*, that means a bilinear form $\langle | \rangle$. This pairing transforms vector g into linear functional $\langle g|$: $\langle g|(x) = \langle g|x \rangle$. Any twice differentiable function $f(x)$ generates a field of pairings: at any point x there exists a second differential of f, a quadratic form $(D_x^2 f)(\Delta x, \Delta x)$. For a convex function these forms are positively definite, and we return to the concept of scalar product. Let us calculate a gradient of f using this scalar product. In coordinate representation the identity $\langle \operatorname{grad} f(x) | y \rangle_x = (D_x f)(y)$ (for any vector y) has a form

$$\sum_{i,j} (\operatorname{grad} f(x))_i \frac{\partial^2 f}{\partial x_i \partial x_j} y_j = \sum_i \frac{\partial f}{\partial x_j} y_j \, , \tag{5.29}$$

hence,

$$(\operatorname{grad} f(x))_i = \sum_j (D_x^2 f)_{ij}^{-1} \frac{\partial f}{\partial x_j} \, . \tag{5.30}$$

As we can see, this $\operatorname{grad} f(x)$ is the *Newtonian direction*, and with this gradient the method of steepest descent transforms into the Newton method of optimization.

Entropy is the concave function and we defined the entropic scalar product through negative second differential of entropy (13.2). Let us define the gradient of entropy by means of this scalar product: $\langle \operatorname{grad}_x S|z \rangle_x = (D_x S)(z)$. The *entropic gradient system* is

$$\frac{dx}{dt} = \varphi(x)\mathrm{grad}_x S , \qquad (5.31)$$

where $\varphi(x) > 0$ is a positive kinetic multiplier.

The system (5.31) is a representative of a family of *model kinetic equations*. One replaces complicated kinetic equations by model equations for simplicity. The main requirements to such models are: they should be as simple as possible and should not violate the basic physical laws. The most known model equation is the BGK model [116] for the collision integral in the Boltzmann equation. There are different models for simplifying kinetics [117, 118]. The entropic gradient models (5.31) possesses all the required properties (if the entropy Hessian is sufficiently simple). It was invented for the lattice Boltzmann kinetics [166]. In many cases it is simpler than the BGK model, because the gradient model is *local* in the sense that it uses only the entropy function and its derivatives at a current state, and it is not necessary to compute the equilibrium (or quasiequilibrium for quasiequilibrium models 2.92 [22, 117]). The entropic gradient model has a one-point relaxation spectrum, because near the equilibrium x^{eq} the gradient vector field (5.31) has an extremely simple linear approximation: $d(\Delta x)/dt = -\varphi(x^{eq})\Delta x$. It corresponds to a well-known fact that the Newton method minimizes a positively defined quadratic form in one step.

A direct computation shows that the thermodynamic projector P (13.4) in a point x onto the tangent space T can be rewritten as

$$P(J) = P^{\perp}(J) + \frac{\mathrm{grad}_x S^{\|}}{\langle \mathrm{grad}_x S^{\|} | \mathrm{grad}_x S^{\|} \rangle_x} \langle \mathrm{grad}_x S^{\perp} | J \rangle_x , \qquad (5.32)$$

where $P^{\perp}$ is the orthogonal projector onto T with respect the entropic scalar product, and the gradient $\mathrm{grad}_x S$ is splitted onto tangent and orthogonal components:

$$\mathrm{grad}_x S = \mathrm{grad}_x S^{\|} + \mathrm{grad}_x S^{\perp} :$$

$$\mathrm{grad}_x S^{\|} = P^{\perp}\mathrm{grad}_x S; \ \mathrm{grad}_x S^{\perp} = (1 - P^{\perp})\mathrm{grad}_x S .$$

From (5.32) it follows that the two properties of an ansatz manifolds are equivalent: orthogonality of the thermodynamic projector and invariance of the manifold with respect to the entropic gradient system (5.31).

Proposition 3. *The thermodynamic projector for an ansatz manifold Ω is orthogonal at any point $x \in \Omega$ if and only if $\mathrm{grad}_x S \in T_x(\Omega)$ at any point $x \in \Omega$.* $\square$

It should be possible to think of gradients as infinitesimal displacements of points x. Usually there are some balances, at least the conservation of the total probability, and the gradient should belong to a given subspace of zero balances change. For example, for the classical Boltzmann-Gibbs-Shannon entropy $(x = \Psi(q))$, $S = -\int \Psi(q)(\ln \Psi(q) - 1)\, dq$, the entropic scalar product is $\langle g(q)|f(q)\rangle_\Psi = \int g(q)f(q)/\Psi(q)\,dq$, and $\operatorname{grad}_\Psi S = -\Psi(q)\ln(\Psi(q)) + c(q)$, where function (vector) $c(q)$ is orthogonal to a given subspace of zero balances. This function have to be founded from the conditions of zero balances for the gradient $\operatorname{grad}_\Psi S$. For example, if the only balance is the conservation of the total probability, $\int \Psi(q)\,dq \equiv 1$, then for the classical Boltzmann-Gibbs-Shannon entropy S

$$\operatorname{grad}_\Psi S = -\Psi(q)\left(\ln(\Psi(q)) - \int \Psi(q')\ln(\Psi(q'))\,dq'\right). \qquad (5.33)$$

For the Kullback-form entropy (i.e. for the negative free energy or the Massieu-Planck function)

$$S = -F/T = -\int \Psi(q)\left(\ln\left(\frac{\Psi(q)}{\Psi^{\mathrm{eq}}(q)}\right) - 1\right)dq,$$

the second differential and the entropic scalar product are the same, as for the classical Boltzmann-Gibbs-Shannon entropy, and

$$\operatorname{grad}_\Psi S = -\Psi(q)\left(\ln\left(\frac{\Psi(q)}{\Psi^{\mathrm{eq}}(q)}\right) - \int \Psi(q')\ln\left(\frac{\Psi(q)}{\Psi^{\mathrm{eq}}(q)}\right)dq'\right). \qquad (5.34)$$

For more complicated system of balances, linear or non-linear, the system of linear equations for $c(q)$ can also be written explicitly.

5.4.4 Violation of the Transversality Condition, Singularity of Thermodynamic Projection, and Steps of Relaxation

The thermodynamic projector transforms the arbitrary vector field equipped with the given Lyapunov function into a vector field with the same Lyapunov function for a given ansatz manifold which is not tangent to the Lyapunov function levels. Sometimes it is useful to create an ansatz which violates this transversality condition. The point of entropy maximum on such an ansatz is not the equilibrium. The usual examples are: the non-correlated approximation $x = \Psi(q_1, \ldots, q_n) = \prod_i f(q_i)$, the Gaussian manifold for a non-quadratic potential, etc. Such manifolds arise often in applications because of simplicity of computations. However, for these manifolds the thermodynamic projector becomes singular in the point of entropy maximum x^* on the ansatz manifold. This is obvious from (5.32): in the neighborhood of x^* it has the form

$$P(J) = P^\perp(J) + \frac{\operatorname{grad}_x S^\|}{\langle \operatorname{grad}_x S^\| | \operatorname{grad}_x S^\|\rangle_x}\langle \operatorname{grad}_x S^\perp | J\rangle_x$$

$$= -\frac{\Delta x}{\langle \Delta x | \Delta x\rangle_{x^*}}\sigma(x^*) + O(1), \qquad (5.35)$$

where $\Delta x = x - x^*$ is the deviation of x from x^*, $\sigma(x^*) = \langle \mathrm{grad}_{x^*} S^\perp | J \rangle_{x^*}$ is the entropy production at the point x^*, $\sigma(x^*) \neq 0$, because the point of entropy maximum x^* is not the equilibrium. In this case the projected system in the neighborhood of x^* reaches the point x^* in finite time t^* as $\sqrt{t^* - t}$ goes to zero. The entropy difference $\Delta S = S(x) - S(x^*) = -\frac{1}{2}\langle \Delta x | \Delta x \rangle_{x^*} + o(\langle \Delta x | \Delta x \rangle_{x^*})$ goes to zero as $-\sigma(x^*)(t^* - t)$ $(t \leq t^*)$.

The singularity of projection has a transparent physical sense. The relaxation along the ansatz manifold to the point x^* is not complete, because this point is not the equilibrium. This motion should be considered as a step of relaxation, and after it was completed, the next step should start. In that sense it is obvious that the motion to the point x^* along the ansatz manifold should take the finite time. The results of this step-by-step relaxation can represent the whole process (with smoothing [26], or without it [27]). The experience of such a step-by-step computing of relaxation trajectories in the initial layer problem for the Boltzmann kinetics demonstrated its efficiency (see [26, 27] and Sect. 9.3).

5.4.5 Thermodynamic Projector, Quasiequilibrium, and Entropy Maximum

The thermodynamic projector projects any vector field which satisfies the second law of thermodynamics into the vector field which satisfies the second law too. Other projectors violate the second law. But what does it mean? Each projector P_x onto tangent space of an ansatz manifold in a point x induces the fast-slow motion splitting: Fast motion is the motion parallel to $\ker P_x$ (on the affine subspace $x + \ker P_x$ in the neighborhood of x), slow motion is the motion on the slow manifold and in the first order it is parallel to the tangent space T_x in the point x (in the first order this slow manifold is the affine subspace $x + \mathrm{im} P_x$, $T_x = \mathrm{im} P_x$), and velocity vector of the slow motion in point x belongs to the image of P_x.

If P_x is the thermodynamic projector, then x is the point of entropy maximum on the affine subspace of fast motion $x + \ker P_x$. It gives the solution to the problem

$$S(z) \to \max, \ z \in x + \ker P_x . \tag{5.36}$$

This is the most important property of thermodynamic projector. It was introduced in [9] as the main thermodynamic condition for model reduction. Let us call it for nonequilibrium points x *the property* **A**:

$$\textbf{A.} \quad \ker P_x \subset \ker D_x S . \tag{5.37}$$

If the projector P_x with the property **A** can be continued to the equilibrium point, x^{eq}, as a smooth function of x, then in this point $\ker P_x \perp \mathrm{im} P_x$. If this is valid for all systems (including systems with additional linear conservation laws), then the following *property* **B** holds:

$$\mathbf{B.} \quad (\ker P_x \cap \ker D_x S) \perp (\operatorname{im} P_x \cap \ker D_x S) \ . \qquad (5.38)$$

Of course, orthogonality in (5.37, 5.38) is considered with respect to the entropic scalar product in the point x.

The property $\mathbf{A}$ means that the value of the entropy production persists for all nonequilibrium points. The sense of the property $\mathbf{B}$ is: each point of the slow manifold can be made an equilibrium point (after a deformation of the system which leads to an additional balance). And for equilibrium points the orthogonality condition (5.38) follows from the property $\mathbf{A}$.

If P_x does not have the property $\mathbf{A}$, then x is not the point of entropy maximum on the affine subspace of fast motion $x + \ker P_x$, so either the fast motion along this subspace does not leads to x (and, hence, the point x does not belong to the slow manifold), or this motion violates the second law, and the entropy decreases. This is the violation of the second law of thermodynamics during the fast motion. If P_x does not have the property $\mathbf{A}$, then such a violation is expected for almost every system.

On the other hand, if P_x is not the thermodynamic projector, then there exists a thermodynamically consistent vector field J, with a non-thermodynamic projection: S is the Lyapunov function for J (it increases), and is not the Lyapunov function for $P_x(J)$ (it decreases in the neighborhood of x). The difference between violation of the second law of thermodynamics in fast and slow motions for a projector without the property $\mathbf{A}$ is: for the fast motion this violation typically exists, for the slow (projected) motion there exist some thermodynamic systems with such a violation. On the other hand, the violation of thermodynamics in the slow motion is worse for applications, if we use the slow dynamics as the answer (and assume that the fast dynamics is relaxed).

If P_x does not have the property $\mathbf{B}$, then there exist systems with violation of the second law of thermodynamics in fast and slow motions. Here we can not claim that the second law is violated for almost every system, but such systems exist.

One particular case of the thermodynamic projector is known during several decades. It is the quasiequilibrium projector (5.15) on the tangent space of the quasiequilibrium (MaxEnt) manifold (5.5) $S(x) \to \max, \Pi(x) = y$. The solution of the problem (5.5) x_y^{qe} parametrized by values of the macroscopic variables y is the quasiequilibrium manifold.

The formula for the quasiequilibrium projector (5.15) was essentially obtained by Robertson [126]. In his dissertation [126] Robertson studied "the equation of motion for the generalized canonical density operator". The generalized canonical density renders entropy a maximum for given statistical expectations of the thermodynamic coordinates. Robertson considered the Liouville equation for a general quantum system. The first main result of Robertson's paper is the explicit expression for splitting of the motion in two components: projection of the motion onto generalized canonical density and the motion in the kernel of this projection. The obtained projector operator is a specific case of the quasiequilibrium projector (5.15). The second result is the exclusion of the motion in the kernel of quasiequilibrium projector from the dynamic equation. This operation is similar to the Zwanzig formalism [125]. It leads to the integro-differential equation with delay in time for the generalized canonical density. The quasiequilibrium projector (5.15) is more general than the projector obtained by Robertson [126] in the following sense: It is derived for any functional S with non-degenerate second differential $D_x^2 S$, for the manifold of conditional maxima of S, and for any (nonlinear) evolution equation. Robertson emphasized that this operator is non-Hermitian with respect to standard L^2 scalar product and in that sense is "not a projector at all". Nevertheless, it is self-adjoint (and, hence, orthogonal), but with respect to another (entropic) scalar product. The general thermodynamic projector (13.4) performs with an arbitrary ansatz manifolds (not obligatory MaxEnt) and in that sense it is much more general.

The thermodynamic projector (5.15) for the quasiequilibrium manifold (5.5) is the orthogonal projector with respect to the entropic scalar product (5.14). In this case both terms in the thermodynamic projector (5.25) are orthogonal projectors with respect to the entropic scalar product (5.14). The first term, P_0, is orthogonal projector by construction. For the second term, $e_g(D_x S)$, it means that the Riesz representation of the linear functional $D_x S$ in the whole space E with respect to the entropic scalar product belongs to the tangent space of the quasiequilibrium manifold. This Riesz representation is the gradient of S with respect to $\langle | \rangle_x$. The following Proposition gives a simple and important condition of orthogonality of the thermodynamic projector (5.25). Let Ω be an ansatz manifold, and let V be some quasiequilibrium manifold, $x \in \Omega \bigcap V$, T_x be the tangent space to the ansatz manifold Ω in the point x. Suppose that there exists a neighborhood of x where $V \subseteq \Omega$. We use the notation $\text{grad}_x S$ for the Riesz representation of the linear functional $D_x S$ in the entropic scalar product $\langle | \rangle_x$: $\langle \text{grad}_x S | f \rangle_x \equiv (D_x S)(f)$ for $f \in E$.

Proposition 4. *Under given assumptions, $\text{grad}_x S \in T_x$, and the thermodynamic projector P_x is the orthogonal projector onto T_x with respect to the entropic scalar product (5.14).* □

So, if a point x on the ansatz manifold Ω belongs to some quasiequilibrium submanifold $V \subseteq \Omega$, then the thermodynamic projector in this point is simply the orthogonal projector with respect to the entropic scalar product (13.2).

Proposition 4 is useful in the following situation. Let the quasiequilibrium approximation be more or less satisfactory, but the "relevant degrees of freedom" depend on the current state of the system. It means that for some changes of the state we should change the list of relevant macroscopic variables (moments of distribution function for generating the quasiequilibrium, for example). Sometimes it can be described as presence of "hidden" degrees of freedom, which are not moments. In these cases the manifold of reduced description should be extended. We have a family of systems of moments $M_\alpha = m_\alpha(x)$, and a family of corresponding quasiequilibrium manifolds Ω_α: The manifold Ω_α consist of solutions of optimization problem $S(x) \to \max$, $m_\alpha(x) = M$ for given α and all admissible values for M. To create a manifold of reduced description it is possible to join all the moments M_α in one family, and construct the corresponding quasiequilibrium manifold. Points on this manifold are parametrized by the family of moments values $\{M_\alpha\}$ for all possible α. It leads to a huge increase of the quasiequilibrium manifold. Another way of extension of the quasiequilibrium manifold is a union of all the manifolds Ω_α for all α. In accordance with the Proposition 4, the thermodynamic projector for this union is simply the orthogonal projector with respect to the entropic scalar product. This kind of manifolds gives a closest generalization of the quasiequilibrium manifolds. Due to (5.36), the thermodynamic projector gives the presentation of almost arbitrary ansatz as the quasiequilibrium manifold. This property opens the natural field for applications of thermodynamic projector: construction of Galerkin approximations with thermodynamic properties.

Of course, there is a "law of the difficulty conservation": for the quasiequilibrium with the moment parameterization the slow manifold is usually not explicitly known, and it can be difficult to calculate it. Thermodynamic projector completely eliminates this difficulty: we can use almost any manifold as appropriate ansatz now. On the other side, on the quasiequilibrium manifold with the moment parameterization (if it is found) it is easy to find the dynamics: simply write $\dot{M} = \Pi(J)$. Building of the thermodynamic projector may require some effort. Finally, if the quasiequilibrium manifold is found, then it is easy to find the projection of any distributions x on the quasiequilibrium manifold: $x \mapsto \Pi(x) \mapsto x^{\mathrm{qe}}_{\Pi(x)}$. It requires just a calculation of the moments $\Pi(x)$. The preimage of the point $x^{\mathrm{qe}}_{\Pi(x)}$ is a set (an affine manifold) of distributions $\{x|\Pi(x - x^{\mathrm{qe}}_{\Pi(x)}) = 0\}$, and $x^{\mathrm{qe}}_{\Pi(x)}$ is the point of entropy maximum on this set. It is possible, but not so easy, to construct such a projector of some neighborhood of the manifold Ω onto Ω for the general thermodynamic projector P_x: for a point z from this neighborhood

$$z \mapsto x \in \Omega, \text{ if } P_x(z - x) = 0 . \tag{5.39}$$

A point $x \in \Omega$ is the point of entropy maximum on the preimage of x, i.e. on the affine manifold $\{z|P_x(z - x) = 0\}$. It is necessary to emphasize that the map (5.39) can be defined only in a neighborhood of the manifold Ω, but not in the whole space, because some of affine subspaces $\{z|P_x(z - x) = 0\}$ for different $x \in \Omega$ can intersect. Let us introduce a special notation for the projection of some neighborhood of the manifold Ω onto Ω, associated with the thermodynamic projector P_x (5.39): $\mathbf{P}_\Omega : z \mapsto x$. The preimage of a point $x \in \Omega$ is:

$$\mathbf{P}_\Omega^{-1} x = x + \ker P_x , \tag{5.40}$$

(or, strictly speaking, a vicinity of x in this affine manifold). Differential of the operator $\mathbf{P}_\Omega$ at a point $x \in \Omega$ from the manifold Ω is simply the projector P_x:

$$\mathbf{P}_\Omega(x + \varepsilon z) = x + \varepsilon P_x z + o(\varepsilon) . \tag{5.41}$$

Generally, differential of $\mathbf{P}_\Omega$ at a point x has not so simple form, if x does not belong Ω.

The "global extension" $\mathbf{P}_\Omega$ of a field of "infinitesimal" projectors P_f ($f \in \Omega$) is needed for a discussion of projector operators technique, memory functions and a short memory approximation.

***

Is it necessary to use the thermodynamic projector everywhere? The persistence of dissipation is necessary, because the violation of the second law may lead to non-physical effects. If one creates a very accurate method for solution of the initial equation (3.1), then it may be possible to expect that the persistence of dissipation will hold without additional effort. But this situation does not appear yet. All methods of model reduction need a special tool to control the persistence of dissipation.

In order to summarize, let us give three reasons to use the thermodynamic projector:

1. It guarantees the persistence of dissipation: all the thermodynamic processes which should produce the entropy conserve this property after projecting, moreover, not only the sign of dissipation conserves, but also the value of entropy production and the reciprocity relations are conserved;
2. The coefficients (and, more generally speaking, the right hand part) of kinetic equations are less known than the thermodynamic functionals, so, the *universality* of the thermodynamic projector (it depends only on thermodynamic data) makes the thermodynamic properties of projected system as reliable, as for the initial system;
3. It is easy (much more easy than the spectral projector, for example).

5.5 Example: Quasiequilibrium Projector and Defect of Invariance for the Local Maxwellians Manifold of the Boltzmann Equation

The Boltzmann equation remains the most inspiring source for the model reduction problems. With this subsection we start a series of examples for the Boltzmann equation.

5.5.1 Difficulties of Classical Methods
of the Boltzmann Equation Theory

The first systematic and (at least partially) successful method of constructing invariant manifolds for dissipative systems was the celebrated *Chapman-
Enskog method* [70] for the Boltzmann kinetic equation (see Chap. 2). The
main difficulty of the Chapman-Enskog method [70] are "nonphysical" properties of high-order approximations. This was stated by a number of authors
and was discussed in detail in [112]. In particular, as it was noted in [72], the
Burnett approximation results in a short-wave instability of the acoustic spectra. This fact contradicts the H-theorem (cf. in [72]). The Hilbert expansion
contains secular terms [112]. The latter contradicts the H-theorem.

The other difficulties of both of these methods are: the restriction upon
the choice of the initial approximation (the local equilibrium approximation),
the requirement for a small parameter, and the usage of slowly converging
Taylor expansion. These difficulties never allow a direct transfer of these
methods on essentially nonequilibrium situations.

The main difficulty of the Grad method [201] is the uncontrollability of
the chosen approximation. An extension of the list of moments can result in a
certain success, but it can also give nothing. Difficulties of moment expansion
in the problems of shock waves and sound propagation are discussed in [112].

Many attempts were made to refine these methods. For the Chapman-
Enskog and Hilbert methods these attempts are based in general on some
better rearrangement of expansions (e.g. neglecting high-order derivatives
[112], reexpanding [112], Pade approximations and partial summing [43, 221,
233], etc.). This type of work with formal series is wide spread in physics.
Sometimes the results are surprisingly good – from the renormalization theory
in quantum fields to the Percus-Yevick equation and the ring-operator in
statistical mechanics. However, one should realize that success cannot be
guaranteed. Moreover, rearrangements never remove the restriction upon the
choice of the initial local equilibrium approximation.

Attempts to improve the Grad method are based on quasiequilibrium approximations [223, 224]. It was found in [224] that the Grad distributions
are linearized versions of appropriate quasiequilibrium approximations (see
also [230, 233, 234]). A method which treats fluxes (e.g. moments with respect to collision integrals) as independent variables in a quasiequilibrium
description was introduced in [233, 234, 246, 248], and will be discussed later
in Example 5.6.

The important feature of quasiequilibrium approximations is that they
are always thermodynamic, i.e. they are consistent with the H-theorem by
construction. However, quasiequilibrium approximations do not remove the
uncontrollability of the Grad method. Dynamic corrections to Grad's approximation will be addressed later in Chap. 6.

5.5.2 Boltzmann Equation

The phase space E consists of distribution functions $f(v, x)$ which depend on the spatial variable x and on velocity variable v. The variable x spans an open domain $\Omega_x^3 \subseteq \mathbf{R}_x$, and the variable v spans the space $\mathbf{R}_v^3$. We require that $f(v, x) \in F$ are nonnegative functions, and also that the following integrals are finite for every $x \in \Omega_x$ (the existence of the moments and of the entropy):

$$I_x^{(i_1 i_2 i_3)}(f) = \int v_1^{i_1} v_2^{i_2} v_3^{i_3} f(v, x) \, \mathrm{d}^3 v, i_1 \geq 0, i_2 \geq 0, i_3 \geq 0 ; \quad (5.42)$$

$$H_x(f) = \int f(v, x)(\ln f(v, x) - 1) \, \mathrm{d}^3 v, H(f) = \int H_x(f) \, \mathrm{d}^3 x . \quad (5.43)$$

Here and below integration in v is done over $\mathbf{R}_v^3$, and it is done over Ω_x in x. For every fixed $x \in \Omega_x$, $I_x^{(\cdots)}$ and H_x might be treated as functionals defined in F.

We write the Boltzmann equation in the form of (3.1) (in the fixed reference system) using standard notation [112]:

$$\frac{\partial f}{\partial t} = J(f), \ J(f) = -v_s \frac{\partial f}{\partial x_s} + Q(f, f) . \quad (5.44)$$

Here and further in this Example summation in two repeated indices is assumed, and $Q(f, f)$ stands for the Boltzmann collision integral. The latter represents the dissipative part of the vector field $J(f)$ (5.44).

In this section we consider the case when boundary conditions for equation (5.44) are relevant to the local with respect to x form of the H-theorem.

For every fixed x, we denote as $H_x^0(f)$ the space of linear functionals

$$\sum_{i=0}^{4} a_i(x) \int \psi_i(v) f(v, x) \, \mathrm{d}^3 v ,$$

where $\psi_i(v)$ represent invariants of a collision ($\psi_0 = 1, \psi_i = v_i, i = 1, 2, 3, \psi_4 = v^2$). We write $(\mathrm{mod} H_x^0(f))$ if an expression is valid within the accuracy of adding a functional from $H_x^0(f)$. The local H-theorem states: for any functional

$$H_x(f) = \int f(v, x)(\ln f(v, x) - 1) \, \mathrm{d}^3 v \ (\mathrm{mod} H_x^0(f)) \quad (5.45)$$

the following inequality is valid:

$$\frac{\mathrm{d} H_x(f)}{\mathrm{d} t} \equiv \int Q(f, f) \big|_{f=f(v, x)} \ln f(v, x) \, \mathrm{d}^3 v \leq 0 . \quad (5.46)$$

Expression (5.46) is equal to zero if and only if $\ln f = \sum_{i=0}^{4} a_i(x) \psi_i(v)$.

Although all functionals (5.45) are equivalent in the sense of the H-theorem, it is convenient to work with the functional

$$H_{\boldsymbol{x}}(f) = \int f(\boldsymbol{v}, \boldsymbol{x})(\ln f(\boldsymbol{v}, \boldsymbol{x}) - 1)\,\mathrm{d}^3\boldsymbol{v}\,.$$

All what was said in this chapter can be applied to the Boltzmann equation (5.44). Now we shall discuss some specific points.

5.5.3 Local Manifolds

Although the general description of manifolds $\Omega \subset F$ holds applies also to the Boltzmann equation, a specific class of manifolds can be defined due to the different character of spatial and velocity dependencies in the Boltzmann equation vector field (5.44). These manifolds will be called **local manifolds**, and they are constructed as follows. Denote as F_{loc} the set of functions $f(\boldsymbol{v})$ with finite integrals

$$a)\,I^{(i_1 i_2 i_3)}(f) = \int v_1^{i_1} v_2^{i_2} v_3^{i_3} f(\boldsymbol{v})\,\mathrm{d}^3\boldsymbol{v}, i_1 \geq 0, i_2 \geq 0, i_3 \geq 0\,;$$

$$b)\,H(f) = \int f(\boldsymbol{v}) \ln f(\boldsymbol{v})\,\mathrm{d}^3\boldsymbol{v}\,. \tag{5.47}$$

In order to construct a local manifold in F, we, first, consider a manifold in F_{loc}. Namely, we define a domain $A \subset B$, where B is a linear space, and consider a smooth immersion $A \to F_{\mathrm{loc}}$: $a \to f(a, \boldsymbol{v})$. The set of functions $f(a, \boldsymbol{v}) \in F_{\mathrm{loc}}$, where a spans a domain A, is a manifold in F_{loc}. Second, we consider *all* bounded and sufficiently smooth functions $a(\boldsymbol{x})$: $\Omega_{\boldsymbol{x}} \to A$, and we define the local manifold in F as the set of functions $f(a(\boldsymbol{x}), \boldsymbol{v})$. Roughly speaking, the local manifold is a set of functions which are parameterized with $\boldsymbol{x}$-dependent functions $a(\boldsymbol{x})$. A local manifold will be called a *locally finite-dimensional* manifold if B is a finite-dimensional linear space.

Locally finite-dimensional manifolds are the natural source of initial approximations for constructing dynamic invariant manifolds in the Boltzmann equation theory. For example, the Tamm–Mott-Smith (TMS) approximation is a locally two-dimensional manifold $\{f(a_-, a_+)\}$ which consists of distributions

$$f(a_-, a_+) = a_- f_- + a_+ f_+\,. \tag{5.48}$$

Here a_- and a_+ (the coordinates on the manifold $\Omega_{\mathrm{TMS}} = \{f(a_-, a_+)\}$) are non-negative real functions of the position vector $\boldsymbol{x}$, and f_- and f_+ are fixed up- and downstream Maxwellians.

The next example is the locally five-dimensional manifold $\{f(n, \boldsymbol{u}, T)\}$ which consists of local Maxwellians (LM). The LM manifold consists of distributions f_0 which are labeled with parameters $n, \boldsymbol{u}$, and T:

$$f_0(n, \boldsymbol{u}, T) = n\left(\frac{2\pi k_{\mathrm{B}} T}{m}\right)^{-3/2} \exp\left(-\frac{m(\boldsymbol{v} - \boldsymbol{u})^2}{2k_{\mathrm{B}} T}\right)\,. \tag{5.49}$$

Parameters n, u, and T in (5.49) are functions of x. In this section we shall not indicate this dependency explicitly.

Distribution $f_0(n, u, T)$ is the unique solution of the variational problem:

$$H(f) = \int f \ln f \, \mathrm{d}^3 v \to \min$$

for:

$$M_0(f) = \int 1 \cdot f \, \mathrm{d}^3 v \; ;$$

$$M_i(f) = \int v_i f \, \mathrm{d}^3 v = n u_i, i = 1, 2, 3 \; ;$$

$$M_4(f) = \int v^2 f \, \mathrm{d}^3 v = \frac{3 n k_{\mathrm{B}} T}{m} + n u^2 \; . \tag{5.50}$$

Hence, the LM manifold is the quasiequilibrium manifold. Considering n, u, and T as five parameters, we see that the LM manifold is parameterized with the values of $M_s(f), s = 0, \ldots, 4$, which are defined in the neighborhood of the LM manifold. It is sometimes convenient to consider the variables $M_s(f_0), s = 0, \ldots, 4$, as a new coordinates on the LM manifold. The relationship between the coordinates $\{M_s(f_0)\}$ and $\{n, u, T\}$ is:

$$n = M_0; u_i = M_0^{-1} M_i, i = 1, 2, 3; T = \frac{m}{3 k_{\mathrm{B}}} M_0^{-1} (M_4 - M_0^{-1} M_i M_i) \; . \tag{5.51}$$

This is the standard moment parametrization of the quasiequilibrium manifold.

5.5.4 Thermodynamic Quasiequilibrium Projector

Thermodynamic quasiequilibrium projector $P_{f_0(n, u, T)}(J)$ onto the tangent space $T_{f_0(n, u, T)}$ is defined as:

$$P_{f_0(n, u, T)}(J) = \sum_{s=0}^{4} \frac{\partial f_0(n, u, T)}{\partial M_s} \int \psi_s J \, \mathrm{d}^3 v \; . \tag{5.52}$$

Here we have assumed that n, u, and T are functions of $M_0, \ldots, M_4$ (see relationship (5.51)), and

$$\psi_0 = 1, \psi_i = v_i, i = 1, 2, 3, \psi_4 = v^2 \; . \tag{5.53}$$

Calculating derivatives in (5.52), and next returning to variables n, u, and T, we obtain:

$$P_{f_0(n, u, T)}(J) = f_0(n, u, T) \tag{5.54}$$

$$\times \left\{ \left[\frac{1}{n} - \frac{m u_i}{n k_{\mathrm{B}} T} (v_i - u_i) + \left(\frac{m u^2}{3 n k_{\mathrm{B}}} - \frac{T}{n} \right) \left(\frac{m (v - u)^2}{2 k_{\mathrm{B}} T^2} - \frac{3}{2T} \right) \right] \int J \, \mathrm{d}^3 v \right.$$

$$+\left[\frac{m}{nk_\mathrm{B}T}(v_i - u_i) - \frac{2mu_i}{3nk_\mathrm{B}}\left(\frac{m(\boldsymbol{v}-\boldsymbol{u})^2}{2k_\mathrm{B}T^2} - \frac{3}{2T}\right)\right]\int v_i J\,\mathrm{d}^3\boldsymbol{v}$$

$$+\frac{m}{3nk_\mathrm{B}}\left(\frac{m(\boldsymbol{v}-\boldsymbol{u})^2}{2k_\mathrm{B}T^2} - \frac{3}{2T}\right)\int v^2 J\,\mathrm{d}^3\boldsymbol{v}\Bigg\}\ .$$

It is sometimes convenient to rewrite (5.55) as

$$P_{f_0(n,\boldsymbol{u},T)}(J) = f_0(n,\boldsymbol{u},T)\sum_{s=0}^{4}\psi^{(s)}_{f_0(n,\boldsymbol{u},T)}\int\psi^{(s)}_{f_0(n,\boldsymbol{u},T)}J\,\mathrm{d}^3\boldsymbol{v}\ . \tag{5.55}$$

Here

$$\psi^{(0)}_{f_0(n,\boldsymbol{u},T)} = n^{-1/2},\ \psi^{(i)}_{f_0(n,\boldsymbol{u},T)} = (2/n)^{1/2}c_i, \tag{5.56}$$

$$\psi^{(4)}_{f_0(n,\boldsymbol{u},T)} = (2/3n)^{1/2}(c^2 - (3/2));\ c_i = (m/2k_\mathrm{B}T)^{1/2}(v_i - u_i),\ i = 1,2,3\ .$$

It is easy to check that

$$\int f_0(n,\boldsymbol{u},T)\psi^{(k)}_{f_0(n,\boldsymbol{u},T)}\psi^{(l)}_{f_0(n,\boldsymbol{u},T)}\,\mathrm{d}^3\boldsymbol{v} = \delta_{kl}\ . \tag{5.57}$$

Here δ_{kl} is the Kronecker delta.

5.5.5 Defect of Invariance for the LM Manifold

The defect of invariance for the LM manifold at the point $f_0(n,\boldsymbol{u},T)$ for the Boltzmann equation vector field in the co-moving reference system is:

$$\Delta(f_0(n,\boldsymbol{u},T)) = P_{f_0(n,\boldsymbol{u},T)}\left(-(v_s - u_s)\frac{\partial f_0(n,\boldsymbol{u},T)}{\partial x_s} + Q(f_0(n,\boldsymbol{u},T))\right)$$

$$-\left(-(v_s - u_s)\frac{\partial f_0(n,\boldsymbol{u},T)}{\partial x_s} + Q(f_0(n,\boldsymbol{u},T))\right)$$

$$= P_{f_0(n,\boldsymbol{u},T)}\left(-(v_s - u_s)\frac{\partial f_0(n,\boldsymbol{u},T)}{\partial x_s}\right) + (v_s - u_s)\frac{\partial f_0(n,\boldsymbol{u},T)}{\partial x_s}\ . \tag{5.58}$$

Substituting (5.55) into (5.58), we obtain:

$$\Delta(f_0(n,\boldsymbol{u},T)) = f_0(n,\boldsymbol{u},T)\left\{\left(\frac{m(\boldsymbol{v}-\boldsymbol{u})^2}{2k_\mathrm{B}T} - \frac{5}{2}\right)(v_i - u_i)\frac{\partial\ln T}{\partial x_i}\right.$$

$$\left.+\frac{m}{k_\mathrm{B}T}(((v_i - u_i)(v_s - u_s) - \frac{1}{3}\delta_{is}(\boldsymbol{v}-\boldsymbol{u})^2)\frac{\partial u_s}{\partial x_i}\right\}\ . \tag{5.59}$$

The LM *manifold is not a dynamic invariant manifold of the Boltzmann equation and the defect (5.59) is not equal to zero. Indeed, inhomogeneity of the temperature and of the flow velocity drives the invariant manifold away from the local equilibrium.*

5.6 Example: Quasiequilibrium Closure Hierarchies for the Boltzmann Equation

Explicit method of constructing approximations (the Triangle Entropy Method [233]) is developed for strongly nonequilibrium problems of Boltzmann's–type kinetics, i.e. when the standard moment variables become insufficient. This method enables one to treat any complicated nonlinear functionals that fit best the physics of a problem (such as, for example, rates of processes) as new independent variables.

The method is applied to the problem of derivation of hydrodynamics from the Boltzmann equation. New macroscopic variables are introduced (moments of the Boltzmann collision integral, or scattering rates). They are treated as independent variables rather than as infinite moment series. This approach gives the complete account of rates of scattering processes. Transport equations for scattering rates are obtained (the second hydrodynamic chain), similar to the usual moment chain (the first hydrodynamic chain). Using the triangle entropy method, three different types of the macroscopic description are considered. The first type involves only moments of distribution functions, and results coincide with those of the Grad method in the Maximum Entropy version. The second type of description involves only scattering rates. Finally, the third type involves both the moments and the scattering rates (the mixed description). The second and the mixed hydrodynamics are sensitive to the choice of the collision model. The second hydrodynamics is equivalent to the first hydrodynamics only for Maxwell molecules, and the mixed hydrodynamics exists for all types of collision models excluding Maxwell molecules. Various examples of the closure of the first, of the second, and of the mixed hydrodynamic chains are considered for the hard spheres model. It is shown, in particular, that the complete account of scattering processes leads to a renormalization of transport coefficients.

5.6.1 Triangle Entropy Method

In the present subsection, which is of introductory character, we shall refer, to be specific, to the Boltzmann kinetic equation for a one-component gas whose state (in the microscopic sense) is described by the one-particle distribution function $f(\boldsymbol{v}, \boldsymbol{x}, t)$ depending on the velocity vector $\boldsymbol{v} = \{v_k\}_{k=1}^3$, the spatial position $\boldsymbol{x} = \{x_k\}_{k=1}^3$ and time t. The the Boltzmann equation describes the evolution of f and in the absence of external forces is

$$\partial_t f + v_k \partial_k f = Q(f, f) , \tag{5.60}$$

where $\partial_t \equiv \partial/\partial t$ is the time partial derivative, $\partial_k \equiv \partial/\partial x_k$ is partial derivative with respect to k-th component of $\boldsymbol{x}$, summation in two repeating indices is assumed, and $Q(f, f)$ is the collision integral (its concrete form is of no importance right now, just note that it is functional-integral operator quadratic with respect to f).

The Boltzmann equation possesses two properties principal for the subsequent reasoning (for the basic properties of the Boltzmann equation see Chap. 2) .

1. There exist five functions $\psi_\alpha(\boldsymbol{v})$ (additive collision invariants),

$$1, \boldsymbol{v}, v^2$$

such that for any their linear combination with coefficients depending on $\boldsymbol{x}, t$ and for arbitrary f the following equality is true:

$$\int \sum_{\alpha=1}^{5} a_\alpha(\boldsymbol{x}, t)\psi_\alpha(\boldsymbol{v})Q(f, f)\,\mathrm{d}\boldsymbol{v} = 0 , \tag{5.61}$$

provided the integrals exist.

2. The equation (5.60) possesses global Lyapunov functional: the H-function,

$$H(t) \equiv H[f] = \int f(\boldsymbol{v}, \boldsymbol{x}, t)\ln f(\boldsymbol{v}, \boldsymbol{x}, t)\,\mathrm{d}\boldsymbol{v}\,\mathrm{d}\boldsymbol{x} , \tag{5.62}$$

the derivative of which by virtue of the equation (5.60) is non-positive under appropriate boundary conditions:

$$\mathrm{d}H(t)/\mathrm{d}t \leq 0 . \tag{5.63}$$

Grad's method [201] and its variants construct closed systems of equations for macroscopic variables when the latter are represented by moments (or, more general, linear functionals) of the distribution function f (hence their alternative name is the "moment methods"). The entropy maximum method for the Boltzmann equation is of particular importance for the subsequent reasoning. It consists in the following. A finite set of moments describing the macroscopic state is chosen. The distribution function of the quasiequilibrium state under given values of the chosen moments is determined, i.e. the problem is solved

$$H[f] \to \min, \text{ for } \hat{M}_i[f] = M_i, \quad i = 1, \ldots, k , \tag{5.64}$$

where $\hat{M}_i[f]$ are linear functionals with respect to f; M_i are the corresponding values of chosen set of k macroscopic variables. The quasiequilibrium distribution function $f^*(\boldsymbol{v}, M(\boldsymbol{x}, t))$, $M = \{M_1, \ldots, M_k\}$, parametrically depends on M_i, its dependence on space $\boldsymbol{x}$ and on time t being represented only by $M(\boldsymbol{x}, t)$. Then the obtained f^* is substituted into the Boltzmann equation (5.60), and operators $\hat{M}_i$ are applied on the latter formal expression.

In the result we have closed systems of equations with respect to $M_i(\boldsymbol{x}, t)$, $i = 1, \ldots, k$:

$$\partial_t M_i + \hat{M}_i[v_k\partial_k f^*(\boldsymbol{v}, M)] = \hat{M}_i[Q(f^*(\boldsymbol{v}, M), f^*(\boldsymbol{v}, M))] . \tag{5.65}$$

The following heuristic explanation can be given to the entropy method. A state of the gas can be described by a finite set of moments on some time scale θ only if all the other moments ("fast") relax on a shorter time scale time $\tau, \tau \ll \theta$, to their values determined by the chosen set of "slow" moments, while the slow ones almost do not change appreciably on the time scale τ. In the process of the fast relaxation the H-function decreases, and in the end of this fast relaxation process a quasiequilibrium state sets in with the distribution function being the solution of the problem (5.64). Then "slow" moments relax to the equilibrium state by virtue of (5.65).

The entropy method has a number of advantages in comparison with the classical Grad's method. First, being not necessarily restricted to any specific system of orthogonal polynomials, and leading to solving an optimization problem, it is more convenient from the technical point of view. Second, and ever more important, the resulting quasiequilibrium H-function, $H^*(M) = H[f^*(v, M)]$, decreases due of the moment equations (5.65).

Let us note one common disadvantage of all the moment methods, and, in particular, of the entropy method. Macroscopic parameters, for which these methods enable to obtain closed systems, must be moments of the distribution function. On the other hand, it is easy to find examples when the interesting macroscopic parameters are nonlinear functionals of the distribution function. In the case of the one-component gas these are the integrals of velocity polynomials with respect to the collision integral $Q(f, f)$ of (5.60) (scattering rates of moments). For chemically reacting mixtures these are the reaction rates, and so on. If the characteristic relaxation time of such nonlinear macroscopic parameters is comparable with that of the "slow" moments, then they should be also included into the list of "slow" variables on the same footing.

In this Example for constructing closed systems of equations for nonlinear (in a general case) macroscopic variables the *triangle entropy method* is used. Let us outline the scheme of this method.

Let a set of macroscopic variables be chosen: linear functionals $\hat{M}[f]$ and nonlinear functionals (in a general case) $\hat{N}[f]$:

$$\hat{M}[f] = \left\{ \hat{M}_1[f], \ldots, \hat{M}_k[f] \right\}, \ \hat{N}[f] = \left\{ \hat{N}_1[f], \ldots, \hat{N}_l[f] \right\} .$$

Then, just as for the problem (5.64), the first quasiequilibrium approximation is constructed under fixed values of the linear macroscopic parameters M:

$$H[f] \to \min \text{ for } \hat{M}_i[f] = M_i, \ i = 1, \ldots, k , \tag{5.66}$$

and the resulting distribution function is $f^*(v, M)$. After that, we seek the true quasiequilibrium distribution function in the form,

$$f = f^*(1 + \varphi) , \tag{5.67}$$

where φ is a deviation from the first quasiequilibrium approximation. In order to determine φ, the second quasiequilibrium approximation is constructed.

Let us denote $\Delta H[f^*, \varphi]$ as the quadratic term in the expansion of the H-function into powers of φ in the neighbourhood of the first quasiequilibrium state f^*. The distribution function of the second quasiequilibrium approximation is the solution to the problem,

$$\Delta H[f^*, \varphi] \to \min \text{ for}$$
$$\hat{M}_i[f^* \varphi] = 0, \quad i = 1, \dots, k \ ,$$
$$\Delta \hat{N}_j[f^*, \varphi] = \Delta N_j, \quad j = 1, \dots, l \ , \tag{5.68}$$

where $\Delta \hat{N}_j$ are linear operators characterizing the linear with respect to φ deviation of (nonlinear) macroscopic parameters N_j from their values, $N_j^* = \hat{N}_j[f^*]$, in the first quasiequilibrium state. Note the importance of the homogeneous constraints $\hat{M}_i[f^* \varphi] = 0$ in the problem (5.68). Physically, it means that the variables ΔN_j are "slow" in the same sense, as the variables M_i, at least in the small neighborhood of the first quasiequilibrium f^*. The obtained distribution function,

$$f = f^*(\boldsymbol{v}, M)(1 + \varphi^{**}(\boldsymbol{v}, M, \Delta N)) \tag{5.69}$$

is used to construct the closed system of equations for the macroparameters M, and ΔN. Because the functional in the problem (5.68) is quadratic, and all constraints in this problem are linear, it is always explicitly solvable.

Further in this section some examples of using the triangle entropy method for the one-component gas are considered. Applications to chemically reacting mixtures were discussed in [246, 247].

5.6.2 Linear Macroscopic Variables

Let us consider the simplest example of using the triangle entropy method, when all the macroscopic variables of the first and of the second quasiequilibrium states are the moments of the distribution function.

Quasiequilibrium Projector

Let $\mu_1(\boldsymbol{v}), \dots, \mu_k(\boldsymbol{v})$ be the microscopic densities of the moments

$$M_1(\boldsymbol{x}, t), \ \dots, M_k(\boldsymbol{x}, t)$$

which determine the first quasiequilibrium state,

$$M_i(\boldsymbol{x}, t) = \int \mu_i(\boldsymbol{v}) f(\boldsymbol{v}, \boldsymbol{x}, t) \, d\boldsymbol{v} \ , \tag{5.70}$$

and let $\nu_1(\boldsymbol{v}), \dots, \nu_l(\boldsymbol{v})$ be the microscopic densities of the moments

$$N_1(\boldsymbol{x}, t), \ \dots, N_l(\boldsymbol{x}, t)$$

determining together with (5.60) the second quasiequilibrium state,

$$N_i(\boldsymbol{x}, t) = \int \nu_i(\boldsymbol{v}) f(\boldsymbol{v}, \boldsymbol{x}, t) \, d\boldsymbol{v} \ . \tag{5.71}$$

The choice of the set of the moments of the first and second quasiequilibrium approximations depends on a specific problem. Further on we assume that the microscopic density $\mu \equiv 1$ corresponding to the normalization condition is always included in the list of microscopic densities of the moments of the first quasiequilibrium state. The distribution function of the first quasiequilibrium state results from solving the optimization problem,

$$H[f] = \int f(\boldsymbol{v}) \ln f(\boldsymbol{v}) \, d\boldsymbol{v} \to \min \tag{5.72}$$

for

$$\int \mu_i(\boldsymbol{v}) f(\boldsymbol{v}) \, d\boldsymbol{v} = M_i, i = 1, \dots, k \ .$$

Let us denote by $M = \{M_1, \dots, M_k\}$ the moments of the first quasiequilibrium state, and by $f^*(\boldsymbol{v}, M)$ let us denote the solution of the problem (5.72).

The distribution function of the second quasiequilibrium state is sought in the form,

$$f = f^*(\boldsymbol{v}, M)(1 + \varphi) \ . \tag{5.73}$$

Expanding the H-function (5.62) in the neighbourhood of $f^*(\boldsymbol{v}, M)$ into powers of φ to second order we obtain,

$$\Delta H(\boldsymbol{x}, t) \equiv \Delta H[f^*, \varphi] = H^*(M) + \int f^*(\boldsymbol{v}, M) \ln f^*(\boldsymbol{v}, M) \varphi(\boldsymbol{v}) \, d\boldsymbol{v}$$

$$+ \frac{1}{2} \int f^*(\boldsymbol{v}, M) \varphi^2(\boldsymbol{v}) \, d\boldsymbol{v} \ , \tag{5.74}$$

where $H^*(M) = H[f^*(\boldsymbol{v}, M)]$ is the value of the H-function in the first quasiequilibrium state.

When searching for the second quasiequilibrium state, it is necessary that the true values of the moments M coincide with their values in the first quasiequilibrium state, i.e.,

$$M_i = \int \mu_i(\boldsymbol{v}) f^*(\boldsymbol{v}, M)(1 + \varphi(\boldsymbol{v})) \, d\boldsymbol{v}$$

$$= \int \mu_i(\boldsymbol{v}) f^*(\boldsymbol{v}, M) \, d\boldsymbol{v} = M_i^*, \ i = 1, \dots, k \ . \tag{5.75}$$

In other words, the set of the homogeneous conditions on φ in the problem (5.68),

$$\int \mu_i(\boldsymbol{v}) f^*(\boldsymbol{v}, M) \varphi(\boldsymbol{v}) \, d\boldsymbol{v} = 0, i = 1, \dots, k \ , \tag{5.76}$$

ensures a shift (change) of the first quasiequilibrium state only due to the new moments $N_1, \ldots, N_l$. In order to take this condition into account automatically, let us introduce the following structure of a Hilbert space:

1. Define the scalar product

$$(\psi_1, \psi_2) = \int f^*(\boldsymbol{v}, M)\psi_1(\boldsymbol{v})\psi_2(\boldsymbol{v})\, d\boldsymbol{v} \,. \qquad (5.77)$$

2. Let E_μ be the linear hull of the set of moment densities

$$\{\mu_1(\boldsymbol{v}), \ldots, \mu_k(\boldsymbol{v})\} \,.$$

Let us construct a basis of E_μ $\{e_1(\boldsymbol{v}), \ldots, e_r(\boldsymbol{v})\}$ that is orthonormal in the sense of the scalar product (5.77):

$$(e_i, e_j) = \delta_{ij} \,, \qquad (5.78)$$

$i, j = 1, \ldots, r$; δ_{ij} is the Kronecker delta.
3. Define a projector $\hat{P}^*$ on the first quasiequilibrium state,

$$\hat{P}^*\psi = \sum_{i=1}^{r} e_i(e_i, \psi) \,. \qquad (5.79)$$

The projector $\hat{P}^*$ is orthogonal: for any pair of functions ψ_1, ψ_2,

$$(\hat{P}^*\psi_1, (\hat{1} - \hat{P}^*)\psi_2) = 0 \,, \qquad (5.80)$$

where $\hat{1}$ is the unit operator. Then the condition (5.76) amounts to

$$\hat{P}^*\varphi = 0 \,, \qquad (5.81)$$

and the expression for the quadratic part of the H-function (5.74) takes the form,

$$\Delta H[f^*, \varphi] = H^*(M) + (\ln f^*, \varphi) + (1/2)(\varphi, \varphi) \,. \qquad (5.82)$$

Now, let us note that the function $\ln f^*$ is invariant with respect to the action of the projector $\hat{P}^*$:

$$\hat{P}^* \ln f^* = \ln f^* \,. \qquad (5.83)$$

This follows directly from the solution of the problem (5.72) using of the method of Lagrange multipliers:

$$f^* = \exp \sum_{i=1}^{k} \lambda_i(M)\mu_i(\boldsymbol{v}) \,,$$

where $\lambda_i(M)$ are Lagrange multipliers. Thus, if the condition (5.81) is satisfied, then from (5.80) and (5.83) it follows that

$$(\ln f^*, \varphi) = (\hat{P}^* \ln f^*, (\hat{1} - \hat{P}^*)\varphi) = 0 \ .$$

Condition (5.81) is satisfied automatically, if ΔN_i are taken as follows:

$$\Delta N_i = ((\hat{1} - \hat{P}^*)\nu_i, \varphi), i = 1, \ldots, l \ . \tag{5.84}$$

Thus, the problem (5.68) of finding the second quasiequilibrium state reduces to

$$\Delta H[f^*, \varphi] - H^*(M) = (1/2)(\varphi, \varphi) \to \min \text{ for}$$
$$((\hat{1} - \hat{P}^*)\nu_i, \varphi) = \Delta N_i, \quad i = 1, \ldots, l \ . \tag{5.85}$$

Note that it is not ultimatively necessary to introduce the structure of the Hilbert space. Moreover that may be impossible, since the "distribution function" and the "microscopic moment densities" are, strictly speaking, elements of different (conjugate one to another) spaces, which may be not reflexive. However, in the examples considered below the mentioned difference is not manifested.

In the remainder of this section we demonstrate how the triangle entropy method is related to Grad's moment method.

Ten-Moment Grad Approximation

Let us take the five additive collision invariants as moment densities of the first quasiequilibrium state:

$$\mu_0 = 1; \ \mu_k = v_k \ (k = 1, 2, 3); \ \mu_4 = \frac{mv^2}{2} \ , \tag{5.86}$$

where v_k are Cartesian components of the velocity, and m is particle's mass. Then the solution to the problem (5.72) is the local Maxwell distribution function $f^{(0)}(\boldsymbol{v}, \boldsymbol{x}, t)$:

$$f^{(0)} = n(\boldsymbol{x}, t) \left(\frac{2\pi k_{\mathrm{B}} T(\boldsymbol{x}, t)}{m} \right)^{-3/2} \exp \left\{ -\frac{m(\boldsymbol{v} - \boldsymbol{u}(\boldsymbol{x}, t))^2}{2 k_{\mathrm{B}} T(\boldsymbol{x}, t)} \right\} \ , \tag{5.87}$$

where
$n(\boldsymbol{x}, t) = \int f(\boldsymbol{v}) \, d\boldsymbol{v}$ is local number density,
$\boldsymbol{u}(\boldsymbol{x}, t) = n^{-1}(\boldsymbol{x}, t) \int f(\boldsymbol{v}) \boldsymbol{v} \, d\boldsymbol{v}$ is the local flow density,
$T(\boldsymbol{x}, t) = \int f(\boldsymbol{v}) \frac{m(\boldsymbol{v} - \boldsymbol{u}(\boldsymbol{x}, t))^2}{3 k_{\mathrm{B}} n(\boldsymbol{x}, t)} \, d\boldsymbol{v}$ is the local temperature,
k_{B} is the Boltzmann constant.

Orthonormalization of the set of moment densities (5.86) with the weight (5.87) gives one of the possible orthonormal basis

$$e_0 = \frac{5k_{\mathrm{B}}T - m(\boldsymbol{v} - \boldsymbol{u})^2}{(10n)^{1/2}k_{\mathrm{B}}T} \ ,$$

$$e_k = \frac{m^{1/2}(v_k - u_k)}{(nk_{\mathrm{B}}T)^{1/2}}, k = 1, 2, 3 \ , \tag{5.88}$$

$$e_4 = \frac{m(\boldsymbol{v} - \boldsymbol{u})^2}{(15n)^{1/2}k_{\mathrm{B}}T} .$$

For the moment densities of the second quasiequilibrium state let us take,

$$\nu_{ik} = mv_iv_k, \ i, k = 1, 2, 3 \ . \tag{5.89}$$

Then

$$(\hat{1} - \hat{P}^{(0)})\nu_{ik} = m(v_i - u_i)(v_k - u_k) - \frac{1}{3}\delta_{ik}m(\boldsymbol{v} - \boldsymbol{u})^2 \ , \tag{5.90}$$

and, since $((\hat{1} - \hat{P}^{(0)})\nu_{ik}, (\hat{1} - \hat{P}^{(0)})\nu_{ls}) = (\delta_{il}\delta_{ks} + \delta_{kl}\delta_{is})Pk_{\mathrm{B}}T/m$, where $P = nk_{\mathrm{B}}T$ is the pressure, and $\sigma_{ik} = (f, (\hat{1} - \hat{P}^{(0)})\nu_{ik})$ is the traceless part of the stress tensor, then from (5.73), (5.86), (5.87), (5.90) we obtain the distribution function of the second quasiequilibrium state in the form

$$f = f^{(0)}\left(1 + \frac{\sigma_{ik}m}{2Pk_{\mathrm{B}}T}\left[(v_i - u_i)(v_k - u_k) - \frac{1}{3}\delta_{ik}(\boldsymbol{v} - \boldsymbol{u})^2\right]\right) \tag{5.91}$$

This is precisely the distribution function of the ten-moment Grad approximation (let us recall that here summation in two repeated indices is assumed).

Thirteen-Moment Grad Approximation

In addition to (5.86), (5.89), let us extend the list of moment densities of the second quasiequilibrium state with the functions

$$\xi_i = \frac{mv_iv^2}{2}, \ i = 1, 2, 3 \ . \tag{5.92}$$

The corresponding orthogonal complements to the projection on the first quasiequilibrium state are

$$(\hat{1} - \hat{P}^{(0)})\xi_i = \frac{m}{2}(v_i - u_i)\left((\boldsymbol{v} - \boldsymbol{u})^2 - \frac{5k_{\mathrm{B}}T}{m}\right) \ . \tag{5.93}$$

The moments corresponding to the densities $(\hat{1} - \hat{P}^{(0)})\xi_i$ are the components of the heat flux vector q_i:

$$q_i = (\varphi, (\hat{1} - \hat{P}^{(0)})\xi_i) \ . \tag{5.94}$$

Since

$$((\hat{1} - \hat{P}^{(0)})\xi_i, (\hat{1} - \hat{P}^{(0)})\nu_{lk}) = 0 ,$$

for any i, k, l, then the constraints

$$((\hat{1} - \hat{P}^{(0)})\nu_{lk}, \varphi) = \sigma_{lk}, ((\hat{1} - \hat{P}^{(0)})\xi_i, \varphi) = q_i$$

in the problem (5.85) are independent, and Lagrange multipliers corresponding to ξ_i are

$$\frac{1}{5n}\left(\frac{k_B T}{m}\right)^2 q_i . \tag{5.95}$$

Finally, taking into account (5.86), (5.91), (5.93), (5.95), we find the distribution function of the second quasiequilibrium state in the form

$$f = f^{(0)}\left(1 + \frac{\sigma_{ik}m}{2Pk_B T}\left((v_i - u_i)(v_k - u_k) - \frac{1}{3}\delta_{ik}(\boldsymbol{v} - \boldsymbol{u})^2\right)\right.$$
$$\left. + \frac{q_i m}{Pk_B T}(v_i - u_i)\left(\frac{m(\boldsymbol{v} - \boldsymbol{u})^2}{5k_B T} - 1\right)\right) , \tag{5.96}$$

which coincides with the thirteen-moment Grad distribution function [201].

Let us remark on the thirteen-moment approximation. From (5.96) it follows that for large enough negative values of $(v_i - u_i)$ the thirteen-moment distribution function becomes negative. This peculiarity of the thirteen-moment approximation is due to the fact that the moment density ξ_i is odd-order polynomial of v_i. In order to eliminate this difficulty, one may consider from the very beginning that in a finite volume the square of velocity of a particle does not exceed a certain value v_{max}^2, which is finite owing to the finiteness of the total energy, and q_i is such that when changing to infinite volume $q_i \to 0$, $v_{max}^2 \to \infty$ and $q_i(v_i - u_i)(\boldsymbol{v} - \boldsymbol{u})^2$ remains finite.

On the other hand, the solution to the optimization problem (5.64) does not exist (is not normalizable), if the highest-order velocity polynomial is odd, as it is for the full 13-moment quasiequilibrium.

Approximation (5.91) yields ΔH (5.82) as follows:

$$\Delta H = H^{(0)} + n\frac{\sigma_{ik}\sigma_{ik}}{4P^2} , \tag{5.97}$$

while ΔH corresponding to (5.96) is,

$$\Delta H = H^{(0)} + n\frac{\sigma_{ik}\sigma_{ik}}{4P^2} + n\frac{q_k q_k \rho}{5P^3} , \tag{5.98}$$

where $\rho = mn$, and $H^{(0)}$ is the local equilibrium value of the H-function

$$H^{(0)} = \frac{5}{2}n\ln n - \frac{3}{2}n\ln P - \frac{3}{2}n\left(1 + \ln\frac{2\pi}{m}\right) . \tag{5.99}$$

These expressions coincide with the corresponding expansions of the quasiequilibrium H-functions obtained by the entropy method, if microscopic moment densities of the first quasiequilibrium approximation are chosen as $1, v_i$, and $v_i v_j$, or as $1, v_i$, $v_i v_j$, and $v_i v^2$. As it was noted in [224], they differs from the H-functions obtained by the Grad method (without the maximum entropy hypothesis), and in contrast to the latter they give proper entropy balance equations.

The transition to the closed system of equations for the moments of the first and of the second quasiequilibrium approximations is accomplished by proceeding from the chain of the Maxwell moment equations, which is equivalent to the Boltzmann equation. Substituting f in the form of $f^{(0)}(1 + \varphi)$ into equation (5.60), and multiplying by $\mu_i(v)$, and integrating over v, we obtain

$$\partial_t(1, \hat{P}^{(0)}\mu_i(v)) + \partial_t(\varphi(v), \mu_i(v)) + \partial_k(v_k\varphi(v), \mu_i(v))$$
$$+\partial_k(v_k, \mu_i(v)) = M_Q[\mu_i, \varphi] \, . \tag{5.100}$$

Here

$$M_Q[\mu_i, \varphi] = \int Q(f^{(0)}(1 + \varphi), f^{(0)}(1 + \varphi))\mu_i(v)\,\mathrm{d}v$$

is a "moment" (corresponding to the microscopic density) $\mu_i(v)$ with respect to the collision integral (further we term M_Q the collision moment or the scattering rate). Now, if one uses f given by equations (5.91), and (5.96) as a closure assumption, then the system (5.100) gives the ten- and thirteen-moment Grad equations, respectively, whereas only linear terms in φ should be kept when calculating M_Q.

Let us note some limitations of truncating the moment hierarchy (5.100) by means of the quasiequilibrium distribution functions (5.91) and (5.96) (or for any other closure which depends on the moments of the distribution functions only). When such closure is used, it is assumed implicitly that the scattering rates in the right hand side of (5.100) "rapidly" relax to their values determined by "slow" (quasiequilibrium) moments. Scattering rates are, generally speaking, independent variables. This peculiarity of the chain (5.100), resulting from the nonlinear character of the Boltzmann equation, distinct it essentially from the other hierarchy equations of statistical mechanics (for example, from the BBGKY chain which follows from the linear Liouville equation). Thus, equations (5.100) are not closed twice: into the left hand side of the equation for the i-th moment enters the $(i+1)$-th moment, and the right hand side contains additional variables – scattering rates.

A consequent way of closure of (5.100) should address both sets of variables (moments and scattering rates) as independent variables. The triangle entropy method enables to do this.

5.6.3 Transport Equations for Scattering Rates in the Neighbourhood of Local Equilibrium. Second and Mixed Hydrodynamic Chains

In this section we derive equations of motion for the scattering rates. It proves convenient to use the following form of the collision integral $Q(f, f)$:

$$Q(f, f)(\boldsymbol{v}) = \int w(\boldsymbol{v}_1', \boldsymbol{v}'|\boldsymbol{v}, \boldsymbol{v}_1)\,(f(\boldsymbol{v}')f(\boldsymbol{v}_1') - f(\boldsymbol{v})f(\boldsymbol{v}_1))\,\mathrm{d}\boldsymbol{v}'\,\mathrm{d}\boldsymbol{v}_1'\,\mathrm{d}\boldsymbol{v}_1 ,$$

(5.101)

where $\boldsymbol{v}$ and $\boldsymbol{v}_1$ are velocities of the two colliding particles before the collision, $\boldsymbol{v}'$ and $\boldsymbol{v}_1'$ are their velocities after the collision, w is a kernel responsible for the post-collision relations $\boldsymbol{v}'(\boldsymbol{v}, \boldsymbol{v}_1)$ and $\boldsymbol{v}_1'(\boldsymbol{v}, \boldsymbol{v}_1)$, momentum and energy conservation laws are taken into account in w by means of corresponding δ-functions. The kernel w has the following symmetry property with respect to its arguments:

$$w(\boldsymbol{v}_1', \boldsymbol{v}'|\boldsymbol{v}, \boldsymbol{v}_1) = w(\boldsymbol{v}_1', \boldsymbol{v}'|\boldsymbol{v}_1, \boldsymbol{v}) = w(\boldsymbol{v}', \boldsymbol{v}_1' \mid \boldsymbol{v}_1, \boldsymbol{v}) = w(\boldsymbol{v}, \boldsymbol{v}_1 \mid \boldsymbol{v}', \boldsymbol{v}_1') .$$

(5.102)

Let $\mu(\boldsymbol{v})$ be the microscopic density of a moment M. The corresponding scattering rate $M_Q[f, \mu]$ is defined as follows:

$$M_Q[f, \mu] = \int Q(f, f)(\boldsymbol{v})\mu(\boldsymbol{v})\,\mathrm{d}\boldsymbol{v} .$$

(5.103)

First, we should obtain transport equations for scattering rates (5.103), analogous to the moment's transport equations. Let us restrict ourselves to the case when f is represented in the form,

$$f = f^{(0)}(1 + \varphi) ,$$

(5.104)

where $f^{(0)}$ is local Maxwell distribution function (5.87), and all the quadratic with respect to φ terms will be neglected below. It is the linear approximation around the local equilibrium.

Since, by detailed balance,

$$f^{(0)}(\boldsymbol{v})f^{(0)}(\boldsymbol{v}_1) = f^{(0)}(\boldsymbol{v}')f^{(0)}(\boldsymbol{v}_1')$$

(5.105)

for all such $(\boldsymbol{v}, \boldsymbol{v}_1)$, $(\boldsymbol{v}', \boldsymbol{v}_1')$ which are related to each other by conservation laws, we have,

$$M_Q[f^{(0)}, \mu] = 0, \quad \text{for any } \mu .$$

(5.106)

Further, by virtue of conservation laws,

$$M_Q[f, \hat{P}^{(0)}\mu] = 0, \quad \text{for any } f .$$

(5.107)

From (5.105)–(5.107) it follows,

$$M_Q[f^{(0)}(1+\varphi),\mu] = M_Q[\varphi,(\hat{1}-\hat{P}^{(0)})\mu] \tag{5.108}$$
$$= -\int w(\boldsymbol{v}',\boldsymbol{v}_1' \mid \boldsymbol{v},\boldsymbol{v}_1)f^{(0)}(\boldsymbol{v})f^{(0)}(\boldsymbol{v}_1)\left\{(1-\hat{P}^{(0)})\mu(\boldsymbol{v})\right\}\,\mathrm{d}\boldsymbol{v}'\,\mathrm{d}\boldsymbol{v}_1'\,\mathrm{d}\boldsymbol{v}_1\,\mathrm{d}\boldsymbol{v}\;.$$

We used notation,

$$\{\psi(\boldsymbol{v})\} = \psi(\boldsymbol{v}) + \psi(\boldsymbol{v}_1) - \psi(\boldsymbol{v}') - \psi(\boldsymbol{v}_1')\;. \tag{5.109}$$

Also, it proves convenient to introduce the microscopic density of the scattering rate, $\mu_Q(\boldsymbol{v})$:

$$\mu_Q(\boldsymbol{v}) = \int w(\boldsymbol{v}',\boldsymbol{v}_1' \mid \boldsymbol{v},\boldsymbol{v}_1)f^{(0)}(\boldsymbol{v}_1)\left\{(1-\hat{P}^{(0)})\mu(\boldsymbol{v})\right\}\,\mathrm{d}\boldsymbol{v}'\,\mathrm{d}\boldsymbol{v}_1'\,\mathrm{d}\boldsymbol{v}_1\;. \tag{5.110}$$

Then,

$$M_Q[\varphi,\mu] = -(\varphi,\mu_Q)\;, \tag{5.111}$$

where $(\cdot,\cdot)$ is the L_2 scalar product with the weight $f^{(0)}$ (5.87). This is a natural scalar product in the space of functions φ (5.104) (multipliers), and it is obviously related to the entropic scalar product in the space of distribution functions at the local equilibrium $f^{(0)}$, which is the L_2 scalar product with the weight $(f^{(0)})^{-1}$.

Now, we obtain transport equations for the scattering rates (5.111). We write down the time derivative of the collision integral due to the Boltzmann equation,

$$\partial_t Q(f,f)(\boldsymbol{v}) = \hat{T}Q(f,f)(\boldsymbol{v}) + \hat{R}Q(f,f)(\boldsymbol{v})\;, \tag{5.112}$$

where

$$\hat{T}Q(f,f)(\boldsymbol{v}) = \int w(\boldsymbol{v}',\boldsymbol{v}_1' \mid \boldsymbol{v},\boldsymbol{v}_1)\left[f(\boldsymbol{v})v_{1k}\partial_k f(\boldsymbol{v}_1) + f(\boldsymbol{v}_1)v_k\partial_k f(\boldsymbol{v})\right.$$
$$\left. - f(\boldsymbol{v}')v_{1k}'\partial_k f(\boldsymbol{v}_1') - f(\boldsymbol{v}_1')v_k'\partial_k f(\boldsymbol{v}')\right]\,\mathrm{d}\boldsymbol{v}'\,\mathrm{d}\boldsymbol{v}_1'\,\mathrm{d}\boldsymbol{v}_1\,\mathrm{d}\boldsymbol{v}\;; \tag{5.113}$$
$$\hat{R}Q(f,f)(\boldsymbol{v}) = \int w(\boldsymbol{v}',\boldsymbol{v}_1' \mid \boldsymbol{v},\boldsymbol{v}_1)\left[Q(f,f)(\boldsymbol{v}')f(\boldsymbol{v}_1') + Q(f,f)(\boldsymbol{v}_1')f(\boldsymbol{v}')\right.$$
$$\left. - Q(f,f)(\boldsymbol{v}_1)f(\boldsymbol{v}) - Q(f,f)(\boldsymbol{v})f(\boldsymbol{v}_1)\right]\,\mathrm{d}\boldsymbol{v}'\,\mathrm{d}\boldsymbol{v}_1'\,\mathrm{d}\boldsymbol{v}_1\,\mathrm{d}\boldsymbol{v}\;. \tag{5.114}$$

Using the representation,

$$\partial_k f^{(0)}(\boldsymbol{v}) = A_k(\boldsymbol{v})f^{(0)}(\boldsymbol{v})\;; \tag{5.115}$$
$$A_k(\boldsymbol{v}) = \partial_k \ln(nT^{-3/2}) + \frac{m}{k_{\mathrm{B}}T}(v_i - u_i)\partial_k u_i + \frac{m(\boldsymbol{v}-\boldsymbol{u})^2}{2k_{\mathrm{B}}T}\partial_k \ln T\;,$$

and after some simple transformations using the relation

$$\{A_k(\boldsymbol{v})\} = 0\;, \tag{5.116}$$

in linear with respect to φ deviation from $f^{(0)}$ (5.104), we obtain in (5.112):

$$\hat{T}Q(f,f)(\boldsymbol{v}) = \partial_k \int w(\boldsymbol{v}',\boldsymbol{v}_1' \mid \boldsymbol{v},\boldsymbol{v}_1)f^{(0)}(\boldsymbol{v}_1)f^{(0)}(\boldsymbol{v})\left\{v_k\varphi(\boldsymbol{v})\right\} d\boldsymbol{v}_1' \, d\boldsymbol{v}' \, d\boldsymbol{v}_1$$

$$+ \int w(\boldsymbol{v}',\boldsymbol{v}_1' \mid \boldsymbol{v},\boldsymbol{v}_1)f^{(0)}(\boldsymbol{v}_1)f^{(0)}(\boldsymbol{v})\left\{v_k A_k(\boldsymbol{v})\right\} d\boldsymbol{v}' \, d\boldsymbol{v}_1' \, d\boldsymbol{v}_1$$

$$+ \int w(\boldsymbol{v}',\boldsymbol{v}_1' \mid \boldsymbol{v},\boldsymbol{v}_1)f^{(0)}(\boldsymbol{v})f^{(0)}(\boldsymbol{v}_1)\left[\varphi(\boldsymbol{v})A_k(\boldsymbol{v}_1)(v_{1k}-v_k)\right.$$

$$+\varphi(\boldsymbol{v}_1)A_k(\boldsymbol{v})(v_k-v_{1k}) + \varphi(\boldsymbol{v}')A_k(\boldsymbol{v}_1')(v_k'-v_{1k}')$$

$$\left. + \varphi(\boldsymbol{v}_1')A_k(\boldsymbol{v}')(v_{1k}'-v_k')\right] d\boldsymbol{v}_1' \, d\boldsymbol{v}' \, d\boldsymbol{v}_1 \; ; \qquad (5.117)$$

$$\hat{R}Q(f,f)(\boldsymbol{v}) = \int w(\boldsymbol{v}',\boldsymbol{v}_1' \mid \boldsymbol{v},\boldsymbol{v}_1)f^{(0)}(\boldsymbol{v})f^{(0)}(\boldsymbol{v}_1)\left\{\xi(\boldsymbol{v})\right\} d\boldsymbol{v}_1' \, d\boldsymbol{v}' \, d\boldsymbol{v}_1 \; ;$$

$$\xi(\boldsymbol{v}) = \int w(\boldsymbol{v}',\boldsymbol{v}_1' \mid \boldsymbol{v},\boldsymbol{v}_1)f^{(0)}(\boldsymbol{v}_1)\left\{\varphi(\boldsymbol{v})\right\} d\boldsymbol{v}_1' \, d\boldsymbol{v}' \, d\boldsymbol{v}_1 \; ; \qquad (5.118)$$

$$\partial_t Q(f,f)(\boldsymbol{v}) \qquad\qquad\qquad\qquad\qquad\qquad (5.119)$$

$$= -\partial_t \int w(\boldsymbol{v}',\boldsymbol{v}_1' \mid \boldsymbol{v},\boldsymbol{v}_1)f^{(0)}(\boldsymbol{v})f^{(0)}(\boldsymbol{v}_1)\left\{\varphi(\boldsymbol{v})\right\} d\boldsymbol{v}' \, d\boldsymbol{v}_1' \, d\boldsymbol{v}_1 \; .$$

Let us use two identities:

1. From the conservation laws it follows

$$\left\{\varphi(\boldsymbol{v})\right\} = \left\{(\hat{1}-\hat{P}^{(0)})\varphi(\boldsymbol{v})\right\} . \qquad (5.120)$$

2. The symmetry property of the kernel w (5.102) which follows from (5.102), (5.105)

$$\int w(\boldsymbol{v}',\boldsymbol{v}_1' \mid \boldsymbol{v},\boldsymbol{v}_1)f^{(0)}(\boldsymbol{v}_1)f^{(0)}(\boldsymbol{v})g_1(\boldsymbol{v})\left\{g_2(\boldsymbol{v})\right\} d\boldsymbol{v}' \, d\boldsymbol{v}_1' \, d\boldsymbol{v}_1 \, d\boldsymbol{v} \quad (5.121)$$

$$= \int w(\boldsymbol{v}',\boldsymbol{v}_1' \mid \boldsymbol{v},\boldsymbol{v}_1)f^{(0)}(\boldsymbol{v}_1)f^{(0)}(\boldsymbol{v})g_2(\boldsymbol{v})\left\{g_1(\boldsymbol{v})\right\} d\boldsymbol{v}' \, d\boldsymbol{v}_1' \, d\boldsymbol{v}_1 \, d\boldsymbol{v} \; .$$

It is valid for any two functions g_1, g_2 ensuring existence of the integrals, and also using the first identity.

Now, multiplying (5.117)–(5.120) by the microscopic moment density $\mu(\boldsymbol{v})$, performing integration over $\boldsymbol{v}$ (and using identities (5.120), (5.122)) we obtain the required transport equation for the scattering rate in the linear neighborhood of the local equilibrium:

$$-\partial_t \Delta M_Q[\varphi,\mu] \equiv -\partial_t(\varphi,\mu_Q)$$

$$= (v_k A_k(\boldsymbol{v}), \mu_Q((\hat{1}-\hat{P}^{(0)})\mu(\boldsymbol{v})))$$

$$+\partial_k(\varphi(\boldsymbol{v})v_k, \mu_Q((\hat{1}-\hat{P}^{(0)})\mu(\boldsymbol{v}))) + \int w(\boldsymbol{v}',\boldsymbol{v}_1' \mid \boldsymbol{v},\boldsymbol{v}_1)f^{(0)}(\boldsymbol{v}_1)f^{(0)}(\boldsymbol{v})$$

$$\times \left\{(\hat{1}-\hat{P}^{(0)})\mu(\boldsymbol{v})\right\} A_k(\boldsymbol{v}_1)(v_{1k}-v_k)\varphi(\boldsymbol{v}) \, d\boldsymbol{v}' \, d\boldsymbol{v}_1' \, d\boldsymbol{v}_1 d\boldsymbol{v}$$

$$+ \left(\xi(\boldsymbol{v}), \mu_Q\left((\hat{1}-\hat{P}^{(0)})\mu(\boldsymbol{v})\right)\right) . \qquad (5.122)$$

The chain of equations (5.122) for scattering rates is a counterpart of the hydrodynamic moment chain (5.100). Below we call (5.122) the *second chain*, and (5.100) – the *first chain*. Equations of the second chain are coupled in the same way as the first one: the last term in the right part of (5.91) $(\xi, \mu_Q((\hat{1} - \hat{P}^{(0)})\mu))$ depends on the whole totality of moments and scattering rates and may be treated as a new variable. Therefore, generally speaking, we have an infinite sequence of chains of increasingly higher orders. Only in the case of a special choice of the collision model – Maxwell potential $U = -\kappa r^{-4}$ – this sequence degenerates: the second and the higher-order chains are equivalent to the first (see below).

Let us restrict our consideration to the first and second hydrodynamic chains. Then a deviation from the local equilibrium state and transition to a closed macroscopic description may be performed in three different ways for the microscopic moment density $\mu(\boldsymbol{v})$. First, one can specify the moment $\hat{M}[\mu]$ and perform a closure of the chain (5.100) by the triangle method given in previous subsections. This leads to Grad's moment method. Second, one can specify scattering rate $\hat{M}_Q[\mu]$ and perform a closure of the second hydrodynamic chain (5.91). Finally, one can consider simultaneously both $\hat{M}[\mu]$ and $\hat{M}_Q[\mu]$ (mixed chain). Quasiequilibrium distribution functions corresponding to the last two variants will be constructed in the following subsection. The hard spheres model (H.S.) and Maxwell's molecules (M.M.) will be considered.

5.6.4 Distribution Functions of the Second Quasiequilibrium Approximation for Scattering Rates

First Five Moments and Collision Stress Tensor

Elsewhere below the local equilibrium $f^{(0)}$ (5.87) is chosen as the first quasiequilibrium approximation.

Let us choose $\nu_{ik} = mv_iv_k$ (5.89) as the microscopic density $\mu(\boldsymbol{v})$ of the second quasiequilibrium state. Let us write down the corresponding scattering rate (collision stress tensor) Δ_{ik} in the form,

$$\Delta_{ik} = -(\varphi, \nu_{Qik}) , \qquad (5.123)$$

where

$$\nu_{Qik}(\boldsymbol{v}) = m \int w(\boldsymbol{v}', \boldsymbol{v}'_1 \mid \boldsymbol{v}_1, \boldsymbol{v}) f^{(0)}(\boldsymbol{v}_1)$$
$$\times \left\{ (v_i - u_i)(v_k - u_k) - \frac{1}{3}\delta_{ik}(\boldsymbol{v} - \boldsymbol{u})^2 \right\} d\boldsymbol{v}' \, d\boldsymbol{v}'_1 \, d\boldsymbol{v}_1 \qquad (5.124)$$

is the microscopic density of the scattering rate Δ_{ik}.

The quasiequilibrium distribution function of the second quasiequilibrium approximation for fixed scattering rates (5.123) is determined as the solution to the problem

$$(\varphi, \varphi) \to \min \text{ for}$$
$$(\varphi, \nu_{Qik}) = -\Delta_{ik} . \tag{5.125}$$

The method of Lagrange multipliers yields

$$\varphi(\boldsymbol{v}) = \lambda_{ik} \nu_{Qik}(\boldsymbol{v}) ,$$
$$\lambda_{ik}(\nu_{Qik}, \nu_{Qls}) = \Delta_{ls} , \tag{5.126}$$

where λ_{ik} are the Lagrange multipliers.

In the examples of collision models considered below (and in general, for centrally symmetric interactions) ν_{Qik} is of the form

$$\nu_{Qik}(\boldsymbol{v}) = (\hat{1} - \hat{P}^{(0)}) \nu_{ik}(\boldsymbol{v}) \Phi((\boldsymbol{v} - \boldsymbol{u})^2) , \tag{5.127}$$

where $(\hat{1} - \hat{P}^{(0)}) \nu_{ik}$ is determined by relationship (5.90) only, and function Φ depends only on the absolute value of the peculiar velocity $(\boldsymbol{v} - \boldsymbol{u})$. Then

$$\lambda_{ik} = r \Delta_{ik} ;$$
$$r^{-1} = (2/15) \left(\Phi^2((\boldsymbol{v} - \boldsymbol{u})^2), (\boldsymbol{v} - \boldsymbol{u})^4 \right) , \tag{5.128}$$

and the distribution function of the second quasiequilibrium approximation for scattering rates (5.123) is given by the expression

$$f = f^{(0)}(1 + r \Delta_{ik} \mu_{Qik}) . \tag{5.129}$$

The form of the function $\Phi((\boldsymbol{v} - \boldsymbol{u})^2)$, and the value of the parameter r are determined by the model of particle's interaction. In the Appendix to this example, they are found for hard spheres and Maxwell molecules models (see (5.187)–(5.192)). The distribution function (5.129) is given by the following expressions:
For Maxwell molecules:

$$f = f^{(0)}$$
$$\times \left\{ 1 + \mu_0^{\text{M.M.}} m (2P^2 k_{\mathrm{B}} T)^{-1} \Delta_{ik} \left((v_i - u_i)(v_k - u_k) - \frac{1}{3} \delta_{ik} (\boldsymbol{v} - \boldsymbol{u})^2 \right) \right\} ,$$
$$\mu_0^{\text{M.M.}} = \frac{k_{\mathrm{B}} T \sqrt{2m}}{3\pi A_2(5) \sqrt{\kappa}} , \tag{5.130}$$

where $\mu_0^{\text{M.M.}}$ is viscosity coefficient in the first approximation of the Chapman-Enskog method (it is exact in the case of Maxwell molecules), κ is a force constant, $A_2(5)$ is a number, $A_2(5) \approx 0.436$ (see [70]);

For the hard spheres model:

$$f = f^{(0)}$$
$$\times \left\{ 1 + \frac{2\sqrt{2}\tilde{r}m\mu_0^{\text{H.S.}}}{5P^2 k_B T} \Delta_{ik} \int_{+1}^{-1} \exp\left\{ -\frac{m(v-u)^2}{2k_B T} y^2 \right\} (1-y^2)(1+y^2) \right.$$
$$\times \left. \left(\frac{m(v-u)^2}{2k_B T}(1-y^2) + 2 \right) dy \left((v_i - u_i)(v_k - u_k) - \frac{1}{3}\delta_{ik}(v-u)^2 \right) \right\},$$
$$\mu_0^{\text{H.S.}} = (5\sqrt{k_B Tm})/(16\sqrt{\pi}\sigma^2), \tag{5.131}$$

where $\tilde{r}$ is a number represented as follows:

$$\tilde{r}^{-1} = \frac{1}{16} \int_{-1}^{+1} \int_{-1}^{+1} \alpha^{-11/2} \beta(y)\beta(z)\gamma(y)\gamma(z)$$
$$\times (16\alpha^2 + 28\alpha(\gamma(y) + \gamma(z)) + 63\gamma(y)\gamma(z)) \, dy \, dz, \tag{5.132}$$
$$\alpha = 1 + y^2 + z^2, \qquad \beta(y) = 1 + y^2, \qquad \gamma(y) = 1 - y^2.$$

Numerical value of $\tilde{r}^{-1}$ is 5.212, to third decimal point accuracy.

In the mixed description, the distribution function of the second quasiequilibrium approximation under fixed values of the moments and of the scattering rates corresponding to the microscopic density (5.89) is determined as a solution of the problem

$$(\varphi, \varphi) \to \min \quad \text{for} \tag{5.133}$$
$$((\hat{1} - \hat{P}^{(0)})\nu_{ik}, \varphi) = \sigma_{ik},$$
$$(\nu_{Qik}, \varphi) = \Delta_{ik}.$$

Taking into account the relation (5.127), we obtain the solution of the problem (5.133) in the form,

$$\varphi(v) = (\lambda_{ik}\Phi((v-u)^2) + \beta_{ik})((v_i - u_i)(v_k - u_k) - (1/3)\delta_{ik}(v-u)^2). \tag{5.134}$$

Lagrange multipliers λ_{ik}, β_{ik} are determined from the system of linear equations,

$$ms^{-1}\lambda_{ik} + 2Pk_B Tm^{-1}\beta_{ik} = \sigma_{ik},$$
$$mr^{-1}\lambda_{ik} + ms^{-1}\beta_{ik} = \Delta_{ik}, \tag{5.135}$$

where
$$s^{-1} = (2/15)(\Phi((v-u)^2), (v-u)^4). \tag{5.136}$$

If the solvability condition of the system (5.135) is satisfied,

$$D = m^2 s^{-2} - 2Pk_B T r^{-1} \neq 0, \tag{5.137}$$

then the distribution function of the second quasiequilibrium approximation exists and takes the form

$$f = f^{(0)} \left\{ 1 + (m^2 s^{-2} - 2Pk_{\mathrm{B}}Tr^{-1})^{-1} \right. \tag{5.138}$$
$$\times [(ms^{-1}\sigma_{ik} - 2Pk_{\mathrm{B}}Tm^{-1}\Delta_{ik})\Phi((\boldsymbol{v} - \boldsymbol{u})^2)$$
$$\left. + (ms^{-1}\Delta_{ik} - mr^{-1}\sigma_{ik})]((v_i - u_i)(v_k - u_k) - (1/3)\delta_{ik}(\boldsymbol{v} - \boldsymbol{u})^2) \right\} \; .$$

The condition (5.137) means independence of the set of moments σ_{ik} from the scattering rates Δ_{ik}. If this condition is not satisfied, then the scattering rates Δ_{ik} can be represented in the form of linear combinations of σ_{ik} (with coefficients depending on the hydrodynamic moments). Then the closed by means of (5.129) equations of the second chain are equivalent to the ten moment Grad equations, while the mixed chain does not exist. This happens only in the case of Maxwell molecules. Indeed, in this case

$$s^{-1} = 2P^2 k_{\mathrm{B}}T(m^2\mu_0^{\mathrm{M.M.}})^{-1}; D = 0 \; .$$

The transformation changing Δ_{ik} to σ_{ik} is

$$\mu_0^{\mathrm{M.M.}}\Delta_{ik}P^{-1} = \sigma_{ik} \; . \tag{5.139}$$

For hard spheres:

$$s^{-1} = \frac{5P^2 k_{\mathrm{B}}T}{4\sqrt{2}\mu_0^{\mathrm{H.S.}}m^2} \cdot \tilde{s}^{-1}, \; \tilde{s}^{-1} = \int_{-1}^{+1} \gamma(y)(\beta(y))^{-7/2} \left(\beta(y) + \frac{7}{4}\gamma(y) \right) \mathrm{d}y \; . \tag{5.140}$$

The numerical value of $\tilde{s}^{-1}$ is 1.115 to third decimal point. The condition (5.136) takes the form,

$$D = \frac{25}{32} \left(\frac{P^2 k_{\mathrm{B}}T}{m\mu_0^{\mathrm{H.S.}}} \right)^2 (\tilde{s}^{-2} - \tilde{r}^{-1}) \neq 0 \; . \tag{5.141}$$

Consequently, for the hard spheres model the distribution function of the second quasiequilibrium approximation of the mixed chain exists and is determined by the expression

$$f = f^{(0)} \left\{ 1 + m(4Pk_{\mathrm{B}}T(\tilde{s}^{-2} - \tilde{r}^{-1}))^{-1} \right.$$
$$\times \left[\left(\sigma_{ik}\tilde{s}^{-1} - \frac{8\sqrt{2}}{5P}\mu_0^{\mathrm{H.S.}}\Delta_{ik} \right) \int_{-1}^{+1} \exp\left(-\frac{m(\boldsymbol{v} - \boldsymbol{u})^2}{2k_{\mathrm{B}}T}y^2 \right) \right.$$
$$\times (1 - y^2)(1 + y^2) \left(\frac{m(\boldsymbol{v} - \boldsymbol{u})^2}{2k_{\mathrm{B}}T}(1 - y^2) + 2 \right) \mathrm{d}y$$
$$\left. + 2 \left(\tilde{s}^{-1} \cdot \frac{8\sqrt{2}}{5P}\mu_0^{\mathrm{H.S.}}\Delta_{ik} - \tilde{r}^{-1}\sigma_{ik} \right) \right]$$
$$\left. \times ((v_i - u_i)(v_k - u_k) - \frac{1}{3}\delta_{ik}(\boldsymbol{v} - \boldsymbol{u})^2) \right\} \; . \tag{5.142}$$

First Five Moments, Collision Stress Tensor, and Collision Heat Flux Vector

Distribution function of the second quasiequilibrium approximation which takes into account the collision heat flux vector Q is constructed in a similar way. The microscopic density ξ_{Qi} is

$$\xi_{Qi}(\boldsymbol{v}) = \int w(\boldsymbol{v}', \boldsymbol{v}_1' \mid \boldsymbol{v}, \boldsymbol{v}_1) f^{(0)}(\boldsymbol{v}_1) \left\{ (\hat{1} - \hat{P}^{(0)}) \frac{v_i^2 v}{2} \right\} \, d\boldsymbol{v}' \, d\boldsymbol{v}_1' \, d\boldsymbol{v}_1 \;. \tag{5.143}$$

The desired distribution functions are the solutions to the following optimization problems: for the second chain it is the solution to the problem (5.125) with the additional constraints,

$$m(\varphi, \xi_{Qi}) = Q_i \;. \tag{5.144}$$

For the mixed chain, the distribution functions is the solution to the problem (5.133) with additional conditions,

$$m(\varphi, \xi_{Qi}) = Q_i \;, \tag{5.145}$$

$$m(\varphi, (\hat{1} - \hat{P}^{(0)})\xi_i) = q_i \;. \tag{5.146}$$

Here $\xi_i = v_i v^2/2$ (see (5.92)). In the Appendix functions ξ_{Qi} are found for Maxwell molecules and hard spheres (see (5.192)–(5.197)). Since

$$(\xi_{Qi}, \nu_{Qkj}) = ((\hat{1} - \hat{P}^{(0)})\xi_i, \nu_{Qkj})$$
$$= (\xi_{Qi}, (\hat{1} - \hat{P}^{(0)})\nu_{kj}) = ((\hat{1} - \hat{P}^{(0)})\xi_i, (\hat{1} - \hat{P}^{(0)})\nu_{kj}) = 0 \;, \tag{5.147}$$

the conditions (5.144) are linearly independent from the constraints of the problem (5.125), and the conditions (5.146) do not depend on the constraints of the problem (5.133).

Distribution function of the second quasiequilibrium approximation of the second chain for fixed Δ_{ik}, Q_i is of the form,

$$f = f^{(0)}(1 + r\Delta_{ik}\nu_{Qik} + \eta Q_i \xi_{Qi}) \;. \tag{5.148}$$

The parameter η is determined by the relation

$$\eta^{-1} = (1/3)(\xi_{Qi}, \xi_{Qi}) \;. \tag{5.149}$$

According to (5.196), for Maxwell molecules

$$\eta = \frac{9m^3(\mu_0^{\text{M.M.}})^2}{10P^3(k_{\text{B}}T)^2} \;, \tag{5.150}$$

and the distribution function (5.148) is

$$f = f^{(0)}$$
$$\times \left\{ 1 + \mu_0^{\text{M.M.}} m (2P^2 k_{\text{B}} T)^{-1} \Delta_{ik} ((v_i - u_i)(v_k - u_k) - (1/3)\delta_{ik}(\boldsymbol{v} - \boldsymbol{u})^2) \right.$$
$$\left. + \mu_0^{\text{M.M.}} m (P^2 k_{\text{B}} T)^{-1} (v_i - u_i) \left(\frac{m(\boldsymbol{v} - \boldsymbol{u})^2}{5 k_{\text{B}} T} - 1 \right) \right\} . \tag{5.151}$$

For hard spheres (see Appendix)

$$\eta = \tilde{\eta} \frac{64 m^3 (\mu_0^{\text{H.S.}})^2}{125 P^3 (k_{\text{B}} T)^2} , \tag{5.152}$$

where η is a number equal to 16.077 to third decimal point accuracy.
The distribution function (5.148) for hard spheres takes the form

$$f = f^{(0)} \left\{ 1 + \frac{2\sqrt{2}\tilde{r} m \mu_0^{\text{H.S.}}}{5 P^2 k_{\text{B}} T} \Delta_{ik} \int_{-1}^{+1} \exp\left(-\frac{m(\boldsymbol{v} - \boldsymbol{u})^2}{2 k_{\text{B}} T} y^2 \right) \beta(y)\gamma(y) \right.$$
$$\times \left(\frac{m(\boldsymbol{v} - \boldsymbol{u})^2}{2 k_{\text{B}} T} \gamma(y) + 2 \right) dy \left((v_i - u_i)(v_k - u_k) - \frac{1}{3}\delta_{ik}(\boldsymbol{v} - \boldsymbol{u})^2 \right)$$
$$+ \frac{2\sqrt{2}\tilde{\eta} m^3 \mu_0^{\text{H.S.}}}{25 P^2 (k_{\text{B}} T)^2} Q_i \left[(v_i - u_i) \left((\boldsymbol{v} - \boldsymbol{u})^2 - \frac{5 k_{\text{B}} T}{m} \right) \right.$$
$$\times \int_{-1}^{+1} \exp\left(-\frac{m(\boldsymbol{v} - \boldsymbol{u})^2}{2 k_{\text{B}} T} y^2 \right) \beta(y)\gamma(y) \left(\frac{m(\boldsymbol{v} - \boldsymbol{u})^2}{2 k_{\text{B}} T} \gamma(y) + 2 \right) dy$$
$$+ (v_i - u_i)(\boldsymbol{v} - \boldsymbol{u})^2 \int_{-1}^{+1} \exp\left(-\frac{m(\boldsymbol{v} - \boldsymbol{u})^2}{2 k_{\text{B}} T} y^2 \right) \beta(y)\gamma(y)$$
$$\left. \times \left(\sigma(y) \frac{m(\boldsymbol{v} - \boldsymbol{u})^2}{2 k_{\text{B}} T} + \delta(y) \right) dy \right] \right\} . \tag{5.153}$$

The functions $\beta(y), \gamma(y), \sigma(y)$ and $\delta(y)$ are

$$\beta(y) = 1 + y^2, \ \gamma(y) = 1 - y^2, \ \sigma(y) = y^2(1 - y^2), \ \delta(y) = 3y^2 - 1 . \tag{5.154}$$

The condition of existence of the second quasiequilibrium approximation of the mixed chain (5.137) should be supplemented with the requirement

$$R = m^2 \tau^{-2} - \frac{5P(k_{\text{B}} T)^2}{2m} \eta^{-1} \neq 0 . \tag{5.155}$$

Here

$$\tau^{-1} = \frac{1}{3} \left((\hat{1} - \hat{P}^{(0)}) \frac{v_i^2 v}{2}, \xi_{Qi}(\boldsymbol{v}) \right) . \tag{5.156}$$

For Maxwell molecules

$$\tau^{-1} = \left(5P^2 k_B^2 T^2 \right) / \left(3\mu_0^{\text{M.M.}} m^3 \right) ,$$

and the solvability condition (5.155) is not satisfied. Distribution function of the second quasiequilibrium approximation of mixed chain does not exist for Maxwell molecules. The variables Q_i are changed to q_i by the transformation

$$3\mu_0^{\text{M.M.}} Q_i = 2Pq_i . \tag{5.157}$$

For hard spheres,

$$\tau^{-1} = \tilde{\tau}^{-1} = \frac{25(Pk_{\text{B}}T)^2}{8\sqrt{2}m^3\mu_0^{\text{H.S.}}} , \tag{5.158}$$

where

$$\tilde{\tau}^{-1} = \frac{1}{8}\int_{-1}^{+1}\beta^{-9/2}(y)\gamma(y)\{63(\gamma(y)+\sigma(y))$$
$$+7\beta(y)(4-10\gamma(y)+2\delta(y)-5\sigma(y))$$
$$+\beta^2(y)(25\gamma(y)-10\delta(y)-40)+20\beta^3(y)\} \, dy . \tag{5.159}$$

The numerical value of $\tilde{\tau}^{-1}$ is about 4.322. Then the condition (5.155) is verified:

$$R \approx 66m^{-4}(Pk_{\text{B}}T)^4(\mu_0^{\text{H.S.}})^2 .$$

Finally, for the fixed values of $\sigma_{ik}, \Delta_{ik}, q_i$ and Q_i the distribution function of the second quasiequilibrium approximation of the second chain for hard spheres is of the form,

$$f = f^{(0)}\left\{1+\frac{m}{4Pk_{\text{B}}T}(\tilde{s}^{-2}-\tilde{r}^{-1})^{-1}\right.$$
$$\times\left[\left(\tilde{s}^{-1}\sigma_{ik}-\frac{8\sqrt{2}}{5P}\mu_0^{\text{H.S.}}\Delta_{ik}\right)\int_{-1}^{+1}\exp\left(-\frac{m(\boldsymbol{v}-\boldsymbol{u})^2}{2k_{\text{B}}T}y^2\right)\right.$$
$$\times\beta(y)\gamma(y)\left(\frac{m(\boldsymbol{v}-\boldsymbol{u})^2}{2k_{\text{B}}T}\gamma(y)+2\right)\,dy+2\left(\tilde{s}^{-1}\frac{8\sqrt{2}}{5P}\mu_0^{\text{H.S.}}\Delta_{ik}-\tilde{r}^{-1}\sigma_{ik}\right)\right]$$
$$\times\left((v_i-u_i)(v_k-u_k)-\frac{1}{3}\delta_{ik}(\boldsymbol{v}-\boldsymbol{u})^2\right)$$
$$+\frac{m^2}{10(Pk_{\text{B}}T)^2}(\tilde{\tau}^{-2}-\tilde{\eta}^{-1})^{-1}\left[\left(\tilde{\tau}^{-1}q_i-\frac{4\sqrt{2}}{5P}\mu_0^{\text{H.S.}}Q_i\right)\right.$$
$$\times\left((v_i-u_i)\left((\boldsymbol{v}-\boldsymbol{u})^2-\frac{5k_{\text{B}}T}{m}\right)\int_{-1}^{+1}\exp\left(-\frac{m(\boldsymbol{v}-\boldsymbol{u})^2}{2k_{\text{B}}T}y^2\right)\right.$$
$$\times\beta(y)\gamma(y)\left(\frac{m(\boldsymbol{v}-\boldsymbol{u})^2}{2k_{\text{B}}T}\gamma(y)+2\right)\,dy+(v_i-u_i)(\boldsymbol{v}-\boldsymbol{u})^2$$
$$\times\int_{-1}^{+1}\exp\left(-\frac{m(\boldsymbol{v}-\boldsymbol{u})^2}{2k_{\text{B}}T}y^2\right)\beta(y)\gamma(y)\left(\frac{m(\boldsymbol{v}-\boldsymbol{u})^2}{2k_{\text{B}}T}\sigma(y)+\delta(y)\right)\,dy\right)$$
$$\left.\left.+2\left(\frac{4\sqrt{2}}{5P}\mu_0^{\text{H.S.}}\tilde{\tau}^{-1}Q_i-\tilde{\eta}^{-1}q_i\right)(v_i-u_i)\left((\boldsymbol{v}-\boldsymbol{u})^2-\frac{5k_{\text{B}}T}{m}\right)\right]\right\} .$$
$$\tag{5.160}$$

Thus, the expressions (5.130), (5.131), (5.142), (5.151), (5.153) and (5.160) give distribution functions of the second quasiequilibrium approximation of the second and mixed hydrodynamic chains for Maxwell molecules and hard spheres. They are analogues of ten- and thirteen-moment Grad approximations (5.91), (5.95).

The next step is to close the second and mixed hydrodynamic chains by means of the found distribution functions.

5.6.5 Closure of the Second and Mixed Hydrodynamic Chains

Second Chain, Maxwell Molecules

The distribution function of the second quasiequilibrium approximation under fixed Δ_{ik} for Maxwell molecules (5.130) presents the simplest example of the closure of the first (5.99) and second (5.122) hydrodynamic chains. With the help of it, we obtain from (5.99) the following transport equations for the moments of the first (local equilibrium) approximation:

$$\partial_t \rho + \partial_i(u_i \rho) = 0 \; ;$$
$$\rho(\partial_t u_k + u_i \partial_i u_k) + \partial_k P + \partial_i(P^{-1} \mu_0^{\text{M.M.}} \Delta_{ik}) = 0 \; ;$$
$$\frac{3}{2}(\partial_t P + u_i \partial_i P) + \frac{5}{2} P \partial_i u_i + P^{-1} \mu_0^{\text{M.M.}} \Delta_{ik} \partial_i u_k = 0 \; . \tag{5.161}$$

Now, let us from the scattering rate transport chain (5.122) find an equation for Δ_{ik} which closes the system (5.123). Substituting (5.130) into (5.122), we obtain after some computation:

$$\partial_t \Delta_{ik} + \partial_s(u_s \Delta_{ik}) + \Delta_{is} \partial_s u_k + \Delta_{ks} \partial_s u_i - \frac{2}{3} \delta_{ik} \Delta_{ls} \partial_s u_l$$
$$+ P^2 (\mu_0^{\text{M.M.}})^{-1} \left(\partial_i u_k + \partial_k u_i - \frac{2}{3} \delta_{ik} \partial_s u_s \right)$$
$$+ P(\mu_0^{\text{M.M.}})^{-1} \Delta_{ik} + \Delta_{ik} \partial_s u_s = 0 \; . \tag{5.162}$$

For comparison, let us give ten-moment Grad equations obtained when closing the chain (5.99) by the distribution functions (5.91):

$$\partial_t \rho + \partial_i(u_i \rho) = 0 \; ;$$
$$\rho(\partial_t u_k + u_i \partial_i u_k) + \partial_k P + \partial_i \sigma_{ik} = 0 \; ;$$
$$\frac{3}{2}(\partial_t P + u_i \partial_i P) + \frac{5}{2} P \partial_i u_i + \sigma_{ik} \partial_i u_k = 0 \; ; \tag{5.163}$$
$$\partial_t \sigma_{ik} + \partial_s(u_s \sigma_{ik}) + P \left(\partial_i u_k + \partial_k u_i - \frac{2}{3} \delta_{ik} \partial_s u_s \right)$$
$$+ \sigma_{is} \partial_s u_k + \sigma_{ks} \partial_s u_i - \frac{2}{3} \delta_{ik} \sigma_{ls} \partial_s u_l + P(\mu_0^{\text{M.M.}})^{-1} \sigma_{ik} = 0 \; . \tag{5.164}$$

Using the explicit form of $\mu_0^{\text{M.M.}}$ (5.130), it is easy to verify that the transformation (5.139) maps the systems (5.161), (5.162) and (5.163) into one another. This is a consequence of the degeneration of the mixed hydrodynamic chain which was already discussed. The systems (5.161), (5.162) and (5.163) are essentially equivalent. These specific properties of Maxwell molecules result from the fact that for them the microscopic densities $(\hat{1} - \hat{P}^{(0)})v_i v_k$ and $(\hat{1} - \hat{P}^{(0)})v_i v^2$ are eigen functions of the linearized collision integral.

Second Chain, Hard Spheres

We now turn our attention to the closure of the second and of the mixed hydrodynamic chains for the hard spheres model. Substituting the distribution function (5.131) into (5.99) and (5.122), we obtain an analogue of the systems (5.161) and (5.162) (second chain, hard spheres):

$$\partial_t \rho + \partial_i(u_i \rho) = 0 \; ; \tag{5.165}$$

$$\rho(\partial_t u_k + u_i \partial_i u_k) + \partial_k P + \tilde{r}\tilde{s}^{-1} \cdot \frac{8\sqrt{2}}{5} \partial_i(\mu_0^{\text{H.S.}} P^{-1} \Delta_{ik}) = 0 \; ;$$

$$\frac{3}{2}(\partial_t P + u_i \partial_i P) + \frac{5}{2} P \partial_i u_i + \tilde{r}\tilde{s}^{-1} \cdot \frac{8\sqrt{2}}{5} \mu_0^{\text{H.S.}} P^{-1} \Delta_{ik} \partial_i u_k = 0 \; ;$$

$$\partial_t \Delta_{ik} + \partial_s(u_s \Delta_{ik}) + \tilde{r}\tilde{a}_1(\partial_s u_s)\Delta_{ik} + \frac{5\tilde{s}^{-1}P^2}{8\sqrt{2}\mu_0^{\text{H.S.}}}\left(\partial_i u_k + \partial_k u_i - \frac{2}{3}\delta_{ik}\partial_s u_s\right)$$

$$+\tilde{r}(\tilde{a}_1 + \tilde{a}_2)\left(\Delta_{is}\partial_s u_k + \Delta_{ks}\partial_s u_i - \frac{2}{3}\delta_{ik}\Delta_{ls}\partial_s u_l\right)$$

$$+\tilde{r}(\tilde{a}_1 + \tilde{a}_3)\left(\Delta_{is}\partial_k u_s + \Delta_{ks}\partial_i u_s - \frac{2}{3}\delta_{ik}\Delta_{ls}\partial_s u_l\right) + (P\tilde{r}\tilde{a}_0/\mu_0^{\text{H.S.}})\Delta_{ik} = 0 \; .$$

The dimensionless parameters $\tilde{a}_0, \tilde{a}_1, \tilde{a}_2$ and $\tilde{a}_3$ are determined by the quadratures

$$\tilde{a}_1 = \frac{1}{16}\int_{-1}^{+1}\int_{-1}^{+1}\beta(y)\beta(z)\gamma^2(z)\gamma(y)\alpha^{-13/2}(y,z)$$
$$\times\{99\gamma(y)\gamma(z)(\gamma(z) - 1) + 18\alpha(y,z)(2\gamma(z)(\gamma(z) - 1)$$
$$+4\gamma(y)(4\gamma(z) - 3)) + 8\alpha^2(y,z)(4\gamma(z) - 3)\}\, dy\, dz \; ;$$

$$\tilde{a}_2 = \frac{1}{16}\int_{-1}^{+1}\int_{-1}^{+1}\beta(y)\beta(z)\gamma(y)\gamma^2(z)\alpha^{-11/2}(y,z)\{63\gamma(y)\gamma(z)$$
$$+14\alpha(y,z)(3\gamma(y) + 2\gamma(z)) + 24\alpha^2(y,z)\}\, dy\, dz \; ;$$

$$\tilde{a}_3 = \frac{1}{16}\int_{-1}^{+1}\int_{-1}^{+1}\alpha^{-11/2}(y,z)\beta(y)\beta(z)\gamma(y)\gamma(z)$$
$$\times\{63\gamma(y)\gamma(z)(\gamma(z) - 1) + 14(2\gamma(z)(\gamma(z) - 1)$$
$$+\gamma(y)(3\gamma(z) - 2))\alpha(y,z) + 8\alpha^2(y,z)(3\gamma(z) - 2)\}dydz \; . \tag{5.166}$$

$$\tilde{a}_0 \approx \frac{1}{1536\sqrt{2}} \int_{-1}^{+1} \int_{-1}^{+1} \int_{-1}^{+1} (\psi(x,y,z))^{-13/2} \beta(x)\beta(y)\beta(z)$$
$$\times \gamma(x)\gamma(y)\gamma(z)\{10395\gamma(x)\gamma(y)\gamma(z) + 3780\psi(x,y,z)$$
$$\times (\gamma(x)\gamma(y) + \gamma(x)\gamma(z) + \gamma(y)\gamma(z)) + 1680\psi^2(x,y,z)$$
$$\times (\gamma(x) + \gamma(y) + \gamma(z)) + 960\psi^3(x,y,z)\}\,dx\,dy\,dz\ ;$$
$$\psi(x,y,z) = 1 + x^2 + y^2 + z^2\ . \tag{5.167}$$

Their numerical values are $\tilde{a}_1 \approx 0.36$, $\tilde{a}_2 \approx 5.59$, $\tilde{a}_3 \approx 0.38$, $\tilde{a}_0 \approx 2.92$ to second decimal point.

Mixed Chain

The closure of the mixed hydrodynamic chain with the functions (5.142) gives the following modification of the system of equations (5.166):

$$\partial_t \rho + \partial_i(u_i \rho) = 0\ ;$$
$$\rho(\partial_t u_k + u_i \partial_i u_k) + \partial_k P + \partial_i \sigma_{ik} = 0\ ;$$
$$\frac{3}{2}(\partial_t P + u_i \partial_i P) + \frac{5}{2} P \partial_i u_i + \sigma_{ik} \partial_i u_k = 0\ ;$$
$$\partial_t \sigma_{ik} + \partial_s(u_s \sigma_{ik}) + P\left(\partial_i u_k + \partial_k u_i - \frac{2}{3}\delta_{ik}\partial_s u_s\right)$$
$$+\sigma_{is}\partial_s u_k + \sigma_{ks}\partial_s u_i - \frac{2}{3}\delta_{ik}\sigma_{ls}\partial_s u_l + \Delta_{ik} = 0\ ;$$
$$\partial_t \Delta_{ik} + \partial_s(u_s \Delta_{ik}) + \frac{5P^2}{\tilde{s}8\sqrt{2}\mu_0^{\text{H.S.}}}\left(\partial_i u_k + \partial_k u_i - \frac{2}{3}\delta_{ik}\partial_s u_s\right)$$
$$+\frac{5P}{4\sqrt{2}\mu_0^{\text{H.S.}}(\tilde{s}^{-2} - \tilde{r}^{-1})}\left\{\frac{\tilde{a}_1}{2}(\partial_s u_s)\alpha_{ik}\right.$$
$$+\frac{1}{2}(\tilde{a}_1 + \tilde{a}_2)\left(\alpha_{is}\partial_s u_k + \alpha_{ks}\partial_s u_i - \frac{2}{3}\delta_{ik}\alpha_{ls}\partial_s u_l\right)$$
$$+\frac{1}{2}(\tilde{a}_1 + \tilde{a}_3)\left(\alpha_{is}\partial_k u_s + \alpha_{ks}\partial_i u_s - \frac{2}{3}\delta_{ik}\alpha_{ls}\partial_s u_l\right)$$
$$+\tilde{b}_1(\partial_s u_s)\beta_{ik} + (\tilde{b}_1 + \tilde{b}_2)\left(\beta_{is}\partial_s u_k + \beta_{ks}\partial_s u_i - \frac{2}{3}\delta_{ik}\beta_{ls}\partial_s u_l\right)$$
$$+(\tilde{b}_1 + \tilde{b}_3)\left(\beta_{is}\partial_k u_s + \beta_{ks}\partial_i u_s - \frac{2}{3}\delta_{ik}\beta_{ls}\partial_s u_l\right)\right\}$$
$$+\frac{5P^2}{8\sqrt{2}(\mu_0^{\text{H.S.}})^2(\tilde{s}^{-2} - \tilde{r}^{-1})}\left\{\frac{5}{8\sqrt{2}\tilde{r}}\beta_{ik} + \tilde{a}_0\alpha_{ik}\right\} = 0\ ; \tag{5.168}$$
$$\alpha_{ik} = \tilde{s}^{-1}\sigma_{ik} - \frac{8\sqrt{2}}{5P}\cdot\mu_0^{\text{H.S.}}\Delta_{ik}\ ;$$
$$\beta_{ik} = \tilde{s}^{-1}\frac{8\sqrt{2}}{5P}\cdot\mu_0^{\text{H.S.}}\Delta_{ik} - \tilde{r}^{-1}\sigma_{ik}\ . \tag{5.169}$$

It is clear from the analysis of distribution functions of the second quasiequilibrium approximations of the second hydrodynamic chain that in the Grad moment method the function $\Phi(c^2)$ is substituted by a constant. Finally, let us note the simplest consequence of the variability of function $\Phi(c^2)$. If μ_0 is multiplied with a small parameter (Knudsen number Kn equal to the ratio of the main free path the to characteristic spatial scale of variations of hydrodynamic values), then the first with respect to Kn approximation of collision stress tensor $\Delta_{ik}^{(0)}$ has the form,

$$\Delta_{ik}^{(0)} = P\left(\partial_i u_k + \partial_k u_i - \frac{2}{3}\delta_{ik}\partial_s u\right) \tag{5.170}$$

for Maxwell molecules, and

$$\Delta_{ik}^{(0)} = \frac{5\tilde{r}}{8\sqrt{2}\tilde{s}\tilde{a}_0}P\left(\partial_i u_k + \partial_k u_i - \frac{2}{3}\delta_{ik}\partial_s u_s\right) \tag{5.171}$$

for hard spheres. Substitution of these expressions into the momentum equations results in the Navier-Stokes equations with effective viscosity coefficients μ_{eff},

$$\mu_{\text{eff}} = \mu_0^{\text{M.M.}} \tag{5.172}$$

for Maxwell molecules and

$$\mu_{\text{eff}} = \tilde{a}_0^{-1}\mu_0^{\text{H.S.}} \tag{5.173}$$

for hard spheres. When using ten-moment Grad approximation which does not distinguish Maxwell molecules and hard spheres, we obtain $\mu_{\text{eff}} = \mu_0^{\text{H.S.}}$. Some consequences of this fact are studied below in Sect. 5.7.

5.6.6 Appendix:
Formulas of the Second Quasiequilibrium Approximation
of the Second and Mixed Hydrodynamic Chains
for Maxwell Molecules and Hard Spheres

Write ν_{Qik} (5.124) in the standard form:

$$\nu_{Qik} = \int f^{(0)} \mid v_1 - v \mid \left\{(v_i - u_i)(v_k - u_k) - \frac{1}{3}\delta_{ik}(v - u)^2\right\} b\,db\,d\epsilon\,dv_1 , \tag{5.174}$$

where b is the impact parameter, ϵ is the angle between the plane containing the trajectory of the particle being scattered in the system of the center of mass and the plane containing the entering asymptote, the trajectory, and a certain fixed direction. It is convenient to switch to the dimensionless velocity $\mathbf{c}$:

$$c_i = \left(\frac{m}{2k_{\text{B}}T}\right)^{1/2}(v_i - u_i) \tag{5.175}$$

and to the dimensionless relative velocity $\mathbf{g}$:

$$g_i = \frac{1}{2}\left(\frac{m}{k_{\mathrm{B}}T}\right)^{1/2}(v_{1i} - u_i) \tag{5.176}$$

After standard transformations and integration with respect to ϵ (see [70]) we obtain in (5.174)

$$\nu_{Qik} = \frac{3P}{m}\pi^{-1/2} \tag{5.177}$$

$$\times \int \exp(-c_1^2)\varphi_1^{(2)}(g)\left((c_{1i} - c_i)(c_{1k} - c_k) - \frac{1}{3}\delta_{ik}(\mathbf{c}_1 - \mathbf{c})^2\right)d\mathbf{c_1}\ .$$

Here

$$\varphi_1^{(2)} = \int (1 - \cos^2\chi)\mid \mathbf{v}_1 - \mathbf{v}\mid b(\chi)\left|\frac{db}{d\chi}\right|d\chi\ , \tag{5.178}$$

and χ is an angle between the vectors g and g'.

The dependence of $\varphi_1^{(2)}$ on the vector g is determined by the choice of the model of particle's interaction.

For Maxwell molecules,

$$\varphi_1^{(2)} = \left(\frac{2\kappa}{m}\right)^{1/2}A_2(5)\ , \tag{5.179}$$

where κ is a force constant, $A_2(5)$ is a number, $A_2(5) \approx 0.436$.

For the model of hard spheres

$$\varphi_1^{(2)} = \frac{\sqrt{2}\sigma^2}{3}\left(\frac{k_{\mathrm{B}}T}{m}\right)^{1/2}\mid \mathbf{c}_1 - \mathbf{c}\mid\ , \tag{5.180}$$

where σ is diameter of the sphere modelling the particle.

Substituting (5.179) and (5.180) into (5.178), we transform the latter to the form:

for Maxwell molecules

$$\nu_{Qik} = \frac{3P}{4m}\left(\frac{2\kappa}{\pi m}\right)^{1/2}A_2(5)\exp(-c^2)\left(\frac{\partial}{\partial c_i}\frac{\partial}{\partial c_k} - \frac{1}{3}\delta_{ik}\frac{\partial}{\partial c_s}\frac{\partial}{\partial c_s}\right)T^{\mathrm{M.M.}}(c^2)\ ;$$

$$T^{\mathrm{M.M.}}(c^2) = \int \exp(-x^2 - 2x_k c_k)\,d\mathbf{x}\ ; \tag{5.181}$$

for hard spheres

$$\nu_{Qik} = \frac{P\sigma^2}{2\sqrt{2}m}\left(\frac{k_{\mathrm{B}}T}{\pi m}\right)^{1/2}\exp(-c^2)\left(\frac{\partial}{\partial c_i}\frac{\partial}{\partial c_k} - \frac{1}{3}\delta_{ik}\frac{\partial}{\partial c_s}\frac{\partial}{\partial c_s}\right)T^{\mathrm{H.S.}}(c^2)\ ;$$

$$T^{\mathrm{H.S.}}(c^2) = \int \mid \mathbf{x}\mid \exp(-x^2 - 2x_k c_k)\,d\mathbf{x}\ . \tag{5.182}$$

It is an easy matter to perform integration in (5.181), the integral is equal to $\pi^{3/2}e^{c^2}$.

Therefore for Maxwell molecules,

$$\nu_{Qik} = \frac{3}{2}n\pi \left(\frac{2\kappa}{m}\right)^{1/2} A_2(5) \left((v_i - u_i)(v_k - u_k) - \frac{1}{3}\delta_{ik}(\boldsymbol{v} - \boldsymbol{u})^2\right) . \quad (5.183)$$

The integral $T^{\text{H.S.}}$ in (5.182) can be transformed as follows:

$$T^{\text{H.S.}}(c^2) = 2\pi + \pi \int_{-1}^{+1} \exp(c^2(1 - y^2))c^2(1 + y^2) \, dy . \quad (5.184)$$

Then for the model of hard spheres,

$$\nu_{Qik} = \sqrt{2\pi}n\sigma^2 \left(\frac{k_{\text{B}}T}{m}\right)^{3/2} \left(c_i c_k - \frac{1}{3}\delta_{ik}c^2\right)$$

$$\times \int_{-1}^{+1} \exp(-c^2 y^2)(1 + y^2)(1 - y^2)(c^2(1 - y^2) + 2) \, dy . \quad (5.185)$$

Let us note a useful relationship:

$$d^n T^{\text{H.S.}}/d(c^2)^n = \pi \int_{-1}^{+1} \exp(c^2(1 - y^2))$$

$$\times (1 + y^2)(1 - y^2)^{n-1}(c^2(1 - y^2) + n) \, dy, n \geq 1 . \quad (5.186)$$

Use the expressions for the viscosity coefficient μ_0 which are obtained in the first approximation of the Chapman-Enskog method:

for Maxwell molecules,

$$\mu_0^{\text{M.M.}} = \left(\frac{2m}{\kappa}\right)^{1/2} \frac{k_{\text{B}}T}{3\pi A_2(5)} ; \quad (5.187)$$

for hard spheres,

$$\mu_0^{\text{H.S.}} = \frac{5(k_{\text{B}}Tm)^{1/2}}{16\pi^{1/2}\sigma^2} . \quad (5.188)$$

Transformation of (5.183), (5.185) to the form of (5.127) gives the following functions $\Phi((\boldsymbol{v} - \boldsymbol{u})^2)$:

for Maxwell molecules,

$$\Phi = P/\mu_0^{\text{M.M.}} ; \quad (5.189)$$

for hard spheres

$$\Phi = \frac{5P}{16\sqrt{2}\mu_0^{\text{H.S.}}} \int_{-1}^{+1} \exp\left(-\frac{m(\boldsymbol{v} - \boldsymbol{u})^2}{2k_{\text{B}}T}y^2\right)$$

$$\times (1 + y^2)(1 - y^2) \left(\frac{m(\boldsymbol{v} - \boldsymbol{u})^2}{2k_{\text{B}}T}(1 - y^2) + 2\right) dy . \quad (5.190)$$

The parameter r from (5.128) is:
for Maxwell molecules:

$$r = \left(m\mu_0^{\text{M.M.}}\right)^2 /(2P^3 k_{\mathrm{B}}T) \; ; \tag{5.191}$$

for hard spheres:

$$r = \tilde{r}\frac{64\left(m\mu_0^{\text{M.M.}}\right)^2}{25P^3 k_{\mathrm{B}}T} \; . \tag{5.192}$$

The dimensionless parameter $\tilde{r}$ is represented as follows:

$$\tilde{r}^{-1} = \frac{1}{16}\int_{-1}^{+1}\int_{-1}^{+1} \alpha^{-11/2}\beta(y)\beta(z)\gamma(y)\gamma(z)$$
$$\times\,(16\alpha^2 + 28\alpha(\gamma(y) + \gamma(z)) + 63\gamma(y)\gamma(z))\,dy\,dz \; . \tag{5.193}$$

Here and below the following notations are used:

$$\beta(y) = 1 + y^2 \; , \qquad \gamma(y) = 1 - y^2 \; , \qquad \alpha = 1 + y^2 + z^2 \; . \tag{5.194}$$

Numerical value of $\tilde{r}^{-1}$ is 5.212 to third decimal point.
    The parameter (5.136) is:
for Maxwell molecules

$$s^{-1} = (2P^2 k_{\mathrm{B}}T)/\left(m^2\mu_0^{\text{M.M.}}\right) \; ; \tag{5.195}$$

for hard spheres

$$s^{-1} = \tilde{s}^{-1}\frac{5\sqrt{2}P^2 k_{\mathrm{B}}T}{8m^2\mu_0^{\text{H.S.}}} \; . \tag{5.196}$$

The dimensionless parameter $\tilde{s}^{-1}$ is of the form

$$\tilde{s}^{-1} = \int_{-1}^{+1}\gamma(y)\beta^{-7/2}(y)\left(\beta(y) + \frac{7}{4}\gamma(y)\right)\,dy \; . \tag{5.197}$$

Numerical value of $\tilde{s}^{-1}$ is 1.115 to third decimal point.
    The scattering rate density (5.143) is of the form,

$$\xi_{Qi} = \sqrt{2}\left(\frac{k_{\mathrm{B}}T}{m}\right)^{3/2}\int f^{(0)}(\boldsymbol{v}_1)\,|\,\boldsymbol{v}_1 - \boldsymbol{v}\,|\left\{c_i\left(c^2 - \frac{5}{2}\right)\right\}b\,db\,d\epsilon\,d\boldsymbol{v}_1 \; . \tag{5.198}$$

Standard transformation of the expression $\{c_i(c^2 - 5/2)\}$ and integration with respect to ϵ change (5.198) to the form,

$$\xi_{Qi} = \frac{P}{\sqrt{2\pi m}}\int \exp(-c_1^2)\varphi_1^{(2)}(3(c_1^2 - c^2)(c_{1i} - c_i) - (\boldsymbol{c}_1 - \boldsymbol{c})^2(c_{1i} + c_i))\,d\boldsymbol{c_1} \; . \tag{5.199}$$

Further, using the expressions (5.179) and (5.180) for $\varphi_1^{(2)}$, we obtain:

for Maxwell molecules:

$$\xi_{Qi} = \frac{P}{m^2} \left(\frac{\kappa k_B T}{\pi}\right)^{1/2} A_2(5) \exp\left(-c^2\right) \hat{D}_i T^{\text{M.M.}}(c^2) ; \qquad (5.200)$$

for hard spheres:

$$\xi_{Qi} = \frac{P k_B T \sigma^2}{\sqrt{\pi} m^2} \exp(-c^2) \hat{D}_i T^{\text{H.S.}}(c^2) . \qquad (5.201)$$

The operator $\hat{D}_i$ is of the form

$$\frac{1}{4}\frac{\partial}{\partial c_i}\frac{\partial}{\partial c_s}\frac{\partial}{\partial c_s} + \frac{3}{2}c_s\frac{\partial}{\partial c_s}\frac{\partial}{\partial c_i} - \frac{1}{2}c_i\frac{\partial}{\partial c_s}\frac{\partial}{\partial c_s} . \qquad (5.202)$$

The operator $\hat{D}_i$ acts on the function $\psi(c^2)$ as follows:

$$\frac{d^2\psi}{d(c^2)^2}2c_i\left(c^2 - \frac{5}{2}\right) + c_i c^2\left(\frac{d^2\psi}{d(c^2)^2} - \frac{d^3\psi}{d(c^2)^3}\right) . \qquad (5.203)$$

From (5.200), (5.201) we obtain:
for Maxwell molecules:

$$\xi_{Qi} = \frac{P}{3\mu_0^{\text{M.M.}}}(v_i - u_i)\left((\boldsymbol{v} - \boldsymbol{u})^2 - \frac{5k_B T}{m}\right) ; \qquad (5.204)$$

for hard spheres:

$$\begin{aligned}
\xi_{Qi} = \frac{5P}{16\sqrt{2}\mu_0^{\text{H.S.}}} &\left\{(v_i - u_i)\left((\boldsymbol{v}-\boldsymbol{u})^2 - \frac{5k_B T}{m}\right)\right.\\
&\times \int_{-1}^{+1}\exp\left(-\frac{m(\boldsymbol{v}-\boldsymbol{u})^2}{2k_B T}y^2\right)\beta(y)\gamma(y)\left(\frac{m(\boldsymbol{v}-\boldsymbol{u})^2}{2k_B T}\gamma(y) + 2\right)dy\\
&+(v_i - u_i)(\boldsymbol{v}-\boldsymbol{u})^2\\
&\left.\times \int_{-1}^{+1}\exp\left(-\frac{m(\boldsymbol{v}-\boldsymbol{u})^2}{2k_B T}y^2\right)\beta(y)\gamma(y)\left(\sigma(y)\frac{m(\boldsymbol{v}-\boldsymbol{u})^2}{2k_B T} + \delta(y)\right)dy\right\} .
\end{aligned} \qquad (5.205)$$

The functions $\sigma(y), \delta(y)$ are of the form

$$\sigma(y) = y^2(1 - y^2) , \qquad \delta(y) = 3y^2 - 1 . \qquad (5.206)$$

The parameter η from (5.149) is:
for Maxwell molecules:

$$\eta = \frac{9m^3\left(\mu_0^{\text{M.M.}}\right)^2}{10P^3(k_B T)^2} ; \qquad (5.207)$$

for hard spheres:

$$\eta = \tilde{\eta} \frac{64m^3 \left(\mu_0^{\text{H.S.}}\right)^2}{125P^3(k_{\text{B}}T)^2} \ . \tag{5.208}$$

The dimensionless parameter $\tilde{\eta}$ is of the form

$$\tilde{\eta}^{-1} = \int_{-1}^{+1} \int_{-1}^{+1} \beta(y)\beta(z)\gamma(y)\gamma(z)\alpha^{-13/2} \left\{ \frac{639}{32}(\gamma(y)\gamma(z) + \sigma(y)\sigma(z) \right.$$

$$+ \sigma(y)\gamma(z) + \sigma(z)\gamma(y)) + \frac{63}{16}\alpha(2\gamma(y) + 2\gamma(z) - 5\gamma(y)\gamma(z)$$

$$+ 2(\sigma(y) + \sigma(z)) + \gamma(z)\delta(y) + \gamma(y)\delta(z) + \sigma(y)\delta(z) + \sigma(z)\delta(y))$$

$$+ \frac{7}{8}\alpha^2(4 - 10\gamma(y) - 10\gamma(z)) + \frac{25}{4}\gamma(y)\gamma(z) + 2\delta(y) \tag{5.209}$$

$$+ 2\delta(z) - 5\sigma(y) - 5\sigma(z) - \frac{5}{2}(\gamma(z)\delta(y) + \gamma(y)\delta(z) + \delta(y)\delta(z))$$

$$+ \frac{1}{4}\alpha^3 \left(-20 + \frac{25}{4}(\gamma(y) + \gamma(z)) - 5(\delta(y) + \delta(z)) \right) + \frac{5}{2}\alpha^4 \right\} dy\, dz \ .$$

Numerical value of $\tilde{\eta}^{-1}$ is 0.622 to second decimal point.

Finally, from (5.204), (5.206) we obtain τ^{-1}(5.156):
for Maxwell molecules

$$\tau^{-1} = \frac{5(Pk_{\text{B}}T)^2}{3\mu_0^{\text{M.M.}}m^3} \ ; \tag{5.210}$$

for hard spheres

$$\tau^{-1} = \tilde{\tau}^{-1} \frac{25P^2(k_{\text{B}}T)^2}{8\sqrt{2}m^3\mu_0^{\text{H.S.}}} \ ;$$

$$\tilde{\tau}^{-1} = \frac{1}{8} \int_{-1}^{+1} \beta^{-9/2}(y)\gamma(y)\{63(\gamma(y) + \sigma(y))$$

$$+ 7\beta(y)(4 - 10\gamma(y) + 2\delta(y) - 5\sigma(y)) + 20\beta^3(y)$$

$$+ \beta^2(y)(25\gamma(y) - 10\delta(y) - 40)\} dy \approx 4.322 \ . \tag{5.211}$$

5.7 Example: Alternative Grad Equations and a "New Determination of Molecular Dimensions" (Revisited)

Here we apply the method developed in the previous section to a classical problem: determination of molecular dimensions (as diameters of equivalent hard spheres) from experimental viscosity data. Scattering rates (moments of collision integral) are treated as new independent variables, and as an alternative to moments of the distribution function, to describe the rarefied gas near local equilibrium. A version of entropy maximum principle is used to derive the Grad-like description in terms of a finite number of scattering rates. New equations are compared to the Grad moment system in the heat non-conductive case. Estimations for hard spheres demonstrate, in particular, some 10% excess of the viscosity coefficient resulting from the scattering rate description, as compared to the Grad moment estimation. All necessary details of the second chain formalism are explained below.

The classical Grad moment method provides an approximate solution to the Boltzmann equation, and leads to a closed system of equations where hydrodynamic variables ρ, $\boldsymbol{u}$, and P (density, mean flux, and pressure) are coupled to a finite set of non-hydrodynamic variables. The latter are usually the stress tensor σ and the heat flux $\boldsymbol{q}$ constituting 10 and 13 moment Grad systems. The Grad method was originally introduced for diluted gases to describe regimes beyond the normal solutions [70], but later it was used, in particular, as a prototype of certain phenomenological schemes in nonequilibrium thermodynamics [235].

However, the moments do not constitute the unique system of non-hydrodynamic variables, and the exact dynamics might be equally expressed in terms of other infinite sets of variables (possibly, of a non-moment nature). Moreover, as long as one shortens the description to only a finite subset of variables, the advantage of the moment description above other systems is not obvious. As we have seen it above, the two sets of variables

5.7.1 Nonlinear Functionals Instead of Moments in the Closure Problem

Here we consider a new system of non-hydrodynamic variables, *scattering rates* $M_Q(f)$:

$$M_{Q\,i_1 i_2 i_3}(f) = \int \mu_{i_1 i_2 i_3} Q(f, f)\, \mathrm{d}\boldsymbol{v} \ ; \tag{5.212}$$

$$\mu_{i_1 i_2 i_3} = m v_1^{i_1} v_2^{i_2} v_3^{i_3} \ ,$$

which, by definition, are the moments of the Boltzmann collision integral $Q(f, f)$:

$$Q(f, f) = \int w(\boldsymbol{v}', \boldsymbol{v}_1', \boldsymbol{v}, \boldsymbol{v}_1)\left\{ f(\boldsymbol{v}') f(\boldsymbol{v}_1') - f(\boldsymbol{v}) f(\boldsymbol{v}_1) \right\} \, \mathrm{d}\boldsymbol{v}'\, \mathrm{d}\boldsymbol{v}_1'\, \mathrm{d}\boldsymbol{v}_1 \ .$$

Here w is the probability density of a change of the velocities, $(\boldsymbol{v}, \boldsymbol{v}_1) \rightarrow (\boldsymbol{v}', \boldsymbol{v}_1')$, of the two particles after their encounter, and w is defined by a model of pair interactions. The description in terms of the scattering rates M_Q (5.212) is alternative to the usually treated description in terms of the moments M: $M_{i_1 i_2 i_3}(f) = \int \mu_{i_1 i_2 i_3} f\, \mathrm{d}\boldsymbol{v}$.

A reason to consider scattering rates instead of the moments is that M_Q (5.212) reflect features of the interactions because of the w incorporated in their definition, while the moments do not. For this reason we can expect that, in general, a description with a *finite* number of scattering rates will be more informative than a description provided by the same number of their moment counterparts.

To come to the Grad-like equations in terms of the scattering rates, we have to complete the following two steps:

(i) To derive a hierarchy of transport equations for ρ, $\boldsymbol{u}$, P, and $M_{Q\,i_1 i_2 i_3}$ in a neighborhood of the local Maxwell states $f_0(\rho, \boldsymbol{u}, P)$.

(ii) To truncate this hierarchy, and to come to a closed set of equations with respect to ρ, $\boldsymbol{u}$, P, and a finite number of scattering rates.

In the step (i), we derive a description with infinite number of variables, which is formally equivalent both to the Boltzmann equation near the local equilibrium, and to the description with an infinite number of moments. The approximation comes into play in the step (ii) where we reduce the description to a finite number of variables. The difference between the moment and the alternative description occurs at this point.

The program (i) and (ii) is similar to what is done in the Grad method [201], with the only exception (and this is important) that we should always use scattering rates as independent variables and not to expand them into series in moments. Consequently, we use a method of a closure in the step (ii) that does not refer to the moment expansions. Major steps of the computation will be presented below.

5.7.2 Linearization

To complete the step (i), we represent f as $f_0(1 + \varphi)$, where f_0 is the local Maxwellian, and we linearize the scattering rates (5.212) with respect to φ:

$$\Delta M_{Q\,i_1 i_2 i_3}(\varphi) = \int \Delta \mu_{Q\,i_1 i_2 i_3} f_0 \varphi \, d\boldsymbol{v} \; ; \tag{5.213}$$

$$\Delta \mu_{Q\,i_1 i_2 i_3} = L_Q(\mu_{i_1 i_2 i_3}) \; .$$

Here L_Q is the usual linearized collision integral, divided by f_0. Though ΔM_Q are linear in φ, they are not moments because their microscopic densities, $\Delta \mu_Q$, are not velocity polynomials for a general case of w.

It is not difficult to derive the corresponding hierarchy of transport equations for variables $\Delta M_{Q\,i_1 i_2 i_3}$, ρ, $\boldsymbol{u}$, and P (we refer further to this hierarchy as to the alternative chain): one has to calculate the time derivative of the scattering rates (5.212) due to the Boltzmann equation, in the linear approximation (5.213), and to complete the system with the five known balance equations for the hydrodynamic moments (scattering rates of the hydrodynamic moments are equal to zero due to conservation laws). The structure of the alternative chain is quite similar to that of the usual moment transport chain, and for this reason we do not reproduce it here (details of calculations can be found in [237]). One should only keep in mind that the stress tensor and the heat flux vector in the balance equations for u and P are no more independent variables, and they are expressed in terms of $\Delta M_{Q\,i_1 i_2 i_3}$, ρ, u, and P.

5.7.3 Truncating the Chain

To truncate the alternative chain (step (ii)), we have, first, to choose a finite set of "essential" scattering rates (5.213), and, second, to obtain the distribution functions which depend parametrically only on ρ, u, P, and on the chosen set of scattering rates. We will restrict our consideration to a single non-hydrodynamic variable, $\sigma_{Q\,ij}$, which is the counterpart of the stress tensor σ_{ij}. This choice corresponds to the polynomial $m v_i v_j$ in the expressions (5.212) and (5.213), and the resulting equations will be alternative to the 10 moment Grad system[2]. For a spherically symmetric interaction, the expression for $\sigma_{Q\,ij}$ may be written:

$$\sigma_{Q\,ij}(\varphi) = \int \Delta\mu_{Q\,ij} f_0 \varphi \, d\boldsymbol{v} ; \tag{5.214}$$

$$\Delta\mu_{Q\,ij} = L_Q(m v_i v_j) = \frac{P}{\eta_{Q\,0}(T)} S_Q(c^2) \left\{ c_i c_j - \frac{1}{3} \delta_{ij} c^2 \right\} .$$

Here $\eta_{Q\,0}(T)$ is the first Sonine polynomial approximation of the Chapman-Enskog viscosity coefficient (VC) [70], and, as usual, $\boldsymbol{c} = \sqrt{\frac{m}{2kT}}(\boldsymbol{v} - \boldsymbol{u})$. The scalar dimensionless function S_Q depends only on c^2, and its form depends on the choice of interaction w.

5.7.4 Entropy Maximization

Next, we find the functions

$$f^*(\rho, \boldsymbol{u}, P, \sigma_{Q\,ij}) = f_0(\rho, \boldsymbol{u}, P)(1 + \varphi^*(\rho, \boldsymbol{u}, P, \sigma_{Q\,ij}))$$

which maximize the Boltzmann entropy $S(f)$ in a neighborhood of f_0 (the quadratic approximation to the entropy is valid within the accuracy of our consideration), for fixed values of $\sigma_{Q\,ij}$. That is, φ^* is a solution to the following conditional variational problem:

$$\Delta S(\varphi) = -\frac{k_B}{2} \int f_0 \varphi^2 \, d\boldsymbol{v} \to \max , \tag{5.215}$$

i) $\int \Delta\mu_{Q\,ij} f_0 \varphi \, d\boldsymbol{v} = \sigma_{Q\,ij};$ ii) $\int \{1, \boldsymbol{v}, v^2\} f_0 \varphi \, d\boldsymbol{v} = 0 .$

---

[2] To get the alternative to the 13 moment Grad equations, one should take into account the scattering counterpart of the heat flux, $q_{Q\,i} = m \int v_i \frac{v^2}{2} Q(f, f) \, d\boldsymbol{v}$.

The second (homogeneous) condition in (5.215) reflects that a deviation φ from the state f_0 is due only to non-hydrodynamic degrees of freedom, and it is straightforwardly satisfied for $\Delta\mu_{Q\,ij}$ (5.214).

Notice, that if we turn to the usual moment description, then condition (i) in (5.215) would fix the stress tensor σ_{ij} instead of its scattering counterpart $\sigma_{Q\,ij}$. Then the resulting function $f^*(\rho, \boldsymbol{u}, P, \sigma_{ij})$ will be exactly the 10 moment Grad approximation. It can be shown that a choice of any finite set of higher moments as the constraint (i) in (5.215) results in the corresponding Grad approximation. In that sense our method of constructing f^* is a direct generalization of the Grad method onto the alternative description.

The Lagrange multipliers method gives straightforwardly the solution to the problem (5.215). After the alternative chain is closed with the functions $f^*(\rho, \boldsymbol{u}, P, \sigma_{Q\,ij})$, the step (ii) is completed, and we arrive at a set of equations with respect to the variables ρ, $\boldsymbol{u}$, P, and $\sigma_{Q\,ij}$. Switching to the variables $\zeta_{ij} = n^{-1}\sigma_{Q\,ij}$, we have:

$$\partial_t n + \partial_i(nu_i) = 0 \; ; \tag{5.216}$$

$$\rho(\partial_t u_k + u_i\partial_i u_k) + \partial_k P + \partial_i\left\{\frac{\eta_{Q\,0}(T)n}{2r_Q P}\zeta_{ik}\right\} = 0 \; ; \tag{5.217}$$

$$\frac{3}{2}(\partial_t P + u_i\partial_i P) + \frac{5}{2}P\partial_i u_i + \left\{\frac{\eta_{Q\,0}(T)n}{2r_Q P}\zeta_{ik}\right\}\partial_i u_k = 0 \; ; \tag{5.218}$$

$$\partial_t\zeta_{ik} + \partial_s(u_s\zeta_{ik}) + \left\{\zeta_{ks}\partial_s u_i + \zeta_{is}\partial_s u_k - \frac{2}{3}\delta_{ik}\zeta_{rs}\partial_s u_r\right\} \tag{5.219}$$

$$+ \left\{\gamma_Q - \frac{2\beta_Q}{r_Q}\right\}\zeta_{ik}\partial_s u_s - \frac{P^2}{\eta_{Q\,0}(T)n}\left(\partial_i u_k + \partial_k u_i - \frac{2}{3}\delta_{ik}\partial_s u_s\right)$$

$$- \frac{\alpha_Q P}{r_Q\eta_{Q\,0}(T)}\zeta_{ik} = 0 \; .$$

Here $\partial_t = \partial/\partial t, \partial_i = \partial/\partial x_i$, summation in two repeated indices is assumed, and the coefficients r_Q, β_Q, and α_Q are defined with the help of the function S_Q (5.214) as follows:

$$r_Q = \frac{8}{15\sqrt{\pi}}\int_0^\infty e^{-c^2}c^6\left(S_Q(c^2)\right)^2 dc \; ;$$

$$\beta_Q = \frac{8}{15\sqrt{\pi}}\int_0^\infty e^{-c^2}c^6 S_Q(c^2)\frac{dS_Q(c^2)}{d(c^2)} dc \; ;$$

$$\alpha_Q = \frac{8}{15\sqrt{\pi}}\int_0^\infty e^{-c^2}c^6 S_Q(c^2)R_Q(c^2) dc \; . \tag{5.220}$$

The function $R_Q(c^2)$ in the last expression is defined due to the action of the operator L_Q on the function $S_Q(c^2)(c_ic_j - \frac{1}{3}\delta_{ij}c^2)$:

$$\frac{P}{\eta_{Q\,0}}R_Q(c^2)(c_ic_j - \frac{1}{3}\delta_{ij}c^2) = L_Q(S_Q(c^2)(c_ic_j - \frac{1}{3}\delta_{ij}c^2)) \; . \tag{5.221}$$

Finally, the parameter γ_Q in (5.216–5.220) reflects the temperature dependence of the VC:

$$\gamma_Q = \frac{2}{3}\left(1 - \frac{T}{\eta_{Q0}(T)}\left(\frac{\mathrm{d}\eta_{Q0}(T)}{\mathrm{d}T}\right)\right).$$

The set of ten equations (5.216–5.220) is alternative to the 10 moment Grad equations.

5.7.5 A New Determination of Molecular Dimensions (Revisited)

The observation already made is that for Maxwell molecules we have: $S^{\mathrm{M.M.}} \equiv 1$, and $\eta_0^{\mathrm{M.M.}} \propto T$; thus $\gamma^{\mathrm{M.M.}} = \beta^{\mathrm{M.M.}} = 0$, $r^{\mathrm{M.M.}} = \alpha^{\mathrm{M.M.}} = \frac{1}{2}$, and (5.216–5.220) becomes the 10 moment Grad system under a simple change of variables $\lambda\zeta_{ij} = \sigma_{ij}$, where λ is the proportionality coefficient in the temperature dependence of $\eta_0^{\mathrm{M.M.}}$.

These properties (the function S_Q is a constant, and the VC is proportional to T) are true only for Maxwell molecules. For all other interactions, the function S_Q is not identical to one, and the VC $\eta_{Q0}(T)$ is not proportional to T. Thus, the shortened alternative description is not equivalent indeed to the Grad moment description. In particular, for hard spheres, the exact expression for the function $S^{\mathrm{H.S.}}$ (5.214) reads:

$$S^{\mathrm{H.S.}} = \frac{5\sqrt{2}}{16}\int_0^1 \exp(-c^2t^2)(1-t^4)\left(c^2(1-t^2)+2\right)\,\mathrm{d}t ; \quad (5.222)$$

$$\eta_0^{\mathrm{H.S.}} \propto \sqrt{T} .$$

Thus, $\gamma^{\mathrm{H.S.}} = \frac{1}{3}$, and $\frac{\beta^{\mathrm{H.S.}}}{r^{\mathrm{H.S.}}} \approx 0.07$, and the equation for the function ζ_{ik} (5.220) contains a nonlinear term,

$$\theta^{\mathrm{H.S.}}\zeta_{ik}\partial_s u_s , \quad (5.223)$$

where $\theta^{\mathrm{H.S.}} \approx 0.19$. This term is missing in the Grad 10 moment equation.

Finally, let us evaluate the VC which results from the alternative description (5.216–5.220). Following Grad's arguments [201], we see that, if the relaxation of ζ_{ik} is fast compared to the hydrodynamic variables, then the two last terms in the equation for ζ_{ik} (5.216–5.220) become dominant, and the equation for u casts into the standard Navier-Stokes form with an effective VC $\eta_{Q\,\mathrm{eff}}$:

$$\eta_{Q\,\mathrm{eff}} = \frac{1}{2\alpha_Q}\eta_{Q0} . \quad (5.224)$$

For Maxwell molecules, we easily derive that the coefficient α_Q in (5.224) is equal to $\frac{1}{2}$. Thus, as one expects, the effective VC (5.224) is equal to the Grad value, which, in turn, is equal to the exact value in the frames of the Chapman-Enskog method for this model.

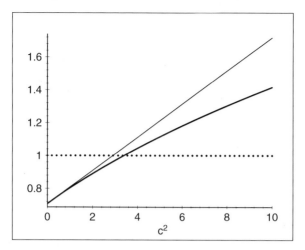

Fig. 5.1. Approximations for hard spheres: *bold* line – function $S^{\text{H.S.}}$, *solid* line – approximation $S_a^{\text{H.S.}}$, *dotted* line – Grad moment approximation

For all interactions different from the Maxwell molecules, the VC $\eta_{Q\,\text{eff}}$ (5.224) is not equal to $\eta_{Q\,0}$. For hard spheres, in particular, a computation of the VC (5.224) requires information about the function $R^{\text{H.S.}}$ (5.221). This is achieved upon a substitution of the function $S^{\text{H.S.}}$ (5.222) into (5.221). Further, we have to compute the action of the operator $L^{\text{H.S.}}$ on the function $S^{\text{H.S.}}(c_i c_j - \frac{1}{3}\delta_{ij}c^2)$, which is rather complicated. However, the VC $\eta_{\text{eff}}^{\text{H.S.}}$ can be relatively easily estimated by using a function $S_a^{\text{H.S.}} = \frac{1}{\sqrt{2}}(1+\frac{1}{7}c^2)$, instead of the function $S^{\text{H.S.}}$, in (5.221). Indeed, the function $S_a^{\text{H.S.}}$ is tangent to the function $S^{\text{H.S.}}$ at $c^2 = 0$, and is its majorant (see Fig. 5.1). Substituting $S_a^{\text{H.S.}}$ into (5.221), and computing the action of the collision integral, we find the approximation $R_a^{\text{H.S.}}$; thereafter we evaluate the integral $\alpha^{\text{H.S.}}$ (5.220), and finally come to the following expression:

$$\eta_{\text{eff}}^{\text{H.S.}} \gtrsim \frac{75264}{67237}\eta_0^{\text{H.S.}} \approx 1.12\eta_0^{\text{H.S.}} \,. \tag{5.225}$$

Thus, for hard spheres, the description in terms of scattering rates results in the VC of more than 10% higher than in the Grad moment description.

A discussion of the results concerns the following two items.

1. Having two not equivalent descriptions which were obtained within one method, we may ask: which is more relevant? A simple test is to compare characteristic times of an approach to hydrodynamic regime. We have $\tau_G \sim \eta_0^{\text{H.S.}}/P$ for 10-moment description, and $\tau_a \sim \eta_{\text{eff}}^{\text{H.S.}}/P$ for alternative description. As $\tau_a > \tau_G$, we see that scattering rate decay slower than corresponding moment, hence, at least for rigid spheres, the alternative description is more relevant. For Maxwell molecules both the descriptions are, of course, equivalent.

2. The VC $\eta_{\text{eff}}^{\text{H.S.}}$ (5.225) has the same temperature dependence as $\eta_0^{\text{H.S.}}$, and also the same dependence on a scaling parameter (a diameter of the sphere). In the classical book [70] (pp. 228–229), "sizes" of molecules are presented, assuming that a molecule is represented with an equivalent sphere and VC is estimated as $\eta_0^{\text{H.S.}}$. Since our estimation of VC differs only by a dimensionless factor from $\eta_0^{\text{H.S.}}$, it is straightforward to conclude that effective sizes of molecules will be reduced by the factor b, where

$$b = \sqrt{\eta_0^{\text{H.S.}}/\eta_{\text{eff}}^{\text{H.S.}}} \approx 0.94 \ .$$

Further, it is well known that sizes of molecules estimated via viscosity in [70] disagree with the estimation via the virial expansion of the equation of state. In particular, in book [238], p. 5, the measured second virial coefficient B_{exp} was compared with the calculated B_0, in which the diameter of the sphere was taken from the viscosity data. The reduction of the diameter by factor b gives $B_{\text{eff}} = b^3 B_0$. The values B_{exp} and B_0 [238] are compared with B_{eff} in the Table 5.1 for three gases at $T = 500\,K$. The results for argon and helium are better for B_{eff}, while for nitrogen B_{eff} is worth than B_0. However, both B_0 and B_{eff} are far from the experimental values.

Table 5.1. Three virial coefficients: experimental B_{exp}, classical B_0 [238], and reduced B_{eff} for three gases at $T = 500\,K$

	B_{exp}	B_0	B_{eff}
Argon	8.4	60.9	50.5
Helium	10.8	21.9	18.2
Nitrogen	168	66.5	55.2

Hard spheres is, of course, an oversimplified model of interaction, and the comparison presented does not allow for a decision between $\eta_0^{\text{H.S.}}$ and $\eta_{\text{eff}}^{\text{H.S.}}$. However, this simple example illustrates to what extend the correction to the VC can affect a comparison with experiment. Indeed, as it is well known, the first-order Sonine polynomial computation for the Lennard-Jones (LJ) potential gives a very good fit of the temperature dependence of the VC for all noble gases [239], subject to a proper choice of the two unknown scaling parameters of the LJ potential[3]. We may expect that a dimensionless correction of the VC for the LJ potential might be of the same order as above for rigid spheres. However, the functional character of the temperature dependence will not be affected, and a fit will be obtained subject to a different choice of the molecular parameters of the LJ potential.

---

[3] A comparison of molecular parameters of the LJ potential, as derived from the viscosity data, to those obtained from independent sources, can be found elsewhere, e.g. in [70], p. 237.

6 Newton Method
with Incomplete Linearization

The Newton method with incomplete linearization is developed for solving the invariance equation. It is the basis of an iterative construction of the manifolds of slow motions.

6.1 The Method

Let us come back to the invariance equation (3.3),

$$\Delta_y = (1 - P_y)J(F(y)) = 0 .$$

One of the most efficient methods to solve this equation is the Newton method with incomplete linearization. Let us linearize the vector field J around $F(y)$:

$$J(F(y) + \delta F(y)) = J(F(y)) + (DJ)_{F(y)}\delta F(y) + o(\delta F(y)) . \tag{6.1}$$

Equation of the Newton method with incomplete linearization makes it possible to determine $\delta F(y)$ from a linear system:

$$\begin{cases} P_y\delta F(y) = 0 , \\ (1 - P_y)(DJ)_{F(y)}\delta F(y) = (1 - P_y)J(F(y)) . \end{cases} \tag{6.2}$$

The crucial point here is that the same projector P_y is used as in the equation (3.3), that is, the variation of the projector δP is not computed (hence, the suggested linearization of equation (3.3) is incomplete). We recall that projector P_y depends on the tangent space $T_y = \mathrm{im}(DF)_y$. If the thermodynamic projector (5.25) is used here, then P_y depends also on $\langle | \rangle_{F(y)}$ and on $g = (DS)_{F(y)}$.

Equations of the Newton method with incomplete linearization (6.2) are not differential equations in y anymore, they do not contain derivatives of the unknown $\delta F(y)$ with respect to y (which would be the case if the variation of the projector δP has been taken into account). The absence of the derivatives in equation (6.2) significantly simplifies its solving. However, even this is not the main advantage of the incomplete linearization. More essential is the fact that iterations of the Newton method with incomplete linearization are

Alexander N. Gorban and Iliya V. Karlin: *Invariant Manifolds for Physical and Chemical Kinetics*, Lect. Notes Phys. **660**, 139–178 (2005)
www.springerlink.com

expected to converge to slow invariant manifolds, unlike the usual Newton method (with "complete linearization").

In order to clarify this feature of the Newton method with incomplete linearization (6.2), let us consider the case of linear manifolds for linear systems. Let a linear evolution equation be given in the real Hilbert space:

$$\dot{\boldsymbol{x}} = \mathbf{A}\boldsymbol{x} ,$$

where $\mathbf{A}$ is negative definite symmetric operator with a simple spectrum. The square of the norm is the Lyapunov function,

$$S(\boldsymbol{x}) = \langle \boldsymbol{x} \mid \boldsymbol{x} \rangle .$$

The manifolds we consider are lines, $\boldsymbol{l}(y) = y\boldsymbol{e}$, where $\boldsymbol{e}$ is the unit vector, and y is a scalar. The invariance equation for such manifolds reads:

$$\boldsymbol{e}\langle \boldsymbol{e} \mid \mathbf{A}\boldsymbol{e} \rangle - \mathbf{A}\boldsymbol{e} = 0 ,$$

and it is simply the eigenvalue problem for the operator $\mathbf{A}$. Solutions to the latter equation are eigenvectors $\boldsymbol{e}_i$, corresponding to eigenvalues λ_i.

Assume that we choose an initial approximation, that is the line $\boldsymbol{l}_0 = y\boldsymbol{e}_0$ defined by the unit vector $\boldsymbol{e}_0$. Let the vector $\boldsymbol{e}_0$ be not an eigenvector of $\mathbf{A}$. We seek another line, $\boldsymbol{l}_1 = a\boldsymbol{e}_1$, where $\boldsymbol{e}_1$ is another unit vector, $\boldsymbol{e}_1 = \boldsymbol{x}_1/\|\boldsymbol{x}_1\|$, $\boldsymbol{x}_1 = \boldsymbol{e}_0 + \delta\boldsymbol{x}$. The additional condition in (6.2) reads: $P_y \delta F(y) = 0$, i.e. $\langle \boldsymbol{e}_0 \mid \delta\boldsymbol{x} \rangle = 0$. Then (6.2) becomes

$$[1 - \boldsymbol{e}_0\langle \boldsymbol{e}_0 \mid \cdot \rangle]\mathbf{A}[\boldsymbol{e}_0 + \delta\boldsymbol{x}] = 0 .$$

Subject to the additional condition, the unique solution is as follows:

$$\boldsymbol{e}_0 + \delta\boldsymbol{x} = \langle \boldsymbol{e}_0 \mid \mathbf{A}^{-1}\boldsymbol{e}_0 \rangle^{-1}\mathbf{A}^{-1}\boldsymbol{e}_0 .$$

Upon rewriting the latter expression in the eigen-basis of $\mathbf{A}$, we have:

$$\boldsymbol{e}_0 + \delta\boldsymbol{y} \propto \sum_i \lambda_i^{-1}\boldsymbol{e}_i\langle \boldsymbol{e}_i \mid \boldsymbol{e}_0 \rangle .$$

The leading term in this sum corresponds to the eigenvalue with the *minimal absolute value*. The example indicates that the method (6.2) seeks the direction of the *slowest relaxation*. For this reason, the Newton method with incomplete linearization (6.2) can be recognized as the basis of iterative construction of the manifolds of slow motions.

In an attempt to simplify computations, the question which always can be asked is as follows: To what extend is the choice of the projector essential in the equation (6.2)? This question is a valid one, because if we accept that iterations converge to a relevant slow manifold, and also that the projection on the true invariant manifold is insensible to the choice of the projector, should one care of the projector on each iteration? In particular, for the moment parameterizations, can one use in equation (6.2) the projector (5.1)? Experience gained from some of the problems studied by this method indicates that this is possible. However, in order to derive physically meaningful equations of motion along the approximate slow manifolds, one has to use the thermodynamic projector (5.25). Otherwise we cannot guarantee the dissipation properties of these equations of motion.

6.2 Example: Two-Step Catalytic Reaction

We consider here a two-step four-component reaction with one catalyst $A_2 = Z$ (2.98):

$$A_1 + A_2 \rightleftharpoons A_3 \rightleftharpoons A_2 + A_4 . \tag{6.3}$$

We assume the Lyapunov function of the form (2.86), $G = \sum_{i=1}^4 c_i[\ln(c_i/c_i^{\mathrm{eq}}) - 1]$. The kinetic equation for the four-component vector of concentrations, $\boldsymbol{c} = (c_1, c_2, c_3, c_4)$, has the form

$$\dot{\boldsymbol{c}} = \boldsymbol{\gamma}_1 W_1 + \boldsymbol{\gamma}_2 W_2 . \tag{6.4}$$

Here $\boldsymbol{\gamma}_{1,2}$ are stoichiometric vectors,

$$\boldsymbol{\gamma}_1 = (-1, -1, 1, 0), \ \boldsymbol{\gamma}_2 = (0, 1, -1, 1) , \tag{6.5}$$

while functions $W_{1,2}$ are reaction rates:

$$W_1 = k_1^+ c_1 c_2 - k_1^- c_3, \ W_2 = k_2^+ c_3 - k_2^- c_2 c_4 . \tag{6.6}$$

Here $k_{1,2}^\pm$ are reaction rate constants. The system under consideration has two conservation laws,

$$c_1 + c_3 + c_4 = B_1, \ c_2 + c_3 = B_2 , \tag{6.7}$$

or $(\boldsymbol{b}_{1,2}, \boldsymbol{c}) = B_{1,2}$, where $\boldsymbol{b}_1 = (1, 0, 1, 1)$ and $\boldsymbol{b}_2 = (0, 1, 1, 0)$. The nonlinear system (6.4) is effectively two-dimensional, and we consider one-dimensional manifolds of reduced description.

We have chosen the concentration of the specie A_1 as the variable of reduced description: $M = c_1$, and $c_1 = (\boldsymbol{m}, \boldsymbol{c})$, where $\boldsymbol{m} = (1, 0, 0, 0)$. The initial manifold $\boldsymbol{c} = \boldsymbol{c}_0(M)$ (i.e. $\boldsymbol{c} = \boldsymbol{c}_0(c_1, B_1, B_2)$) was taken as the quasi-equilibrium approximation, i.e. the vector function $\boldsymbol{c}_0$ is the solution to the problem:

$$G \to \min \text{ for } (\boldsymbol{m}, \boldsymbol{c}) = c_1, \ (\boldsymbol{b}_1, \boldsymbol{c}) = B_1, \ (\boldsymbol{b}_2, \boldsymbol{c}) = B_2 . \tag{6.8}$$

The solution to the problem (6.8) can be computed explicitly:

$$c_{01} = c_1 \,, \tag{6.9}$$
$$c_{02} = B_2 - \phi(c_1) \,,$$
$$c_{03} = \phi(c_1) \,,$$
$$c_{04} = B_1 - c_1 - \phi(c_1) \,,$$
$$\phi(M) = A(c_1) - \sqrt{A^2(c_1) - B_2(B_1 - c_1)} \,,$$
$$A(c_1) = \frac{B_2(B_1 - c_1^{\mathrm{eq}}) + c_3^{\mathrm{eq}}(c_1^{\mathrm{eq}} + c_3^{\mathrm{eq}} - c_1)}{2c_3^{\mathrm{eq}}} \,.$$

The thermodynamic projector associated with the manifold (6.9) reads:

$$\boldsymbol{P}_0 \boldsymbol{x} = \frac{\partial \boldsymbol{c}_0}{\partial c_1}(\boldsymbol{m}, \boldsymbol{x}) + \frac{\partial \boldsymbol{c}_0}{\partial B_1}(\boldsymbol{b}_1, \boldsymbol{x}) + \frac{\partial \boldsymbol{c}_0}{\partial B_2}(\boldsymbol{b}_2, \boldsymbol{x}) \,. \tag{6.10}$$

Computing $\boldsymbol{\Delta}_0 = (1 - \boldsymbol{P}_0)\boldsymbol{J}(\boldsymbol{c}_0)$ we find that it is not equal to zero, and thus the quasiequilibrium manifold $\boldsymbol{c}_0$ is not invariant. The first correction, $\boldsymbol{c}_1 = \boldsymbol{c}_0 + \delta\boldsymbol{c}$, is found from the linear algebraic system (6.2)

$$(1 - \boldsymbol{P}_0)\boldsymbol{L}_0'\delta\boldsymbol{c} = -[1 - \boldsymbol{P}_0]\boldsymbol{J}(\boldsymbol{c}_0) \,, \tag{6.11}$$
$$\delta c_1 = 0$$
$$\delta c_1 + \delta c_3 + \delta c_4 = 0$$
$$\delta c_3 + \delta c_2 = 0 \,, \tag{6.12}$$

where the symmetric 4×4 matrix $\boldsymbol{L}_0'$ has the form (we write 0 instead of $\boldsymbol{c}_0$ in the subscript in order to simplify notations):

$$L_{0,kl}' = -\gamma_{1k}\frac{W_1^+(\boldsymbol{c}_0) + W_1^-(\boldsymbol{c}_0)}{2}\frac{\gamma_{1l}}{c_{0l}} - \gamma_{2k}\frac{W_2^+(\boldsymbol{c}_0) + W_2^-(\boldsymbol{c}_0)}{2}\frac{\gamma_{2l}}{c_{0l}} \tag{6.13}$$

Here we use the self-adjoint linearization[1].

The explicit solution $\boldsymbol{c}_1(c_1, B_1, B_2)$ to the linear system (6.11) is easily found, and we do not reproduce it here. The process was iterated. On the $k+1$ iteration, the following projector $\boldsymbol{P}_k$ was used:

$$\boldsymbol{P}_k \boldsymbol{x} = \frac{\partial \boldsymbol{c}_k}{\partial c_1}(\boldsymbol{m}, \boldsymbol{x}) + \frac{\partial \boldsymbol{c}_k}{\partial B_1}(\boldsymbol{b}_1, \boldsymbol{x}) + \frac{\partial \boldsymbol{c}_k}{\partial B_2}(\boldsymbol{b}_2, \boldsymbol{x}) \,. \tag{6.14}$$

Note that projector $\boldsymbol{P}_k$ (6.14) is thermodynamic only if $k = 0$. In the process of finding the corrections to the manifold, the non-thermodynamic projectors are allowed (we should return to the thermodynamic projector for projection of the vector field onto ansatz manifold). The linear equation at the $k+1$ iteration is thus obtained by replacing $\boldsymbol{c}_0$, $\boldsymbol{P}_0$, and $\boldsymbol{L}_0'$ with $\boldsymbol{c}_k$, $\boldsymbol{P}_k$, and $\boldsymbol{L}_k'$ in all the entries of (6.11) and (6.13).

---

[1] The self-adjoint linearization was introduced in Chap. 2 (2.33), more detailed discussion follows in Chap. 7 (7.15)

Once the manifold c_k was obtained on the kth iteration, we derived the corresponding dynamics by introducing the corresponding thermodynamic projector. The resulting dynamic equation for the variable c_1 in the kth approximation has the form:

$$\left(\boldsymbol{\nabla} G \big|_{\boldsymbol{c}_k}, \partial \boldsymbol{c}_k/\partial c_1\right)\dot{c}_1 = \left(\boldsymbol{\nabla} G \big|_{\boldsymbol{c}_k}, \boldsymbol{J}(\boldsymbol{c}_k)\right). \tag{6.15}$$

Here $[\boldsymbol{\nabla} G \big|_{\boldsymbol{c}_k}]_i = \ln[c_{ki}/c_i^{\mathrm{eq}}]$.

Analytic results were compared with the results of the numerical integration of the system (6.4). The following set of parameters was used:

$$k_1^+ = 1.0, \; k_1^- = 0.5, \; k_2^+ = 0.4, \; k_2^- = 1.0\,;$$
$$c_1^{\mathrm{eq}} = 0.5, \; c_2^{\mathrm{eq}} = 0.1, \; c_3^{\mathrm{eq}} = 0.1, \; c_4^{\mathrm{eq}} = 0.4\,,$$
$$B_1 = 1.0, \;\; B_2 = 0.2\,.$$

Figure 6.1 demonstrates the quasi-equilibrium manifold (6.9) and the first two corrections. It should be stressed that we spent no special effort on the construction of the initial approximation, that is, of the quasi-equilibrium manifold, have not used any information about the Jacobian field (unlike, for example, the ILDM [93] or CSP [90] methods) etc. The initial quasi-equilibrium approximation is in a rather poor agreement with the reduced description. Therefore, it should be appreciated that the further corrections

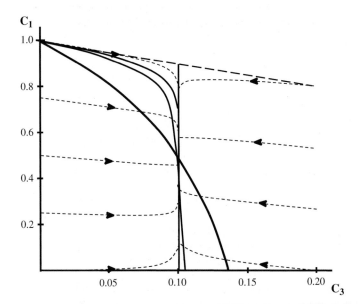

Fig. 6.1. Images of the initial quasi-equilibrium manifold (*bold* line) and the first two corrections (*solid* normal lines) in the phase plane $[c_1, c_3]$ for two-step catalytic reaction (6.3). *Dashed* lines are individual trajectories

rapidly improve the situation while no small parameter considerations were used. This confirms our expectation of the advantage of using the iteration methods instead of methods based on a small parameter expansions for model reduction problems.

6.3 Example: Non-Perturbative Correction of Local Maxvellian Manifold and Derivation of Nonlinear Hydrodynamics from Boltzmann Equation (1D)

We apply here the method of invariant manifold to a particularly important situation when the initial manifold consists of local Maxwellians (5.49) (the LM manifold). This manifold and its corrections play the central role in the problem of derivation of hydrodynamics from the Boltzmann equation. Hence, any method of approximate investigation of the Boltzmann equation should be tested with the LM manifold. Classical methods (the Chapman-Enskog and Hilbert methods) use Taylor-type expansions into powers of a small parameter (the Knudsen number expansion). However, as we have mentioned above, the method of invariant manifold, generally speaking, assumes no small parameters, at least in its formal part where convergency properties are not discussed. We shall develop an appropriate technique to consider the invariance equation of the first iteration. This technique involves ideas of the parametrix expansion of the theory of pseudodifferential and Fourier integral operators [249, 250]. This approach will make it possible to avoid using small parameters.

We seek a correction to the LM manifold in the form (dependence of velocity $\boldsymbol{v}$ will be not displayed whenever possible):

$$f_1(n, \boldsymbol{u}, T) = f_0(n, \boldsymbol{u}, T) + \delta f_1(n, \boldsymbol{u}, T) . \qquad (6.16)$$

We use the Newton method with incomplete linearization for obtaining the correction $\delta f_1(n, \boldsymbol{u}, T)$, because we are interested in a manifold of slow (hydrodynamic) motions. We introduce the representation:

$$\delta f_1(n, \boldsymbol{u}, T) = f_0(n, \boldsymbol{u}, T)\varphi(n, \boldsymbol{u}, T) . \qquad (6.17)$$

6.3.1 Positivity and Normalization

When seeking corrections, we should be ready to face two problems that are typical for any method of successive approximations in the Boltzmann equation theory. Namely, the first of this problems is that the correction

$$f_{\Omega_{k+1}} = f_{\Omega_k} + \delta f_{\Omega_{k+1}}$$

obtained from the linearized invariance equation of the $k+1$-th iteration may be not a non-negatively defined function and thus it cannot be used directly in order to define the thermodynamic projector for the $k+1$-th approximation. In order to overcome this difficulty, we can treat the procedure as a process of correcting the dual variable $\mu_f = D_f H(f)$ rather than the process of immediate correcting the distribution functions.

The dual variable μ_f is:

$$\mu_f\big|_{f=f(\boldsymbol{x},\,\boldsymbol{v})} = D_f H(f)\big|_{f=f(\boldsymbol{x},\,\boldsymbol{v})} = D_f H_{\boldsymbol{x}}(f)\big|_{f=f(\boldsymbol{x},\,\boldsymbol{v})} = \ln f(\boldsymbol{v},\boldsymbol{x})\,. \quad (6.18)$$

Then, at the $k+1$-th iteration, we obtain a new dual variable $\mu_f\big|_{\Omega_{k+1}}$:

$$\mu_f\big|_{\Omega_{k+1}} = \mu_f\big|_{\Omega_k} + \delta\mu_f\big|_{\Omega_{k+1}}\,. \quad (6.19)$$

Due to the relationship $\mu_f \longleftrightarrow f$, we have:

$$\delta\mu_f\big|_{\Omega_{k+1}} = \varphi_{\Omega_{k+1}} + O(\delta f_{\Omega_{k+1}}^2), \varphi_{\Omega_{k+1}} = f_{\Omega_k}^{-1}\delta f_{\Omega_{k+1}}\,. \quad (6.20)$$

Thus, solving the linear invariance equation of the k-th iteration with respect to the unknown function $\delta f_{\Omega_{k+1}}$, we find a correction to the dual variable $\varphi_{\Omega_{k+1}}$ (6.20), and we derive the corrected distributions $f_{\Omega_{k+1}}$ as

$$f_{\Omega_{k+1}} = \exp(\mu_f\big|_{\Omega_k} + \varphi_{\Omega_{k+1}}) = f_{\Omega_k}\exp(\varphi_{\Omega_{k+1}})\,. \quad (6.21)$$

Functions (6.21) are positive, and they satisfy the invariance equation and the additional conditions within the accuracy of $\varphi_{\Omega_{k+1}}$.

However, the second difficulty which might occur is that functions (6.21) might have no finite moments (5.43). In particular, this difficulty can be a result of some approximations used in solving equations. Hence, we have to "regularize" the functions (6.21) in some way. A sketch of an approach to do this regularization is as follows: instead of $f_{\Omega_{k+1}}$ (6.21), we consider functions:

$$f_{\Omega_{k+1}}^{(\beta)} = f_{\Omega_k}\exp(\varphi_{\Omega_{k+1}} + \varphi^{\text{reg}}(\beta))\,. \quad (6.22)$$

Here $\varphi^{\text{reg}}(\beta)$ is a function labeled with $\beta \in B$, and B is a linear space. Then we derive β_* from the condition of matching the macroscopic variables.

For example, corrections to the LM distribution in the Chapman-Enskog method [70] and the thirteen-moment Grad approximation [201] are not non-negatively defined functions, while the thirteen-moment quasiequilibrium approximation [224] has no finite integrals (5.42) and (5.43).

6.3.2 Galilean Invariance of Invariance Equation

In some cases, it is convenient to consider the Boltzmann equation vector field in a reference system which moves with the flow velocity. In this reference system, we define the Boltzmann equation vector field as:

$$\frac{\mathrm{d}f}{\mathrm{d}t} = J_u(f), \frac{\mathrm{d}f}{\mathrm{d}t} = \frac{\partial f}{\partial t} + u_{\boldsymbol{x},s}(f)\frac{\partial f}{\partial x_s} ;$$

$$J_u(f) = -(v_s - u_{\boldsymbol{x},s}(f))\frac{\partial f}{\partial x_s} + Q(f,f) . \tag{6.23}$$

Here $u_{\boldsymbol{x},s}(f)$ stands for the s-th component of the flow velocity:

$$u_{\boldsymbol{x},s}(f) = n_{\boldsymbol{x}}^{-1}(f) \int v_s f(\boldsymbol{v},\boldsymbol{x}) \,\mathrm{d}^3\boldsymbol{v}; \ n_{\boldsymbol{x}}(f) = \int f(\boldsymbol{v},\boldsymbol{x}) \,\mathrm{d}^3\boldsymbol{v} . \tag{6.24}$$

In particular, this form of the Boltzmann equation vector field is convenient when the initial manifold Ω_0 consists of functions f_{Ω_0} which depend explicitly on $(\boldsymbol{v} - u_{\boldsymbol{x}}(f))$ (i.e., if functions $f_{\Omega_0} \in \Omega_0$ do not change under velocity shifts: $\boldsymbol{v} \to \boldsymbol{v} + \boldsymbol{c}$, where $\boldsymbol{c}$ is a constant vector). This is also the case of the LM manifold.

Substituting $J_u(f)$ (6.23) instead of $J(f)$ (5.44) into all expressions which depend on the Boltzmann equation vector field, we transfer all procedures developed above into the moving reference system. In particular, we obtain the following invariance equation of the first iteration for a general locally finite dimensional initial approximation $f_0(a(\boldsymbol{x}),\boldsymbol{v})$:

$$(P_{a(\boldsymbol{x})}^{0*}(\cdot) - 1)J_{u,\mathrm{lin},a(\boldsymbol{x})}^{0}(\delta f_1(a(\boldsymbol{x}),\boldsymbol{v})) + \Delta(f_0(a(\boldsymbol{x}),\boldsymbol{v})) = 0 ; \tag{6.25}$$

where

$$J_{u,\mathrm{lin},a(\boldsymbol{x})}^{0}(g) = \left\{ n_{\boldsymbol{x}}^{-1}(f_0(a(\boldsymbol{x}))) \int v_s g \,\mathrm{d}^3\boldsymbol{v} \right.$$

$$\left. + u_{\boldsymbol{x},s}(f_0(a(\boldsymbol{x})))n_{\boldsymbol{x}}^{-1}(f_0(a(\boldsymbol{x}))) \int g \,\mathrm{d}^3\boldsymbol{v} \right\} \frac{\partial f_0(a(\boldsymbol{x}),\boldsymbol{v})}{\partial x_s}$$

$$- (v_s - u_{\boldsymbol{x},s}(f_0(a(\boldsymbol{x})))) \frac{\partial g}{\partial x_s} + L_{f_0(a(\boldsymbol{x}),\boldsymbol{v})}(g) ;$$

$$\Delta(f_0(a(\boldsymbol{x}),\boldsymbol{v})) = (P_{a(\boldsymbol{x})}^{*}(\cdot) - 1)J_u(f_0(a(\boldsymbol{x}),\boldsymbol{v})) .$$

Here $a(\boldsymbol{x})$ are coordinates on the manifold at the given space point $\boldsymbol{x}$, $P_{a(\boldsymbol{x})}^{*}$ is the corresponding thermodynamic projector. Additional conditions do not depend on the vector field, and thus they remain valid for equation (6.25).

6.3.3 Equation of the First Iteration

The equation of the first iteration in the form of (6.20) for the correction $\varphi(n,\boldsymbol{u},T)$ is:

$$\left\{ P_{f_0(n,\boldsymbol{u},T)}(\cdot) - 1 \right\} \left\{ -(v_s - u_s)\frac{\partial f_0(n,\boldsymbol{u},T)}{\partial x_s} + f_0(n,\boldsymbol{u},T)L_{f_0(n,\boldsymbol{u},T)}(\varphi) \right.$$

$$-(v_s - u_s)\frac{\partial (f_0(n,\boldsymbol{u},T)\varphi)}{\partial x_s} - n^{-1}(f_0(n,\boldsymbol{u},T)) \left(\int v_s f_0(n,\boldsymbol{u},T)\varphi \,\mathrm{d}^3\boldsymbol{v} \right.$$

$$\left. +u_s(f_0(n,\boldsymbol{u},T)) \int f_0(n,\boldsymbol{u},T)\varphi \,\mathrm{d}^3\boldsymbol{v} \right) \frac{\partial f_0(n,\boldsymbol{u},T)}{\partial x_s} \right\} = 0 . \tag{6.26}$$

Here $P_{f_0(n,\boldsymbol{u},T)}$ is the thermodynamic projector on the LM manifold and $f_0(n,\boldsymbol{u},T)L_{f_0(n,\boldsymbol{u},T)}(\varphi)$ is the linearized Boltzmann collision integral:

$$
f_0(n,\boldsymbol{u},T)L_{f_0(n,\boldsymbol{u},T)}(\varphi) = \int w(\boldsymbol{v}',\boldsymbol{v}_1'|\boldsymbol{v},\boldsymbol{v}_1)f_0(n,\boldsymbol{u},T)
$$
$$
\times \{\varphi' + \varphi_1' - \varphi_1 - \varphi\}\, \mathrm{d}^3\boldsymbol{v}'\, \mathrm{d}^3\boldsymbol{v}_1'\, \mathrm{d}^3\boldsymbol{v}_1 . \tag{6.27}
$$

Additional condition for equation (6.26) has the form:

$$
P_{f_0(n,\boldsymbol{u},T)}(f_0(n,\boldsymbol{u},T)\varphi) = 0 . \tag{6.28}
$$

In detail notation:

$$
\int 1 \cdot f_0(n,\boldsymbol{u},T)\varphi\, \mathrm{d}^3\boldsymbol{v} = 0, \quad \int v_i f_0(n,\boldsymbol{u},T)\varphi\, \mathrm{d}^3\boldsymbol{v} = 0, \; i = 1,2,3 ,
$$
$$
\int v^2 f_0(n,\boldsymbol{u},T)\varphi\, \mathrm{d}^3\boldsymbol{v} = 0 . \tag{6.29}
$$

Eliminating in (6.26) the terms containing

$$
\int v_s f_0(n,\boldsymbol{u},T)\varphi\, \mathrm{d}^3\boldsymbol{v} \text{ and } \int f_0(n,\boldsymbol{u},T)\varphi\, \mathrm{d}^3\boldsymbol{v}
$$

with the use of (6.29), we obtain the following form of equation (6.26):

$$
\{P_{f_0(n,\boldsymbol{u},T)}(\cdot) - 1\}\left(-(v_s - u_s)\frac{\partial f_0(n,\boldsymbol{u},T)}{\partial x_s}\right. \tag{6.30}
$$
$$
\left. + f_0(n,\boldsymbol{u},T)L_{f_0(n,\boldsymbol{u},T)}(\varphi) - (v_s - u_s)\frac{\partial(f_0(n,\boldsymbol{u},T)\varphi)}{\partial x_s}\right) = 0 .
$$

In order to address the properties of equation (6.30), it proves useful to introduce real Hilbert spaces $G_{f_0(n,\boldsymbol{u},T)}$ with scalar products:

$$
(\varphi,\psi)_{f_0(n,\boldsymbol{u},T)} = \int f_0(n,\boldsymbol{u},T)\varphi\psi\, \mathrm{d}^3\boldsymbol{v} . \tag{6.31}
$$

Each Hilbert space is associated with the corresponding LM distribution $f_0(n,\boldsymbol{u},T)$.

The projector $P_{f_0(n,\boldsymbol{u},T)}$ (5.55) is associated with a projector $\Pi_{f_0(n,\boldsymbol{u},T)}$ which acts in the space $G_{f_0(n,\boldsymbol{u},T)}$:

$$
\Pi_{f_0(n,\boldsymbol{u},T)}(\varphi) = f_0^{-1}(n,\boldsymbol{u},T)P_{f_0(n,\boldsymbol{u},T)}(f_0(n,\boldsymbol{u},T)\varphi) . \tag{6.32}
$$

It is an orthogonal projector, because

$$
\Pi_{f_0(n,\boldsymbol{u},T)}(\varphi) = \sum_{s=0}^{4}\psi_{f_0(n,\boldsymbol{u},T)}^{(s)}(\psi_{f_0(n,\boldsymbol{u},T)}^{(s)},\varphi)_{f_0(n,\boldsymbol{u},T)} . \tag{6.33}
$$

Here $\psi^{(s)}_{f_0(n,\boldsymbol{u},T)}$ are given by the expression (5.57).

We can rewrite the equation of the first iteration (6.30) in the form:

$$L_{f_0(n,\boldsymbol{u},T)}(\varphi) + K_{f_0(n,\boldsymbol{u},T)}(\varphi) = D_{f_0(n,\boldsymbol{u},T)} \ . \tag{6.34}$$

Notations used here are:

$$D_{f_0(n,\boldsymbol{u},T)} = f_0^{-1}(n,\boldsymbol{u},T)\Delta(f_0(n,\boldsymbol{u},T)) \ ; \tag{6.35}$$

$$K_{f_0(n,\boldsymbol{u},T)}(\varphi) = \{\Pi_{f_0(n,\boldsymbol{u},T)}(\cdot) - 1\} f_0^{-1}(n,\boldsymbol{u},T)(v_s - u_s)\frac{\partial(f_0(n,\boldsymbol{u},T)\varphi)}{\partial x_s} \ .$$

The additional condition for equation (6.34) is:

$$(\psi^{(s)}_{f_0(n,\boldsymbol{u},T)},\varphi)_{f_0(n,\boldsymbol{u},T)} = 0, s = 0,\dots,4 \ . \tag{6.36}$$

We list now the properties of the equation (6.34) for usual collision models [70]:

(a) The linear integral operator $L_{f_0(n,\boldsymbol{u},T)}$ is self-adjoint with respect to the scalar product $(\cdot,\cdot)_{f_0(n,\boldsymbol{u},T)}$, and the quadratic form $(\varphi, L_{f_0(n,\boldsymbol{u},T)}(\varphi))$ is negatively definite in $\mathrm{im} L_{f_0(n,\boldsymbol{u},T)}$.

(b) The kernel of $L_{f_0(n,\boldsymbol{u},T)}$ does not depend on $f_0(n,\boldsymbol{u},T)$, and it is the linear hull of the polynomials $\psi_0 = 1, \psi_i = v_i, i = 1,2,3$, and $\psi_4 = v^2$.

(c) The right hand side $D_{f_0(n,\boldsymbol{u},T)}$ is orthogonal to $\ker L_{f_0(n,\boldsymbol{u},T)}$ in the sense of the scalar product $(\cdot,\cdot)_{f_0(n,\boldsymbol{u},T)}$.

(d) The projection operator $\Pi_{f_0(n,\boldsymbol{u},T)}$ is the self-adjoint projector onto $\ker L_{f_0(n,\boldsymbol{u},T)}$:

$$\Pi_{f_0(n,\boldsymbol{u},T)}(\varphi) \in \ker L_{f_0(n,\boldsymbol{u},T)} \tag{6.37}$$

Projector $\Pi_{f_0(n,\boldsymbol{u},T)}$ projects orthogonally.

(e) The image of the operator $K_{f_0(n,\boldsymbol{u},T)}$ is orthogonal to $\ker L_{f_0(n,\boldsymbol{u},T)}$.

(f) Additional condition (6.36) requires the solution of equation (6.34) to be orthogonal to $\ker L_{f_0(n,\boldsymbol{u},T)}$.

These properties result in the *necessary condition* for solving the equation (6.34) with the additional constraint (6.36). This means the following: equation (6.34), provided with constraint (6.36), satisfies the condition which is necessary to have the unique solution in $\mathrm{im} L_{f_0(n,\boldsymbol{u},T)}$.

Remark. Because of the *differential* part of the operator $K_{f_0(n,\boldsymbol{u},T)}$, we are not able to apply the Fredholm alternative to obtain the *necessary and sufficient* conditions for solvability of equation (6.36). Thus, the condition mentioned here is, rigorously speaking, only the necessary condition. Nevertheless, we shall continue to develop a formal procedure for solving the equation (6.34).

To this end, we paid no attention to the dependence of functions, spaces, operators, etc, on the space variable $\boldsymbol{x}$. It is useful to rewrite once again the equation (6.34) in order to separate the local in $\boldsymbol{x}$ operators from the differential operators. Furthermore, we shall replace the subscript $f_0(n, \boldsymbol{u}, T)$ with the subscript $\boldsymbol{x}$ in all the expressions. We represent (6.34) as:

$$A_{\text{loc}}(\boldsymbol{x}, \boldsymbol{v})\varphi - A_{\text{diff}}\left(\boldsymbol{x}, \frac{\partial}{\partial \boldsymbol{x}}, \boldsymbol{v}\right)\varphi = -D(\boldsymbol{x}, \boldsymbol{v}) ;$$

$$A_{\text{loc}}(\boldsymbol{x}, \boldsymbol{v})\varphi = -\left\{L_{\boldsymbol{x}}(\boldsymbol{v})\varphi + (\Pi_{\boldsymbol{x}}(\boldsymbol{v}) - 1)r_{\boldsymbol{x}}\varphi\right\} ;$$

$$A_{\text{diff}}\left(\boldsymbol{x}, \frac{\partial}{\partial \boldsymbol{x}}, \boldsymbol{v}\right)\varphi = (\Pi_{\boldsymbol{x}}(\cdot) - 1)\left((v_s - u_s)\frac{\partial}{\partial x_s}\varphi\right) ;$$

$$\Pi_{\boldsymbol{x}}(\boldsymbol{v})g = \sum_{s=0}^{4} \psi_{\boldsymbol{x}}^{(s)}(\psi_{\boldsymbol{x}}^{(s)}, g) ;$$

$$\psi_{\boldsymbol{x}}^{(0)} = n^{-1/2}, \ \psi_{\boldsymbol{x}}^{(s)} = (2/n)^{1/2}c_s(\boldsymbol{x}, \boldsymbol{v}), \ s = 1, 2, 3 ,$$

$$\psi_{\boldsymbol{x}}^{(4)} = (2/3n)^{1/2}(c^2(\boldsymbol{x}, \boldsymbol{v}) - 3/2); \ c_i(\boldsymbol{x}, \boldsymbol{v}) = (m/2k_BT(\boldsymbol{x}))^{1/2}(v_i - u_i(\boldsymbol{x})) ,$$

$$r_{\boldsymbol{x}} = (v_s - u_s)\left(\frac{\partial \ln n}{\partial x_s} + \frac{m}{k_BT}(v_i - u_i)\frac{\partial u_i}{\partial x_s} + \left(\frac{m(\boldsymbol{v} - \boldsymbol{u})^2}{2k_BT} - \frac{3}{2}\right)\frac{\partial \ln T}{\partial x_s}\right) ;$$

$$D(\boldsymbol{x}, \boldsymbol{v}) = \left\{\left(\frac{m(\boldsymbol{v} - \boldsymbol{u})^2}{2k_BT} - \frac{5}{2}\right)(v_i - u_i)\frac{\partial \ln T}{\partial x_i}\right.$$
$$\left. + \frac{m}{k_BT}\left(((v_i - u_i)(v_s - u_s) - \frac{1}{3}\delta_{is}(\boldsymbol{v} - \boldsymbol{u})^2\right)\frac{\partial u_s}{\partial x_i}\right\} . \tag{6.38}$$

Here we have omitted the dependence on $\boldsymbol{x}$ in the functions $n(\boldsymbol{x})$, $u_i(\boldsymbol{x})$, and $T(\boldsymbol{x})$. Further, if no confusion might occur, we always assume this dependence, and we shall not indicate it explicitly.

The additional condition for this equation is:

$$\Pi_{\boldsymbol{x}}(\varphi) = 0 . \tag{6.39}$$

Equation (6.38) is linear in φ. However, the main difficulty in solving this equation is caused by the differential in $\boldsymbol{x}$ operator A_{diff} which does not commute with the local in $\boldsymbol{x}$ operator A_{loc}.

6.3.4 Parametrix Expansion

In this subsection we introduce a method to construct approximate solutions of equation (6.37). This procedure involves an expansion similar to the parametrix expansion in the theory of pseudo-differential (PDO) and Fourier integral operators (FIO).

Considering $\varphi \in \text{im}L_{\boldsymbol{x}}$, we write a formal solution of equation (6.38) as:

$$\varphi(\boldsymbol{x}, \boldsymbol{v}) = \left(A_{\text{loc}}(\boldsymbol{x}, \boldsymbol{v}) - A_{\text{diff}}\left(\boldsymbol{x}, \frac{\partial}{\partial \boldsymbol{x}}, \boldsymbol{v}\right)\right)^{-1}(-D(\boldsymbol{x}, \boldsymbol{v})) \tag{6.40}$$

It is useful to extract the differential operator $\frac{\partial}{\partial \boldsymbol{x}}$ from the operator $A_{\text{diff}}(\boldsymbol{x}, \frac{\partial}{\partial \boldsymbol{x}}, \boldsymbol{v})$:

$$\varphi(\boldsymbol{x}, \boldsymbol{v}) = \left(1 - B_s(\boldsymbol{x}, \boldsymbol{v})\frac{\partial}{\partial x_s}\right)^{-1} \varphi_{\mathrm{loc}}(\boldsymbol{x}, \boldsymbol{v}) . \qquad (6.41)$$

Notations used here are:

$$\begin{aligned}
\varphi_{\mathrm{loc}}(\boldsymbol{x}, \boldsymbol{v}) &= A_{\mathrm{loc}}^{-1}(\boldsymbol{x}, \boldsymbol{v})(-D(\boldsymbol{x}, \boldsymbol{v})) \\
&= [-L_{\boldsymbol{x}}(\boldsymbol{v}) - (\Pi_{\boldsymbol{x}}(\boldsymbol{v}) - 1)r_{\boldsymbol{x}}]^{-1}(-D(\boldsymbol{x}, \boldsymbol{v})) ; \\
B_s(\boldsymbol{x}, \boldsymbol{v}) &= A_{\mathrm{loc}}^{-1}(\boldsymbol{x}, \boldsymbol{v})(\Pi_{\boldsymbol{x}}(\boldsymbol{v}) - 1)(v_s - u_s) \\
&= [-L_{\boldsymbol{x}}(\boldsymbol{v}) - (\Pi_{\boldsymbol{x}}(\boldsymbol{v}) - 1)r_{\boldsymbol{x}}]^{-1}(\Pi_{\boldsymbol{x}}(\boldsymbol{v}) - 1)(v_s - u_s) .
\end{aligned} \qquad (6.42)$$

We shall now discuss in more details the properties of the terms in (6.42). For every $\boldsymbol{x}$, the function $\varphi_{\mathrm{loc}}(\boldsymbol{x}, \boldsymbol{v})$, considered as a function of $\boldsymbol{v}$, is an element of the Hilbert space $G_{\boldsymbol{x}}$. It gives a solution to the integral equation:

$$- L_{\boldsymbol{x}}(\boldsymbol{v})\varphi_{\mathrm{loc}} - (\Pi_{\boldsymbol{x}}(\boldsymbol{v}) - 1)(r_{\boldsymbol{x}}\varphi_{\mathrm{loc}}) = (-D(\boldsymbol{x}, \boldsymbol{v})) \qquad (6.43)$$

This latter linear integral equation has the unique solution in $\mathrm{im}L_{\boldsymbol{x}}(\boldsymbol{v})$. Indeed,

$$\begin{aligned}
\ker A_{\mathrm{loc}}^+(\boldsymbol{x}, \boldsymbol{v}) &= \ker(L_{\boldsymbol{x}}(\boldsymbol{v}) + (\Pi_{\boldsymbol{x}}(\boldsymbol{v}) - 1)r_{\boldsymbol{x}})^+ \\
&= \ker(L_{\boldsymbol{x}}(\boldsymbol{v}))^+ \bigcap \ker((\Pi_{\boldsymbol{x}}(\boldsymbol{v}) - 1)r_{\boldsymbol{x}})^+ \\
&= \ker(L_{\boldsymbol{x}}(\boldsymbol{v}))^+ \bigcap \ker(r_{\boldsymbol{x}}(\Pi_{\boldsymbol{x}}(\boldsymbol{v}) - 1)) ,
\end{aligned}$$

$$\text{and } G_{\boldsymbol{x}} \bigcap \Pi_{\boldsymbol{x}}(\boldsymbol{v})G_{\boldsymbol{x}} = \{0\} . \qquad (6.44)$$

Thus, the existence of the unique solution of equation (6.43) follows from the Fredholm alternative.

Let us consider the operator $R(\boldsymbol{x}, \frac{\partial}{\partial \boldsymbol{x}}, \boldsymbol{v})$:

$$R\left(\boldsymbol{x}, \frac{\partial}{\partial \boldsymbol{x}}, \boldsymbol{v}\right) = \left(1 - B_s(\boldsymbol{x}, \boldsymbol{v})\frac{\partial}{\partial x_s}\right)^{-1} . \qquad (6.45)$$

One can represent it as a formal series:

$$R\left(\boldsymbol{x}, \frac{\partial}{\partial \boldsymbol{x}}, \boldsymbol{v}\right) = \sum_{m=0}^{\infty} \left[B_s(\boldsymbol{x}, \boldsymbol{v})\frac{\partial}{\partial x_s}\right]^m . \qquad (6.46)$$

Here

$$\left[B_s(\boldsymbol{x}, \boldsymbol{v})\frac{\partial}{\partial x_s}\right]^m = B_{s_1}(\boldsymbol{x}, \boldsymbol{v})\frac{\partial}{\partial x_{s_1}} \ldots B_{s_m}(\boldsymbol{x}, \boldsymbol{v})\frac{\partial}{\partial x_{s_m}} . \qquad (6.47)$$

Every term of the type (6.47) can be represented as a finite sum of operators which are superpositions of the following two operations: of the integral in $\boldsymbol{v}$ operations with kernels depending on $\boldsymbol{x}$, and of differential in $\boldsymbol{x}$ operations.

Our goal is to obtain explicit representation of the operator $R(\boldsymbol{x}, \frac{\partial}{\partial \boldsymbol{x}}, \boldsymbol{v})$ (6.45) as an integral operator. If the operator $B_s(\boldsymbol{x}, \boldsymbol{v})$ would not depend on $\boldsymbol{x}$ i.e., if no dependence on spatial variables would occur in kernels of integral operators, in $B_s(\boldsymbol{x}, \boldsymbol{v})$), then we could reach our goal via the usual Fourier transform. However, operators $B_s(\boldsymbol{x}, \boldsymbol{v})$ and $\frac{\partial}{\partial x_k}$ do not commute, and thus this elementary approach does not work. We shall develop a method to obtain the required explicit representation using the ideas of PDO and IOF technique.

We start with the representation (6.46). Our strategy is to transform every summand (6.47) in order to place integral in $\boldsymbol{v}$ operators $B_s(\boldsymbol{x}, \boldsymbol{v})$ on the left of the differential operators $\frac{\partial}{\partial x_k}$. The commutation of every pair $\frac{\partial}{\partial x_k} B_s(\boldsymbol{x}, \boldsymbol{v})$ yields an elementary transform:

$$\frac{\partial}{\partial x_k} B_s(\boldsymbol{x}, \boldsymbol{v}) \rightarrow B_s(\boldsymbol{x}, \boldsymbol{v}) \frac{\partial}{\partial x_k} - \left[B_s(\boldsymbol{x}, \boldsymbol{v}), \frac{\partial}{\partial x_k} \right] . \qquad (6.48)$$

Here $[M, N] = MN - NM$ denotes the commutator of operators M and N. We can represent (6.47) as:

$$\left[B_s(\boldsymbol{x}, \boldsymbol{v}) \frac{\partial}{\partial x_s} \right]^m = B_{s_1}(\boldsymbol{x}, \boldsymbol{v}) \ldots B_{s_m}(\boldsymbol{x}, \boldsymbol{v}) \frac{\partial}{\partial x_{s_1}} \ldots \frac{\partial}{\partial x_{s_m}}$$
$$+ O\left(\left[B_{s_i}(\boldsymbol{x}, \boldsymbol{v}), \frac{\partial}{\partial x_{s_k}} \right] \right) . \qquad (6.49)$$

Here $O([B_{s_i}(\boldsymbol{x}, \boldsymbol{v}), \frac{\partial}{\partial x_{s_k}}])$ denotes the terms which contain one or more pairs of brackets $[\cdot, \cdot]$. The first term in (6.49) contains no brackets. We can continue this process of selection and extract the first-order in the number of pairs of brackets terms, the second-order terms, etc. Thus, we arrive at the *expansion into powers of commutator* of the expressions (6.47).

In this section we consider explicitly the zeroth-order term of this commutator expansion. Neglecting all the terms with brackets in (6.49), we write:

$$\left[B_s(\boldsymbol{x}, \boldsymbol{v}) \frac{\partial}{\partial x_s} \right]^m_0 = B_{s_1}(\boldsymbol{x}, \boldsymbol{v}) \ldots B_{s_m}(\boldsymbol{x}, \boldsymbol{v}) \frac{\partial}{\partial x_{s_1}} \ldots \frac{\partial}{\partial x_{s_m}} . \qquad (6.50)$$

Here the subscript zero indicates the zeroth order with respect to the number of brackets.

We should now substitute expressions $[B_s(\boldsymbol{x}, \boldsymbol{v}) \frac{\partial}{\partial x_s}]^m_0$ (6.50) instead of expressions $[B_s(\boldsymbol{x}, \boldsymbol{v}) \frac{\partial}{\partial x_s}]^m$ (6.47) into the series (6.46):

$$R_0\left(\boldsymbol{x}, \frac{\partial}{\partial \boldsymbol{x}}, \boldsymbol{v} \right) = \sum_{m=0}^{\infty} \left[B_s(\boldsymbol{x}, \boldsymbol{v}) \frac{\partial}{\partial x_s} \right]^m_0 . \qquad (6.51)$$

The action of every summand (6.50) might be defined via the Fourier transform with respect to spatial variables.

Denote as F the direct Fourier transform of a function $g(\boldsymbol{x}, \boldsymbol{v})$:

$$Fg(\boldsymbol{x}, \boldsymbol{v}) \equiv \hat{g}(\boldsymbol{k}, \boldsymbol{v}) = \int g(\boldsymbol{x}, \boldsymbol{v}) \exp(-ik_s x_s) \, \mathrm{d}^p \boldsymbol{x} . \tag{6.52}$$

Here p is the spatial dimension. Then the inverse Fourier transform is:

$$g(x, \boldsymbol{v}) \equiv F^{-1} \hat{g}(\boldsymbol{k}, \boldsymbol{v}) = (2\pi)^{-p} \int \hat{g}(\boldsymbol{k}, \boldsymbol{v}) \exp(ik_s x_s) \, \mathrm{d}^p \boldsymbol{k} . \tag{6.53}$$

The action of the operator (6.50) on a function $g(\boldsymbol{x}, \boldsymbol{v})$ is defined as:

$$\left[B_s(\boldsymbol{x}, \boldsymbol{v}) \frac{\partial}{\partial x_s} \right]_0^m g(\boldsymbol{x}, \boldsymbol{v})$$

$$= \left(B_{s_1}(\boldsymbol{x}, \boldsymbol{v}) \dots B_{s_m}(\boldsymbol{x}, \boldsymbol{v}) \frac{\partial}{\partial x_{s_1}} \dots \frac{\partial}{\partial x_{s_m}} \right) (2\pi)^{-p} \int \hat{g}(\boldsymbol{k}, \boldsymbol{v}) e^{ik_s x_s} \, \mathrm{d}^p \boldsymbol{k}$$

$$= (2\pi)^{-p} \int \exp(ik_s x_s) [ik_l B_l(\boldsymbol{x}, \boldsymbol{v})]^m \hat{g}(\boldsymbol{k}, \boldsymbol{v}) \, \mathrm{d}^p \boldsymbol{k} . \tag{6.54}$$

Taking into account (6.54) in (6.51) yields the following definition of the operator R_0:

$$R_0 g(\boldsymbol{x}, \boldsymbol{v}) = (2\pi)^{-p} \int e^{ik_s x_s} (1 - ik_l B_l(\boldsymbol{x}, \boldsymbol{v}))^{-1} \hat{g}(\boldsymbol{k}, \boldsymbol{v}) \, \mathrm{d}^p \boldsymbol{k} . \tag{6.55}$$

This is the *Fourier integral operator* (note that the kernel of this integral operator depends on $\boldsymbol{k}$ and on $\boldsymbol{x}$). The commutator expansion introduced above is a version of the *parametrix expansion* [249, 250], while expression (6.55) is the leading term of this expansion. The kernel $(1 - ik_l B_l(\boldsymbol{x}, \boldsymbol{v}))^{-1}$ is called the *main symbol of the parametrix*.

The account of (6.55) in the formula (6.41) yields the zeroth-order term of parametrix expansion $\varphi_0(\boldsymbol{x}, \boldsymbol{v})$:

$$\varphi_0(\boldsymbol{x}, \boldsymbol{v}) = F^{-1}(1 - ik_l B_l(\boldsymbol{x}, \boldsymbol{v}))^{-1} F \varphi_{\mathrm{loc}} . \tag{6.56}$$

In detail notation:

$$\varphi_0(\boldsymbol{x}, \boldsymbol{v}) = (2\pi)^{-p} \int \int \exp(ik_s(x_s - y_s))$$

$$\times (1 - ik_s[-L_{\boldsymbol{x}}(\boldsymbol{v}) - (\Pi_{\boldsymbol{x}}(\boldsymbol{v}) - 1)r_{\boldsymbol{x}}]^{-1}(\Pi_{\boldsymbol{x}}(\boldsymbol{v}) - 1)(v_s - u_s(\boldsymbol{x})))^{-1}$$

$$\times [-L_{\boldsymbol{y}}(\boldsymbol{v}) - (\Pi_{\boldsymbol{y}}(\boldsymbol{v}) - 1)r_{\boldsymbol{y}}]^{-1}(-D(\boldsymbol{y}, \boldsymbol{v})) \, \mathrm{d}^p \boldsymbol{y} \, \mathrm{d}^p \boldsymbol{k} . \tag{6.57}$$

We shall now list the steps to calculate the function $\varphi_0(\boldsymbol{x}, \boldsymbol{v})$ (6.57).
Step 1. Solve the linear integral equation

$$[-L_{\boldsymbol{x}}(\boldsymbol{v}) - (\Pi_{\boldsymbol{x}}(\boldsymbol{v}) - 1)r_{\boldsymbol{x}}] \varphi_{\mathrm{loc}}(\boldsymbol{x}, \boldsymbol{v}) = -D(\boldsymbol{x}, \boldsymbol{v}) . \tag{6.58}$$

and obtain the function $\varphi_{\mathrm{loc}}(\boldsymbol{x}, \boldsymbol{v})$.

Step 2. Calculate the Fourier transform $\hat{\varphi}_{\mathrm{loc}}(\boldsymbol{k}, \boldsymbol{v})$:

$$\hat{\varphi}_{\mathrm{loc}}(\boldsymbol{k}, \boldsymbol{v}) = \int \varphi_{\mathrm{loc}}(\boldsymbol{y}, \boldsymbol{v}) \exp(-ik_s y_s) \, \mathrm{d}^p \boldsymbol{y} \ . \tag{6.59}$$

Step 3. Solve the linear integral equation

$$\begin{aligned}
[-L_{\boldsymbol{x}}(\boldsymbol{v}) - (\Pi_{\boldsymbol{x}}(\boldsymbol{v}) - 1)(r_{\boldsymbol{x}} + ik_s(v_s - u_s(\boldsymbol{x}))]\hat{\varphi}_0(\boldsymbol{x}, \boldsymbol{k}, \boldsymbol{v}) &= -\hat{D}(\boldsymbol{x}, \boldsymbol{k}, \boldsymbol{v}) \ ; \\
-\hat{D}(\boldsymbol{x}, \boldsymbol{k}, \boldsymbol{v}) &= [-L_{\boldsymbol{x}}(\boldsymbol{v}) - (\Pi_{\boldsymbol{x}}(\boldsymbol{v}) - 1)r_{\boldsymbol{x}}]\hat{\varphi}_{\mathrm{loc}}(\boldsymbol{k}, \boldsymbol{v}) \ .
\end{aligned} \tag{6.60}$$

and obtain the function $\hat{\varphi}_0(\boldsymbol{x}, \boldsymbol{k}, \boldsymbol{v})$.

Step 4. Calculate the inverse Fourier transform $\varphi_0(\boldsymbol{x}, \boldsymbol{v})$:

$$\varphi_0(\boldsymbol{x}, \boldsymbol{v}) = (2\pi)^{-p} \int \hat{\varphi}_0(\boldsymbol{x}, \boldsymbol{k}, \boldsymbol{v}) \exp(ik_s x_s) \, \mathrm{d}^p \boldsymbol{k} \ . \tag{6.61}$$

Completing these four steps, we obtain an explicit expression for the zeroth-order term of parametrix expansion $\varphi_0(\boldsymbol{x}, \boldsymbol{v})$ (6.56).

As we have already mentioned it above, equation (6.58) of Step 1 has the unique solution in $\mathrm{im}L_{\boldsymbol{x}}(\boldsymbol{v})$. Equation (6.60) of Step 3 has the same property. Indeed, for every $\boldsymbol{k}$, the right hand side $-\hat{D}(\boldsymbol{x}, \boldsymbol{k}, \boldsymbol{v})$ is orthogonal to $\mathrm{im}\Pi_{\boldsymbol{x}}(\boldsymbol{v})$, and thus the existence and the uniqueness of the formal solution $\hat{\varphi}_0(\boldsymbol{x}, \boldsymbol{k}, \boldsymbol{v})$ follows again from the Fredholm alternative.

Thus, in Step 3, we obtain the unique solution $\hat{\varphi}_0(\boldsymbol{x}, \boldsymbol{k}, \boldsymbol{v})$. For every $\boldsymbol{k}$, this is a function which belongs to $\mathrm{im}L_{\boldsymbol{x}}(\boldsymbol{v})$. Because the LM distribution $f_0(\boldsymbol{x}, \boldsymbol{v}) = f_0(n(\boldsymbol{x}), \boldsymbol{u}(\boldsymbol{x}), T(\boldsymbol{x}), \boldsymbol{v})$ has no explicit dependency on $\boldsymbol{x}$, we see that the inverse Fourier transform of Step 4 gives $\varphi_0(\boldsymbol{x}, \boldsymbol{v}) \in \mathrm{im}L_{\boldsymbol{x}}(\boldsymbol{v})$.

Equations (6.58)–(6.61) provide us with the scheme of constructing the zeroth-order term of parametrix expansion. Closing this section, we outline briefly the way to calculate the first-order term of this expansion.

Consider a formal operator $R = (1 - AB)^{-1}$. Operator R is defined by a formal series:

$$R = \sum_{m=0}^{\infty} (AB)^m \ . \tag{6.62}$$

In every term of this series, we want to place operators A on the left to operators B. In order to do this, we have to commute B with A from left to right. The commutation of every pair BA yields the elementary transform, $BA \to AB - [A, B]$, where $[A, B] = AB - BA$. Extracting the terms with no commutators $[A, B]$ and with a single commutator $[A, B]$, we arrive at the following representation:

$$R = R_0 + R_1 + (\text{terms with more than two brackets}) \ . \tag{6.63}$$

Here

$$R_0 = \sum_{m=0}^{\infty} A^m B^m \; ; \tag{6.64}$$

$$R_1 = - \sum_{m=2}^{\infty} \sum_{i=2}^{\infty} i A^{m-i}[A, B] A^{i-1} B^{i-1} B^{m-i} \; . \tag{6.65}$$

Operator R_0 (6.64) is the zeroth-order term of parametrix expansion derived above. Operator R_1 (the *first-order term of parametrix expansion*) can be represented as follows:

$$R_1 = - \sum_{m=1}^{\infty} m A^m [A, B] \left(\sum_{i=0}^{\infty} A^i B^i \right) B^m = - \sum_{m=1}^{\infty} m A^m C B^m \; ,$$
$$C = [A, B] R_0 \; . \tag{6.66}$$

This expression can be considered as an *ansatz* for the formal series (6.62), and it gives the most convenient way to calculate R_1. Its structure is similar to that of R_0. Continuing in this manner, we can derive the second-order term R_2, etc.

In the next subsection we shall consider in more detail the zero-order term of parametrix expansion.

6.3.5 Finite-Dimensional Approximations to Integral Equations

Dealing further only with the zeroth-order term of parametrix expansion (6.57), we have to solve two linear integral equations, (6.58) and (6.60). These equations satisfy the Fredholm alternative, and thus they have unique solutions. After the problem is reduced to solving linear integral equations, we are at the same level of complexity as in the Chapman-Enskog method. The usual approach is to replace integral operators with some appropriate finite-dimensional operators.

First we remind some standard objectives of finite-dimensional approximations, considering equation (6.58). Let $p_i(\boldsymbol{x}, \boldsymbol{v})$, where $i = 1, 2, \ldots$, be a basis in $\mathrm{im} L_{\boldsymbol{x}}(\boldsymbol{v})$. Every function $\varphi(\boldsymbol{x}, \boldsymbol{v}) \in \mathrm{im} L_{\boldsymbol{x}}(\boldsymbol{v})$ can be represented in this basis as:

$$\varphi(\boldsymbol{x}, \boldsymbol{v}) = \sum_{i=1}^{\infty} a_i(\boldsymbol{x}) p_i(\boldsymbol{x}, \boldsymbol{v}); a_i(\boldsymbol{x}) = (\varphi(\boldsymbol{x}, \boldsymbol{v}), p_i(\boldsymbol{x}, \boldsymbol{v}))_{\boldsymbol{x}} \; . \tag{6.67}$$

Equation (6.58) is equivalent to an infinite set of linear algebraic equations with respect to unknowns $a_i(\boldsymbol{x})$:

$$\sum_{i=1}^{\infty} m_{ki}(\boldsymbol{x}) a_i(\boldsymbol{x}) = d_k(\boldsymbol{x}), \quad k = 1, 2, \ldots \; . \tag{6.68}$$

Here

$$m_{ki}(\boldsymbol{x}) = (p_k(\boldsymbol{x}, \boldsymbol{v}), A_{\text{loc}}(\boldsymbol{x}, \boldsymbol{v})p_i(\boldsymbol{x}, \boldsymbol{v}))_{\boldsymbol{x}} \ ;$$
$$d_k(\boldsymbol{x}) = -(p_k(\boldsymbol{x}, \boldsymbol{v}), D(\boldsymbol{x}, \boldsymbol{v}))_{\boldsymbol{x}} \ . \tag{6.69}$$

For a finite-dimensional approximation of equation (6.68) we use a projection onto a finite number of basis functions $p_i(\boldsymbol{x}, \boldsymbol{v}), i = i_1, \ldots, i_n$. Then, instead of (6.67), we search for the function φ_{fin}:

$$\varphi_{\text{fin}}(\boldsymbol{x}, \boldsymbol{v}) = \sum_{s=1}^{n} a_{i_s}(\boldsymbol{x})p_{i_s}(\boldsymbol{x}, \boldsymbol{v}) \ . \tag{6.70}$$

Infinite set of equations (6.68) is replaced with a finite set of linear algebraic equations with respect to $a_{i_s}(\boldsymbol{x})$, where $s = 1, \ldots, n$:

$$\sum_{l=1}^{n} m_{i_s i_l}(\boldsymbol{x})a_{i_l}(\boldsymbol{x}) = d_{i_s}(\boldsymbol{x}), \quad s = 1, \ldots, n \ . \tag{6.71}$$

There are no a priori restrictions upon the choice of the basis, as well as upon the choice of its finite-dimensional approximations. Here we use the standard basis of irreducible Hermite tensors (see, for example, [112,201]). The simplest finite-dimensional approximation occurs if the finite set of Hermite tensors is chosen as:

$$p_k(\boldsymbol{x}, \boldsymbol{v}) = c_k(\boldsymbol{x}, \boldsymbol{v})(c^2(\boldsymbol{x}, \boldsymbol{v}) - (5/2)), k = 1, 2, 3 \ ;$$
$$p_{ij}(\boldsymbol{x}, \boldsymbol{v}) = c_i(\boldsymbol{x}, \boldsymbol{v})c_j(\boldsymbol{x}, \boldsymbol{v}) - \frac{1}{3}\delta_{ij}c^2(\boldsymbol{x}, \boldsymbol{v}), \quad i, j = 1, 2, 3 \ ;$$
$$c_i(\boldsymbol{x}, \boldsymbol{v}) = v_T^{-1}(\boldsymbol{x})(v_i - u_i(\boldsymbol{x})), \quad v_T(\boldsymbol{x}) = (2k_B T(\boldsymbol{x})/m)^{1/2} \ . \tag{6.72}$$

It is important to stress here that "good" properties of orthogonality of Hermite tensors, as well as of other similar polynomial systems in the Boltzmann equation theory, have the local in $\boldsymbol{x}$ character, i.e. when these functions are treated as polynomials in $c(\boldsymbol{x}, \boldsymbol{v})$ rather than polynomials in $\boldsymbol{v}$. For example, functions $p_k(\boldsymbol{x}, \boldsymbol{v})$ and $p_{ij}(\boldsymbol{x}, \boldsymbol{v})$(6.72) are orthogonal in the sense of the scalar product $(\cdot, \cdot)_{\boldsymbol{x}}$:

$$(p_k(\boldsymbol{x}, \boldsymbol{v}), p_{ij}(\boldsymbol{x}, \boldsymbol{v}))_{\boldsymbol{x}} \propto \int e^{-c^2(\boldsymbol{x}, \boldsymbol{v})}p_k(\boldsymbol{x}, \boldsymbol{v})p_{ij}(\boldsymbol{x}, \boldsymbol{v}) \, \mathrm{d}^3 c(\boldsymbol{x}, \boldsymbol{v}) = 0 \ . \tag{6.73}$$

On the contrary, functions $p_k(\boldsymbol{y}, \boldsymbol{v})$ and $p_{ij}(\boldsymbol{x}, \boldsymbol{v})$ are not orthogonal neither in the sense of the scalar product $(\cdot, \cdot)_{\boldsymbol{y}}$, nor in the sense of the scalar product $(\cdot, \cdot)_{\boldsymbol{x}}$, if $\boldsymbol{y} \neq \boldsymbol{x}$. This distinction is important for constructing the parametrix expansion. Further, we omit the dependencies on $\boldsymbol{x}$ and $\boldsymbol{v}$ in the dimensionless velocity $c_i(\boldsymbol{x}, \boldsymbol{v})$(6.72) if no confusion might occur.

In this section we consider the case of one-dimensional in $\boldsymbol{x}$ equations. We assume that:

$$u_1(x) = u(x_1) , \quad u_2 = u_3 = 0 , \quad T(x) = T(x_1), \quad n(x) = n(x_1) . \quad (6.74)$$

We write x instead of x_1 below. Finite-dimensional approximation (6.72) requires only two functions:

$$p_3(x, v) = c_1^2(x, v) - \frac{1}{3}c^2(x, v) , \quad p_4(x, v) = c_1(x, v)(c^2(x, v) - (5/2)) ,$$

$$c_1(x, v) = v_T^{-1}(x)(v_1 - u(x)) , \quad c_{2,3}(x, v) = v_T^{-1}(x)v_{2,3} . \quad (6.75)$$

We shall now perform a step-by-step calculation of the zeroth-order term of the parametrix expansion, in the one-dimensional case, for the finite-dimensional approximation (6.75).

Step 1. *Calculation of $\varphi_{\text{loc}}(x, v)$ from equation (6.58).*

We seek the function $\varphi_{\text{loc}}(x, v)$ in the approximation (6.75) as:

$$\varphi_{\text{loc}}(x, v) = a_{\text{loc}}(x)(c_1^2 - (1/3)c^2) + b_{\text{loc}}(x)c_1(c^2 - (5/2)) . \quad (6.76)$$

Finite-dimensional approximation (6.71) of integral equation (6.58) in the basis (6.75) yields:

$$m_{33}(x)a_{\text{loc}}(x) + m_{34}(x)b_{\text{loc}}(x) = \alpha_{\text{loc}}(x) ;$$

$$m_{43}(x)a_{\text{loc}}(x) + m_{44}(x)b_{\text{loc}}(x) = \beta_{\text{loc}}(x) . \quad (6.77)$$

Notations used are:

$$m_{33}(x) = n(x)\lambda_3(x) + \frac{11}{9}\frac{\partial u}{\partial x} ; \qquad m_{44}(x) = n(x)\lambda_4(x) + \frac{27}{4}\frac{\partial u}{\partial x} ;$$

$$m_{34}(x) = m_{43}(x) = \frac{v_T(x)}{3}\left(\frac{\partial \ln n}{\partial x} + \frac{11}{2}\frac{\partial \ln T}{\partial x}\right) ;$$

$$\lambda_{3,4}(x) = -\frac{1}{\pi^{3/2}}\int e^{-c^2(x, v)}p_{3,4}(x, v)L_x(v)p_{3,4}(x, v)\, \mathrm{d}^3c(x, v) > 0 ;$$

$$\alpha_{\text{loc}}(x) = -\frac{2}{3}\frac{\partial u}{\partial x} ; \qquad \beta_{\text{loc}}(x) = -\frac{5}{4}v_T(x)\frac{\partial \ln T}{\partial x} . \quad (6.78)$$

Parameters $\lambda_3(x)$ and $\lambda_4(x)$ are easily expressed via the so-called Enskog integral brackets, and they are calculated in [70] for a wide class of molecular models.

Solving equation (6.77), we obtain coefficients $a_{\text{loc}}(x)$ and $b_{\text{loc}}(x)$ in the expression (6.76):

$$a_{\text{loc}} = \frac{A_{\text{loc}}(x)}{Z(x,0)} ; \quad b_{\text{loc}} = \frac{B_{\text{loc}}(x)}{Z(x,0)} ; \quad Z(x,0) = m_{33}(x)m_{44}(x) - m_{34}^2(x) ;$$

$$A_{\text{loc}}(x) = \alpha_{\text{loc}}(x)m_{44}(x) - \beta_{\text{loc}}(x)m_{34}(x) ;$$

$$B_{\text{loc}}(x) = \beta_{\text{loc}}(x)m_{33}(x) - \alpha_{\text{loc}}(x)m_{34}(x) ;$$

$$a_{\text{loc}} = \frac{-\dfrac{2}{3}\dfrac{\partial u}{\partial x}\left(n\lambda_4 + \dfrac{27}{4}\dfrac{\partial u}{\partial x}\right) + \dfrac{5}{12}v_T^2\dfrac{\partial \ln T}{\partial x}\left(\dfrac{\partial \ln n}{\partial x} + \dfrac{11}{2}\dfrac{\partial \ln T}{\partial x}\right)}{\left(n\lambda_3 + \dfrac{11}{9}\dfrac{\partial u}{\partial x}\right)\left(n\lambda_4 + \dfrac{27}{4}\dfrac{\partial u}{\partial x}\right) - \dfrac{v_T^2}{9}\left(\dfrac{\partial \ln n}{\partial x} + \dfrac{11}{2}\dfrac{\partial \ln T}{\partial x}\right)^2} ;$$

$$b_{\text{loc}} = \frac{-\dfrac{5}{4}v_T\dfrac{\partial \ln T}{\partial x}\left(n\lambda_3 + \dfrac{11}{9}\dfrac{\partial u}{\partial x}\right) + \dfrac{2}{9}v_T\dfrac{\partial u}{\partial x}\left(\dfrac{\partial \ln n}{\partial x} + \dfrac{11}{2}\dfrac{\partial \ln T}{\partial x}\right)}{\left(n\lambda_3 + \dfrac{11}{9}\dfrac{\partial u}{x}\right)\left(n\lambda_4 + \dfrac{27}{4}\dfrac{\partial u}{\partial x}\right) - \dfrac{v_T^2}{9}\left(\dfrac{\partial \ln n}{x} + \dfrac{11}{2}\dfrac{\partial \ln T}{x}\right)^2} .$$
$$(6.79)$$

These expressions complete Step 1.

Step 2. *Calculation of Fourier transform of* $\varphi_{loc}(x, \boldsymbol{v})$ *and its expression in the local basis.*

In this step we make two operations:

(i) The Fourier transformation of the function $\varphi_{\text{loc}}(x, \boldsymbol{v})$:

$$\hat{\varphi}_{\text{loc}}(k, \boldsymbol{v}) = \int_{-\infty}^{+\infty} \exp(-iky)\varphi_{\text{loc}}(y, \boldsymbol{v})\,dy .$$
$$(6.80)$$

(ii) The representation of $\hat{\varphi}_{\text{loc}}(k, \boldsymbol{v})$ in the local basis $\{p_0(x, \boldsymbol{v}), \dots, p_4(x, \boldsymbol{v})\}$:

$$p_0(x, \boldsymbol{v}) = 1, p_1(x, \boldsymbol{v}) = c_1(x, \boldsymbol{v}), p_2(x, \boldsymbol{v}) = c^2(x, \boldsymbol{v}) - (3/2) , \qquad (6.81)$$
$$p_3(x, \boldsymbol{v}) = c_1^2(x, \boldsymbol{v}) - (1/3)c^2(x, \boldsymbol{v}), p_4(x, \boldsymbol{v}) = c_1(x, \boldsymbol{v})(c^2(x, \boldsymbol{v}) - (5/2)) .$$

Operation (ii) is necessary for completing Step 3 because there we deal with x-dependent operators. Obviously, the function $\hat{\varphi}_{\text{loc}}(k, \boldsymbol{v})$ (6.80) is a finite-order polynomial in $\boldsymbol{v}$, and thus representation (ii) is exact.

We obtain in (ii):

$$\hat{\varphi}_{\text{loc}}(x, k, \boldsymbol{v}) \equiv \hat{\varphi}_{\text{loc}}(x, k, c(x, \boldsymbol{v})) = \sum_{i=0}^{4} \hat{h}_i(x, k)p_i(x, \boldsymbol{v}) .$$
$$(6.82)$$

Here

$$\hat{h}_i(x, k) = (p_i(x, \boldsymbol{v}), p_i(x, \boldsymbol{v}))_x^{-2}(\hat{\varphi}_{\text{loc}}(k, \boldsymbol{v}), p_i(x, \boldsymbol{v}))_x .$$
$$(6.83)$$

Let us introduce notations:

$$\vartheta \equiv \vartheta(x, y) = (T(x)/T(y))^{1/2} , \quad \gamma \equiv \gamma(x, y) = \frac{u(x) - u(y)}{v_T(y)} .$$
$$(6.84)$$

Coefficients $\hat{h}_i(x, k)$ (6.83) have the following explicit form:

$$\hat{h}_i(x, k) = \int_{-\infty}^{+\infty} \exp(-iky)h_i(x, y)\,dy; h_i(x, y) = Z^{-1}(y, 0)g_i(x, y)$$

$$g_0(x, y) = B_{\text{loc}}(y)(\gamma^3 + \frac{5}{2}\gamma(\vartheta^2 - 1)) + \frac{2}{3}A_{\text{loc}}(y)\gamma^2 ;$$

$$g_1(x, y) = B_{\text{loc}}(y)(3\vartheta\gamma^2 + \frac{5}{2}\vartheta(\vartheta^2 - 1)) + \frac{4}{3}A_{\text{loc}}(y)\vartheta\gamma ;$$

$$g_2(x, y) = \frac{5}{3}B_{\text{loc}}(y)\vartheta^2\gamma ;$$

$$g_3(x, y) = B_{\text{loc}}(y)2\vartheta\gamma + A_{\text{loc}}(y)\vartheta^2 ;$$

$$g_4(x, y) = B_{\text{loc}}(y)\vartheta^3 .$$
$$(6.85)$$

Here $Z(y, 0), B_{\text{loc}}(y)$ and $A_{\text{loc}}(y)$ are the functions defined in (6.79)

Step 3. *Calculation of the function $\hat{\varphi}_0(x, k, \boldsymbol{v})$ from equation* (6.60).

Linear integral equation (6.60) is similar to equation (6.58). We search for the function $\hat{\varphi}_0(x, k, \boldsymbol{v})$ in the basis (6.75) as:

$$\hat{\varphi}_0(x, k, \boldsymbol{v}) = \hat{a}_0(x, k)p_3(x, \boldsymbol{v}) + \hat{b}_0(x, k)p_4(x, \boldsymbol{v}) \,. \tag{6.86}$$

Finite-dimensional approximation of the integral equation (6.60) in the basis (6.75) yields the following equations for unknowns $\hat{a}_0(x, k)$ and $\hat{b}_0(x, k)$:

$$m_{33}(x)\hat{a}_0(x, k) + \left[m_{34}(x) + \frac{1}{3}ikv_T(x) \right] \hat{b}_0(x, k) = \hat{\alpha}_0(x, k) \,;$$

$$\left[m_{43}(x) + \frac{1}{3}ikv_T(x) \right] \hat{a}_0(x, k) + m_{44}(x)\hat{b}_0(x, k) = \hat{\beta}_0(x, k) \,. \tag{6.87}$$

Notations used here are:

$$\hat{\alpha}_0(x, k) = m_{33}(x)\hat{h}_3(x, k) + m_{34}(x)\hat{h}_4(x, k) + \hat{s}_\alpha(x, k) \,; \tag{6.88}$$

$$\hat{\beta}_0(x, k) = m_{43}(x)\hat{h}_3(x, k) + m_{44}(x)\hat{h}_4(x, k) + \hat{s}_\beta(x, k) \,;$$

$$\hat{s}_{\alpha,\beta}(x, k) = \int_{-\infty}^{+\infty} \exp(-iky)s_{\alpha,\beta}(x, y) \, dy \,;$$

$$s_\alpha(x, y) = \frac{1}{3}v_T(x) \left(\frac{\partial \ln n}{\partial x} + 2\frac{\partial \ln T}{\partial x} \right) h_1(x, y) \tag{6.89}$$

$$+ \frac{2}{3}\frac{\partial u}{\partial x}(h_0(x, y) + 2h_2(x, y)) \,;$$

$$s_\beta(x, y) = \frac{5}{4}v_T(x) \left(\frac{\partial \ln n}{\partial x}h_2(x, y) + \frac{\partial \ln T}{\partial x}(3h_2(x, y) + h_0(x, y)) \right)$$

$$+ \frac{2\partial u}{3\partial x}h_1(x, y) \,.$$

Solving equations (6.87), we obtain functions $\hat{a}_0(x, k)$ and $\hat{b}_0(x, k)$ in (6.86):

$$\hat{a}_0(x, k) = \frac{\hat{\alpha}_0(x, k)m_{44}(x) - \hat{\beta}_0(x, k)(m_{34}(x) + \frac{1}{3}ikv_T(x))}{Z(x, \frac{1}{3}ikv_T(x))} \,;$$

$$\hat{b}_0(x, k) = \frac{\hat{\beta}_0(x, k)m_{33}(x) - \hat{\alpha}_0(x, k)(m_{34}(x) + \frac{1}{3}ikv_T(x))}{Z(x, \frac{1}{3}ikv_T(x))} \,. \tag{6.90}$$

Here

$$Z(x, \frac{1}{3}ikv_T(x)) = Z(x, 0) + \frac{k^2v_T^2(x)}{9} + \frac{2}{3}ikv_T(x)m_{34}(x)$$

$$= \left(n\lambda_3 + \frac{11\partial u}{9\partial x} \right) \left(n\lambda_4 + \frac{27\partial u}{4\partial x} \right)$$

$$-\frac{v_T^2(x)}{9}\left(\frac{\partial \ln n}{\partial x}+\frac{11\partial \ln T}{2\partial x}\right)^2$$

$$+\frac{k^2 v_T^2(x)}{9}+\frac{2}{9}ikv_T^2(x)\left(\frac{\partial \ln n}{\partial x}+\frac{11\partial \ln T}{2\partial x}\right). \quad (6.91)$$

Step 4. *Calculation of the inverse Fourier transform of the function* $\hat{\varphi}_0(x,k,\boldsymbol{v})$.

The inverse Fourier transform of the function $\hat{\varphi}_0(x,k,\boldsymbol{v})$ (6.86) yields:

$$\varphi_0(x,\boldsymbol{v})=a_0(x)p_3(x,\boldsymbol{v})+b_0(x)p_4(x,\boldsymbol{v}). \quad (6.92)$$

Here

$$a_0(x)=\frac{1}{2\pi}\int_{-\infty}^{+\infty}\exp(ikx)\hat{a}_0(x,k)\,dk,$$

$$b_0(x)=\frac{1}{2\pi}\int_{-\infty}^{+\infty}\exp(ikx)\hat{b}_0(x,k)\,dk. \quad (6.93)$$

Taking into account expressions (6.79), (6.90)–(6.91), and (6.85), we obtain finally the explicit expression for the *finite-dimensional approximation of the zeroth-order term of parametrix expansion* (6.92):

$$a_0(x)=\frac{1}{2\pi}\int_{-\infty}^{+\infty}dy\int_{-\infty}^{+\infty}dk\exp(ik(x-y))Z^{-1}(x,\tfrac{1}{3}ikv_T(x))$$
$$\times\{Z(x,0)h_3(x,y)+[s_\alpha(x,y)m_{44}(x)-s_\beta(x,y)m_{34}(x)]$$
$$-\frac{1}{3}ikv_T(x)[m_{34}(x)h_3(x,y)+m_{44}(x)h_4(x,y)+s_\beta(x,y)]\};$$

$$b_0(x)=\frac{1}{2\pi}\int_{-\infty}^{+\infty}dy\int_{-\infty}^{+\infty}dk\exp(ik(x-y))Z^{-1}(x,\tfrac{1}{3}ikv_T(x))$$
$$\times\{Z(x,0)h_4(x,y)+[s_\beta(x,y)m_{33}(x)-s_\alpha(x,y)m_{34}(x)]$$
$$-\frac{1}{3}ikv_T(x)[m_{34}(x)h_4(x,y)+m_{33}(x)h_3(x,y)+s_\alpha(x,y)]\}. \quad (6.94)$$

6.3.6 Hydrodynamic Equations

Now we discuss the utility of obtained results for hydrodynamics.

The correction to the LM manifold $f_0(n,\boldsymbol{u},T)$(5.49) has the form:

$$f_1(n,\boldsymbol{u},T)=f_0(n,\boldsymbol{u},T)(1+\varphi_0(n,\boldsymbol{u},T)) \quad (6.95)$$

Here the function $\varphi_0(n,\boldsymbol{u},T)$ is given explicitly by expressions (6.92)–(6.94).

The usual form of closed hydrodynamic equations for $n,\boldsymbol{u}$, and T, where the traceless stress tensor σ_{ik} and the heat flux vector q_i are expressed via hydrodynamic variables, will be obtained if we substitute the function (6.95) into balance equations of the density, of the momentum, and of the energy. For the LM approximation, these balance equations result in the Euler equation of the nonviscid liquid (i.e. $\sigma_{ik}(f_0) \equiv 0$, and $q_i(f_0) \equiv 0$). For the correction f_1 (6.95), we obtain the following expressions of $\sigma = \sigma_{xx}(f_1)$ and $q = q_x(f_1)$ (all other components are equal to zero in the one-dimensional situation under consideration):

$$\sigma = \frac{1}{3}na_0 , \qquad q = \frac{5}{4}nb_0 . \tag{6.96}$$

Here a_0 and b_0 are given by expression (6.94).

From the geometrical viewpoint, hydrodynamic equations with the stress tensor and the heat flux vector (6.96) have the following interpretation: we take the corrected manifold Ω_1 which consists of functions f_1 (6.95), and we project the Boltzmann equation vectors $J_u(f_1)$ onto the tangent spaces T_{f_1} using the quasiequilibrium projector P_{f_0} (5.55).

6.3.7 Nonlocality

Expressions (6.94) include nonlocal spatial dependence, and, hence, the corresponding hydrodynamic equations are nonlocal. This nonlocality enters in two different ways. The first source of nonlocality might be called a *frequency-response nonlocality*, and it enters through explicit non-polynomial k-dependence of integrands in (6.94). This latter dependence has the form:

$$\int_{-\infty}^{+\infty} \frac{A(x,y) + ikB(x,y)}{C(x,y) + ikD(x,y) + k^2 E(x,y)} \exp(ik(x-y)) \, dk . \tag{6.97}$$

Integration over k in (6.97) can be completed via auxiliary functions.

The second type of nonlocal contributions might be called *correlative nonlocality*, and it is due to the terms $(u(x) - u(y))$ (the difference of flow velocities in points x and y) and via $T(x)/T(y)$ (the ratio of temperatures in distant points x and y).

6.3.8 Acoustic Spectra

The frequency-response nonlocality in hydrodynamic equations is relevant to small perturbations of the uniform equilibrium. The stress tensor σ and the heat flux q(6.96) are:

$$\sigma = -(2/3)n_0 T_0 R \left(2\varepsilon \frac{\partial u'}{\partial \xi} - 3\varepsilon^2 \frac{\partial^2 T}{\partial \xi^2} \right) ;$$

$$q = -(5/4)T_0^{3/2} n_0 R \left(3\varepsilon \frac{\partial T'}{\partial \xi} - (8/5)\varepsilon^2 \frac{\partial^2 u}{\partial \xi^2} \right) . \tag{6.98}$$

Here

$$R = \left(1 - (2/5)\varepsilon^2 \frac{\partial^2}{\partial \xi^2}\right) - 1 \ . \tag{6.99}$$

In (6.98), we have expressed parameters λ_3 and λ_4 via the viscosity coefficient μ of the Chapman-Enskog method [70] (it is easy to see from (6.78) that $\lambda_3 = \lambda_4 \propto \mu^{-1}$ for spherically symmetric models of a collision), and we have used the following notations: T_0 and n_0 are the equilibrium temperature and density, $\xi = (\eta T_0^{1/2})^{-1} n_0 x$ is the dimensionless coordinate, $\eta = \mu(T_0)/T_0, u' = T_0^{-1/2} \delta u, T' = \delta T/T_0, n' = \delta n/n_0$, and $\delta u, \delta T, \delta n$ are the deviations of the flux velocity, of the temperature and of the density from their equilibrium values $u = 0, T = T_0$ and $n = n_0$. We also used the system of units with $k_B = m = 1$.

In the linear case, the parametrix expansion degenerates, and its zeroth-order term (6.61) gives the exact solution to equation (6.38).

The dispersion relationship for the approximation (6.98) is:

$$\omega^3 + (23k^2/6D)\omega^2 + \left\{k^2 + (2k^4/D^2) + (8k^6/5D^2)\right\}\omega + (5k^4/2D) = 0 \ ;$$
$$D = 1 + (4/5)k^2 \ . \tag{6.100}$$

Here k is the wave vector.

The acoustic spectrum given by the dispersion relationship (6.100) contains no nonphysical short-wave instability, unlike the Burnett approximation (Fig. 6.2). The regularization of the Burnett approximation [43, 44] gives a similar result. Both of these approximations predict a limit of the decrement $\text{Re}\,\omega$ for short waves. These issues will be addressed in more detail in Chap. 8.

6.3.9 Nonlinearity

Nonlinear dependence on $\frac{\partial u}{\partial x}$, on $\frac{\partial \ln T}{\partial x}$, and on $\frac{\partial \ln n}{\partial x}$ appears already in the local approximation φ_{loc} (6.79). In order to outline some features of this nonlinearity, we represent the zeroth-order term of the expansion of a_{loc} (6.79) into powers of $\frac{\partial \ln T}{\partial x}$ and $\frac{\partial \ln n}{\partial x}$:

$$a_{\text{loc}} = -\frac{2}{3}\frac{\partial u}{\partial x}\left(n\lambda_3 + \frac{11}{9}\frac{\partial u}{\partial x}\right)^{-1} + O\left(\frac{\partial \ln T}{\partial x}, \frac{\partial \ln n}{\partial x}\right) \ . \tag{6.101}$$

This expression describes the asymptotic of the "purely nonlinear" contribution to the stress tensor σ (6.96) for a strong divergency of a flow. The account of nonlocality yields instead of (6.98):

$$a_0(x) = -\frac{1}{2\pi}\int_{-\infty}^{+\infty} dy \int_{-\infty}^{+\infty} dk \exp(ik(x-y))\frac{2}{3}\frac{\partial u}{\partial y}\left(n\lambda_3 + \frac{11}{9}\frac{\partial u}{\partial y}\right)^{-1}$$
$$\times \left[\left(n\lambda_3 + \frac{11}{9}\frac{\partial u}{\partial x}\right)\left(n\lambda_4 + \frac{27}{4}\frac{\partial u}{\partial x}\right) + \frac{k^2 v_T^2}{9}\right]^{-1}$$

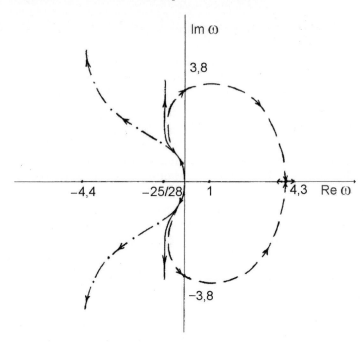

Fig. 6.2. Acoustic dispersion curves for approximation (6.98) (*solid* line), for second (the Burnett) approximation of the Chapman-Enskog expansion [72] (*dashed* line) and for the regularization of the Burnett approximation via partial summing of the Chapman-Enskog expansion [43, 44] (punctuated *dashed* line). Arrows indicate the direction of increase of k^2

$$
\times \left[\left(n\lambda_3 + \frac{11}{9}\frac{\partial u}{\partial x} \right) \left(n\lambda_4 + \frac{27}{4}\frac{\partial u}{\partial x} \right) \right.
$$
$$
\left. + \frac{4}{9}\left(n\lambda_4 + \frac{27}{4}\frac{\partial u}{dy} \right)\frac{\partial u}{\partial x}v_T^{-2}(u(x)-u(y))^2 - \frac{2}{3}ik\frac{\partial u}{\partial x}(u(x)-u(y)) \right]
$$
$$
+ O\left(\frac{\partial \ln T}{\partial x}, \frac{\partial \ln n}{\partial x} \right). \tag{6.102}
$$

Both expressions, (6.101) and (6.102) become singular when

$$
\frac{\partial u}{\partial y} \rightarrow \left(\frac{\partial u}{\partial y} \right)^{*} = -\frac{9n\lambda_3}{11}. \tag{6.103}
$$

Hence, the stress tensor (6.97) becomes infinite if $\frac{\partial u}{\partial y}$ tends to $\frac{\partial u}{\partial y}^{*}$ in any point y. In other words, the flow becomes "infinitely viscous" when $\frac{\partial u}{\partial y}$ approaches the negative value $-\frac{9n\lambda_3}{11}$. This infinite viscosity threshold prevents a transfer of the flow into nonphysical region of negative viscosity if $\frac{\partial u}{\partial y} > \frac{\partial u}{\partial y}^{*}$ because of the "infinitely strong damping" at $\frac{\partial u}{\partial y}^{*}$. This peculiarity was detected in

[43, 44] as a result of partial summation of the Chapman-Enskog expansion. In particular, partial summing for the simplest nonlinear situation [45, 233] yields the following expression for the stress tensor σ:

$$\sigma = \sigma_{IR} + \sigma_{IIR} \; ; \quad \sigma_{IR} = -\frac{4}{3}\left(1 - \frac{5}{3}\varepsilon^2\frac{\partial^2}{\partial\xi^2}\right)^{-1}\left(\varepsilon\frac{\partial u'}{\partial\xi} + \varepsilon^2\frac{\partial^2\theta'}{\partial\xi^2}\right) \; ;$$

$$\theta' = T' + n' \; ; \quad \sigma_{IIR} = \frac{28}{9}\left(1 + \frac{7}{3}\varepsilon\frac{\partial u'}{\partial\xi}\right)^{-1}\frac{\partial^2 u'}{\partial\xi^2} \; . \tag{6.104}$$

Notations here follow (6.98) and (6.99). Expression (6.104) might be considered as a scetch of the "full" stress tensor defined by a_0(6.94). It takes into account both the frequency-response and the nonlinear contributions (σ_{IR} and σ_{IIR}, respectively) in a simple form of a sum. However, the superposition of these contributions in (6.94) is more complicated. Moreover, the explicit correlative nonlocality of expression (6.94) was detected neither in [45], nor in numerous examples of partial summation [233].

Nevertheless, approximation (6.104) contains the peculiarity of viscosity similar to that in (6.101) and (6.102). In dimensionless variables and $\varepsilon = 1$, expression (6.104) predicts the infinite threshold at velocity divergency equal to $-(3/7)$, rather than $-(9/11)$ in (6.101) and (6.102). Viscosity tends to zero as the divergency tends to positive infinity in both approximations. A physical interpretation of these phenomena was given in [45]: large *positive* values of $\frac{\partial u}{\partial x}$ means that the gas diverges rapidly, and the flow becomes nonviscid because the particles retard to exchange their momentum. On contrary, its *negative* values (such as $-(3/7)$ for (6.104) and $-(9/11)$) for (6.101) and (6.102)) describe a strong compression of the flow. Strong deceleration results in a "solid fluid" limit with an infinite viscosity (Fig. 6.3).

Thus, hydrodynamic equations for approximation (6.95) are both nonlinear and nonlocal. This result is not surprising, accounting for the integro-differential nature of equation (6.38).

It is important that no small parameters were used neither when we were deriving equation (6.38) nor when we were obtaining the correction (6.95).

6.4 Example: Non-Perturbative Derivation of Linear Hydrodynamics from the Boltzmann Equation (3D)

In this example we shall discuss a bit more about the linear hydrodynamics obtained by the Newtom method with incomplete linearization. Using the Newton method instead of power series, a model of linear hydrodynamics is derived from the Boltzmann equation for regimes where the Knudsen number is of order unity. The model demonstrates no violation of stability of acoustic spectra in contrast to the Burnett hydrodynamics.

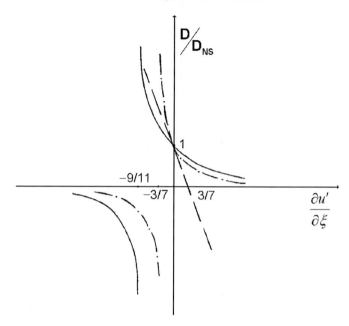

Fig. 6.3. Dependency of viscosity on compression for approximation (6.101) (*solid* line), for partial summing (6.104) (punctuated *dashed* line), and for the Burnett approximation [45,233] (*dashed* line). The latter changes the sign at a regular point and, hence, nothing prevents the flow to transfer into the nonphysical region

The Knudsen number ε (a ratio between the mean free path, l_c, and a scale of hydrodynamic flows, l_h) is a smalness parameter when hydrodynamics is derived from the Boltzmann equation [239]. The Chapman–Enskog method [70] derives the Navier-Stokes hydrodynamic equations as the first-order correction to the Euler hydrodynamics at $\varepsilon \to 0$, and it also derives formal corrections of order ε^2, ε^3, ... (known as the Burnett and super-Burnett corrections). These corrections are important outside the strictly hydrodynamic domain $\varepsilon \ll 1$, and has to be considered for an exension of hydrodynamic description into a highly nonequilidrium domain $\varepsilon \lesssim 1$. Not much is known about high-order in ε hydrodynamics, especially in nonlinear case. Nonetheless, in linear case, some definite information can be obtained. On the one hand, experiments on sound propagation in noble gases are considerably better explained with the Burnett and super-Burnett hydrodynamics rather than with the Navier-Stokes approximation alone [241]. On the other hand, direct calculation shows non-physical behavior of the Burnett hydrodynamics for ultra-short waves: acoustic waves increase instead of decay [72]. The latter failure of the Burnett approximation cannot be ignored. For the Navier-Stokes approximation no such violation is observed.

These two results indicate that, at least in a linear regime, it makes sense to consider hydrodynamics at $\varepsilon \sim 1$, but the Chapman-Enskog method of deriving such hydrodynamics is problematic. The problem of constructing solutions to the Boltzmann equation valid when ε is of order one is one of the main open problems of classical kinetic theory [239].

The main idea of the present example is to formulate the problem of a finding a correction to the Euler hydrodynamics in such a fashion that expansions in ε do not appear as a necessary element of analysis. This will be possible by using the Newton method instead of Taylor expansions to get such correction. Resulting hydrodynamic equations do not exhibit the mentioned violation.

The starting point is the set of local Maxwell distribution functions (LM) $f_0(n, \boldsymbol{u}, T; \boldsymbol{v})$, where $\boldsymbol{v}$ is the particle's velocity, and n, $\boldsymbol{u}$, and T are local number density, average velocity, and temperature. We write the Boltzmann equation as before in the co-moving reference frame (6.23):

$$\frac{\mathrm{d}f}{\mathrm{d}t} = J(f), \quad J(f) = -(v - u)_i \cdot \partial_i f + Q(f, f) , \tag{6.105}$$

where $\mathrm{d}/\mathrm{d}t = \partial/\partial t + u_i \cdot \partial_i$ is the material derivative, $\partial_i = \partial/\partial x_i$, while Q is the Boltzmann collision integral.

On the one hand, calculating right hand site of (6.105) in the LM-states, we obtain $J(f_0)$, a time derivative of the LM-states due to the Boltzmann equation. On the other hand, calculating a time derivative of the LM-states due to the Euler dynamics, we obtain $P_0 J(f_0)$, where P_0 is the thermodynamic projector operator onto the LM manifold (see [11] and (5.55)):

$$P_0 J = \frac{f_0}{n} \left\{ \int J \, \mathrm{d}\boldsymbol{c} + 2c_i \cdot \int c_i J \, \mathrm{d}\boldsymbol{c} + \frac{2}{3} \left(c^2 - \frac{3}{2} \right) \int \left(c^2 - \frac{3}{2} \right) J \, \mathrm{d}\boldsymbol{c} \right\} , \tag{6.106}$$

Since the LM functions are not solutions to the Boltzmann equation (6.105) (except for constant n, $\boldsymbol{u}$, and T), a difference $\Delta(f_0)$ between $J(f_0)$ and $P_0 J(f_0)$ is not equal to zero (5.59):

$$\Delta(f_0) = J(f_0) - P_0 J(f_0) \tag{6.107}$$

$$= -f_0 \left\{ 2(\partial_i u_k) \left(c_i c_k - \frac{1}{3} \delta_{ik} c^2 \right) + v_T \frac{\partial_i T}{T} c_i \left(c^2 - \frac{5}{2} \right) \right\} .$$

here $\boldsymbol{c} = v_T^{-1}(\boldsymbol{v} - \boldsymbol{u})$, and $v_T = \sqrt{2k_B T/m}$ is the thermal velocity. Note that the latter expression gives the complete invariance defect of the linearized local Maxwell approximation, and it is neither big nor small by itself. An unknown hydrodynamic solution of (6.105), $f_\infty(n, \boldsymbol{u}, T; \boldsymbol{v})$, satisfies the following invariance equation:

$$\Delta(f_\infty) = J(f_\infty) - P_\infty J(f_\infty) = 0 , \tag{6.108}$$

where P_∞ is an unknown projecting operator. Both P_∞ and f_∞ are unknown in (6.108), but, nontheless, one is able to consider a sequence of corrections

$\{f_1, f_2, \ldots\}$, $\{P_1, P_2, \ldots\}$ to the initial approximation f_0 and P_0. Above it was shown, how to ensure the H-theorem on every step of approximations by choosing appropriate projecting operators P_n. In the present illustrative example we do not consider projectors other than P_0.

Let us apply the Newton method with incomplete linearization to (6.108) with f_0 as initial approximation for f_∞ and with P_0 as an initial approximation for P_∞. Writing $f_1 = f_0 + \delta f$, we get the first iteration:

$$L(\delta f/f_0) + (P_0 - 1)(v - u)_i \partial_i \delta f + \Delta(f_0) = 0 , \qquad (6.109)$$

where L is a linearized collision integral.

$$
\begin{aligned}
&L(g)\\
&= f_0(\boldsymbol{v}) \int w(\boldsymbol{v}_1', \boldsymbol{v}'; \boldsymbol{v}_1, \boldsymbol{v}) f_0(\boldsymbol{v}_1) \{g(\boldsymbol{v}_1') + g(\boldsymbol{v}') - g(\boldsymbol{v}_1) - g(\boldsymbol{v})\} \, \mathrm{d}\boldsymbol{v}_1' \, \mathrm{d}\boldsymbol{v}' \, \mathrm{d}\boldsymbol{v}_1 .
\end{aligned}
$$
$$(6.110)$$

Here w is a probability density of velocities change, $(\boldsymbol{v}, \boldsymbol{v}_1) \leftrightarrow (\boldsymbol{v}', \boldsymbol{v}_1')$, of a pair of molecules after their encounter. When deriving (6.109), we have accounted $P_0 L = 0$, and an additional condition which fixes the same values of n, $\boldsymbol{u}$, and T in states f_1 as in LM states f_0:

$$P_0 \delta f = 0 . \qquad (6.111)$$

Equation (6.109) is basic in what follows. Note that it contains no Knudsen number explicitly. Our strategy will be to treat equation (6.109) in such a way that the Knudsen number will appear explicitly only at the latest stage of computations.

The two further approximations will be adopted. The first concerns a linearization of (6.109) about the global equilibria F_0. The second concerns a finite-dimensional approximation of integral operator in (6.109) in velocity space. It is worthwhile noting here that none of these approximations concerns an assumption about the smallness of the Knudsen number.

Following the first of the approximations mentioned, denote as δn, δu, and δT deviations of hydrodynamic variables from their equilibrium values n_0, $\boldsymbol{u}_0 = 0$, and T_0. Introduce also dimensionless variables $\Delta n = \delta n/n_0$, $\Delta u = \delta u/v_T^0$, and $\Delta T = \delta T/T_0$, where v_T^0 is a heat velocity in equilibria, and a dimensionless relative velocity $\xi = \boldsymbol{v}/v_T^0$. Correction f_1 in the approximation, linear in deviations from F_0, reads:

$$f_1 = F_0(1 + \varphi_0 + \varphi_1) ,$$

where

$$\varphi_0 = \Delta n + 2\Delta u_i \xi_i + \Delta T(\xi^2 - 3/2)$$

is a linearized deviation of LM from F_0, and φ_1 is an unknown function. The latter is to be obtained from a linearized version of (6.109).

Following the second approximation, we seek φ_1 in a form:

$$\varphi_1 = A_i(\boldsymbol{x})\xi_i\left(\xi^2 - \frac{5}{2}\right) + B_{ik}(\boldsymbol{x})\left(\xi_i\xi_k - \frac{1}{3}\delta_{ik}\xi^2\right) + \dots \qquad (6.112)$$

where dots denote terms of an expansion of φ_1 in velocity polynomials, orthogonal to $\xi_i(\xi^2 - 5/2)$ and $\xi_i\xi_k - 1/3\delta_{ik}\xi^2$, as well as to 1, to $\boldsymbol{\xi}$, and to ξ^2. These terms do not contribute to shear stress tensor and heat flux vector in hydrodynamic equations. Independency of functions A and B from ξ^2 amounts to the first Sonine polynomial approximation of viscosity and heat transfer coefficients. Thus, we consider a projection onto a finite-dimensional subspace spanned by $\xi_i(\xi^2 - 5/2)$ and $\xi_i\xi_k - 1/3\delta_{ik}\xi^2$. Our goal is to derive functions A and B from a linearized version of (6.109). Knowing A and B, we get the following expressions for shear stress tensor $\boldsymbol{\sigma}$ and heat flux vector $\boldsymbol{q}$:

$$\sigma = p_0 B, \quad \boldsymbol{q} = \frac{5}{4}p_0 v_T^0 A, \qquad (6.113)$$

where p_0 is equilibrium pressure of ideal gas.

Linearizing (6.109) near F_0, using an ansatz for φ_1 cited above, and turning to Fourier transform in space, we derive:

$$\frac{5p_0}{3\eta_0}a_i(\boldsymbol{k}) + iv_T^0 b_{ij}(\boldsymbol{k})k_j = -\frac{5}{2}iv_T^0 k_i\tau(\boldsymbol{k}) ; \qquad (6.114)$$

$$\frac{p_0}{\eta_0}b_{ij}(\boldsymbol{k}) + iv_T^0\overline{k_i a_j(\boldsymbol{k})} = -2iv_T^0\overline{k_i\gamma_j(\boldsymbol{k})} ,$$

where $i = \sqrt{-1}$, $\boldsymbol{k}$ is the wave vector, η_0 is the first Sonine polynomial approximation of shear viscosity coefficient, $\boldsymbol{a}(\boldsymbol{k})$, $\boldsymbol{b}(\boldsymbol{k})$, $\tau(\boldsymbol{k})$ and $\boldsymbol{\gamma}(\boldsymbol{k})$ are Fourier transforms of $\boldsymbol{A}(\boldsymbol{x})$, $\boldsymbol{B}(\boldsymbol{x})$, $\Delta T(\boldsymbol{x})$, and $\Delta \boldsymbol{u}(\boldsymbol{x})$, respectively, and the over-bar denotes a symmetric traceless dyad:

$$\overline{a_i b_j} = 2a_i b_j - \frac{2}{3}\delta_{ij}a_s b_s .$$

Introducing a dimensionless wave vector $\boldsymbol{f} = [(v_T^0\eta_0)/(p_0)]\boldsymbol{k}$, solution to (6.114) may be written:

$$b_{lj}(\boldsymbol{k}) = -\frac{10}{3}i\overline{\gamma_l(\boldsymbol{k})f_j}[(5/3) + (1/2)f^2]^{-1} \qquad (6.115)$$

$$+\frac{5}{3}i(\gamma_s(\boldsymbol{k})f_s)\overline{f_l f_j}[(5/3)+(1/2)f^2]^{-1}[5 + 2f^2]^{-1} - \frac{15}{2}\tau(\boldsymbol{k})\overline{f_l f_j}[5 + 2f^2]^{-1} ;$$

$$a_l(\boldsymbol{k}) = -\frac{15}{2}if_l\tau(\boldsymbol{k})[5 + 2f^2]^{-1}$$

$$-[5 + 2f^2]^{-1}[(5/3) + (1/2)f^2]^{-1}[(5/3)f_l(\gamma_s(\boldsymbol{k})f_s) + \gamma_l(\boldsymbol{k})f^2(5 + 2f^2)] .$$

Considering z-axis as a direction of propagation and denoting k_z as k, γ as γ_z, we obtain from (6.114) the k-dependence of $a = a_z$ and $b = b_{zz}$:

$$a(k) = -\frac{\frac{3}{2}p_0^{-1}\eta_0 v_T^0 ik\tau(k) + \frac{4}{5}p_0^{-2}\eta_0^2(v_T^0)^2 k^2\gamma(k)}{1 + \frac{2}{5}p_0^{-2}\eta_0^2(v_T^0)^2 k^2} , \qquad (6.116)$$

$$b(k) = -\frac{\frac{4}{3}p_0^{-1}\eta_0 v_T^0 ik\gamma(k) + p_0^{-2}\eta_0^2(v_T^0)^2 k^2\tau(k)}{1 + \frac{2}{5}p_0^{-2}\eta_0^2(v_T^0)^2 k^2} .$$

Using expressions for σ and q cited above, and also using (6.116), it is an easy matter to close the linearized balance equations (given in Fourier terms):

$$\frac{1}{v_T^0}\partial_t\nu(k) + ik\gamma_k = 0 , \qquad (6.117)$$

$$\frac{2}{v_T^0}\partial_t\gamma(k) + ik(\tau(k) + \nu(k)) + ikb(k) = 0 ,$$

$$\frac{3}{2v_T^0}\partial\tau + ik\gamma(k) + \frac{5}{4}ika(k) = 0 .$$

The equations (6.117), together with expressions (6.116), complete our derivation of hydrodynamic equations.

To this end, the Knudsen number was not penetrating our derivations. Now it is worthwhile to introduce it. The Knudsen number will appear most naturally if we turn to dimensionless form of (6.116). Taking $l_c = v_T^0\eta_0/p_0$ (l_c is of order of a mean free path), and introducing a hydrodynamic scale l_h, so that $k = \kappa/l_h$, where κ is a not-dimensional wave vector, we obtain in (6.116):

$$a(\kappa) = -\frac{\frac{3}{2}i\varepsilon\kappa\tau(\kappa) + \frac{4}{5}\varepsilon^2\kappa^2\gamma_\kappa}{1 + \frac{2}{5}\varepsilon^2\kappa^2} , \qquad (6.118)$$

$$b(\kappa) = -\frac{\frac{4}{3}i\varepsilon\kappa\gamma(\kappa) + \varepsilon^2\kappa^2\tau(\kappa)}{1 + \frac{2}{5}\varepsilon^2\kappa^2} ,$$

where $\varepsilon = l_c/l_h$. Considering the limit $\varepsilon \to 0$ in (6.118), we come back to the familiar Navier-Stokes expressions: $\sigma_{zz}^{NS} = -\frac{4}{3}\eta_0\partial_z\delta u_z$, $q_z^{NS} = -\lambda_0\partial_z\delta T$, where $\lambda_0 = 15k_B\eta_0/4m$ is the first Sonine polynomial approximation of heat conductivity coefficient.

Since we were not assuming smallness of the Knudsen number ε while deriving (6.118), we can write $\varepsilon = 1$. With all the approximations mentioned above, (6.117) and (6.116) (or, equivalently, (6.117) and (6.118)) may be considered as a model of a linear hydrodynamics at ε of order one. The most interesting feature of this model is a non-polynomial dependence on κ. This amounts to that share stress tensor and heat flux vector depend on spatial derivatives of δu and of δT to arbitrary high order.

To find out a result of the non-polynomial behavior (6.118), it is most informative to calculate a dispersion relation for plane waves. Let us introduce a dimensionless frequency $\lambda = \omega l_h/v_T^0$, where ω is a complex frequency of a

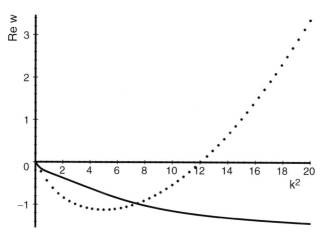

Fig. 6.4. Attenuation rate of sound waves. *Dotts*: the Burnett approximation. Bobylev's instability occurs when the curve intersects the horizontal axis. *Solid*: First iteration of the Newton method on the invariance equation

wave $\sim\exp(\omega t + ikz)$ ($\mathrm{Re}\omega$ is a damping rate, and $\mathrm{Im}\omega$ is a circular frequency). Making use of (6.117) and (6.118), writing $\varepsilon = 1$, we obtain the following dispersion relation $\lambda(\kappa)$:

$$12(1+\tfrac{2}{5}\kappa^2)^2\lambda^3 + 23\kappa^2(1+\tfrac{2}{5}\kappa^2)\lambda^2 + 2\kappa^2(5+5\kappa^2+\tfrac{6}{5}\kappa^4)\lambda + \tfrac{15}{2}\kappa^4(1+\tfrac{2}{5}\kappa^2) = 0 \;.$$
$$(6.119)$$

Figure 6.4 presents a dependence $\mathrm{Re}\lambda(\kappa^2)$ for acoustic waves obtained from (6.119) and for the Burnett approximation [72]. The violation in the latter occurs when the curve crosses the horizontal axis. In contrast to the Burnett approximation [72], the acoustic spectrum (6.119) is stable for all κ. Moreover, $\mathrm{Re}\lambda(\kappa^2)$ demonstrates a finite limit, as $\kappa^2 \to \infty$.

A discussion of results concerns the following two items:

1. The approach used avoids expansion into powers of the Knudsen number, and thus we obtain a hydrodynamics valid (at least formally) for moderate Knudsen numbers as an immediate correction to the Euler hydrodynamics. This is in a contrast to the usual treatment of high-order hydrodynamics as "(the well established) Navier-Stokes approximation + high-order terms". The Navier-Stokes hydrodynamics is recovered a posteriori, as a limiting case, but not as a necessary intermediate step of computations.
2. Linear hydrodynamics derived is stable for all k, same as the Navier-Stokes hydrodynamics alone. The $(1+\alpha k^2)^{-1}$ "cut-off", as in (6.116) and (6.118), was earlier found in a "partial summing" of Enskog series [42, 43].

Thus, we come to the following two conclusions:

1. A positive answer is given to the question of whether is it possible to construct solutions of the Boltzmann equation valid for the Knudsen number of order one.
2. Linear hydrodynamics derived can be used as a model for $\varepsilon = 1$ without a violation of acoustic spectra at large k.

6.5 Example: Dynamic Correction to Moment Approximations

6.5.1 Dynamic Correction or Extension of the List of Variables?

Considering the Grad moment ansatz as a suitable first approximation to a closed finite-moment dynamics, the correction is derived from the Boltzmann equation. The correction consists of two parts, local and nonlocal. Locally corrected thirteen-moment equations are demonstrated to contain exact transport coefficients. Equations resulting from the nonlocal correction give a microscopic justification to some phenomenological theories of extended hydrodynamics.

A considerable part of the modern development of nonequilibrium thermodynamics is based on the idea of extension of the list of relevant variables. Various phenomenological and semi-phenomenological theories in this domain are known under the common title of the extended irreversible thermodynamics (EIT) [235]. With this, the question of a microscopic justification of the EIT becomes important. Recall that a justification for some of the versions of the EIT was found witin the well known Grad moment method [201].

Originally, the Grad moment approximation was introduced for the purpose of solving the Boltzmann-like equations of the classical kinetic theory. The Grad method is used in various kinetic problems, e.g., in plasma and in phonon transport. We mention also that Grad equations assist in understanding asymptotic features of gradient expansions, both in linear and nonlinear domains [40, 42, 205, 219, 233].

The essence of the Grad method is to introduce an approximation to the one-particle distribution function f which would depend only on a finite number N of moments, and, subsequently, to use this approximation to derive a closed system of N moment equations from the kinetic equation. The number N (the level at which the moment transport hierarchy is truncated) is not specified in the Grad method. One particular way to choose N is to obtain an estimation of the transport coefficients (viscosity and heat conductivity) sufficiently close to their exact values provided by the Chapman–Enskog method (CE) [70]. In particular, for the thirteen-moment Grad approximation it is well known that transport coefficients are equal to the first Sonine polynomial approximation to the exact CE values. Accounting for higher moments with

$N > 13$ can improve this approximation (good for neutral gases but poor for plasmas [231]). However, what should be done, starting with the thirteen-moment approximation, to come to the exact CE transport coefficients is an open question. It is also well known [204] that the Grad method provides a poorly converging approximation when applied to strongly nonequilibrium problems (such as shock and kinetic layers).

Another question comes from the approximate character of the Grad equations, and is discussed in frames of the EIT: while the Grad equations are strictly hyperbolic at any level N (i.e., predicting a finite speed of propagation), whether this feature will be preserved in the further corrections.

These two questions are special cases of a more general one, namely, how to derive a closed description with a given number of moments? Such a description is sometimes called mesoscopic [251] since it occupies an intermediate level between the hydrodynamic (macroscopic) and the kinetic (microscopic) levels of description.

Here we aim at deriving the mesoscopic dynamics of thirteen moments [21] in the simplest case when the kinetic description satisfies the linearized Boltzmann equation. Our approach will be based on the two assumptions:

(i)   The mesoscopic dynamics of thirteen moments exists, and is invariant with respect to the microscopic dynamics,
(ii) The thirteen-moment Grad approximation is a suitable first approximation to this mesoscopic dynamics.

The assumption (i) is realized as the invariance equation for the (unknown) mesoscopic distribution function. Following the assumption (ii), we solve the invariance equation iteratively, taking the Grad approximation for the input approximation, and consider the first iteration (further we refer to this as to the dynamic correction, to distinguish from constructing another ansatz). We demonstrate that the correction results in the exact CE transport coefficients. We also demonstrate how the dynamic correction modifies the hyperbolicity of the Grad equations. A similar viewpoint on derivation of hydrodynamics was earlier developed in [11] (see previous examples). We shall return to a comparison below.

6.5.2 Invariance Equation
for Thirteen-Moment Parameterization

We denote as n_0, $\boldsymbol{u}_0 = 0$, and p_0 the equilibrium values of the hydrodynamic parameters (n is the number density, $\boldsymbol{u}$ is the average velocity, and $p = n k_B T$ is the pressure). The global Maxwell distribution function F is

$$F = n_0 (v_T)^{-3} \pi^{-3/2} \exp(-c^2) \,,$$

where $v_T = \sqrt{2 k_B T_0 m^{-1}}$ is the equilibrium thermal velocity, and $\boldsymbol{c} = \boldsymbol{v}/v_T$ is the peculiar velocity of a particle. The near-equilibrium dynamics of the distribution function, $f = F(1 + \varphi)$, is due to the linearized Boltzmann equation:

$$\partial_t \varphi = \hat{J}\varphi \equiv -v_T c_i \partial_i \varphi + \hat{L}\varphi ,$$

$$\hat{L}\varphi = \int w F(\boldsymbol{v}_1)[\varphi(\boldsymbol{v}_1') + \varphi(\boldsymbol{v}') - \varphi(\boldsymbol{v}_1) - \varphi(\boldsymbol{v})] \, d\boldsymbol{v}_1' \, d\boldsymbol{v}' \, d\boldsymbol{v}_1 ,$$

where $\hat{L}$ is the linearized collision operator, and w is the probability density of pair encounters. Furthermore, $\partial_i = \partial/\partial x_i$, and summation convention in two repeated indices is assumed.

Let $n = \delta n/n_0$, $\boldsymbol{u} = \delta \boldsymbol{u}/v_T$, $p = \delta p/p_0$ ($p = n + T$, $T = \delta T/T_0$), be dimensionless deviations of the hydrodynamic variables, while $\boldsymbol{\sigma} = \delta \boldsymbol{\sigma}/p_0$ and $\boldsymbol{q} = \delta \boldsymbol{q}/(p_0 v_T)$ are dimensionless deviations of the stress tensor $\boldsymbol{\sigma}$, and of the heat flux $\boldsymbol{q}$. The linearized thirteen-moment Grad distribution function is $f_0 = F(\boldsymbol{c})\,[1 + \varphi_0]$, where

$$\varphi_0 = \varphi_1 + \varphi_2 , \tag{6.120}$$
$$\varphi_1 = n + 2u_i c_i + T\left[c^2 - (3/2)\right] ,$$
$$\varphi_2 = \sigma_{ik}\overline{c_i c_k} + (4/5)q_i c_i \left[c^2 - (5/2)\right] .$$

The overline denotes a symmetric traceless dyad. We use the following convention:

$$\overline{a_i b_k} = a_i b_k + a_k b_i - \frac{2}{3}\delta_{ik} a_l b_l ,$$

$$\overline{\partial_i f_k} = \partial_i f_k + \partial_k f_i - \frac{2}{3}\delta_{ik}\partial_l f_l .$$

The thirteen-moment Grad's equations are derived in two steps: first, the Grad's distribution function (6.120) is inserted into the linearized Boltzmann equation to give a formal expression, $\partial_t \varphi_0 = \hat{J}\varphi_0$, second, projector P_0 is applied to this expression, where $P_0 = P_1 + P_2$, and operators P_1 and P_2 act as follows:

$$P_1 J = \frac{F}{n_0}\left\{ X_0 \int X_0 J \, d\boldsymbol{v} + X_i \int X_i J \, d\boldsymbol{v} + X_4 \int X_4 J \, d\boldsymbol{v} \right\} , \tag{6.121}$$

$$P_2 J = \frac{F}{n_0}\left\{ Y_{ik} \int Y_{ik} J d\boldsymbol{v} + Z_i \int Z_i J d\boldsymbol{v} \right\} .$$

Here $X_0 = 1$, $X_i = \sqrt{2}c_i$, where $i = 1, 2, 3$, $X_4 = \sqrt{2/3}\left(c^2 - \frac{3}{2}\right)$, $Y_{ik} = \sqrt{2\overline{c_i c_k}}$, and $Z_i = \frac{2}{\sqrt{5}}c_i\left(c^2 - \frac{5}{2}\right)$. The resulting equation,

$$P_0[F\partial_t \varphi_0] = P_0[F\hat{J}\varphi_0] ,$$

is a compressed representation for the thirteen-moment Grad equations for the macroscopic variables $M_{13} = \{n, \boldsymbol{u}, T, \boldsymbol{\sigma}, \boldsymbol{q}\}$.

Now we turn our attention to the main purpose of this example, and derive the dynamic invariance correction to the thirteen-moment distribution function (6.120). The assumption (i) [existence of closed dynamics of thirteen moments] implies the invariance equation for the true mesoscopic distribution function, $\tilde{f}(M_{13}, \boldsymbol{c}) = F[1 + \tilde{\varphi}(M_{13}, \boldsymbol{c})]$, where we have stressed that this function depends parametrically on the same thirteen macroscopic parameters, as the original Grad approximation. The invariance condition for $\tilde{f}(M_{13}, \boldsymbol{c})$ reads [11]:

$$(1 - \tilde{P})[F\hat{J}\tilde{\varphi}] = 0 , \qquad (6.122)$$

where $\tilde{P}$ is the projector associated with $\tilde{f}$. Generally speaking, the projector $\tilde{P}$ depends on the distribution function $\tilde{f}$ [11,231]. In the following, we use the projector P_0 (6.121) which will be consistent with our approximate treatment of (6.122).

Following the assumption (ii) [Grad's distribution function (6.120) is a good initial approximation], the Grad's function f_0, and the projector P_0, are chosen as the input data for solving the equation (6.122) iteratively. The dynamic correction amounts to the first iterate. Let us consider these steps in a more detail.

Substituting φ_0 (6.120) and P_0 (6.121) instead of φ and P in the equation (6.122), we get: $(1 - P_0)[F\hat{J}\varphi_0] \equiv \Delta_0 \neq 0$, which demonstrates that (6.120) is not a solution to the equation (6.122). Moreover, Δ_0 splits in two natural pieces: $\Delta_0 = \Delta_0^{\mathrm{loc}} + \Delta_0^{\mathrm{nloc}}$, where

$$\Delta_0^{\mathrm{loc}} = (1 - P_2)[F\hat{L}\varphi_2] , \qquad (6.123)$$
$$\Delta_0^{\mathrm{nloc}} = (1 - P_0)[-v_T F c_i \partial_i \varphi_0] .$$

Here we have accounted for $P_1[F\hat{L}\varphi] = 0$, and $\hat{L}\varphi_1 = 0$. The first piece of (6.123), Δ_0^{loc}, can be termed *local* because it does not account for spatial gradients. Its origin is twofold. In the first place, recall that we are performing our analysis in a non-local-equilibrium state (the thirteen-moment Grad's approximation is not a zero point of the Boltzmann collision integral, hence $\hat{L}\varphi_0 \neq 0$). In the second place, specializing to the linearized case under consideration, functions $\overline{\boldsymbol{cc}}$ and $\boldsymbol{c}[c^2 - (5/2)]$, in general, are not the eigenfunctions of the linearized collision integral, and hence $P_2[F\hat{L}\varphi_0] \neq F\hat{L}\varphi_0$, resulting in $\Delta_0^{\mathrm{loc}} \neq 0^2$.

The nonlocal part may be written as:

$$\Delta_0^{\mathrm{nloc}} = -v_T F(\Pi_{1|krs}\partial_k \sigma_{rs} + \Pi_{2|ik}\overline{\partial_k q_i} + \Pi_3 \partial_k q_k) , \qquad (6.124)$$

where Π are velocity polynomials:

---

[2] Except for Maxwell molecules (interaction potential $U \sim r^{-4}$) for which $\hat{L}\varphi_0 \neq 0$ but $P_2[F\hat{L}\varphi_{\mathrm{G}}] = F\hat{L}\varphi_0$. Same goes for the relaxation time approximation of the collision integral ($\hat{L} = -\tau^{-1}$).

$$\Pi_{1|krs} = c_k \left[c_r c_s - (1/3)\delta_{rs}c^2 \right] - (2/5)\delta_{ks}c_r c^2 \;,$$
$$\Pi_{2|ik} = (4/5)\left[c^2 - (7/2) \right] \left[c_i c_k - (1/3)\delta_{ik}c^2 \right] \;,$$
$$\Pi_3 = (4/5)\left[c^2 - (5/2) \right] \left[c^2 - (3/2) \right] - c^2 \;.$$

We seek the dynamic correction of the form:

$$f = F[1 + \varphi_0 + \phi] \;.$$

Substituting $\varphi = \varphi_0 + \phi$, and $P = P_0$, into (6.122), we derive an equation for the correction ϕ:

$$(1 - P_2)[F\hat{L}(\varphi_2 + \phi)] = (1 - P_0)[v_T F c_i \partial_i(\varphi_0 + \phi)] \;. \tag{6.125}$$

The equation (6.125) should be supplied with the additional condition, $P_0[F\phi] = 0$.

6.5.3 Solution of the Invariance Equation

Let us apply the usual ordering to solve (6.125), introducing a small parameter ϵ, multiplying the collision integral $\hat{L}$ with ϵ^{-1}, and expanding $\phi = \sum_n \epsilon^n \phi^{(n)}$. Subject to the additional condition, the resulting sequence of linear integral equations is uniquely soluble. Let us consider the first two orders in ϵ.

Because $\Delta_0^{loc} \neq 0$, the leading correction is of the order ϵ^0, i.e. of the same order as the initial approximation φ_0. The function $\phi^{(0)}$ is due the following equation:

$$(1 - P_2)[F\hat{L}(\varphi_2 + \phi^{(0)})] = 0 \;, \tag{6.126}$$

subject to the condition, $P_0[F\phi^{(0)}] = 0$. The equation (6.126) has the unique solution: $\varphi_2 + \phi^{(0)} = \sigma_{ik} Y_{ik}^{(0)} + q_i Z_i^{(0)}$, where functions, $Y_{ik}^{(0)}$ and $Z_i^{(0)}$, are solutions to the integral equations:

$$\hat{L} Y_{ik}^{(0)} = b Y_{ik} \;, \quad \hat{L} Z_i^{(0)} = a Z_i \;, \tag{6.127}$$

subject to the conditions, $P_1[F\mathbf{Y}^{(0)}] = 0$ and $P_1[F\mathbf{Z}^{(0)}] = 0$. Factors a and b are:

$$a = \pi^{-3/2} \int e^{-c^2} Z_i^{(0)} \hat{L} Z_i^{(0)} \, d\mathbf{c} \;,$$

$$b = \pi^{-3/2} \int e^{-c^2} Y_{ik}^{(0)} \hat{L} Y_{ik}^{(0)} \, d\mathbf{c} \;.$$

Now we are able to notice that the equation (6.127) coincides with the CE equations [70] for the *exact transport coefficients* (viscosity and temperature conductivity). Emergency of these well known equations in the present context is important and rather unexpected: *when the moment transport equations are closed with the locally corrected function* $f^{loc} = F(1 + \varphi_0 + \phi^{(0)})$, *we come to a closed set of thirteen equations containing the exact CE transport coefficients.*

Let us analyze the next order (ϵ^1), where Δ_0^{nloc} comes into play. To simplify matters, we neglect the difference between the exact and the approximate CE transport coefficients. The correction $\phi^{(1)}$ is due to the equation,

$$(1 - P_2)[F\hat{L}\phi^{(1)}] + \Delta_0^{\text{nloc}} = 0 , \qquad (6.128)$$

the additional condition is: $P_0[F\phi^{(1)}] = 0$. The problem (6.128) reduces to three integral equations of a familiar form:

$$\hat{L}\Psi_{1|krs} = \Pi_{1|krs} , \quad \hat{L}\Psi_{2|ik} = \Pi_{2|ik} , \quad \hat{L}\Psi_3 = \Pi_3 , \qquad (6.129)$$

subject to conditions: $P_1[F\Psi_{1|krs}] = 0$, $P_1[F\Psi_{2|ik}] = 0$, and $P_1[F\Psi_3] = 0$. Integral equations (6.129) are of the same structure as are the integral equations appearing in the CE method, and the methods to handle them are well developed [70]. In particular, a reasonable and simple approximation is to take $\Psi_{\alpha|...} = -A_\alpha \Pi_{\alpha|...}$. Then

$$\phi^{(1)} = -v_T(A_1\Pi_{1|krs}\partial_k\sigma_{rs} + A_2\Pi_{2|ik}\overline{\partial_k q_i} + A_3\Pi_3\partial_k q_k) , \qquad (6.130)$$

where A_α are the approximate values of the kinetic coefficients, and which are expressed via matrix elements of the linearized collision integral:

$$A_\alpha^{-1} \propto -\int \exp(-c^2)\Pi_{\alpha|...}\hat{L}\Pi_{\alpha|...} \, d\mathbf{c} > 0 . \qquad (6.131)$$

The evaluation can be extended to a computational scheme for any given molecular model (e.g., for the Lennard-Jones potential), in the manner of the transport coefficients computations in the classical Chapman–Enskog method.

6.5.4 Corrected Thirteen-Moment Equations

To summarize the results of the dynamic correction, we quote first the unclosed equations for the variables $M_{13} = M_{13} = \{n, \mathbf{u}, T, \boldsymbol{\sigma}, \mathbf{q}\}$:

$$(1/v_T^0)\partial_t n + \partial_i u_i = 0 , \qquad (6.132)$$
$$(2/v_T^0)\partial_t u_i + \partial_i(T + n) + \partial_k\sigma_{ik} = 0 , \qquad (6.133)$$
$$(1/v_T^0)\partial_t T + (2/3)\partial_i u_i + (2/3)\partial_i q_i = 0 , \qquad (6.134)$$
$$(1/v_T^0)\partial_t\sigma_{ik} + 2\overline{\partial_i u_k} - (2/3)\overline{\partial_i q_k} + \partial_l h_{ikl} = R_{ik} , \qquad (6.135)$$
$$(2/v_T)\partial_t q_i - (5/2)\partial_i p - (5/2)\partial_k\sigma_{ik} + \partial_k g_{ik} = R_i . \qquad (6.136)$$

Terms spoiling the closure are: the higher moments of the distribution function,

$$h_{ikl} = 2\pi^{-3/2} \int e^{-c^2} \varphi c_i c_k c_l d\mathbf{c} \,,$$

$$g_{ik} = 2\pi^{-3/2} \int e^{-c^2} \varphi c_i c_k c^2 \, d\mathbf{c} \,,$$

and the scattering rates,

$$R_{ik} = \frac{2}{v_T} \pi^{-3/2} \int e^{-c^2} c_i c_k \hat{L} \varphi \, d\mathbf{c} \,,$$

$$R_i = \frac{2}{v_T} \pi^{-3/2} \int e^{-c^2} c_i c^2 \hat{L} \varphi \, d\mathbf{c} \,.$$

Grad's distribution function (6.120) provides the zeroth-order closure approximation to both the higher-order moments and the scattering rates:

$$R_{ik}^{(0)} = -\mu_0^{-1}\sigma_{ik}, \quad R_i^{(0)} = -\lambda_0^{-1} q_i \,, \tag{6.137}$$

$$\partial_l h_{ikl}^{(0)} = (2/3)\delta_{ik}\partial_l q_l + (4/5)\overline{\partial_i q_k} \,,$$

$$\partial_l g_{lk}^{(0)} = (5/2)\partial_k(p+T) + (7/2)\partial_l \sigma_{lk} \,,$$

where μ_0 and λ_0 are the first Sonine polynomial approximations to the viscosity and the temperature conductivity coefficients [70], respectively.

The local correction improves the closure of the scattering rates:

$$R_{ik} = -\mu_{\mathrm{CE}}^{-1}\sigma_{ik}, \quad R_i = -\lambda_{\mathrm{CE}}^{-1} q_i \,, \tag{6.138}$$

where the subscript CE corresponds to the *exact* Chapman–Enskog values of the transport coefficients.

The nonlocal correction adds the following terms to the higher-order moments:

$$\partial_l g_{lk} = \partial_l g_{lk}^{(0)} - A_3 \partial_k \partial_l q_l - A_2 \partial_l \overline{\partial_l q_k} \,, \tag{6.139}$$

$$\partial_l h_{ikl} = \partial_l h_{ikl}^{(0)} - A_1 \partial_l \partial_l \sigma_{ik} \,,$$

where A_i are the kinetic coefficients derived above.

In order to illustrate what changes in Grad equations with the nonlocal correction, let us consider a model with two scalar variables, $T(x,t)$ and $q(x,t)$ (a simplified case of the one-dimensional corrected thirteen-moment system where one retains only the variables responsible for heat conduction):

$$\partial_t T + \partial_x q = 0, \quad \partial_t q + \partial_x T - a\partial_x^2 q + q = 0 \,. \tag{6.140}$$

Parameter $a \geq 0$ controls "turning on" the nonlocal correction. Using $\{q(k,\omega), T(k,\omega)\} \exp(\omega t + ikx)$, we come to a dispersion relation for the two roots $\omega_{1,2}(k)$. Without the correction ($a = 0$), there are two domains of k: for $0 \leq k < k_-$, dispersion is diffusion-like ($\mathrm{Re}\,\omega_{1,2}(k) \leq 0$, $\mathrm{Im}\,\omega_{1,2}(k) = 0$), while as $k \geq k_-$, dispersion is wave-like ($\omega_1(k) = \omega_2^*(k)$, $\mathrm{Im}\,\omega_1(k) \neq 0$). For

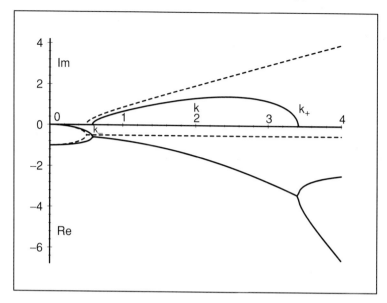

Fig. 6.5. Attenuation $\mathrm{Re}\,\omega_{1,2}(k)$ (*lower* pair of curves), frequency $\mathrm{Im}\,\omega_{1,2}(k)$ (*upper* pair of curves). *Dashed* lines – Grad case ($a = 0$), drawn lines – dynamic correction ($a = 0.5$)

a between 0 and 1, the dispersion modifies in the following way: The wave-like domain becomes bounded, and exists for $k \in]k_-(a), k_+(a)[$, while the diffusion-like domain consists of two pieces, $k < k_-(a)$ and $k > k_+(a)$.

The dispersion relation for $a = 1/2$ is shown in Fig. 6.5. As a increases to 1, the boundaries of the wave-like domain, $k_-(a)$ and $k_+(a)$, move towards each other, and collapse at $a = 1$. For $a > 1$, the dispersion relation becomes purely diffusive ($\mathrm{Im}\,\omega_{1,2} = 0$) for all k.

6.5.5 Discussion: Transport Coefficients, Destroying the Hyperbolicity, etc.

1. Considering the thirteen-moment Grad's ansatz as a suitable approxima-tion to the closed dynamics of thirteen moments, we have found that the first correction leads to the exact Chapman–Enskog transport coefficients. Further, the nonlocal part of this correction extends the Grad equations with terms containing spatial gradients of the heat flux and of the stress tensor, destroying the hyperbolic nature of Grad's moment system. Cor-responding kinetic coefficients are explicitly derived for the Boltzmann equation.
2. Extension of Grad equations with terms like in (6.139) was mentioned in the EIT [252]. These derivations were based on phenomenological and semi-phenomenological argument. In particular, the extension of the heat

flux with appealing to nonlocality effects in *dense* fluids. Here we have derived the similar contribution from the *simplest* (i.e. dilute gas) kinetics, in fact, from the assumption about existence of the mesoscopic dynamics. The advantage of using the simplest kinetics is that corresponding kinetic coefficients (6.131) become a matter of a *computation* for any molecular model.

3. When the invariance principle is applied to derive hydrodynamics (closed equations for the variables n, $\boldsymbol{u}$ and T) then [11] the local Maxwellian f_{lm} is chosen as the input distribution function for the invariance equation. In the linear domain, $f_{lm} = F[1 + \varphi_1]$, and the projector is $P_{lm} = P_1$, see (6.120) and (6.121). When the latter expressions are substituted into the invariance equation (6.122), we obtain $\Delta_{lm} = \Delta_{lm}^{\mathrm{nloc}} = -v_T F\{2\partial_i u_k \overline{c_i c_k} + \partial_i T c_i [c^2 - (5/2)]\}$, while $\Delta_{lm}^{\mathrm{loc}} \equiv 0$ because the local Maxwellians are zero points of the Boltzmann collision integral. Consequently, the dynamic correction begins with the order ϵ, and the analog of the equation (6.128) reads:

$$\hat{L}\phi_{lm}^{(1)} = v_T\{2\partial_i u_k \overline{c_i c_k} + \partial_i T c_i [c^2 - (5/2)]\} \, ,$$

subject to a condition, $P_1[F\phi_{lm}^{(1)}] = 0$. The latter is the familiar Chapman-Enskog equation, resulting in the Navier-Stokes correction to the Euler equations [70]. Thus, *the nonlocal dynamic correction is related to the thirteen-moment Grad equations entirely in the same way as the Navier-Stokes are related to the Euler equations.*

4. Let us discuss briefly the further corrections. The first local correction (the functions Y_1 and Z_1 in (6.127)) is not the limiting point of our iterational procedure. When the latter is continued, the subsequent local corrections are found from integral equations, $\hat{L}Y_{n+1} = b_{n+1}Y_n$, and $\hat{L}Z_{n+1} = a_{n+1}Z_n$. Thus, we are led to the following two eigenvalue problems: $\hat{L}Y_\infty = b_\infty Y_\infty$, and $\hat{L}Z_\infty = a_\infty Z_\infty$, where a_∞ and b_∞ are the closest to zero eigenvalues among all the eigenvalue problems with the given tensorial structure [248].

5. Approach of this example [21] can be extended to derive dynamic corrections to other (non-moment) approximations of interest in the kinetic theory. The above analysis has demonstrated, in particular, the importance of the local correction, generically relevant to an approximation which is not a zero point of the collision integral. Very recently, this approach was successfully applied to improve the nonlinear Grad's thirteen-moment equations [253].

7 Quasi-Chemical Representation

7.1 Decomposition of Motions, Non-Uniqueness of Selection of Fast Motions, Self-Adjoint Linearization, Onsager Filter, and Quasi-Chemical Representation

In Chap. 5 we have used the second law of thermodynamics, the existence of the entropy, in order to equip the problem of constructing the slow invariant manifolds with a geometric structure. The requirement of the entropy growth (universally, for all reduced models) restricts significantly the form of the projectors (5.25).

In this chapter we introduce a different but equally important argument – the micro-reversibility (T-invariance), and its macroscopic consequences, the reciprocity relations. As first discussed by Onsager in 1931 [187], the implication of the micro-reversibility is the self-adjointness of the linear approximation of the system (3.1) in the equilibrium x^*: for any z and p,

$$\langle (D_x J)_{x^*} z | p \rangle_{x^*} \equiv \langle z | (D_x J)_{x^*} p \rangle_{x^*} \ . \tag{7.1}$$

The main idea in the present chapter is to use the *reciprocity relations* (7.1) *for the fast motions*. In order to appreciate this idea, we should mention that the decomposition of motions into fast and slow is not unique. Requirement (7.1) for any equilibrium point of fast motions means a selection (filtration) of the fast motions. We term this the *Onsager filter*. Equilibrium points of fast motions are all the points on manifolds of slow motions. Application of the Onsager filter amounts to a distinguished symmetrization of the linearized vector field $(D_x J)_x$ in the points x of the slow manifolds.

To begin with, let us remind the standard way of symmetrization: the linear operator A is decomposed into the symmetric and the skew-symmetric parts, $A = \frac{1}{2}(A + A^\dagger) + \frac{1}{2}(A - A^\dagger)$. Here $A^\dagger$ is adjoint to A with respect to a fixed scalar product (entropic scalar product in the present context). However, a replacement of an operator with its symmetric part can lead to catastrophic (from the physical standpoint) consequences such as, for example, loss of stability. In order to construct a sensible Onsager filter, we shall use the *quasi-chemical representation*.

The formalism of the quasi-chemical representation is one of the most developed means of modelling, it makes it possible to "assemble" complex

Alexander N. Gorban and Iliya V. Karlin: *Invariant Manifolds for Physical and Chemical Kinetics*, Lect. Notes Phys. **660**, 179–187 (2005)
www.springerlink.com

processes out of elementary processes. There exist various presentations of the quasi-chemical formalism. Our presentation here is a generalization of the approach suggested first by Feinberg [243] (see also [81, 242, 244]).

Symbol A_i ("quasi-substance") is put into correspondence to each variable x_i. The *elementary reaction* is defined according to the *stoichiometric equation*,

$$\sum_i \alpha_i A_i \rightleftharpoons \sum_i \beta_i A_i , \qquad (7.2)$$

where α_i (*the loss stoichiometric coefficients*) and β_i (*the gain stoichiometric coefficients*) are real numbers. Apart from the entropy, one has to specify a monotonic function of one variable, $\Psi(a)$, $\Psi'(a) > 0$. In particular, the function $\Psi(a) = \exp(\lambda a)$, $\lambda = \text{const}$, is encountered oft in applications.

Given the elementary reaction (7.2), one defines the rates of the direct and of the reverse reactions:

$$W^+ = w^* \Psi \left(\sum_i \alpha_i \mu_i \right) ,$$

$$W^- = w^* \Psi \left(\sum_i \beta_i \mu_i \right) , \qquad (7.3)$$

where $\mu_i = \frac{\partial S}{\partial x_i}$, $x^* = \text{const}$, $x^* > 0$. The rate of the elementary reaction is then defined as $W = W^+ - W^-$.

The equilibrium of the elementary reaction (7.2) is given by the following equation:

$$W^+ = W^- . \qquad (7.4)$$

Thanks to the strict monotonicity of the function Ψ, equilibrium of the elementary reaction is reached when the arguments of the functions coincide in equation (7.3), that is, whenever

$$\sum_i (\beta_i - \alpha_i)\mu_i = 0 . \qquad (7.5)$$

The vector with the components $\gamma_i = \beta_i - \alpha_i$ is termed the *stoichiometric vector of the elementary reaction*.

Let x^0 be a point of equilibrium of the reaction (7.2). The linear approximation of the reaction rate has a particularly simple form:

$$W(x^0 + \delta) = -w^* \Psi'(a(x^0)) \langle \gamma | \delta \rangle_{x^0} + o(\delta) , \qquad (7.6)$$

where $a(x^0) = \sum_i \alpha_i \mu_i(x^0) = \sum_i \beta_i \mu_i(x^0)$, and $\langle | \rangle_{x^0}$ is the entropic scalar product in the equilibrium. In other words,

$$(D_x W)_{x^0} = -w^* \Psi'(a(x^0)) \langle \gamma | . \qquad (7.7)$$

Let us write down the kinetic equation for the elementary reaction:

$$\frac{\mathrm{d}x}{\mathrm{d}t} = \gamma W(x) \, . \tag{7.8}$$

Linearization of this equation at the equilibrium x^0 has the following form:

$$\frac{\mathrm{d}\delta}{\mathrm{d}t} = -w^* \Psi'(a(x^0)) \gamma \langle \gamma | \delta \rangle_{x^0} \, . \tag{7.9}$$

That is, the matrix of the linear approximation has the form,

$$K = -k^* |\gamma\rangle\langle\gamma| \, , \tag{7.10}$$

where

$$k^* = w^* \Psi'(a(x^0)) > 0 \, ,$$

while the entropic scalar product is taken at the equilibrium point x^0.

If there are several elementary reactions, then the stoichiometric vectors γ^r and the reaction rates $W_r(x)$ are specified for each individual reaction, while the kinetic equation is obtained by summing the right hand sides of equation (7.8) for individual elementary reactions,

$$\frac{\mathrm{d}x}{\mathrm{d}t} = \sum_r \gamma^r W_r(x) \, . \tag{7.11}$$

Let us assume that under the reversion of the motions, the direct reaction transforms into the reverse reaction. Thus, the T-invariance of the equilibrium means that it is reached in the point of the *detailed balance*, where all the elementary reaction equilibrate simultaneously:

$$W_r^+(x^*) = W_r^-(x^*) \, . \tag{7.12}$$

This assumption is nontrivial if vectors γ^r are linearly dependent (for example, if the number of reactions is greater than the number of species minus the number of conservation laws).

One can call the equations of detailed balance (7.12) the "nonlinear Onsager relations". These equations give us the restrictions on the reaction rates not only near the equilibrium, in the linear approximation, but also far away from the equilibrium. The representation (7.3) is crucial for this continuation of the usual linear Onsager relations from the neighbourhood of the equilibrium point to the whole phase space. The problem of a rigorous foundation of nonlinear Onsager relations [188, 189] remains open, but a recent attempt made by Berdichevsky [190] seems to be promising.

In the detailed balance case, the linearization of equation (7.11) in the neighborhood of x^* has the following form ($x = x^* + \delta$):

$$\frac{\mathrm{d}\delta}{\mathrm{d}t} = -\sum_r k_r^* \gamma^r \langle \gamma^r | \delta \rangle_{x^*} \, , \tag{7.13}$$

where

$$k_r^* = w_r^* \Psi_r'(a_r^*) > 0 \ ,$$

$$a_r^* = \sum_i \alpha_i^r \mu_i(x^*) = \sum_i \beta_i^r \mu_i(x^*) \ .$$

The following matrix of the linear approximation is obviously self-adjoint and stable:

$$K = -\sum_r k_r^* |\gamma^r\rangle\langle\gamma^r| \ . \tag{7.14}$$

Note that matrix K is the sum of matrices of rank one.

Let us now extract the self-adjoint part of the form (7.14) in the *arbitrary* point x. Linearizing the reaction rate about x, we obtain:

$$W(x+\delta) = w^* \left(\Psi'(a(x))\langle\alpha|\delta\rangle_x - \Psi'(b(x))\langle\beta|\delta\rangle_x\right) + o(\delta) \ , \tag{7.15}$$

where

$$a(x) = \sum_i \alpha_i \mu_i(x) \ ,$$

$$b(x) = \sum_i \beta_i \mu_i(x) \ .$$

Let us introduce notation,

$$k^{\mathrm{SYM}}(x) = \frac{1}{2} w^* \left(\Psi'(a(x)) + \Psi'(b(x))\right) > 0 \ ,$$

$$k^{\mathrm{A}}(x) = \frac{1}{2} w^* \left(\Psi'(a(x)) - \Psi'(b(x))\right) \ .$$

In terms of this notation, equation (7.15) may be rewritten,

$$W(x+\delta) = -k^{\mathrm{SYM}}(x)\langle\gamma|\delta\rangle_x + k^{\mathrm{A}}(x)\langle\alpha+\beta|\delta\rangle_x + o(\delta) \ . \tag{7.16}$$

The second term vanishes in the equilibrium ($k^{\mathrm{A}}(x^*) = 0$, due to the detailed balance).

The *symmetric linearization* (Onsager filter) amounts to keeping only the first term in the linearized vector field (7.16) when studying the fast motion towards the (approximate) slow manifolds, instead of the full expression (7.15). Matrix $K(x)$ of the linear approximation becomes then similar to (7.14):

$$K(x) = -\sum_r k_r^{\mathrm{SYM}}(x)|\gamma^r\rangle\langle\gamma^r| \ , \tag{7.17}$$

where

$$k_r^{\mathrm{SYM}}(x) = \frac{1}{2} w_r^* \left(\Psi_r'(a(x)) + \Psi_r'(b(x))\right) > 0 \ ,$$

$$a_r(x) = \sum_i \alpha_i^r \mu_i(x) \ ,$$

$$b_r(x) = \sum_i \beta_i^r \mu_i(x) \ ,$$

while the entropic scalar product $\langle|\rangle_x$ is taken at the point x. For each label of the elementary reaction r, the function $k_r^{\mathrm{SYM}}(x)$ is positive. Thus, the stability of the symmetric matrix (7.17) is elicit.

Symmetric linearization (7.17) is distinguished also by the fact that it preserves the rank of the elementary processes contributing to the complex mechanism. Same as in the equilibrium case, the matrix $K(x)$ is the sum of rank one operators corresponding to each individual process. This is not so for the standard symmetrization.

Using the symmetric operator (7.17) in the above Newton method with incomplete linearization can be considered as a version of a heuristic strategy of "we act in such a way as if the manifolds $F(W)$ were already slow invariant manifolds". If this were the case, then, in particular, the fast motions towards the were described by the self-adjoint linear approximation.

We have described the quasi-chemical formalism for finite-dimensional systems. Infinite-dimensional generalizations are almost straightforrwad in many important cases, and are achieved by a mere replacement of summation by integration. The best known example is the Boltzmann collision integral: each velocity v corresponds to a quasi-substance A_v, and a collision is described by a stoichiometric equation:

$$A_v + A_w \rightleftharpoons A_{v'} + A_{w'} .$$

In the Example to this chapter we consider the Boltzmann collision integral from this standpoint in a more detail.

7.2 Example: Quasi-Chemical Representation and Self-Adjoint Linearization of the Boltzmann Collision Operator

A decomposition of motions near a thermodynamically nonequilibrium states results in a linear relaxation towards this state. In this Example, the linear operator of this relaxation is explicitly constructed in the case of the Boltzmann equation.

Let us remind that the entropy-related specification of the equilibrium state is due to the two points of view. From the first, thermodynamic viewpoint, equilibrium is a state in which the entropy is maximal. From the second, kinetic viewpoint, a quadratic form of the entropy increases in a course of linear regression towards this state. If the underlying microscopic dynamics is time-reversible, the kinetic viewpoint is realized due to the well-known symmetry properties of the linearized kinetic operator.

In a majority of near-equilibrium studies, a principle of a decomposition of motions into fast and slow occupies a distinct place. In some special cases, decomposition of motions is taken into account explicitly, by introducing a small parameter into dynamic equations. More frequently, however, it comes into play implicitly, for example, through an assumption of a fast decay of memory in the projection operator formalism [194]. Even in presence of long-living dynamic effects (mode coupling), an assumption about decomposition of motions is required as a final instance to obtain a closed set of equations for slow variables.

However, for closed systems, there remains a question: whether and to what extend the two aforementioned entropy-related points of view are applicable to non-equilibrium states? Further, if an answer is affirmative, then how to make explicitly the corresponding specification?

This Example is aimed at answering the questions just mentioned, and it is a straightforward continuation of results [11, 14]. Namely, in [11, 14], it was demonstrated that the principle of motions decomposition alone constitutes a necessary and sufficient condition for the thermodynamic specification of a non-equilibrium state. However, in a general situation, one deals with states f other than f_0. A question is, whether these two ideas can be applied to $f \neq f_0$ (at least approximately), and if so, then how to make the presentation explicit.

The positive answer to this question was partially given in the framework of the method of invariant manifolds [9, 11, 14]. Objects studied in [9, 11, 14] were manifolds in the space of distribution functions, and the goal was to construct iteratively a manifold that is tangent in all its points to a vector field of a dissipative system (an invariant manifold), beginning with some initial manifold with no such property. It was natural to employ methods of KAM-theory (Newton-type linear iterations to improve the initial manifold). However, additional idea of the decomposition of motions into fast and slow near the manifold was required to adapt KAM-theory to dissipative systems. The geometrical formulation of this idea [9, 11, 14] results in a definition of a plane of fast motion, Γ_f, associated with the state f, and orthogonal to the gradient of the entropy in f. The physical interpretation of Γ_f is that contains all those states from a neighborhood of f, which come into f in the course of fast relaxation (as if f were the final state of fast processes occuring in its neighborhood). Usually, Γ_f contains more states than can come into f in a fast relaxation because of the conservation of certain macroscopic quantities (e.g. density, momentum, and energy, as well as, possibly, higher moments of f which practically do not vary during the fast processes). The redundant states are eliminated by imposing additional restrictions which cut out "thinner" linear manifolds, planes of fast motions P_f, inside Γ_f. Extremal property of f on Γ_f is preserved also on P_f (cf. [9, 11, 14]).

Thus, the decomposition of motions near a manifold results in the thermo-dynamical viewpoint: the states f on the manifold are described as the unique points of the entropy maximum of corresponding planes of fast motions Γ_f. This formulation defines a slow dynamics on the manifolds in agreement with the H-theorem for the Boltzmann equation, or with its analogs for other systems (see [9, 11, 14] for details). As it was demonstrated in [9, 11, 14], the decomposition of motions in a neighborhood of f is a criterion (a necessary and sufficient condition) of the existence of the thermodynamic description of f.

The Newton iteration improves the states of a non-invariant manifold $(f + \delta f)$, while δf is thought on Γ_f. Equation for δf involves a linearization of the collision integral in the state f. Here, if $f \neq f_0$, where f_0 is the local equilibrium, we face a problem of how to perform the linearization of the collision integral in concordance with the H-theorem (corrections to the manifold of local equilibrium states were studied in detail in [11]).

Here we show that the aforementioned assumption about the decomposition of motions results in the kinetic description of states on manifolds of slow motions, and that Onsager's principle can be applied in a natural way to linearize the Boltzmann collision integral.

As it follows from the definition to definition of Γ_f, the state f is the unique point of minimum of the H-function on Γ_f. In the first non-vanishing approximation, we have the following expression for the H-function in the states on Γ_f:

$$H(f + \delta f) \approx H(f) + \frac{1}{2}\langle \delta f | \delta f \rangle_f$$

Here $\langle \cdot | \cdot \rangle_f$ denotes the scalar product generated by the second derivative of H in the state f: $\langle g_1 | g_2 \rangle_f = \int f^{-1} g_1 g_2 \, d\boldsymbol{v}$.

Decomposition of motions means that the quadratic form $\langle \delta f | \delta f \rangle_f$ decays monotonically in the course of the linear relaxation towards the state f. It is natural, therefore, to impose the requirement that this linear relaxation should obey Onsager's principle. Namely, the corresponding linear operator should be symmetric (formally self-adjoint) and non-positively definite with respect to the scalar product $\langle \cdot | \cdot \rangle_f$, and furthermore, the kernel of this operator should consist of linear combinations of conserved quantities (1, $\boldsymbol{v}$, and v^2). In other words, the decomposition of motions should portray the pattern of the linear relaxation in the vicinity of f similar to that in a small neighborhood of f_0. Following this idea, we shall now decompose the linearized collision integral L_f in two parts: L_f^{SYM} (satisfying Onsager's principle), and L_f^{A} (the non-thermodynamic part).

In the state f, each direct encounter, $(\boldsymbol{v}, \boldsymbol{v}_1) \to (\boldsymbol{v}', \boldsymbol{v}_1')$, together with the reverse encounter, $(\boldsymbol{v}', \boldsymbol{v}_1') \to (\boldsymbol{v}, \boldsymbol{v}_1)$, contribute a rate, $G^+(f) - L^-(f)$ ("gain$-$loss"), to the collision integral, where (see Chap. 2):

$$W(f) = W(\boldsymbol{v}', \boldsymbol{v}_1'; \boldsymbol{v}, \boldsymbol{v}_1) \exp\left\{ D_f H|_{f=f(\boldsymbol{v})} + D_f H|_{f=f(\boldsymbol{v}_1)} \right\};$$

$$W'(f) = W(\boldsymbol{v}', \boldsymbol{v}_1'; \boldsymbol{v}, \boldsymbol{v}_1) \exp\left\{ D_f H|_{f=f(\boldsymbol{v}')} + D_f H|_{f=f(\boldsymbol{v}_1')} \right\};$$

A deviation δf from the state f will change the rates of both the direct and the reverse processes. Resulting deviations of the rates are:

$$\delta W = W(f)\left\{ D_f^2 H|_{f=f(\boldsymbol{v})} \cdot \delta f(\boldsymbol{v}) + D_f^2 H|_{f=f(\boldsymbol{v}_1)} \cdot \delta f(\boldsymbol{v}_1) \right\};$$

$$\delta W' = W'(f)\left\{D_f^2 H|_{f=f(\boldsymbol{v}')} \cdot \delta f(\boldsymbol{v}') + D_f^2 H|_{f=f(\boldsymbol{v}_1')} \cdot \delta f(\boldsymbol{v}_1')\right\};$$

Symmetrization with respect to the direct and the reverse encounters will give a term proportional to a balanced rate, $W^{\mathrm{SYM}}(f) = \frac{1}{2}(W(f) + W'(f))$, in both of the expressions δW and $\delta W'$. Thus, we come to the decomposition of the linearized collision integral, $L_f = L_f^{\mathrm{SYM}} + L_f^{\mathrm{A}}$, where

$$L_f^{\mathrm{SYM}}\delta f = \int w\frac{f'f_1' + ff_1}{2}\left\{\frac{\delta f'}{f'} + \frac{\delta f_1'}{f_1'} - \frac{\delta f_1}{f_1} - \frac{\delta f}{f}\right\}\,d\boldsymbol{v}_1'\,d\boldsymbol{v}'\,d\boldsymbol{v}_1 \, ; \quad (7.18)$$

$$L_f^{\mathrm{A}}\delta f = \int w\frac{f'f_1' - ff_1}{2}\left\{\frac{\delta f'}{f'} + \frac{\delta f_1'}{f_1'} + \frac{\delta f_1}{f_1} + \frac{\delta f}{f}\right\}\,d\boldsymbol{v}_1'\,d\boldsymbol{v}'\,d\boldsymbol{v}_1 \, ; \quad (7.19)$$

$f = f(\boldsymbol{v}), f_1 = f(\boldsymbol{v}_1), f' = f(\boldsymbol{v}'), f_1' = f(\boldsymbol{v}_1'), \delta f = \delta f(\boldsymbol{v}), \delta f_1 = \delta f(\boldsymbol{v}_1), \delta f' = \delta f(\boldsymbol{v}'), \delta f_1' = \delta f(\boldsymbol{v}_1')$.

Operator L_f^{SYM} (7.18) satisfies all the aforementioned requirements pertinent to Onsager's principle, namely:

(i) $\langle g_1|L_f^{\mathrm{SYM}}|g_2\rangle_f = \langle g_2|L_f^{\mathrm{SYM}}|g_1\rangle_f$ (symmetry);
(ii) $\langle g|L_f^{\mathrm{SYM}}|g\rangle_f \le 0$ (local entropy production inequality);
(iii) $f, \boldsymbol{v}f, v^2 f \in \ker L_f^{\mathrm{SYM}}$ (conservation laws).

For an unspecified f, the non-thermodynamic operator L_f^{A} (7.19) has none of these properties. If $f = f_0$, then the part (7.19) vanishes, while operator $L_{f_0}^{\mathrm{SYM}}$ becomes the usual linearized collision integral due to the balance $W(f_0) = W'(f_0)$.

The non-negative definite form $\langle \delta f|\delta f\rangle_f$ decays monotonically due to the equation of linear relaxation, $\partial_t \delta f = L_f^{\mathrm{SYM}}\delta f$, and the unique point of minimum, $\delta f = 0$, of $\langle \delta f|\delta f\rangle_f$ corresponds to the equilibrium point of the vector field $L_f^{\mathrm{SYM}}\delta f$.

Operator L_f^{SYM} describes the state f as the equilibrium state of the linear relaxation. Note that the method of extracting the symmetric part (7.18) is strongly based on the representation of the direct and the reverse processes, and it is not a simple procedure like, e.g., $\frac{1}{2}(L_f + L_f^+)$. The latter expression cannot be used as a basis for Onsager's principle since it would violate conditions (ii) and (iii).

Thus, if motions do decompose into a fast motion towards the manifold and a slow motion along the manifold, then states on this manifold can be described from both the thermodynamic and the kinetic points of view. Our consideration results in the explicit construction of the operator L_f^{SYM} (7.18) responsible for the fast relaxation towards the state f. It can be used, in particular, for obtaining corrections to such approximations as the Grad moment approximations and the Tamm–Mott-Smith approximation, in the framework of the method of invariant manifold [9, 14, 21]. The non-thermodynamic part (7.19) is always present in L_f, when $f \ne f_0$, but if trajectories of an equation $\partial_t \delta f = L_f \delta f$ are close to the trajectories of the equation $\partial_t \delta f = L_f^{\mathrm{SYM}}\delta f$, then L_f^{SYM} is a good approximation to L_f. Statements about closeness of trajectories depend on specific features of f, and typically they can be claimed when a small parameter is present. On the other hand, the explicit thermodynamic and kinetic presentation of states on a manifold of slow motions (the extraction of L_f^{SYM} as above and construction of planes Γ_f [9, 11, 14]) is based just on the assumption about the decomposition of motions, and can be used avoiding a consideration of a small parameter.

8 Hydrodynamics From Grad's Equations: What Can We Learn From Exact Solutions?

A detailed treatment of the classical Chapman-Enskog derivation of hydrodynamics is given in the framework of Grad's moment equations. Grad's systems are considered as the minimal kinetic models where the Chapman-Enskog method can be studied exactly, thereby providing the basis to compare various approximations in extending the hydrodynamic description beyond the Navier-Stokes approximation. Various techniques, such as the method of partial summation, Padé approximants, and invariance principle are compared both in linear and nonlinear situations.

8.1 The "Ultra-Violet Catastrophe" of the Chapman-Enskog Expansion

Most of the interesting expansions in non-equilibrium statistical physics are divergent. This paraphrase of the well known folklore "Dorfman's theorem" conveys the intrinsic problem of many-body systems: A number of systematic (at the first glance) methods has led to

- An excellent but already known on the phenomenological grounds first approximation;
- Already the next correction, not known phenomenologically and hence of interest, does not exist because of divergence.

There are many examples of this situations: Cluster expansion of the exact collision integral for dense gases leads to divergent approximations of transport coefficients, non-convergent long tails of correlation functions in the Green–Kubo formulae etc.

The derivation of the hydrodynamic equations from a microscopic description is the classical problem of physical kinetics. As is well known, the famous Chapman–Enskog method [70] provides an opportunity to compute a solution from the Boltzmann kinetic equation as a formal series in powers of the Knudsen number ϵ. The parameter ϵ reflects the ratio between the mean free path of a particle, and the scale of variations of the hydrodynamic fields (density, mean flux, and temperature). If the Chapman–Enskog expansion is truncated at a certain order, we obtain subsequently: the Euler hydrodynamics (ϵ^0), the Navier–Stokes hydrodynamics (ϵ^1), the Burnett hydrodynamics (ϵ^2), the

Alexander N. Gorban and Iliya V. Karlin: *Invariant Manifolds for Physical and Chemical Kinetics*, Lect. Notes Phys. **660**, 189–246 (2005)
www.springerlink.com

super-Burnett hydrodynamics (ϵ^3), etc. The post-Navier–Stokes terms extend the hydrodynamic description beyond the strictly hydrodynamic limit $\epsilon \ll 1$.

However, as it has been first demonstrated by Bobylev [72], even in the simplest regime (one-dimensional linear deviations around the global equilibrium), the Burnett hydrodynamic equations violate the basic physics behind the Boltzmann equation. Namely, sufficiently short acoustic waves are amplified with time instead of decaying. This contradicts the H-theorem, since all near-equilibrium perturbations must decay. The situation does not improve in the next, super-Burnett approximation.

This "ultra-violet catastrophe" which occurs in the lower-order truncations of the Chapman–Enskog expansion creates therefore very serious difficulties in the problem of an extension of the hydrodynamic description into a highly non-equilibrium domain (see [112] for a discussion of other difficulties of the post-Navier–Stokes terms of the Chapman–Enskog expansion). The Euler and the Navier–Stokes approximations remain basic in the hydrodynamic description, while the problem of their extension is one of the central open problems of kinetic theory. The study of approximate solutions based on the Chapman–Enskog method still continues [74].

All this begs for a question: *What is wrong with the Chapman–Enskog method?* At first glance, the failure of the Burnett and of the super-Burnett hydrodynamics may be accounted in favor of a frequently used argument about the asymptotic character of the Chapman–Enskog expansion. However, it is worthwhile to notice here that divergences in the low-order terms of formal expansions are not too surprising. In many occasions, in particular, in quantum field theory [198] and in statistical physics [199], the situation is often improved if one takes into account the very remote terms of the corresponding expansions. Thus, a more constructive viewpoint on the Chapman–Enskog expansion could be to proceed along these lines, and to try to *sum up* the Chapman–Enskog series, at least formally and approximately.

An attempt of this kind of working with the Chapman–Enskog expansion is undertaken in this chapter. The formalities are known to be rather awkward for the Boltzmann equation, and untill now, exact summations of the Chapman–Enskog expansion are known in a very limited number of cases [202]. In this chapter, we shall concentrate on the Chapman–Enskog method as applied to the well known Grad moment equations [201].

The use of the Grad equations for our purpose brings, of course, considerable technical simplifications as compared to the case of the Boltzmann equation but it does not make the problem trivial. Indeed, the Chapman–Enskog method amounts to a nonlinear recurrence procedure even when applied to the simplest, linearized Grad equations. Moreover, as we shall see soon, the Chapman–Enskog expansion for moment systems inherits Bobylev's instability in the low-order approximations. Still, the advantage of our approach is that many explicit results can be obtained and analyzed. In order to summarize, in this chapter we consider Grad's moment equations as finitely-coupled kinetic models where the problem of reduced description is meaningful, rather than as models of extended hydrodynamics. The latter viewpoint is well known as a microscopic background of the extended irreversible thermodynamics [236, 252].

The outline of this chapter is as follows: after an introduction of the Chapman–Enskog procedure for the linearized Grad equations (Subsect. 8.2), we shall start the discussion with two examples (the linearized one- and three-dimensional 10 moment Grad equations) where the Chapman–Enskog series is summed up exactly in closed form (Sects. 8.3.1 and 8.3.2). These results makes it possible to discuss the features of the Chapman–Enskog solution in the short-wave domain in the framework of the model, and will serve the purpose of testing various approximate methods thereafter. We shall see, in particular, that the "smallness" of the Knudsen number ϵ used to develop the Chapman–Enskog method has no direct meaning in the exact result. Also, it will become clear that finite-order truncations, even provided they are stable, give less opportunities to approximate the solution in a whole, and especially in the short-wave domain.

The exact solutions are, of course, the lucky exceptions, and even for the Grad moment equations the complexity of the Chapman–Enskog method increases rapidly with an increase of the number of the moments taken into account. Further (Sect. 8.4.1) we shall review a technique of summing the Chapman–Enskog expansion *partially*. This technique is heuristic (as are the methods of partial summing in general), but it still removes the Bobylev instability, as well as it qualitatively reproduces the features of the exact solutions in the short-wave limit.

The approach of working in the sections mentioned so far falls into the paradigm of the Taylor-like expansions into powers of the Knudsen number. This viewpoint on the problem of the derivation of the hydrodynamics will be *altered* beginning with Sect. 8.4.2. There we demonstrate that a condition of a *dynamic invariance* which can be realized directly and with no restrictions of the Knudsen number brings us to the same result as the exact summation of the Chapman–Enskog expansion. The Chapman–Enskog method thereafter can be regarded as *one* possibility to solve the resulting invariance equations. Further, we demonstrate that iterative methods provide a reasonable alternative to the Taylor expansion in this problem. Namely, we show that the Newton method has certain advantages over the Chapman–Enskog method (Sect. 8.4.3). We also establish a relationship between the method of partial summation and the Newton method.

The material of further sections serves for an illustrative introduction how the pair "invariance equation + Newton method" can be applied to problems of kinetic theory. The remaining sections of this chapter are devoted to further examples of this approach on the level of the Grad equations. In Sects. 8.4.4 and 8.4.5 we derive and discuss the invariance equations for the linearized thirteen-moment Grad equations. Section 8.4.6 is devoted to kinetic equations of the Grad type, arising in problems of phonon transport in massive solids at low temperatures. In particular, we demonstrate that the onset of the second sound regime of phonon propagation corresponds to a branching point of the exact sum of the relevant Chapman–Enskog expansion.

In Sect. 8.4.7 we apply the invariance principle to nonlinear Grad equations. We sum up exactly a *subseries* of the Chapman–Enskog expansion, namely, the dominant contribution in the limit of high average velocities. This type of contribution is therefore important for an extension of the hydrodynamic description into the domain of strong shock waves. We present a relevant analysis of the corresponding invariance equation, and, in particular, discuss the nature of singular points of this equation. A brief discussion concludes this chapter. Some of the results presented below were published earlier in [17, 40, 41, 43–45, 205, 237], and summarized in [42].

8.2 The Chapman–Enskog Method for Linearized Grad's Equations

In this section, for the sake of completeness, we introduce linearized Grad's equations and the Chapman–Enskog method for them in the form that will used in the rest of this chapter. Since the Chapman–Enskog method is extensively discussed in a number of books, especially, in the classical monograph [70], our presentation will be brief.

The notation will follow that of the papers [43, 72]. We denote ρ_0, T_0 and $\boldsymbol{u} = 0$ the fixed equilibrium values of density, temperature and averaged velocity (in the appropriate Galilean reference frame), while $\delta\rho$, δT and $\delta\boldsymbol{u}$ are small deviations of the hydrodynamic quantities from their equilibrium values. Grad's moment equations [201] which will appear below, contain the temperature-dependent viscosity coefficient, $\mu(T)$. It is convenient to write $\mu(T) = \eta(T)T$. The functional form of $\eta(T)$ is dictated by the choice of the model for particle interaction. In particular, we have $\eta = \text{const}$ for Maxwell's molecules, and $\eta \sim \sqrt{T}$ for hard spheres.

We use the system of units in which Boltzmann's constant k_B and the particle mass m are equal to one. Let us introduce the following system of dimensionless variables:

$$\boldsymbol{u} = \frac{\delta\boldsymbol{u}}{\sqrt{T_0}} , \quad \rho = \frac{\delta\rho}{\rho_0} , \quad T = \frac{\delta T}{T_0} , \tag{8.1}$$

$$\boldsymbol{x} = \frac{\rho_0}{\eta(T_0)\sqrt{T_0}}\boldsymbol{x}' , \quad t = \frac{\rho_0}{\eta(T_0)}t' ,$$

where $\boldsymbol{x}'$ are spatial coordinates, and t' is time. Three-dimensional thirteen moment Grad's equations, linearized near the equilibrium, take the following form when written in terms of the dimensionless variables (8.1):

$$\partial_t \rho = -\nabla \cdot \boldsymbol{u} \,, \tag{8.2}$$

$$\partial_t \boldsymbol{u} = -\nabla \rho - \nabla T - \nabla \cdot \boldsymbol{\sigma} \,,$$

$$\partial_t T = -\frac{2}{3}(\nabla \cdot \boldsymbol{u} + \nabla \cdot \boldsymbol{q}) \,,$$

$$\partial_t \boldsymbol{\sigma} = -\overline{\nabla \boldsymbol{u}} - \frac{2}{5}\overline{\nabla \boldsymbol{q}} - \boldsymbol{\sigma} \,, \tag{8.3}$$

$$\partial_t \boldsymbol{q} = -\frac{5}{2}\nabla T - \nabla \cdot \boldsymbol{\sigma} - \frac{2}{3}\boldsymbol{q} \,.$$

In these equations, $\boldsymbol{\sigma}(\boldsymbol{x},t)$ and $\boldsymbol{q}(\boldsymbol{x},t)$ are dimensionless quantities corresponding to the stress tensor and to the heat flux, respectively. Further, the gradient ∇ stands for the vector of spatial derivatives $\partial/\partial\boldsymbol{x}$. The dot denotes the standard scalar product, while the overline stands for a symmetric traceless dyad. In particular,

$$\overline{\nabla \boldsymbol{u}} = \nabla \boldsymbol{u} + (\nabla \boldsymbol{u})^T - \frac{2}{3}I\nabla \cdot \boldsymbol{u} \,,$$

where I is unit matrix.

Grad's equations (8.2) and (8.3) is the simplest model of a coupling of the hydrodynamic variables, $\rho(\boldsymbol{x},t)$, $T(\boldsymbol{x},t)$ and $\boldsymbol{u}(\boldsymbol{x},t)$, to the non-hydrodynamic variables $\boldsymbol{\sigma}(\boldsymbol{x},t)$ and $\boldsymbol{q}(\boldsymbol{x},t)$. The problem of reduced description is to close the first three equations (8.2), and to get an autonomous system for the hydrodynamic variables alone. In other words, the non-hydrodynamic variables $\boldsymbol{\sigma}(\boldsymbol{x},t)$ and $\boldsymbol{q}(\boldsymbol{x},t)$ should be expressed in terms of $\rho(\boldsymbol{x},t)$, $T(\boldsymbol{x},t)$ and $\boldsymbol{u}(\boldsymbol{x},t)$. The Chapman–Enskog method, as applied for this purpose to Grad's system (8.2) and (8.3), involves the following steps:

First, we introduce a formal parameter ϵ, and write instead of equations (8.3):

$$\partial_t \boldsymbol{\sigma} = -\overline{\nabla \boldsymbol{u}} - \frac{2}{5}\overline{\nabla \boldsymbol{q}} - \frac{1}{\epsilon}\boldsymbol{\sigma} \,, \tag{8.4}$$

$$\partial_t \boldsymbol{q} = -\frac{5}{2}\nabla T - \nabla \cdot \boldsymbol{\sigma} - \frac{2}{3\epsilon}\boldsymbol{q} \,.$$

Second, the Chapman–Enskog solution is found as a formal expansions of the stress tensor and of the heat flux vector:

$$\boldsymbol{\sigma} = \sum_{n=0}^{\infty} \epsilon^{n+1}\boldsymbol{\sigma}^{(n)} \,; \tag{8.5}$$

$$\boldsymbol{q} = \sum_{n=0}^{\infty} \epsilon^{n+1}\boldsymbol{q}^{(n)} \,.$$

The zero-order coefficients, $\boldsymbol{\sigma}^{(0)}$ and $\boldsymbol{q}^{(0)}$, are:

$$\boldsymbol{\sigma}^{(0)} = -\overline{\nabla \boldsymbol{u}} \,, \quad \boldsymbol{q}^{(0)} = -\frac{15}{4}\nabla T \,. \tag{8.6}$$

Coefficients of order $n \geq 1$ are found from the recurrence procedure:

$$\sigma^{(n)} = -\left\{ \sum_{m=0}^{n-1} \partial_t^{(m)} \sigma^{(n-1-m)} + \frac{2}{5} \overline{\nabla q^{(n-1)}} \right\}, \qquad (8.7)$$

$$q^{(n)} = -\frac{3}{2} \left\{ \sum_{m=0}^{n-1} \partial_t^{(m)} q^{(n-1-m)} + \nabla \cdot \sigma^{(n-1)} \right\},$$

where $\partial_t^{(m)}$ are recurrently defined *Chapman–Enskog operators*. They act on functions $\rho(\boldsymbol{x}, t)$, $T(\boldsymbol{x}, t)$ and $\boldsymbol{u}(\boldsymbol{x}, t)$, and on their spatial derivatives, according to the following rule:

$$\partial_t^{(m)} D\rho = \begin{cases} -D\nabla \cdot \boldsymbol{u} & m = 0 \\ 0 & m \geq 1 \end{cases} ; \qquad (8.8)$$

$$\partial_t^{(m)} DT = \begin{cases} -\frac{2}{3} D\nabla \cdot \boldsymbol{u} & m = 0 \\ -\frac{2}{3} D\nabla \cdot q^{(m-1)} & m \geq 1 \end{cases} ;$$

$$\partial_t^{(m)} D\boldsymbol{u} = \begin{cases} -D\nabla(\rho + T) & m = 0 \\ -D\nabla \cdot \sigma^{(m-1)} & m \geq 1 \end{cases} .$$

Here D is an arbitrary differential operator with constant coefficients.

Given the initial condition (8.6), the Chapman–Enskog equations (8.7) and (8.8) are recurrently solvable. Finally, by terminating the computation at the order $N \geq 0$, we obtain the Nth order approximations to the expansions (8.5), σ_N and q_N:

$$\sigma_N = \sum_{n=0}^{N} \epsilon^{n+1} \sigma^{(n)} , \quad q_N = \sum_{n=0}^{N} \epsilon^{n+1} q^{(n)} . \qquad (8.9)$$

Substituting these expressions instead of the functions σ and q in (8.2), we close the latter to give the hydrodynamic equations of the order N. In particular, $N = 0$ results in the Navier–Stokes approximation, $N = 1$ and $N = 2$ give the Burnett and the super-Burnett approximations, respectively, and so on.

Though the "microscopic" features of Grad's moment equations are, of course, much simpler in comparison to the Boltzmann equation, the Chapman–Enskog procedure just described is not trivial. Our purpose is to study explicitly the features of the gradient expansions like (8.5) in the highly non-equilibrium domain, and, in particular, to find out to what extend the finite-order truncations (8.9) approximate the solution, and what kind of alternative strategies to find approximations are possible. In the following, when referring to Grad's equations, we use the notation $mDnM$, where m is the spatial dimension of the corresponding fields, and n is the number of these fields. For example, the above system is the $3D13M$ Grad's system.

8.3 Exact Summation of the Chapman–Enskog Expansion

8.3.1 The $1D10M$ Grad Equations

In this section, we start the discussion with the exact summation of the Chapman–Enskog series for the simplest Grad's system, the one-dimensional linearized ten-moment equations. Throughout the section we use the hydrodynamic variables $p(x,t) = \rho(x,t) + T(x,t)$ and $u(x,t)$, representing the dimensionless deviations of the pressure and of the average velocity from their equilibrium values (see (8.1)). The starting point is the linearized Grad's equations for p, u, and σ, where σ is the dimensionless xx-component of the stress tensor:

$$\partial_t p = -\frac{5}{3}\partial_x u \;, \tag{8.10}$$

$$\partial_t u = -\partial_x p - \partial_x \sigma \;,$$

$$\partial_t \sigma = -\frac{4}{3}\partial_x u - \frac{1}{\epsilon}\sigma \;.$$

The system of equations for three functions is derived from the ten-moment Grad's system (see (8.38) below). Equations (8.10) provides the simplest model of a coupling of the hydrodynamic variables, u and p, to the single non-hydrodynamic variable σ, and corresponds to a heat non-conductive case.

Our goal here is to reduce the description, and to get a closed set of equations with respect to variables p and u only. That is, we have to express the function σ in the terms of spatial derivatives of p and u. The Chapman–Enskog method, as applied to (8.10) results in the following series representation:

$$\sigma = \sum_{n=0}^{\infty} \epsilon^{n+1}\sigma^{(n)} \;. \tag{8.11}$$

The coefficients $\sigma^{(n)}$ are obtained from the following recurrence procedure [43]:

$$\sigma^{(n)} = -\sum_{m=0}^{n-1} \partial_t^{(m)}\sigma^{(n-1-m)} \;, \tag{8.12}$$

where the Chapman–Enskog operators $\partial_t^{(m)}$ act on p, u, and their spatial derivatives as follows:

$$\partial_t^{(m)}\partial_x^l u = \begin{cases} -\partial_x^{l+1}p, & m = 0 \\ -\partial_x^{l+1}\sigma^{(m-1)}, & m \geq 1 \end{cases} \;, \tag{8.13}$$

$$\partial_t^{(m)}\partial_x^l p = \begin{cases} -\frac{5}{3}\partial_x^{l+1}u, & m = 0 \\ 0, & m \geq 1 \end{cases} \;.$$

Here $l \geq 0$ is an arbitrary integer, and $\partial_x^0 = 1$. Finally,

$$\sigma^{(0)} = -\frac{4}{3}\partial_x u, \tag{8.14}$$

which leads to the Navier–Stokes approximation of the stress tensor: $\sigma_{NS} = \epsilon\sigma^{(0)}$.

Because of the somewhat involved structure of the recurrence procedure (8.12) and (8.13), the Chapman–Enskog method is a nonlinear operation even in the simplest model (8.10). Moreover, the Bobylev instability is again present.

Indeed, computing the coefficients $\sigma^{(1)}$ and $\sigma^{(2)}$ on the basis of (8.12), we obtain:

$$\sigma_B = \epsilon\sigma^{(0)} + \epsilon^2\sigma^{(1)} = -\frac{4}{3}\left(\epsilon\partial_x u + \epsilon^2\partial_x^2 p\right), \tag{8.15}$$

and

$$\sigma_{SB} = \epsilon\sigma^{(0)} + \epsilon^2\sigma^{(1)} + \epsilon^3\sigma^{(2)} = -\frac{4}{3}\left(\epsilon\partial_x u + \epsilon^2\partial_x^2 p + \frac{1}{3}\epsilon^3\partial_x^3 u\right), \tag{8.16}$$

for the Burnett and the super-Burnett approximations, respectively. Now we can substitute each of the approximations, σ_{NS}, σ_B, and σ_{SB} for σ in the second equation of the set (8.10). The equations thus obtained, together with the equation for density ρ, form the closed systems of the hydrodynamic equations of the Navier–Stokes, Burnett, and super-Burnett levels. To see the properties of the resulting equations, we compute the dispersion relation for the hydrodynamic modes. Using a new space-time scale, $x' = \epsilon^{-1}x$, and $t' = \epsilon^{-1}t$, and representing $u = u_k\varphi(x',t')$, and $p = p_k\varphi(x',t')$, where $\varphi(x',t') = \exp(\omega t' + ikx')$, and k is a real-valued wave vector, we obtain the following dispersion relations $\omega(k)$ from the condition of a non-trivial solvability of the corresponding linear system with respect to u_k and p_k:

$$\omega_\pm = -\frac{2}{3}k^2 \pm \frac{1}{3}i|k|\sqrt{4k^2 - 15}, \tag{8.17}$$

for the Navier–Stokes approximation,

$$\omega_\pm = -\frac{2}{3}k^2 \pm \frac{1}{3}i|k|\sqrt{8k^2 + 15}, \tag{8.18}$$

for the Burnett approximation (8.15), and

$$\omega_\pm = \frac{2}{9}k^2(k^2 - 3) \pm \frac{1}{9}i|k|\sqrt{4k^6 - 24k^4 - 72k^2 - 135}, \tag{8.19}$$

for the super-Burnett approximation (8.16).

These examples demonstrate that the real part $\text{Re}(\omega_\pm(k)) \le 0$ for the Navier–Stokes (8.17) and for the Burnett (8.18) approximations, for all wave vectors. Thus, these approximations describe attenuating acoustic waves. However, for the super-Burnett approximation, the function $\text{Re}(\omega_\pm(k))$ (8.19) becomes positive as soon as $|k| > \sqrt{3}$. That is, the equilibrium point is stable within the Navier–Stokes and the Burnett approximation, and it becomes unstable within the super-Burnett approximation for sufficiently short waves. Similar to the case of the Bobylev instability of the Burnett hydrodynamics for the Boltzmann equation, the latter result contradicts the dissipative properties of the Grad system (8.10): the spectrum of the full $1D10M$ system (8.10) is stable for arbitrary k.

Our goal now is to sum up the series (8.11) in closed form. Firstly, we should make some preparations.

As demonstrated in [43] (see also below), the functions $\sigma^{(n)}$ in (8.11) and (8.12) have the following explicit structure to arbitrary order $n \geq 0$:

$$\sigma^{(2n)} = a_n \partial_x^{2n+1} u \,, \tag{8.20}$$
$$\sigma^{(2n+1)} = b_n \partial_x^{2(n+1)} p \,,$$

where the coefficients a_n and b_n are determined through the recurrence procedure (8.12), and (8.13). The Chapman–Enskog procedure (8.12) and (8.13) can be represented in terms of the real-valued coefficients a_n and b_n (8.20).

Knowing the structure (8.20) of the coefficients of the Chapman–Enskog expansion (8.11), we can write down its formal sum. It is convenient to use the Fourier variables introduced above which amounts essentially to the change $\epsilon \partial_x \to ik$. Substituting expression (8.20) into the Chapman–Enskog series (8.11), we obtain the following formal expression for the Fourier image of the sum:

$$\sigma_k = ikA(k^2)u_k - k^2 B(k^2)p_k \,, \tag{8.21}$$

where the functions $A(k^2)$ and $B(k^2)$ are formal power series with the coefficients (8.20):

$$A(k^2) = \sum_{n=0}^{\infty} a_n(-k^2)^n \,, \tag{8.22}$$

$$B(k^2) = \sum_{n=0}^{\infty} b_n(-k^2)^n \,.$$

Thus, the question of the summation of the Chapman–Enskog series (8.11) amounts to finding the two functions, $A(k^2)$ and $B(k^2)$ (8.22). Knowing them, the dispersion relation for the hydrodynamic modes can be derived:

$$\omega_{\pm} = \frac{k^2 A}{2} \pm \frac{|k|}{2}\sqrt{k^2 A^2 - \frac{20}{3}(1 - k^2 B)} \,. \tag{8.23}$$

We shall concentrate now on the problem of deriving $A(k^2)$ and $B(k^2)$ (8.22) in closed form. For this purpose, we shall first express the Chapman–Enskog procedure (8.12) and (8.13) in terms of the coefficients a_n and b_n (8.20). At the same time, our derivation will constitute proof for the structure (8.20).

It is convenient to start with the Fourier representation of (8.12) and (8.13). Writing $u = u_k \exp(ikx)$, $p = p_k \exp(ikx)$, and $\sigma = \sigma_k \exp(ikx)$, we obtain:

$$\partial_t^{(m)} u_k = \begin{cases} -ikp_k, & m = 0 \\ -ik\sigma_k^{(m-1)}, & m \geq 1 \end{cases}, \tag{8.24}$$

$$\partial_t^{(m)} p_k = \begin{cases} -\frac{5}{3}iku_k, & m = 0 \\ 0, & m \geq 1 \end{cases},$$

while

$$\sigma_k^{(n)} = -\sum_{m=0}^{n-1} \partial_t^{(m)} \sigma_k^{(n-1-m)}, \tag{8.25}$$

and

$$\sigma_k^{(2n)} = a_n(-k^2)^n iku_k, \tag{8.26}$$
$$\sigma_k^{(2n+1)} = b_n(-k^2)^n(-k^2)p_k.$$

The Navier–Stokes and the Burnett approximations give $a_0 = -\frac{4}{3}$, and $b_0 = -\frac{4}{3}$. Thus, the structure (8.26) is proved for $n = 0$.

The further derivation relies on induction. Let us assume that the ansatz (8.26) is proven up to the order n. Computing the coefficient $\sigma_k^{(2(n+1))}$ from (8.25), we have:

$$\sigma_k^{(2(n+1))} = -\partial_t^{(0)} \sigma_k^{(2n+1)} - \sum_{m=0}^{n} \partial_t^{(2m+1)} \sigma_k^{(2(n-m))} - \sum_{m=1}^{n} \partial_t^{(2m)} \sigma_k^{(2(n-m)+1)}. \tag{8.27}$$

Due to the assumption of the induction, we can adopt the form of the coefficients $\sigma_k^{(j)}$ (8.26) in all the terms on the right hand side of (8.27). On the basis of (8.26) and (8.24), we conclude that each term in the last sum of (8.27) is equal to zero. Further, the term $\partial_t^{(0)} \sigma_k^{(2n+1)}$ gives the linear contribution:

$$\partial_t^{(0)} \sigma_k^{(2n+1)} = \partial_t^{(0)} b_n(-k^2)^n(-k^2)p_k = -\frac{5}{3}b_n(-k^2)^{n+1}iku_k,$$

while the terms in the remaining sum contribute nonlinearly:

$$\partial_t^{(2m+1)} \sigma_k^{(2(n-m))} = a_{n-m}(-k^2)^{n-m}ik\partial_t^{(2m+1)}u_k = -a_{n-m}a_m(-k^2)^{n+1}iku_k.$$

Substituting the last two expressions into (8.27), we see that it has just the same structure as the coefficient $\sigma_k^{(2(n+1))}$ in (8.26). Thus, we obtain the first recurrence equation:

$$a_{n+1} = \frac{5}{3}b_n + \sum_{m=0}^{n} a_{n-m}a_m.$$

Computing the coefficient $\sigma_k^{(2(n+1)+1)}$ by the same pattern, we come to the second recurrence equation, and the Chapman–Enskog procedure (8.12) and (8.13) can be reformulated in terms of the coefficients a_n and b_n (8.20):

$$a_{n+1} = \frac{5}{3}b_n + \sum_{m=0}^{n} a_{n-m}a_m , \qquad (8.28)$$

$$b_{n+1} = a_{n+1} + \sum_{m=0}^{n} a_{n-m}b_m .$$

The initial condition for this set of equations is dictated by the Navier–Stokes and the Burnett terms:

$$a_0 = -\frac{4}{3} , \quad b_0 = -\frac{4}{3} \qquad (8.29)$$

Our goal now is to compute the functions A and B (8.22) on the basis of the recurrence equations (8.28). At this point, it is worthwhile to notice that the usual way of dealing with the recurrence system (8.28) would be either to truncate it at a certain n, or to calculate all the coefficients explicitly, and substitute the result into the power series (8.22). Both approaches are not successful here. Indeed, retaining the coefficients a_0, b_0, and a_1 gives the super-Burnett approximation (8.16) which has the Bobylev short-wave instability, and there is no guarantee that the same failure will not occur in the higher-order truncation. On the other hand, a term-by-term computation of the whole set of coefficients a_n and b_n is a nontrivial task due to the nonlinearity in (8.28).

Fortunately, another route is possible. Multiplying both the equations in (8.28) with $(-k^2)^{n+1}$, and performing a formal summation in n from zero to infinity, we arrive at the following expressions:

$$A - a_0 = -k^2 \left\{ \frac{5}{3}B + \sum_{n=0}^{\infty}\sum_{m=0}^{n} a_{n-m}(-k^2)^{n-m}a_m(-k^2)^m \right\} , \qquad (8.30)$$

$$B - b_0 = A - a_0 - k^2 \sum_{n=0}^{\infty}\sum_{m=0}^{n} a_{n-m}(-k^2)^{n-m}b_m(-k^2)^m .$$

Now we notice that

$$\lim_{N\to\infty} \sum_{n=0}^{N}\sum_{m=0}^{n} a_{n-m}(-k^2)^{n-m}a_m(-k^2)^m = A^2 , \qquad (8.31)$$

$$\lim_{N\to\infty} \sum_{n=0}^{N}\sum_{m=0}^{n} a_{n-m}(-k^2)^{n-m}b_m(-k^2)^m = AB .$$

Taking into account the initial condition (8.29), equation (8.30) yields a pair of coupled quadratic equations for the functions A and B:

$$A = -\frac{4}{3} - k^2 \left(\frac{5}{3} B + A^2 \right) , \qquad (8.32)$$

$$B = A(1 - k^2 B) .$$

The result (8.32) concludes essentially the question of the computation of functions A and B (8.22). Still, further simplifications are possible. In particular, it is convenient to reduce the consideration to a single function. Solving system (8.32) for B, and introducing a new function, $X(k^2) = k^2 B(k^2)$, we obtain an equivalent cubic equation:

$$-\frac{5}{3}(X - 1)^2 \left(X + \frac{4}{5} \right) = \frac{X}{k^2} . \qquad (8.33)$$

Since A and B (8.22) are real-valued, we are only interested in the real-valued roots of (8.33).

An elementary analysis of this equation brings the following result: *the real-valued root $X(k^2)$ of (8.33) is unique and negative for all finite values k^2. Moreover, the function $X(k^2)$ is a monotonic function of k^2* (Fig. 8.1). The limiting values are:

$$\lim_{|k| \to 0} X(k^2) = 0 , \qquad \lim_{|k| \to \infty} X(k^2) = -0.8 . \qquad (8.34)$$

Under the conditions just mentioned, the function under the root in (8.23) is negative for all values of the wave vector k, including the limits, and we come to the following dispersion law:

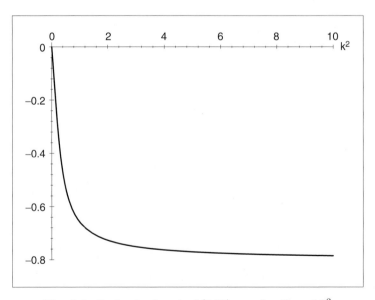

Fig. 8.1. Real-valued root of (8.33) as a function of k^2

$$\omega_{\pm} = \frac{X}{2(1-X)} \pm i\frac{|k|}{2}\sqrt{\frac{5X^2 - 16X + 20}{3}} , \tag{8.35}$$

where $X = X(k^2)$ is the real-valued root of equation (8.33). Since $X(k^2)$ is negative for all $|k| > 0$, the attenuation rate, $\mathrm{Re}(\omega_{\pm})$, is negative for all $|k| > 0$, and the exact acoustic spectrum of the Chapman–Enskog procedure *is stable for arbitrary wave lengths*. In the short-wave limit, from (8.35) we obtain:

$$\lim_{|k|\to\infty} \omega_{\pm} = -\frac{2}{9} \pm i|k|\sqrt{3} . \tag{8.36}$$

The characteristic equation of the original Grad equations (8.10) reads:

$$3\omega^3 + 3\omega^2 + 9k^2\omega + 5k^2 = 0 . \tag{8.37}$$

The two complex-conjugate roots of this equation correspond to the hydrodynamic modes, while for the non-hydrodynamic real mode, $\omega_{nh}(k)$, $\omega_{nh}(0) = -1$, and $\omega_{nh} \to -0.5$ as $|k| \to \infty$. Recall that the non-hydrodynamic modes of the Grad equations are characterized by the common property that for them $\omega(0) \neq 0$. These modes are irrelevant to the Chapman–Enskog method. As the final comment here, (8.36) demonstrates that the exact attenuation rate, $\mathrm{Re}(\omega_{\pm})$, tends to a finite value, $-\frac{2}{9} \approx -0.22$ as $|k| \to \infty$. This asymptotic behavior is in a complete agreement with the data for the hydrodynamic branch of the spectrum (8.37) of the original Grad equations (8.10). The attenuation rates (real parts of the dispersion relations $\omega_{\pm}$ for the Burnett (8.18), the super-Burnett (8.19), the exact Chapman–Enskog solution (8.35), are compared to each other in Fig. 8.2. In this figure, we also represent the attenuation rates of the hydrodynamic and non-hydrodynamic mode of the Grad equations (8.37). The results of this section lead to the following conclusion:

(i) The proposed approach provides a way to deal with the problem of *summation* of the Chapman–Enskog expansion. The exact dispersion relation (8.35) of the Chapman–Enskog procedure is demonstrated to be stable for all wave lengths, while the Bobylev instability is present on the level of the super-Burnett approximation. Moreover, it can be demonstrated that the function X (the real root of (8.33)) is a real-valued analytic function of k. Thus, the treatment of the formal expansions performed above is justified.

(ii) The exact result of the Chapman–Enskog procedure has a clear non-polynomial character. Indeed, this follows directly from (8.34): the function $X(k^2)$ cannot be a polynomial because it maps the axis k into a segment $[0, -0.8]$. As a conjecture here, the resulting exact hydrodynamics is *essentially* nonlocal in space. For this reason, even if the hydrodynamic equations of a certain level of the approximation *is* stable, it cannot reproduce the non-polynomial behavior for sufficiently short waves.

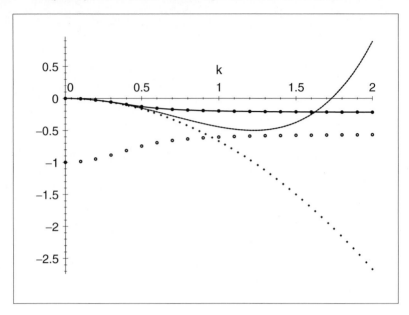

Fig. 8.2. Attenuation rates for the $1D10M$ Grad system. *Solid*: Exact summation of the Chapman–Enskog Expansion. *Dots*: The Navier–Stokes approximation. *Dash*: The super–Burnett approximation. *Circles*: Hydrodynamic and non-hydrodynamic modes of the $1D10M$ Grad system

(iii) The result of this section demonstrates that, at least in some cases, the sum of the Chapman–Enskog series amounts to a quite regular function, and the "smallness" of the Knudsen number ϵ used to develop the Chapman–Enskog procedure (8.12) *is no longer necessary.*

8.3.2 The $3D10M$ Grad Equations

In this section we generalize our considerations of the Chapman–Enskog method to the three-dimensional linearized 10-moment Grad equations [201]. The Chapman–Enskog series for the stress tensor, which is again due to a nonlinear procedure, will be summed up in closed form. The method used follows essentially the one discussed above, though the computations are slightly more extensive. The reason to consider this example is that we would like to know what happens to the diffusive hydrodynamic mode in the short-wave domain.

Throughout this section, we use the variables (8.1), and p and $\boldsymbol{u}$ are dimensionless deviations of pressure and of mean flux from their equilibrium values, respectively. The point of departure is the set of the three-dimensional linearized Grad equations for the p, $\boldsymbol{u}$, and $\boldsymbol{\sigma}$, where $\boldsymbol{\sigma}$ is a dimensionless stress tensor:

$$\partial_t p = -\frac{5}{3}\nabla \cdot \boldsymbol{u} \ , \tag{8.38}$$

$$\partial_t \boldsymbol{u} = -\nabla p - \nabla \cdot \boldsymbol{\sigma} \ ,$$

$$\partial_t \boldsymbol{\sigma} = -\overline{\nabla \boldsymbol{u}} - \frac{1}{\epsilon}\boldsymbol{\sigma} \ .$$

Equation (8.38) provides a simple model of a coupling of the hydro-dynamic variables, $\boldsymbol{u}$ and p, to the non-hydrodynamic variable $\boldsymbol{\sigma}$. These equations are suitable for an application of the Chapman–Enskog procedure. Therefore, our goal here is not to investigate the properties of (8.38) as they are, but to reduce the description, and to get a closed set of equations with respect to the variables p and $\boldsymbol{u}$ only. That is, we have to express $\boldsymbol{\sigma}$ in terms of spatial derivatives of p and of $\boldsymbol{u}$. The Chapman–Enskog method, as applied to (8.38) results in the following:

$$\boldsymbol{\sigma} = \sum_{n=0}^{\infty} \epsilon^{n+1} \boldsymbol{\sigma}^{(n)} \ . \tag{8.39}$$

The coefficients $\boldsymbol{\sigma}^{(n)}$ are due to the following recurrence procedure:

$$\boldsymbol{\sigma}^{(n)} = -\sum_{m=0}^{n-1} \partial_t^{(m)} \boldsymbol{\sigma}^{(n-1-m)} \ , \tag{8.40}$$

where the Chapman–Enskog operators $\partial_t^{(m)}$ act on the functions p and $\boldsymbol{u}$, and on their derivatives, as follows:

$$\partial_t^{(m)} D\boldsymbol{u} = \begin{cases} -D\nabla p, & m = 0 \\ -D\nabla \cdot \boldsymbol{\sigma}^{(m-1)}, & m \geq 1 \end{cases} \ , \tag{8.41}$$

$$\partial_t^{(m)} Dp = \begin{cases} -\frac{5}{3}D\nabla \cdot \boldsymbol{u}, & m = 0 \\ 0, & m \geq 1 \end{cases} \ .$$

Here D is an arbitrary differential operator $D = \prod_{i=1}^{3} \partial_i^{l_i}$, while l_i is an arbitrary integer, and $\partial_i^0 = 1$. Finally, $\boldsymbol{\sigma}^{(0)} = -\overline{\nabla \boldsymbol{u}}$, which leads to the Navier–Stokes approximation.

Our goal is to sum up the series (8.39) in closed form.

The terms $\boldsymbol{\sigma}^{(n)}$ in equations (8.39), (8.40), and (8.41), have the following explicit structure for arbitrary order $n \geq 0$ (a generalization of (8.20) to the three-dimensional case):

$$\boldsymbol{\sigma}^{(2n)} = a_n \Delta^n \overline{\nabla \boldsymbol{u}} + b_n \Delta^{n-1} G\nabla \cdot \boldsymbol{u} \ , \tag{8.42}$$

$$\boldsymbol{\sigma}^{(2n+1)} = c_n \Delta^n Gp \ ,$$

where $\Delta = \nabla \cdot \nabla$ is the Laplace operator, and the operator G has the form:

$$G = \nabla\nabla - \frac{1}{3}I\Delta = \frac{1}{2}\overline{\nabla\nabla} \ . \tag{8.43}$$

The real-valued and yet unknown coefficients a_n, b_n, and c_n in (8.42) are due to the recurrence procedure (8.40), and (8.41). Knowing the structure of the coefficients of the Chapman–Enskog series (8.42), we can reformulate the Chapman–Enskog solution in terms of a self-consistent recurrence procedure for the coefficients a_n, b_n, and c_n. Let us consider this derivation in more detail.

The point of departure is the Fourier representation of the recurrence equations (8.40), (8.41), and (8.42). Writing

$$\boldsymbol{u} = \boldsymbol{u}_k \exp(i\boldsymbol{k} \cdot \boldsymbol{x}) \, ,$$
$$p = p_k \exp(i\boldsymbol{k} \cdot \boldsymbol{x}) \, ,$$
$$\boldsymbol{\sigma}^{(n)} = \boldsymbol{\sigma}_k^{(n)} \exp(i\boldsymbol{k} \cdot \boldsymbol{x}) \, ,$$

and introducing the unit vector $\boldsymbol{e}_k$ directed along $\boldsymbol{k}$ ($\boldsymbol{k} = k\boldsymbol{e}_k$), equations (8.40), (8.41), and (8.42) can be rewritten as:

$$\boldsymbol{\sigma}_k^{(n)} = - \sum_{m=0}^{n-1} \partial_t^{(m)} \boldsymbol{\sigma}_k^{(n-1-m)} \, , \tag{8.44}$$

$$\partial_t^{(m)} D_k \boldsymbol{u}_k = \begin{cases} -D_k i\boldsymbol{k} p_k, & m = 0 \\ -D_k i\boldsymbol{k} \cdot \boldsymbol{\sigma}_k^{(m-1)}, & m \geq 1 \end{cases} \, , \tag{8.45}$$

$$\partial_t^{(m)} D_k p_k = \begin{cases} -\frac{5}{3} D_k i\boldsymbol{k} \cdot \boldsymbol{u}_k, & m = 0 \\ 0, & m \geq 1 \end{cases} \, .$$

where D_k is an arbitrary tensor $D_k = \prod_{s=1}^{3} (ik_s)^{l_s}$, and

$$\boldsymbol{\sigma}_k^{(2n)} = (-k^2)^n (a_n \overline{i k \boldsymbol{u}} + b_n i\boldsymbol{g}_k (\boldsymbol{k} \cdot \boldsymbol{u})) \, , \tag{8.46}$$
$$\boldsymbol{\sigma}_k^{(2n+1)} = c_n (-k^2)^{n+1} \boldsymbol{g}_k p_k \, ,$$

where

$$\boldsymbol{g}_k = \left(\boldsymbol{e}_k \boldsymbol{e}_k - \frac{1}{3} I \right) = \frac{1}{2} \overline{\boldsymbol{e}_k \boldsymbol{e}_k} \, . \tag{8.47}$$

From the form of the Navier–Stokes approximation, $\boldsymbol{\sigma}_k^{(0)}$, it follows that $a_0 = -1$ and $b_0 = 0$, while a direct computation of the Burnett approximation leads to:

$$\boldsymbol{\sigma}_k^{(1)} = \frac{1}{2} k^2 \boldsymbol{g}_k p_k \, . \tag{8.48}$$

Thus, we have $c_0 = -\frac{1}{2}$ which proves the ansatz (8.42) for $n = 0$ in both the even and the odd orders.

The rest of the proof relies on induction. Let the structure (8.46) be proven up to the order n. The computation of the next, $n+1$ order coefficient $\boldsymbol{\sigma}_k^{(2(n+1))}$, involves only terms of lower order. From (8.44) we obtain:

$$\sigma_k^{(2(n+1))} = -\partial_t^{(0)}\sigma_k^{(2n+1)} - \sum_{m=1}^{2n+1}\partial_t^{(m)}\sigma_k^{(2n+1-m)} \ . \tag{8.49}$$

The first term in the right hand side depends linearly on the coefficients c_n:

$$-\partial_t^{(0)}\sigma_k^{(2n+1)} = -c_n(-k^2)^{n+1}\boldsymbol{g}_k\partial_t^{(0)}p_k \tag{8.50}$$

$$= \frac{5}{3}c_n(-k^2)^{n+1}i\boldsymbol{g}_k\boldsymbol{k}\cdot\boldsymbol{u}_k \ .$$

The remaining terms on the right hand side of (8.49) contribute nonlinearly. Splitting the even and the odd orders of the Chapman–Enskog operators $\partial_t^{(m)}$, we rewrite the sum in (8.49):

$$-\sum_{m=1}^{2n+1}\partial_t^{(m)}\sigma_k^{(2n+1-m)} = -\sum_{l=1}^{n}\partial_t^{(2l)}\sigma_k^{(2(n-l)+1)} - \sum_{l=0}^{n}\partial_t^{(2l+1)}\sigma_k^{(2(n-l))} \ . \tag{8.51}$$

Due to (8.46) and (8.45), each term in the first sum is equal to zero, and we are left only with the second sum:

$$\partial_t^{(2l+1)}\sigma_k^{(2(n-l))} = (-k^2)^{n-l}(a_{n-l}i\overline{\boldsymbol{k}\partial_t^{(2l+1)}\boldsymbol{u}_k}+b_{n-l}i\boldsymbol{g}_k\boldsymbol{k}\cdot\partial_t^{(2l+1)}\boldsymbol{u}_k) \ , \tag{8.52}$$

while

$$\partial_t^{(2l+1)}\boldsymbol{u}_k = -(-k^2)^{l+1}\left(a_l\boldsymbol{u}_k + \frac{1}{3}(a_l+2b_l)\boldsymbol{e}_k(\boldsymbol{e}_k\cdot\boldsymbol{u}_k)\right) \ . \tag{8.53}$$

In the last expression, use of the following identities was made:

$$\boldsymbol{k}\cdot\overline{\boldsymbol{k}\boldsymbol{u}_k} = k^2\left(\boldsymbol{u}_k + \frac{1}{3}\boldsymbol{e}_k(\boldsymbol{e}_k\cdot\boldsymbol{u}_k)\right) \ , \tag{8.54}$$

$$\boldsymbol{k}\cdot\boldsymbol{g}_k = \frac{2}{3}\boldsymbol{k} \ .$$

Substituting (8.53) into the right hand side of (8.52), and thereafter substituting the result into the right hand side of (8.51), we obtain the following in the right hand side of (8.49):

$$\sigma_k^{(2(n+1))} = (-k^2)^{n+1}\left(\sum_{m=0}^{n}a_{n-m}a_m\right)i\overline{\boldsymbol{k}\boldsymbol{u}_k} + (-k^2)^{n+1}\left(\frac{5}{3}c_n\right. \tag{8.55}$$

$$+ \sum_{m=0}^{n}\left\{\frac{1}{3}(2a_{n-m}+b_{n-m})(a_m+2b_m)+a_{n-m}b_m\right\}\right)i\boldsymbol{g}_k(\boldsymbol{k}\cdot\boldsymbol{u}_k) \ .$$

The functional structure of the right hand side of this expression is the same as that of the first equation in the set (8.46), and thus we obtain the first recurrence equation:

$$a_{n+1}\overline{ku_k} + b_{n+1}g_k(k \cdot u_k) = \left(\sum_{m=0}^{n} a_{n-m}a_m\right)\overline{ku_k} \tag{8.56}$$

$$+ \left(\frac{5}{3}c_n + \sum_{m=0}^{n}\left\{\frac{1}{3}(2a_{n-m} + b_{n-m})(a_m + 2b_m) + a_{n-m}b_m\right\}\right)g_k(k \cdot u_k) \; .$$

Considering in the same way the coefficient $\sigma_k^{(2(n+1)+1)}$, we come to the second recurrence equation,

$$c_{n+1} = 2a_{n+1} + b_{n+1} + \frac{2}{3}\sum_{m=0}^{n}(2a_{n-m} + b_{n-m})c_m \; . \tag{8.57}$$

Thus, the complete set of the recurrence equations is given by (8.56) and (8.57). Equation (8.56) is equivalent to a pair of scalar equations. Indeed, introducing new variables,

$$r_n = \frac{2}{3}c_n \; , \tag{8.58}$$

$$q_n = \frac{2}{3}(2a_n + b_n) \; ,$$

and using the identity,

$$\overline{ku_k} = (\overline{ku_k} - 2g_k(k \cdot u_k)) + 2g_k(k \cdot u_k) \; ,$$

and also noticing that

$$g_k : (\overline{ku_k} - 2g_k(k \cdot u_k)) = 0 \; ,$$

where : denotes the double contraction of tensors, we arrive in (8.56) and (8.57) at the following three scalar recurrence relations in terms the coefficients r_n, q_n, and a_n:

$$r_{n+1} = q_{n+1} + \sum_{m=0}^{n} q_{n-m}r_m \tag{8.59}$$

$$q_{n+1} = \frac{5}{3}r_n + \sum_{m=0}^{n} q_{n-m}q_m$$

$$a_{n+1} = \sum_{m=0}^{n} a_{n-m}a_m$$

The initial condition for this system is provided by the explicit form of the Navier–Stokes and the Burnett approximations, and reads:

$$r_0 = -4/3, \quad q_0 = -4/3, \quad a_0 = -1 \; . \tag{8.60}$$

The recurrence relations (8.59) are completely equivalent to the original Chapman–Enskog procedure (8.40) and (8.41). In the one-dimensional case, the recurrence system (8.59) reduces to the first two equations for r_n and q_n. In this case, the system of recurrence equations is identical (up to the notations) to the recurrence system (8.28), considered in the preceding section. For what follows, it is important to notice that the recurrence equation for the coefficients a_n is decoupled from the equations for the coefficients r_n and q_n.

Now we shall express the Chapman–Enskog series of the stress tensor (8.39) in terms of r_n, q_n, and a_n. Using again the Fourier transform, and substituting (8.42) into the right hand side of (8.39), we derive:

$$\sigma_k = A(k^2)(\overline{\boldsymbol{k}\boldsymbol{u}_k} - 2\boldsymbol{g}_k(\boldsymbol{k}\cdot\boldsymbol{u}_k)) + \frac{3}{2}Q(k^2)\boldsymbol{g}_k(\boldsymbol{k}\cdot\boldsymbol{u}_k) - \frac{3}{2}k^2 R(k^2)\boldsymbol{g}_k p_k \ , \quad (8.61)$$

From here on, we use a new spatial scale which amounts to $\boldsymbol{k}' = \epsilon\boldsymbol{k}$, and drop the prime. The functions $A(k^2)$, $Q(k^2)$, and $R(k^2)$ in (8.61) are defined by the power series with the coefficients due to (8.59):

$$A(k^2) = \sum_{n=0}^{\infty} a_n(-k^2)^n \ , \quad (8.62)$$

$$Q(k^2) = \sum_{n=0}^{\infty} q_n(-k^2)^n \ ,$$

$$R(k^2) = \sum_{n=0}^{\infty} r_n(-k^2)^n \ .$$

Thus, the question of summation of the Chapman–Enskog series (8.39) amounts to finding the three functions, $A = A(k^2)$, $Q = Q(k^2)$, and $R = R(k^2)$ (8.62) in the three- and two-dimensional cases, or to the two functions, $Q(k^2)$, and $R(k^2)$ in the one-dimensional case.

Now we shall focus on computing the functions (8.62) from the recurrence equations (8.59). At this point, it is worthwhile to notice again that a truncation at a certain n is not successful. Indeed, already in the one-dimensional case, retaining the coefficients q_0, r_0, and q_1 leads to the super-Burnett approximation (8.16) which has the short-wave instability for $k^2 > 3$, as it was demonstrated in the preceding section, and there is no guarantee that the same will not occur in a higher-order truncation.

Fortunately, the approach introduced in the preceding section works again. Multiplying each of the equations in (8.62) with $(-k^2)^{n+1}$, and performing a summation in n from zero to infinity, we derive:

$$Q - q_0 = -k^2 \left\{ \frac{5}{3}R + \sum_{n=0}^{\infty}\sum_{m=0}^{n} q_{n-m}(-k^2)^{n-m}q_m(-k^2)^m \right\} \ , \quad (8.63)$$

$$R - r_0 = Q - q_0 - k^2 \sum_{n=0}^{\infty} \sum_{m=0}^{n} q_{n-m}(-k^2)^{n-m} r_m(-k^2)^m ,$$

$$A - a_0 = -k^2 \sum_{n=0}^{\infty} \sum_{m=0}^{n} a_{n-m}(-k^2)^{n-m} a_m(-k^2)^m .$$

Now we notice that

$$\lim_{N \to \infty} \sum_{n=0}^{N} \sum_{m=0}^{n} a_{n-m}(-k^2)^{n-m} a_m(-k^2)^m = A^2 , \qquad (8.64)$$

$$\lim_{N \to \infty} \sum_{n=0}^{N} \sum_{m=0}^{n} q_{n-m}(-k^2)^{n-m} r_m(-k^2)^m = QR ,$$

$$\lim_{N \to \infty} \sum_{n=0}^{N} \sum_{m=0}^{n} q_{n-m}(-k^2)^{n-m} q_m(-k^2)^m = Q^2 .$$

Taking into account the initial conditions (8.60), and also using (8.64), we derive from (8.63) the following three quadratic equations for the functions A, R, and Q:

$$Q = -\frac{4}{3} - k^2 \left(\frac{5}{3} R + Q^2 \right) , \qquad (8.65)$$

$$R = Q(1 - k^2 R) ,$$

$$A = -(1 + k^2 A^2) .$$

The result (8.65) concludes essentially the question of computation of functions (8.62) in closed form. Still, further simplifications are possible. In particular, it is convenient to use a single unknown function, $X(k^2) = k^2 R(k^2)$, in the first two equations in the system (8.65). We again obtain an equivalent cubic equation:

$$-\frac{5}{3}(X - 1)^2 \left(X + \frac{4}{5} \right) = \frac{X}{k^2} , \qquad (8.66)$$

which coincides with (8.66) of the previous section. We shall also rewrite the third equation of (8.65) using a function $Y(k^2) = k^2 A(k^2)$:

$$Y(1 + Y) = -k^2 . \qquad (8.67)$$

The functions in (8.62) can now be straightforwardly expressed in terms of the relevant solutions to (8.66) and (8.67). Since all functions in (8.62) are real-valued functions, we are interested only in the real-valued roots of the algebraic equations (8.66) and (8.67).

The relevant analysis of the cubic equation (8.66) was already performed above: the real-valued root $X(k^2)$ is unique and negative for all finite values of k^2. Limiting values of the function $X(k^2)$ at $k \to 0$ and at $k \to \infty$ are given by (8.34):

$$\lim_{k\to 0} X(k^2) = 0 , \quad \lim_{k\to\infty} X(k^2) = -\frac{4}{5} .$$

The quadratic equation (8.67) has no real-valued solutions for $k^2 > \frac{1}{4}$, and it has two real-valued solution for each k^2, where $k^2 < \frac{1}{4}$. We denote $k_c = \frac{1}{2}$ the corresponding critical value of the wave vector. For $k = 0$, one of these roots is equal to zero, while the other is equal to one. The asymptotics $Y \to 0$, as $k \to 0$, answers the question which of these two roots of (8.67) is relevant to the Chapman–Enskog solution, and we derive:

$$Y = \begin{cases} -\frac{1}{2}\left(1 - \sqrt{1-4k^2}\right) & k < k_c \\ \text{none} & k > k_c \end{cases} \tag{8.68}$$

The function Y (8.68) is negative for $k \leq k_c$.

From now on, X and Y will denote the relevant roots of (8.66) and (8.67) just discussed. The Fourier image of the expression $\nabla \cdot \boldsymbol{\sigma}$ follows from (8.61):

$$i\boldsymbol{k} \cdot \boldsymbol{\sigma}_k = Y((\boldsymbol{e}_k \cdot \boldsymbol{u}_k)\boldsymbol{e}_k - \boldsymbol{u}_k) - \frac{X}{1-X}(\boldsymbol{e}_k \cdot \boldsymbol{u}_k)\boldsymbol{e}_k - iX\boldsymbol{k}p_k . \tag{8.69}$$

The latter expression contributes to the right-hand side of the second of equations in the Grad system (8.38) (more specifically, it contributes to the corresponding Fourier transform of this equation). Knowing (8.69), we can calculate the dispersion $\omega(\boldsymbol{k})$ of the plane waves $\sim\exp\{\omega t + i\boldsymbol{k} \cdot \boldsymbol{x}\}$ which now follows from the exact solution of the Chapman–Enskog procedure. The calculation of the dispersion relation amounts to an evaluation of the determinant of a $(d+1) \times (d+1)$ matrix, and is quite standard (see, e.g. [240]). We therefore provide only the final result. The exact dispersion relation of the hydrodynamic modes reads:

$$(\omega - Y)^{d-1}\left(\omega^2 - \frac{X}{1-X}\omega + \frac{5}{3}k^2(1-X)\right) = 0 . \tag{8.70}$$

Here, d is the spatial dimension.

From the dispersion relation (8.70), we easily derive the following classification of the hydrodynamic modes:

(i) For $d = 1$, the spectrum of the hydrodynamic modes is purely acoustic with the dispersion ω_a which is given by (8.35):

$$\omega_a = \frac{X}{2(1-X)} \pm i\frac{k}{2}\sqrt{\frac{5X^2 - 16X + 20}{3}} , \tag{8.71}$$

where $X = X(k^2)$ is the real-valued root of (8.66). Since X is a negative function for all $k > 0$, the attenuation rate of the acoustic modes, $\mathrm{Re}(\omega_a)$, is negative for all $k > 0$, and the exact acoustic spectrum of the Chapman–Enskog procedure is free of the Bobylev instability for arbitrary wave lengths.

(ii) For $d > 1$, the dispersion of the acoustic modes is given by (8.71). As follows from the Chapman–Enskog procedure, the diffusion-like (real-valued) mode has the dispersion ω_d:

$$\omega_d = \begin{cases} -\frac{1}{2}\left(1 - \sqrt{1 - 4k^2}\right) & k < k_c \\ \text{none} & k > k_c \end{cases} \qquad (8.72)$$

The diffusion mode is $(d - 1)$ times degenerated, the corresponding attenuation rate is negative for $k < k_c$, and this mode *cannot be extended beyond the critical value* $k_c = \frac{1}{2}$ *within the Chapman–Enskog method*.

The reason why this rather remarkable peculiarity of the Chapman–Enskog procedure occurs can be found upon closer investigation of the spectrum of the underlying Grad moment system (8.38).

Indeed, in the original system (8.38), besides the hydrodynamic modes, there exist several non-hydrodynamic modes which are irrelevant to the Chapman–Enskog solution. All these non-hydrodynamic modes are characterized by the property that the corresponding dispersion relations $\omega(\boldsymbol{k})$ do not go to zero, as $k \to 0$. At the point $k_c = \frac{1}{2}$, the diffusion branch (8.72) intersects with one of the non-hydrodynamic branches of (8.38). For larger values of the wave vector k, these two branches produce a pair of complex conjugate solutions with the real part equal to $-\frac{1}{2}$. Thus, though the spectrum of the original equations (8.38) indeed continues past k_c, *the Chapman–Enskog method does not recognize this extension as part of the hydrodynamic branch.* It is also interesting to notice that if we would accept all the roots of (8.67), including the complex-values for $k > k_c$, and not only the real-valued root as suggested by the asymptotics of the Chapman–Enskog solution (see the explanations preceding (8.68)), then we would come in (8.70) to the structure of the dispersion relation just mentioned.

The attenuation rates (the functions $\text{Re}(\omega_a)$ and $\text{Re}(\omega_d)$) are plotted in Fig. 8.3, together with the relevant dependencies for the approximations of the Chapman–Enskog method. The non-hydrodynamic branch of (8.38) which causes the breakdown of the Chapman–Enskog solution is also represented in Fig. 8.3. It is rather remarkable that while the exact hydrodynamic description becomes inapplicable for the diffusion branch at $k \geq k_c$, the usual Navier–Stokes description still provides a good approximation to the acoustic mode around this point.

The analysis of this section leads to the following additional remarks to the conclusions made at the end of Sect. 8.3.1:

(i) The developed approach provides an understanding of the features of Chapman–Enskog solutions and the problem of extending the hydrodynamic modes into a highly non-equilibrium domain on the exact basis and in the full spatial dimension. The exact acoustic mode in the framework of the Chapman–Enskog procedure is demonstrated to be stable for all wave lengths, while the diffusion-like mode can be regarded for the hydrodynamic mode only in a bounded domain $k < k_c$. It is remarkable that the result of the

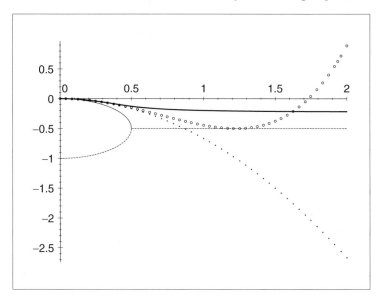

Fig. 8.3. Attenuation rates for the $3D10M$ Grad system as functions of $|k|$. *Bold*: The acoustic branch, exact summation. *Dots*: The acoustic branch, Navier–Stokes approximation. *Circles*: The acoustic branch, super-Burnett approximation. *Solid*: The diffusion branch, exact summation. *Dash*: The critical mode of the $3D10M$ Grad system

Chapman–Enskog procedure has a clear non-polynomial character. As a conjecture here, the resulting hydrodynamics is *essentially* nonlocal in space. It is also clear that *any* polynomial approximation to the Chapman–Enskog series will fail to reproduce the peculiarity of the diffusion mode demonstrated in the framework of the exact solution.

(ii) Concerning the extension of hydrodynamics into a highly non-equilibrium domain on the basis of the Boltzmann equations, the question remains open in the sense of an exact summation as above. In this respect, results for simplified models can serve either for testing approximate procedures or at least as guide. In particular, the mechanism of the singularity of the diffusion-like mode through a coupling to the non-hydrodynamic mode might be a rather general mechanism of limiting of the hydrodynamic description, and not just a feature of the Grad systems.

(iii) The result of this section demonstrates that the sum of the Chapman–Enskog series amounts to either a quite regular function (as is the function X), or to a function with a singularity at finite k_c. In both cases, however, the "smallness" of the Knudsen number ϵ used to develop the Chapman–Enskog procedure plays no role in the result of the Chapman–Enskog procedure.

8.4 The Dynamic Invariance Principle

8.4.1 Partial Summation of the Chapman–Enskog Expansion

The examples considered above demonstrate that it makes sense to speak about the sum of the Chapman–Enskog expansion, at least when the Chapman–Enskog method is applied to the (linearized) Grad equations. However, even in this case, the possibility to perform the summation exactly seems to be the lucky exception rather than the rule. Indeed, computations become more bulky with the increase of the number of the moments included in the Grad equations. Therefore, we arrive at the question: how can we approximate the recurrence equations of the Chapman–Enskog method to account for all the orders in the Knudsen number? Any such method amounts to some "partial" summation of the Chapman–Enskog expansion, and this type of working with formal series is widely spread in various fields of physics.

In this section we shall discuss a method of approximating the Chapman–Enskog expansion as a whole. As we now have the exact expressions for the Chapman–Enskog solution for the linearized 10 moment Grad equations, it is natural to start with this example for comparison purposes.

Let us come back to the originating one-dimensional Grad equations (8.10), and to the corresponding formulas of the Chapman–Enskog method (8.12) and (8.13). Instead of using the exact equations (8.12) in each order n, we introduce the following approximate equations:

Let $N \geq 1$ be some fixed integer. Then, instead of equations (8.12), we write:

$$\sigma^{(n)} = - \sum_{m=0}^{n-1} \partial_t^{(m)} \sigma^{(n-1-m)}, \quad n \leq N, \tag{8.73}$$

$$\sigma^{(n)} = - \sum_{m=0}^{N-1} \partial_t^{(m)} \sigma^{(n-1-m)}, \quad n > N. \tag{8.74}$$

This approximation amounts to the following: up to order N, the Chapman–Enskog procedure (8.12) is taken exactly (equation (8.73)), while in the computation of higher orders (equation (8.74)) we restrict the set of the Chapman–Enskog operators (8.13) only up to order N. Thus, the Chapman–Enskog coefficients $\sigma^{(n)}$ of order higher than N are taken into account only "partially". As N tends to infinity, the recurrence procedure (8.73) and (8.74) tends formally to the exact Chapman–Enskog procedure (8.12). We shall further refer to (8.73) and (8.74) as the *regularization* of the $N - th$ order. In particular, taking $N = 1$, we come to the regularization of the Burnett approximation, taking $N = 2$ we come to the regularization of the super-Burnett approximation, etc.

It can be demonstrated that the approximate procedure just described does not alter the structure of the functions $\sigma^{(2n)}$ and $\sigma^{(2n+1)}$ (8.20), while the recurrence equations for the coefficients a_n and b_n (8.20) will differ from the exact result of the full Chapman–Enskog procedure (8.28). The advantage of the regularization procedure (8.73) and (8.74) over the exact Chapman–Enskog recurrence procedure (8.12) is that the resulting equations for the coefficients a_n and b_n are always linear, as they result from (8.73) and (8.74). This feature enables one to sum up the corresponding series exactly, even if the originating nonlinear procedure leads to a too difficult analysis. The number N can be called the "depth" of the approximation: the large N is, the more low-order terms of the Chapman–Enskog expansion are taken into account exactly due to (8.73).

For the first example, let us take $N = 1$ in (8.73) and (8.74). The regularization of the Burnett approximation then reads:

$$\sigma^{(n)} = -\partial_t^{(0)} \sigma^{(n-1)} , \qquad (8.75)$$

where $n \geq 1$, and $\sigma^{(0)} = -(4/3)\partial_x u$. Turning to the Fourier variables, we derive:

$$\sigma_k^{(2n)} = a_n(-k^2)^n i k u_k , \qquad (8.76)$$
$$\sigma_k^{(2n+1)} = b_n(-k^2)^{n+1} p_k ,$$

where the coefficients a_n and b_n are due to the following recurrence procedure:

$$a_{n+1} = \frac{5}{3} b_n, \quad b_n = a_n, \quad a_0 = -\frac{4}{3} , \qquad (8.77)$$

whereupon

$$a_n = b_n = \left(\frac{5}{3}\right)^n a_0 . \qquad (8.78)$$

Thus, denoting as σ_{1k}^R the Fourier transform of the regularized Burnett approximation, we obtain:

$$\sigma_{1k}^R = -\frac{4}{3 + 5k^2} \left(i k u_k - k^2 p_k\right) . \qquad (8.79)$$

It should be noted that the recurrence equations (8.77) can also be obtained from the exact recurrence equations (8.28) by neglecting the nonlinear terms. Thus, the approximation adopted within the regularization procedure (8.75) amounts to the following rational approximation of the functions A and B (8.22):

$$A_1^R = B_1^R = -\frac{4}{3 + 5k^2} . \qquad (8.80)$$

Substituting the latter expressions instead of the functions A and B in the dispersion formula (8.23), we come to the dispersion relation of the hydrodynamic modes within the regularized Burnett approximation:

$$\omega_\pm = -\frac{2k^2}{3+5k^2} \pm i|k| \sqrt{\frac{75k^2k^2 + 66k^2 + 15}{25k^2k^2 + 30k^2 + 9}} \ . \tag{8.81}$$

The dispersion relation (8.81) is stable for all wave vectors, and in the short-wave limit we have:

$$\lim_{|k|\to\infty} \omega_\pm = -0.4 \pm i|k|\sqrt{3} \ . \tag{8.82}$$

Thus, the regularized Burnett approximation leads qualitatively to the same behavior of the dispersion relation, as the exact result (8.36), with the limiting value of the attenuation rate equal to -0.4 instead of the exact value $-2/9$.

Consider now the regularization of the super-Burnett approximation. This amounts to setting $N = 2$ in the recurrence equations (8.73) and (8.74). Then, instead of (8.75), we have:

$$\sigma^{(1)} = -\partial_t^{(0)} \sigma^{(0)} \ , \tag{8.83}$$
$$\sigma^{(2+n)} = -\partial_t^{(0)} \sigma^{(n+1)} - \partial_t^{(1)} \sigma^{(n)} \ ,$$

where $n \geq 0$. The corresponding recurrence equations for the coefficients a_n and b_n now become:

$$a_{n+1} = \frac{1}{3}b_n, \quad a_n = b_n, \quad a_0 = -\frac{4}{3} \ . \tag{8.84}$$

Thus, instead of (8.80), we obtain:

$$A_2^R = B_2^R = -\frac{4}{3+k^2} \ . \tag{8.85}$$

The corresponding dispersion relation of the regularized super-Burnett approximation reads:

$$\omega_\pm = -\frac{2k^2}{3+k^2} \pm i|k| \sqrt{\frac{25k^2k^2 + 78k^2 + 45}{3k^2k^2 + 18k^2 + 27}} \ , \tag{8.86}$$

while in the short-wave limit the asymptotic behavior becomes:

$$\lim_{|k|\to\infty} \omega_\pm = -2 \pm i|k| \sqrt{\frac{25}{3}} \ . \tag{8.87}$$

The Bobylev instability is removed again within the regularization of the super-Burnett approximation, and the lower-order terms of the Chapman–Enskog expansion are taken into account more precisely in comparison to the regularized Burnett approximation. However, the approximation in a whole has not improved (see Fig. 8.4). *Thus, we can conclude that although the partial summation method (8.73) and (8.74) is capable of removing the Bobylev instability, and reproducing qualitatively the exact Chapman–Enskog solution*

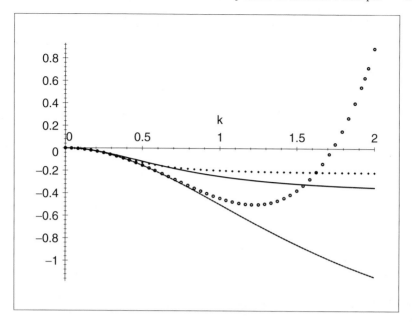

Fig. 8.4. Attenuation rates for the partial summing. *Solid*: The regularized Burnett approximation. *Dash*: The regularized super-Burnett approximation. *Circles*: The super-Burnett approximation. *Dots*: The exact summation

in the short-wave domain, the exactness does not increase monotonically with the depth of the approximation N. This drawback of the regularization procedure indicates once again that an attempt to capture the lower-order terms of the Chapman–Enskog procedure does not succeed in a better approximation as a whole.

8.4.2 The Dynamic Invariance

The starting points of all the approaches considered so far (exact or approximate) is the Chapman–Enskog expansion. However, the result of the summation does not involve the Knudsen number ϵ explicitly and does not require the "smallness" of this parameter. Therefore, it makes sense to reformulate the problem of the reduced description (for the Grad equations (8.10) this amounts to the problem of constructing a function $\sigma_k(u_k, p_k, k)$) in a way where the parameter ϵ does not appear at all. Further, in the framework of such an approach, we can seek a method of explicit construction of the function $\sigma_k(u_k, p_k, k)$, which does not rely upon the Taylor-like expansions as above.

In this section we introduce such an approach, considering again the illustrative example (8.10). These ideas will be extensively used in the sequel, and they also constitute the basis of the so-called method of invariant manifold for dissipative systems [11].

Let us rewrite here (8.10) in the Fourier variables, and cancel the parameter ϵ:

$$\partial_t p_k = -\frac{5}{3} i k u_k , \tag{8.88}$$

$$\partial_t u_k = -i k p_k - i k \sigma_k ,$$

$$\partial_t \sigma_k = -\frac{4}{3} i k u_k - \sigma_k .$$

The result of the reduction in the system (8.88) amounts to a function $\sigma_k(u_k, p_k, k)$, which depends parametrically on the hydrodynamic variables u_k and p_k, and also on the wave vector k. Due to the linearity of the problem under consideration, this function depends linearly on u_k and p_k, and we can start with the form given by (8.21):

$$\sigma_k(u_k, p_k, k) = i k A u_k - k^2 B p_k , \tag{8.89}$$

where A and B are undetermined functions of k. Now, however, we do not refer to a power series representation of these functions as in (8.22).

Given the form of the function $\sigma_k(u_k, p_k, k)$ (8.89), we can compute its time derivative in *two* different ways. On one hand, substituting (8.89) into the right hand side of the third equation in the set (8.88), we derive:

$$\partial_t^{\mathrm{micro}} \sigma_k = -i k \left(\frac{4}{3} + A \right) u_k + k^2 B p_k . \tag{8.90}$$

On the other hand, computing the time derivative and using the first two equations (8.88), we obtain:

$$\partial_t^{\mathrm{macro}} \sigma_k = \frac{\partial \sigma_k}{\partial u_k} \partial_t u_k + \frac{\partial \sigma_k}{\partial p_k} \partial_t p_k \tag{8.91}$$

$$= i k A \left(-i k p_k - i k \sigma_k \right) - k^2 B \left(-\frac{5}{3} i k u_k \right)$$

$$= i k \left(\frac{5}{3} k^2 B + k^2 A \right) u_k + k^2 \left(A - k^2 B \right) p_k .$$

Equating the expressions in the right hand sides of (8.90) and (8.91), and requiring that the resulting equality holds for any values of the variables u_k and p_k, we derive the following two algebraic equations:

$$F(A, B, k) = -A - \frac{4}{3} - k^2 \left(\frac{5}{3} B + A^2 \right) = 0 , \tag{8.92}$$

$$G(A, B, k) = -B + A \left(1 - k^2 B \right) = 0 .$$

These are exactly the equations (8.32), which were obtained after summation of the Chapman–Enskog expansion. Now, however, we have reached the same result without using the expansion. Thus, (8.92) (or, equivalently, (8.32)) can be used as a starting point for the construction of the function (8.89).

It is important to comment on the somewhat formal manipulations which have led to (8.92). First of all, by the very sense of the reduced description problem, we are looking for a set of functions σ_k which depend on time only through the time dependence of the hydrodynamic variables u_k and p_k. That is, we are looking for a set (8.89), which is parameterized with the values of the hydrodynamic variables. Further, the two time derivatives, (8.90) and (8.91), are relevant to the "microscopic" and the "macroscopic" evolution within the set (8.89), respectively. Indeed, the expression in the right hand side of (8.90) is just the value of the vector field of the original Grad equations at the points of the set (8.89). On the other hand, (8.91) expresses the time derivative in terms of the reduced (macroscopic) dynamics, which, in turn, is self-consistently defined by the form (8.89). Equations (8.92) provide, therefore, the *dynamic invariance condition of the reduced description* for the set (8.89): the function $\sigma_k(u_k(t), p_k(t), k)$ is a solution to both the full Grad system (8.88) and to the reduced system which consists of the first two (hydrodynamic) equations. For this reason, equations (8.92) and their analogs which will be obtained on similar reasoning, will be called *the invariance equations*.

8.4.3 The Newton Method

Let us concentrate on the problem of solving the invariance equations (8.92). Clearly, if we are going to expand the functions A and B into power series (8.22), we shall return to the Chapman–Enskog procedure. Now, however, we see that the Chapman–Enskog expansion is just a method to solve the invariance equations (8.92), and maybe not even the optimal one.

Another possibility is to use *iterative* methods. Indeed, we shall apply Newton's method. The algorithm is as follows: Let A_0 and B_0 are some initial approximations chosen for the procedure. The correction, $A_1 = A_0 + \delta A_1$ and $B_1 = B_0 + \delta B_1$, due to the Newton iteration is obtained upon a linearization (8.92) around the approximation A_0 and B_0. Computing the derivatives, we can represent the equation of the Newton iteration in matrix form:

$$
\begin{pmatrix} \frac{\partial F(A,B,k)}{\partial A}\Big|_{A=A_0,B=B_0} & \frac{\partial F(A,B,k)}{\partial B}\Big|_{A=A_0,B=B_0} \\ \frac{\partial G(A,B,k)}{\partial A}\Big|_{A=A_0,B=B_0} & \frac{\partial G(A,B,k)}{\partial B}\Big|_{A=A_0,B=B_0} \end{pmatrix} \begin{pmatrix} \delta A_1 \\ \delta B_1 \end{pmatrix}
$$
$$
+ \begin{pmatrix} F(A_0, B_0, k) \\ G(A_0, B_0, k) \end{pmatrix} = 0 . \qquad (8.93)
$$

where

$$
\frac{\partial F(A, B, k)}{\partial A}\Big|_{A=A_0,B=B_0} = -\left(1 + 2k^2 A_0\right) , \qquad (8.94)
$$
$$
\frac{\partial F(A, B, k)}{\partial B}\Big|_{A=A_0,B=B_0} = -\frac{5}{3}k^2 ,
$$

$$\frac{\partial G(A, B, k)}{\partial A}\Big|_{A=A_0, B=B_0} = 1 - k^2 B_0 \,,$$

$$\frac{\partial G(A, B, k)}{\partial B}\Big|_{A=A_0, B=B_0} = -\left(1 + k^2 A_0\right) \,.$$

Solving the system of linear algebraic equations, we come to the first correction δA_1 and δB_1. Further corrections are found iteratively:

$$A_{n+1} = A_n + \delta A_{n+1} \,, \tag{8.95}$$
$$B_{n+1} = B_n + \delta B_{n+1} \,,$$

where $n \geq 0$, and

$$\begin{pmatrix} -(1 + 2k^2 A_n) & -\frac{5}{3}k^2 \\ 1 - k^2 B_n & -(1 + k^2 A_n) \end{pmatrix} \begin{pmatrix} \delta A_{n+1} \\ \delta B_{n+1} \end{pmatrix} + \begin{pmatrix} F(A_n, B_n, k) \\ G(A_n, B_n, k) \end{pmatrix} = 0 \,. \tag{8.96}$$

Within the algorithm just presented, the problem is how to choose the initial approximation A_0 and B_0. The recursion (8.95) and (8.96) is applicable formally to any initial approximation. However, the convergence (if at all) might be sensitive to the choice.

For the first experiment let us take the Navier–Stokes approximation of the functions A and B:

$$A_0 = B_0 = -\frac{4}{3}$$

The outcome of the first two Newton iterations (the attenuation rates as they follow from the first and second Newton iteration) are presented in Fig. 8.5. It is clearly seen that the Newton iterations converge rapidly to the exact solution for moderate k, but the asymptotic behavior in the short-wave domain does not improve.

Another possibility is to take the result of the regularization procedure as presented above. Let the regularized Burnett approximation (8.80) be taken for the initial approximation, that is:

$$A_0 = A_1^R = -\frac{4}{3 + 5k^2} \,, \qquad B_0 = B_1^R = -\frac{4}{3 + 5k^2} \,. \tag{8.97}$$

Substituting (8.97) into (8.95) and (8.96) for $n = 0$ we obtain, after some algebra, the following first correction:

$$A_1 = -\frac{4(27 + 63k^2 + 153k^2 k^2 + 125k^2 k^2 k^2)}{3(3 + 5k^2)(9 + 9k^2 + 67k^2 k^2 + 75k^2 k^2 k^2)} \,, \tag{8.98}$$

$$B_1 = -\frac{4(9 + 33k^2 + 115k^2 k^2 + 75k^2 k^2 k^2)}{(3 + 5k^2)(9 + 9k^2 + 67k^2 k^2 + 75k^2 k^2 k^2)}$$

Functions (8.98) are not yet the exact solution to (8.92) (that is, the functions $F(A_1, B_1, k)$ and $G(A_1, B_1, k)$ are not equal to zero for all k). However,

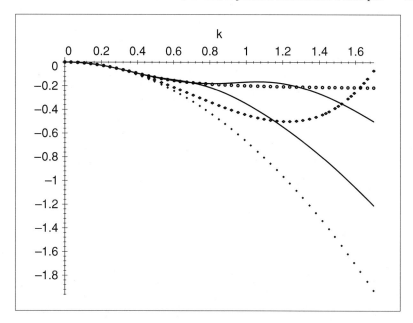

Fig. 8.5. Attenuation rates for the Newton method with the Navier–Stokes approximation as the initial condition. *Dots*: The Navier–Stokes approximation. *Solid*: The first and the second iterations of the invariance equation. *Circles*: The exact solution to the invariance equation. *Diamonds*: The super-Burnett approximation

substituting A_1 and B_1 instead of A and B into the dispersion relation (8.23), we derive in the short-wave limit:

$$\lim_{|k|\to\infty} \omega_\pm = -\frac{2}{9} \pm i|k|\sqrt{3} . \tag{8.99}$$

That is, already the first Newton iteration, as applied to the regularized Burnett approximation, leads to the exact expression in the short-wave domain. Since the first Newton iteration appears to be asymptotically exact, the next iterations improve the solution only for the intermediate values of k, whereas the asymptotic behaviour remains exact in all iterations. The attenuation rates for the first and second Newton iterations with the initial approximation (8.97) are plotted in Fig. 8.6. The agreement with the exact solution is excellent.

One more test is to take the result of the super-Burnett approximation (8.85) as an initial condition in the Newton procedure (8.96). As we know, the regularization of the super-Burnett approximation provides a poorer approximation in comparison to (8.97), particularly in the short-wave domain. Nevertheless, the Newton iterations do converge though less rapidly (see Fig. 8.7).

The examples considered so far demonstrate that the Newton method, as applied to the invariance equations (8.92) is a more powerful tool in

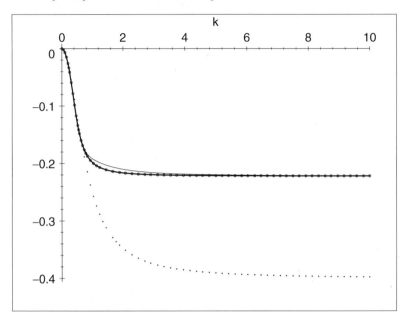

Fig. 8.6. Attenuation rates with the regularized Burnett approximation as the initial condition for the Newton method. *Dots*: The regularized Burnett approximation, or the first Newton iteration with the Euler initial condition (see text). *Solid*: The first and the second Newton iterations with the regularized Burnett approximation as the initial condition. *Circles*: The exact solution to the invariance equation

comparison to the Chapman–Enskog procedure. It is also important that the initial approximation should be "properly chosen", and that it should reproduce, at least qualitatively, the features of the solution not only in the long-wave limit, but over the whole range of wavenumbers.

The best from the initial approximations considered so far is the regularized Burnett approximation (8.97). We have already commented on the relation of this approximation to the invariance equations, as well as on its relation to the Chapman–Enskog procedure. The further important observation is as follows:

Let us choose the *Euler* approximation for the functions A and B, that is:

$$A_0 = B_0 = 0 \tag{8.100}$$

The equation of the first Newton iteration (8.96) is very simple:

$$\begin{pmatrix} -1 & -\frac{5}{3}k^2 \\ 1 & -1 \end{pmatrix} \begin{pmatrix} \delta A_1 \\ \delta B_1 \end{pmatrix} + \begin{pmatrix} -\frac{4}{3} \\ 0 \end{pmatrix} = 0 , \tag{8.101}$$

and

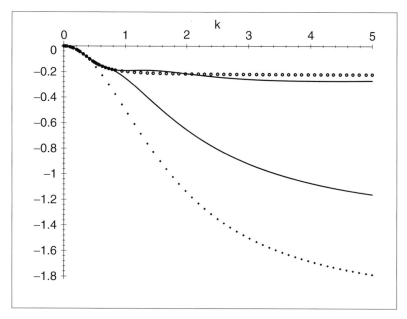

Fig. 8.7. Attenuation rates with the regularized super-Burnett approximation as the initial condition for the Newton method. *Dots*: The regularized super-Burnett approximation. *Solid*: The first and the second Newton iterations. *Circles*: The exact solution to the invariance equation

$$A_1 = B_1 = -\frac{4}{3 + 5k^2} \ . \tag{8.102}$$

Thus, *the regularized Burnett approximation is at the same time the first Newton correction as applied to the Euler initial approximation.* This property distinguishes the regularization of the Burnett approximation from other regularizations. Now the functions (8.98) can be regarded as the *second* Newton correction as applied to the Euler initial approximation (8.100).

Finally, let us examine what Newton's method does in the case of singularities. As we have demonstrated in the previous section, the singularity of the diffusion-like mode occurs when this mode couples to a non-hydrodynamic mode of the 10 moment Grad system if the spatial dimension is greater that one.

Without proving it here, the invariance equation method as applied to the 10 moment Grad system (8.38) leads to the system of equations (8.65). We have already demonstrated what the outcome of the Newton method is when it is applied to the first two equations of this system (responsible for the acoustic mode and containing no singularities). The Newton method, as applied to (8.67), reads:

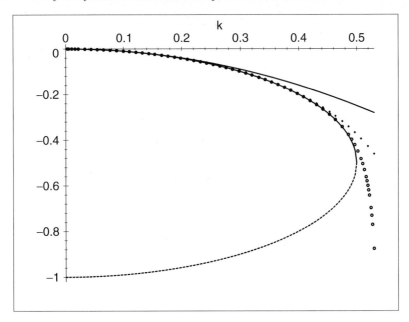

Fig. 8.8. The diffusion mode with the Euler initial approximation for the invariance equation. *Upper solid*: The the first iteration. *Dots*: The second iteration. *Circles*: The third iteration. *Lower solid*: The exact solution. *Dash*: The critical mode

$$Y_{n+1} = Y_n + \delta Y_{n+1} \ , \tag{8.103}$$
$$(1 + 2Y_n)\delta Y_{n+1} + \{Y_n(1 + Y_n) + k^2\} = 0 \ ,$$

where $n \geq 0$, and Y_0 is a chosen initial approximation. Taking the Euler approximation ($Y_0 = 0$), we derive:

$$Y_1 = -k^2 \ , \tag{8.104}$$
$$Y_2 = -\frac{k^2(1 + k^2)}{1 - 2k^2} \ .$$

The second approximation, Y_2, is singular at $k_2 = \sqrt{1/2}$, and it can be demonstrated that all further corrections also have the first singularity at points k_n, and the sequence $k_2, \ldots, k_n$ tends to the actual branching point of the invariance equation (8.67) $k_c = 1/2$. The analysis of further corrections demonstrates that the convergence is very rapid (see Fig. 8.8).

The expressions (8.104) demonstrate that unlike polynomial approximations, Newton's method is capable of detecting the actual singularities of the hydrodynamic spectrum. Formally, the function Y_2 becomes positive as k becomes larger than k_2, and thus the attenuation rate, $\omega_d = Y_2$ becomes positive after this point. However, unlike the super-Burnett approximation for the acoustic mode, this transition occurs now at a singular point. Indeed, the attenuation rate Y_2 tends to "minus infinity", as k tends to k_2 from the left. Thus, as described with the Newton procedure, the non-physical domain is separated from the physical one with an "infinitely viscid" threshold. The occurrence of the poles in the Newton iterations is, of course, quite clear. Indeed, the Newton method involves the derivative of the function $R(Y) = Y(Y+1)$ which appears on the left hand side of (8.67). The derivative $dR(Y)/dY$ becomes zero at the singularity point $Y_c = -1/2$. The results of this section bring us to the following conclusion:

(i) Exact summation of the Chapman–Enskog procedure results in the same system of equations as the principle of dynamic invariance. This was demonstrated above for a specific situation but it holds for any (linearized) Grad system. The resulting equations are always *nonlinear* (even for the simplest linearized kinetic systems, such as Grad equations).

(ii) Now we are able to *alter the viewpoint*: the invariance equations can be considered as basic in the theory, while the Chapman–Enskog method is a way to solve it via an expansion in powers of k. The method of power series expansion is neither the only method to solve equations, nor the optimal. Alternative iteration methods might be better suited to the problem of constructing the reduced description.

(iii) An opportunity to derive the invariance equation in closed form, and next to solve it this or that way is, of course, rather exotic. The situation becomes complicated already for the nonlinear Grad equations, and we should not expect anything simple in the case of the Boltzmann equation. Therefore, if we are willing to proceed along these lines in other problems, attention should be drawn towards approximate procedures. With this, the question arises: what amount of information is required to execute the procedures? Indeed, the Navier–Stokes approximation can be obtained without any knowledge of the whole nonlinear system of invariance equations. It is important that the Newton method, as applied to our problem, does not require any global information as well. This was demonstrated above by a relation between the first iteration as applied to the Euler approximation and the regularization of the Burnett approximation.

8.4.4 Invariance Equation for the $1D13M$ Grad System

Let us consider as the next example the problem of the reduced description for the one-dimensional thirteen moment Grad system. Using the dimensionless variables as above, we write the one-dimensional version of the Grad equations (8.2) and (8.3) in the k-representation:

$$\partial_t \rho_k = -iku_k \,, \tag{8.105}$$
$$\partial_t u_k = -ik\rho_k - ikT_k - ik\sigma_k \,,$$
$$\partial_t T_k = -\frac{2}{3}iku_k - \frac{2}{3}ikq_k \,,$$

$$\partial_t \sigma_k = -\frac{4}{3} i k u_k - \frac{8}{15} i k q_k - \sigma_k ,$$

$$\partial_t q_k = -\frac{5}{2} i k T_k - i k \sigma_k - \frac{2}{3} q_k .$$

The Grad system (8.105) provides the simplest coupling of the hydrodynamic variables ρ_k, u_k, and T_k to the non-hydrodynamic variables, σ_k and q_k, the latter corresponding to the heat flux. As above, our goal is to reduce the description of the Grad system (8.105) to the three hydrodynamic equations with respect to the variables ρ_k, u_k, and T_k. That is, we have to express the functions σ_k and q_k in terms of ρ_k, u_k, and T_k:

$$\sigma_k = \sigma_k(\rho_k, u_k, T_k, k) ,$$

$$q_k = q_k(\rho_k, u_k, T_k, k) .$$

Application of the Chapman–Enskog method in these cases, results in the following algebraic scheme (we omit the Knudsen number ϵ):

$$\sigma_k^{(n)} = -\left\{ \sum_{m=0}^{n-1} \partial_t^{(m)} \sigma_k^{(n-1-m)} + \frac{8}{15} i k q_k^{(n-1)} \right\} , \tag{8.106}$$

$$q_k^{(n)} = -\left\{ \sum_{m=0}^{n-1} \partial_t^{(m)} q_k^{(n-1-m)} + i k \sigma_k^{(n-1)} \right\} ,$$

where the Chapman–Enskog operators act as follows:

$$\partial_t^{(m)} \rho_k = \begin{cases} -i k u_k, & m = 0 \\ 0, & m \geq 1 \end{cases} , \tag{8.107}$$

$$\partial_t^{(m)} u_k = \begin{cases} -i k (\rho_k + T_k), & m = 0 \\ -i k \sigma_k^{(m-1)}, & m \geq 1 \end{cases} ,$$

$$\partial_t^{(m)} T_k = \begin{cases} -\frac{2}{3} i k u_k, & m = 0 \\ -\frac{2}{3} i k q_k^{(m-1)}, & m \geq 1 \end{cases} .$$

The initial condition for the recurrence procedure (8.106) reads: $\sigma_k^{(0)} = -\frac{4}{3} i k u_k$, and $q_k^{(0)} = -\frac{15}{4} i k T_k$, which leads to the Navier–Stokes-Fourier hydrodynamic equations.

Computing the coefficients $\sigma_k^{(1)}$ and $q_k^{(1)}$, obtain the Burnett approximation:

$$\sigma_{1k} = -\frac{4}{3} i k u_k + \frac{4}{3} k^2 \rho_k - \frac{2}{3} k^2 T_k , \tag{8.108}$$

$$q_{1k} = -\frac{15}{4} i k T_k + \frac{7}{4} k^2 u_k .$$

The Burnett approximation (8.108) coincides with that obtained from the Boltzmann equation, and it is precisely the case where the instability was first demonstrated in the paper [72].

The structure of the terms $\sigma_k^{(n)}$ and $q_k^{(n)}$ (an analog of (8.20) and (8.42)) is as follows:

$$\sigma_k^{(2n)} = a_n(-k^2)^n iku_k ,$$ (8.109)

$$\sigma_k^{(2n+1)} = b_n(-k^2)^{n+1}\rho_n + c_n(-k^2)^{n+1}T_k ,$$

$$q_k^{(2n)} = \beta_n(-k^2)^n ik\rho_k + \gamma_n(-k^2)^n ikiT_k ,$$

$$q_k^{(2n+1)} = \alpha_n(-k^2)^{n+1}u_k .$$

The derivation of the invariance equation for the system (8.105) goes along the same lines as in the previous section. We seek the functions of the reduced description in the form:

$$\sigma_k = ikAu_k - k^2B\rho_k - k^2CT_k ,$$ (8.110)

$$q_k = ikX\rho_k + ikYT_k - k^2Zu_k ,$$

where the functions $A, \ldots, Z$ will be determinated in the process.

The invariance condition results in a closed system of equations for the functions A, B, C, X, Y, and Z. As above, computing the microscopic time derivative of the functions (8.110), due to the two last equations of the Grad system (8.105) we derive:

$$\partial_t^{\text{micro}}\sigma_k = -ik\left(\frac{4}{3} - \frac{8}{15}k^2Z + A\right)u_k$$ (8.111)

$$+k^2\left(\frac{8}{15}X + B\right)\rho_k + k^2\left(\frac{8}{15}Y + C\right)T_k ,$$

$$\partial_t^{\text{micro}}q_k = k^2\left(A + \frac{2}{3}Z\right)u_k + ik\left(k^2B - \frac{2}{3}X\right)\rho_k$$

$$-ik\left(\frac{5}{2} - k^2C - \frac{2}{3}Y\right)T_k .$$

On the other hand, computing the macroscopic time derivative due to the first three equations of the system (8.105), we obtain:

$$\partial_t^{\text{macro}}\sigma_k = \frac{\partial\sigma_k}{\partial u_k}\partial_t u_k + \frac{\partial\sigma_k}{\partial\rho_k}\partial_t\rho + \frac{\partial\sigma_k}{\partial T_k}\partial_t T_k$$ (8.112)

$$= ik\left(k^2A^2 + k^2B + \frac{2}{3}k^2C - \frac{2}{3}k^2k^2CZ\right)u_k$$

$$+\left(k^2A - k^2k^2AB - \frac{2}{3}k^2k^2CX\right)\rho_k$$

$$+\left(k^2A - k^2k^2AC - \frac{2}{3}k^2k^2CY\right)T_k ;$$

$$\partial_t^{\text{macro}}q_k = \frac{\partial q_k}{\partial u_k}\partial_t u_k + \frac{\partial q_k}{\partial\rho_k}\partial_t\rho u_k + \frac{\partial q_k}{\partial T_k}\partial_t T_k$$

$$= \left(-k^2 k^2 Z A + k^2 X + \frac{2}{3} k^2 Y - \frac{2}{3} k^2 k^2 Y Z \right) u_k$$

$$+ ik \left(k^2 Z - k^2 k^2 Z B + \frac{2}{3} k^2 Y X \right) \rho_k$$

$$+ ik \left(k^2 Z - k^2 k^2 Z C + \frac{2}{3} k^2 Y^2 \right) T_k .$$

Equating the corresponding expressions in the formulas (8.111) and (8.112), we derive the following system of coupled equations:

$$F_1 = -\frac{4}{3} + \frac{8}{15} k^2 Z - A - k^2 A^2 - k^2 B - \frac{2}{3} k^2 C + \frac{2}{3} k^2 k^2 C Z = 0 ,$$

$$F_2 = \frac{8}{15} X + B - A + k^2 A B + \frac{2}{3} k^2 C X = 0 ,$$

$$F_3 = \frac{8}{15} Y + C - A + k^2 A C + \frac{2}{3} k^2 C Y = 0 ,$$

$$F_4 = A + \frac{2}{3} Z + k^2 Z A - X - \frac{2}{3} Y + \frac{2}{3} k^2 Y Z = 0 ,$$

$$F_5 = k^2 B - \frac{2}{3} X - k^2 Z + k^2 k^2 Z B - \frac{2}{3} k^2 Y X = 0 ,$$

$$F_6 = -\frac{5}{2} + k^2 C - \frac{2}{3} Y - k^2 Z + k^2 k^2 Z C - \frac{2}{3} k^2 Y^2 = 0 . \tag{8.113}$$

As above, the invariance equations (8.113) can also be obtained upon summation of the Chapman–Enskog expansion, after the Chapman–Enskog procedure is casted into a recurrence relations for the coefficients $a_n, \ldots, \alpha_n$ (8.109). This route is less straightforward than the one just presented, and we omit the proof.

The Newton method, as applied to the system (8.113), results in the following algorithm:

Denote as $\boldsymbol{A}$ the six-component vector function $\boldsymbol{A} = (A, B, C, X, Y, Z)$. Let $\boldsymbol{A}_0$ is the initial approximation, then:

$$\boldsymbol{A}_{n+1} = \boldsymbol{A}_n + \delta \boldsymbol{A}_{n+1} , \tag{8.114}$$

where $n \geq 0$, and the vector function $\delta \boldsymbol{A}_{n+1}$ is a solution to the linear system of equations:

$$\boldsymbol{N}_n \delta \boldsymbol{A}_{n+1} + \boldsymbol{F}_n = 0 . \tag{8.115}$$

Here $\boldsymbol{F}_n$ is the vector function with the components $F_i(\boldsymbol{A}_n)$, and $\boldsymbol{N}_n$ is a 6×6 matrix:

$$\begin{pmatrix} -(1 + 2k^2 A_n) & -k^2 & -2/3 k^2 (1 - k^2 Z_n) \\ k^2 B_n - 1 & 1 + k^2 & 2/3 k^2 X_n \\ k^2 C_n - 1 & 0 & 1 + 2/3 k^2 Y_n + k^2 A_n \\ 1 + k^2 Z_n & 0 & 0 \\ 0 & k^2 (1 + k^2 Z_n) & 0 \\ 0 & 0 & k^2 (1 + k^2 Z_n) \end{pmatrix} \tag{8.116}$$

$$\left. \begin{array}{ccc} 0 & 0 & 2/3k^2(4/5 + k^2 C_n) \\ 2/3(4/5 + k^2 C_n) & 0 & 0 \\ 0 & 2/3(4/5 + k^2 C_n) & 0 \\ -1 & -2/3(1 - k^2 Z_n) & 2/3 + k^2 A_n + 2/3k^2 Y_n \\ -2/3(1 + k^2 Y_n) & -2/3k^2 X_n & -k^2(1 - k^2 B_n) \\ 0 & -2/3(1 + 2k^2 Y_n) & -k^2(1 - k^2 C_n) \end{array} \right) .$$

The Euler approximation gives: $A_0 = \ldots = Z_0 = 0$, while $F_1 = -4/3$, $F_6 = -5/2$, and $F_2 = \ldots = F_5 = 0$. The first Newton iteration (8.115) as applied to this initial approximation, leads again to a simple algebraic problem, and we have finally obtained:

$$A_1 = -20\frac{141k^2 + 20}{867k^4 + 2105k^2 + 300} , \qquad (8.117)$$

$$B_1 = -20\frac{459k^2 k^2 + 810k^2 + 100}{3468k^2 k^2 k^2 + 12755k^2 k^2 + 11725k^2 + 1500} ,$$

$$C_1 = -10\frac{51k^2 k^2 - 485k^2 - 100}{3468k^2 k^2 k^2 + 12755k^2 k^2 + 11725k^2 + 1500} ,$$

$$X_1 = -\frac{375k^2(21k^2 - 5)}{2(3468k^2 k^2 k^2 + 12755k^2 k^2 + 11725k^2 + 1500)} ,$$

$$Y_1 = -\frac{225(394k^2 k^2 + 685k^2 + 100)}{4(3468k^2 k^2 k^2 + 12755k^2 k^2 + 11725k^2 + 1500)} ,$$

$$Z_1 = -15\frac{153k^2 + 35}{867k^4 + 2105k^2 + 300} .$$

Substituting (8.109) into the first three equations of the Grad system (8.105), and proceeding with the dispersion relation as above, we derive the latter in terms of the functions $A, \ldots, Z$:

$$\omega^3 - k^2 \left(\frac{2}{3}Y + A\right)\omega^2 \qquad (8.118)$$

$$+ k^2 \left(\frac{5}{3} - \frac{2}{3}k^2 Z - \frac{2}{3}k^2 C - k^2 B + \frac{2}{3}k^2 AY + \frac{2}{3}k^2 k^2 CZ\right)\omega$$

$$+ \frac{2}{3}k^2(k^2 X - k^2 Y + k^2 k^2 BY - k^2 k^2 XC) = 0 .$$

When the functions $A_1, \ldots, Z_1$ (8.117) are substituted instead of $A, \ldots, Z$ into (8.118), the dispersion relation of the first Newton iteration, as applied to the invariance equations (8.113) with the Euler initial approximation, is obtained. This result coincides with the regularization of the Burnett approximation, which was considered in [43]. There it was demonstrate that the equilibrium is stable within this approximation for arbitrary wave lengths. The dispersion relation for the Burnett approximation, in turn, is due to the approximation

$$A = -4/3, \quad B = -4/3, \quad C = 2/3, \quad X = 0, \quad Y = -15/4, \quad Z = -7/4,$$

as it follows from a comparison of (8.108) and (8.110). The dispersion relation for the Burnett approximation coincides with the one obtained in [72] from the Boltzmann equation.

8.4.5 Invariance Equation for the $3D13M$ Grad System

The final example to be considered is the 13 moment Grad system in three spatial dimensions, (8.2) and (8.3). Let us rewrite here the original system in terms of Fourier variables:

$$\partial_t \rho_k = -ike_k \cdot u_k, \qquad (8.119)$$
$$\partial_t u_k = -ike_k \rho_k - ike_k T_k - ike_k \cdot \sigma_k,$$
$$\partial_t T_k = -\frac{2}{3} ik(e_k \cdot u_k + e_k \cdot q_k),$$
$$\partial_t \sigma_k = -ik\overline{e_k u_k} - \frac{2}{5} ik\overline{e_k q_k} - \sigma_k,$$
$$\partial_t q_k = -\frac{5}{2} ike_k T_k - ike_k \cdot \sigma_k - \frac{2}{3} q_k.$$

Here we have represented the wave vector k as $k = ke_k$, and e_k is the unit vector.

The structure of the even and odd Chapman–Enskog coefficients, $\sigma_k^{(n)}$ and $q_k^{(n)}$, turns out to be as follows:

$$\sigma_k^{(2n)} = (-k^2)^n ik \{a_n(\overline{e_k u_k} - 2g_k(e_k \cdot u_k)) + b_n g_k(e_k \cdot u_k)\},$$
$$\sigma_k^{(2n+1)} = (-k^2)^{n+1} g_k \{c_n T_k + d_n \rho_k\},$$
$$q_k^{(2n)} = (-k^2)^n ike_k \{\gamma_n T_k + \delta_n \rho_k\},$$
$$q_k^{(2n+1)} = (-k^2)^{n+1} \{\alpha_n e_k(e_k \cdot u_k) + \beta_n(u_k - e_k(e_k \cdot u_k))\}, \qquad (8.120)$$

where $g_k = 1/2\overline{e_k e_k}$, and the real-valued coefficients $a_n, \dots, \beta_n$ are due to the Chapman–Enskog procedure (8.7) and (8.8).

The expressions just presented suggest that the dynamic invariant form of the stress tensor and of the heat flux reads:

$$\sigma_k = ikA(\overline{e_k u_k} - 2g_k(e_k \cdot u_k)) + 2ikBg_k(e_k \cdot u_k) \qquad (8.121)$$
$$\qquad -2k^2 C g_k T_k - 2k^2 D g_k \rho_k,$$
$$q_k = ikZe_k T_k + ikUe_k \rho_k$$
$$\qquad -k^2 X(u_k - e_k(e_k \cdot u_k)) - k^2 Y e_k(e_k \cdot u_k),$$

where the functions $A, \dots, Y$ depend on k. The dynamic invariance condition results in the following two closed systems for these functions:

$$\frac{2}{5}U + D - B + \frac{2}{3}k^2CU + \frac{4}{3}k^2BD = 0 , \quad (8.122)$$

$$\frac{2}{5}Z + C - B + \frac{2}{3}k^2CZ + \frac{4}{3}k^2BC = 0 ,$$

$$-1 + \frac{2}{5}k^2Y - B - \frac{2}{3}k^2C - k^2D - \frac{4}{3}k^2B^2 + \frac{2}{3}k^2k^2CY = 0 ,$$

$$\frac{4}{3}k^2D - \frac{2}{3}U - k^2Y - \frac{2}{3}k^2ZU + \frac{4}{3}k^2k^2YD = 0 ,$$

$$-\frac{5}{2} + \frac{4}{3}k^2C - \frac{2}{3}Z - k^2Y - \frac{2}{3}k^2Z^2 + \frac{4}{3}k^2k^2YC = 0 ,$$

$$\frac{4}{3}B + \frac{2}{3}Y - U - \frac{2}{3}Z + \frac{2}{3}k^2ZY + \frac{4}{3}k^2YB = 0 ,$$

and

$$-1 - A + \frac{2}{5}k^2X - k^2A^2 = 0 , \quad (8.123)$$

$$A + \frac{2}{3}X + k^2AX = 0$$

The method of summation of the Chapman–Enskog expansion can also be developed, starting with the structure of the Chapman–Enskog coefficients (8.120), in the same manner as in Sect. 8.3. Simple but rather extensive computations in this case lead, of course, to the invariance equations (8.122) and (8.123).

The Newton method, as applied to the systems (8.122) and (8.123) with the initial Euler approximation, leads in the first iteration to the regularization of the Burnett approximation reported earlier in [43].

Introducing the functions $\bar{A} = k^2A$ and $\bar{X} = k^2X$ in (8.123) we obtain:

$$R(\bar{A}) = \frac{5\bar{A}(3\bar{A}^2 + 5\bar{A} + 2)}{4(6\bar{A} + 5)} = -k^2 , \quad (8.124)$$

while

$$\bar{X} = -\frac{3\bar{A}}{2 + 3\bar{A}} .$$

The derivative, $dR(\bar{A})/d\bar{A}$, becomes equal to zero for $\bar{A}_c \approx -0.364$, which gives the critical wave vector $k_c = \sqrt{-R(\bar{A}_c)} \approx 0.305$. The Newton method, as applied to (8.124) with the initial Euler condition $\bar{A} = 0$, gives the following: the results of the first and of the second iterations are regular functions, while the third and the further iterations bring a singularity which converges to the point k_c (see Fig. 8.9). These singularities (the real poles) of the Newton corrections are of the same nature as discussed above.

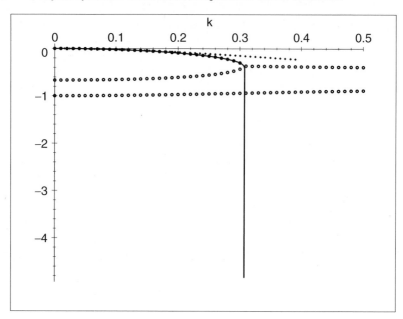

Fig. 8.9. Solutions to (8.123). *Circles*: Numerical solutions. *Dots*: The first Newton iteration. *Solid*: The 4th Newton iteration

8.4.6 Gradient Expansions in Kinetic Theory of Phonons

Exact Chapman–Enskog Solution and Onset of Second Sound

In this section, we close our discussion of linearized Grad systems with an application to simple models for phonon transport in rigid insulators. It is demonstrated that the extended diffusion mode transforms into a second sound mode due to its coupling to a non-hydrodynamic mode at some critical value of the wave vector. This criticality shows up as a branching point of the extension of the diffusion mode within the Chapman–Enskog method. Although the analysis is essentially similar to the examples considered above, it is presented in some details for the sake of completeness.

Experiments on heat pulse propagation through crystalline media [206, 207] confirmed the existence of a temperature window (the Guyer-Krumhansl window [208–210]) with respect to which the features of heat propagation are qualitatively different: At temperatures exceeding the high-temperature edge of the window, the heat propagates in a diffusion-like way. Below the low-temperature edge of the window, the propagation goes in a ballistic way, with the constant speed of sound. Within the window, the propagation becomes wave-like. This latter regime is called second sound (see [211] for a review).

This problem has drawn some renewed attention in the last years. Models relevant for a unified description of diffusion, second sound, and ballistic regimes of heat propagation are intensively discussed (see [212, 251] and references therein). To be specific, recall the simplest and typical model of the phonon transport [212]. Let $e(\boldsymbol{x}, t)$ and $\boldsymbol{p}(\boldsymbol{x}, t)$ be small deviations of the energy density and energy flux of the phonon field from their equilibrium values, respectively. Then

$$\partial_t e = -c^2 \nabla \cdot \boldsymbol{p} , \tag{8.125}$$

$$\partial_t \boldsymbol{p} = -\frac{1}{3}\nabla e - \frac{1}{\tau_R}\boldsymbol{p} . \tag{8.126}$$

Here c is the Debye velocity of phonons, and τ_R is the characteristic time of resistive processes. Equations (8.125) can be derived from the Boltzmann–Peierls kinetic equation, within the relaxation time approximation, by a method similar to Grad's method [212]. Equations (8.125), (8.126) provide the simplest model of coupling between the hydrodynamic variable e and the non-hydrodynamic variable $\boldsymbol{p}$, allowing for a qualitative description of both the diffusion and the second sound. Following the standard argumentation [212], we observe the two limiting cases:

1. As $\tau_R \to 0$, equation (8.126) yields the Fourier relation $\boldsymbol{p} = -\frac{1}{3}\tau_R \nabla e$ which closes (8.125) to give the diffusion equation:

$$\partial_t e + \frac{1}{3}\tau_R c^2 \Delta e = 0 . \tag{8.127}$$

2. As $\tau_R \to \infty$, (8.126) yields $\partial_t \boldsymbol{p} = -\frac{1}{3}\nabla e$, and (8.125) closes to give the wave equation:

$$\partial_t^2 e + \frac{1}{3}c^2 \Delta e = 0 . \tag{8.128}$$

Equation (8.127) describes the usual diffusive regime of heat propagation, while (8.128) is relevant to the (undamped) second sound regime with the velocity $u_2 = c/\sqrt{3}$; both are closed with respect to the variable e.

However, even within the simplest model (8.125), (8.126), the problem of closure remains unsolved in a systematic way when τ_R is finite. The natural way of doing so is provided by the Chapman–Enskog method. In the situation under consideration, the Chapman–Enskog method yields an extension of the diffusive transport to finite values of the parameter τ_R, and leads to an expansion of the non-hydrodynamic variable $\boldsymbol{p}$ in terms of the hydrodynamic variable e. With this, if we are able to make this extension of the diffusive mode exactly, we could learn more about the transition between the diffusion and second sound (within the framework of the model).

The Chapman–Enskog method, as applied to (8.125), (8.126), results in the following series representation:

$$\boldsymbol{p} = \sum_{n=0}^{\infty} \boldsymbol{p}^{(n)} , \tag{8.129}$$

where the coefficients $p^{(n)}$ are due to the Chapman–Enskog recurrence procedure,

$$p^{(n)} = -\tau_R \sum_{m=0}^{n-1} \partial_t^{(m)} p^{(n-1-m)} , \qquad (8.130)$$

while the Chapman–Enskog operators $\partial_t^{(m)}$ act on e as follows:

$$\partial_t^{(m)} e = -c^2 \nabla \cdot p^{(m)} . \qquad (8.131)$$

Finally, the zero order term reads: $p^{(0)} = -\frac{1}{3}\tau_R \nabla e$, and leads to the Fourier approximation of the energy flux.

To sum up the series (8.129) in closed form, we shall specify the non-linearity appearing in equations (8.130) and (8.131). The coefficients $p^{(n)}$ in equations (8.129) and (8.130) have the following explicit structure for arbitrary order $n \geq 0$:

$$p^{(n)} = a_n \Delta^n \nabla e , \qquad (8.132)$$

where the real-valued and yet unknown coefficients a_n are due to the recurrence procedure (8.130), and (8.131). Indeed, the form (8.132) is true for $n = 0$ ($a_0 = -\frac{1}{3}\tau_R$). Let us assume that (8.132) is proven up to order $n - 1$. Then, computing the nth order coefficient $p^{(n)}$, we derive:

$$p^{(n)} = -\tau_R \sum_{m=0}^{n-1} \partial_t^{(m)} a_{n-1-m} \Delta^{(n-1-m)} \nabla e \qquad (8.133)$$

$$= -\tau_R \sum_{m=0}^{n-1} a_{n-1-m} \Delta^{(n-1-m)} \nabla \left(-c^2 a_m \nabla \cdot \nabla \Delta^m e \right)$$

$$= \tau_R c^2 \left\{ \sum_{m=0}^{n-1} a_{n-1-m} a_m \right\} \Delta^n \nabla e .$$

The last expression has the same form as (8.132). Thus, the Chapman–Enskog procedure for the model (8.125), (8.126) is equivalent to the following nonlinear recurrence relation in terms of the coefficients a_n:

$$a_n = \tau_R c^2 \sum_{m=0}^{n-1} a_{n-1-m} a_m , \qquad (8.134)$$

subject to the initial condition $a_0 = -\frac{1}{3}\tau_R$. Further, it is convenient to make the Fourier transform. Using $p = p_k \exp\{ik \cdot x\}$ and $e = e_k \exp\{ik \cdot x\}$, where k is the real-valued wave vector, we derive in (8.132): $p_k^{(n)} = a_n ik(-k^2)^n e_k$, and

$$p_k = ikA(k^2)e_k , \qquad (8.135)$$

where

$$A(k^2) = \sum_{n=0}^{\infty} a_n(-k^2)^n . \qquad (8.136)$$

Thus, the the Chapman–Enskog solution (8.129) amounts to finding the function $A(k^2)$ represented by the power series (8.136). If the function A is known, the exact Chapman–Enskog closure of the system (8.125), (8.126) amounts to the following dispersion relation of plane waves $\sim \exp\{\omega_k t + i\boldsymbol{k} \cdot \boldsymbol{x}\}$:

$$\omega_k = c^2 k^2 A(k^2) . \qquad (8.137)$$

Here ω_k is a complex-valued function of the real-valued vector $\boldsymbol{k}$: $\mathrm{Re}(\omega_k)$ is the attenuation rate, $\mathrm{Im}(\omega_k)$ is the frequency.

Multiplying both equations in (8.134) with $(-k^2)^n$, and performing a summation in n from 1 to infinity, we get:

$$A - a_0 = -\tau_R c^2 k^2 \sum_{n=0}^{\infty} \sum_{m=0}^{n} a_{n-m}(-k^2)^{n-m} a_m(-k^2)^m ,$$

Now we notice that

$$\lim_{N \to \infty} \sum_{n=0}^{N} \sum_{m=0}^{n} a_{n-m}(-k^2)^{n-m} a_m(-k^2)^m = A^2 ,$$

Setting $a_0 = -\frac{1}{3}\tau_R$, we derive a quadratic equation for the function A:

$$\tau_R c^2 k^2 A^2 + A + \frac{1}{3}\tau_R = 0 . \qquad (8.138)$$

Further, a selection procedure is required to choose the relevant root of (8.138). Firstly, recall that all the coefficients a_n (8.132) are real-valued by the sense of the Chapman–Enskog method (8.130) and (8.131), hence the function A (8.136) is real-valued. Therefore, only the real-valued root of (8.138) is relevant to the Chapman–Enskog solution. The first observation is that (8.138) has no real-valued solutions as soon as k becomes bigger than the critical value k_c, where

$$k_c = \frac{\sqrt{3}}{2\tau_R c} . \qquad (8.139)$$

Secondly, there are two real-valued solutions to (8.138) at $k < k_c$. However, only one of them satisfies the Chapman–Enskog asymptotic $\lim_{k \to 0} A(k^2) = -\frac{1}{3}\tau_R$.

With these two remarks, we finally derive the following exact Chapman–Enskog dispersion relation (8.137):

$$\omega_k = \begin{cases} -(2\tau_R)^{-1}\left(1 - \sqrt{1 - (k^2)/(k_c^2)}\right) & k < k_c \\ \text{none} & k > k_c \end{cases} . \qquad (8.140)$$

This dispersion relation corresponds to the extended diffusion transport, and it comes back to the standard Fourier approximation in the limit of long waves $k/k_c \ll 1$. The Chapman–Enskog solution does not exist as soon as $k/k_c > 1$. For $k = k_c$, the extended diffusion branch crosses one of the non-hydrodynamic branches of (8.125), (8.126). For larger k, the extended diffusion mode and the critical non-hydrodynamic mode produce a pair of complex conjugate solutions with the real part equal to $-\frac{1}{2\tau_R}$. The imaginary part of this extension after k_c attains the asymptotic value $\pm i u_2 k$, as $k \to \infty$, where $u_2 = c/\sqrt{3}$ is the (undamped) second sound velocity in the model (8.125), (8.126) (see equation (8.128)). Although the spectrum of the original (8.125), (8.126) continues indeed after k_c, the Chapman–Enskog method does not recognize this extension as part of the hydrodynamic branch, *while the second sound regime is born from the extended diffusion after coupling with the critical non-hydrodynamic mode.*

Finally, let us consider the opportunities provided by the Newton method as applied to the invariance equation. First, the invariance equation can be easily obtained in closed form here. Consider again the expression for the heat flux in terms of the energy density (8.135), $p_k = ikA(k^2)e_k$, where now the function A is not thought as the Chapman–Enskog series (8.136). The invariance equation is a constraint on the function A, expressing the form-invariance of the heat flux (8.135) under both the dynamic equations (8.125) and (8.126). Computing the time derivative of function (8.135) due to equation (8.125), we obtain:

$$\partial_t^{\text{macro}} p_k = ikA(k^2)\partial_t e_k = c^2 k^2 A^2 ike_k . \tag{8.141}$$

On the other hand, computing the time derivative of the same function due to equation (8.126), we have:

$$\partial_t^{\text{micro}} p_k = -\frac{1}{3}ike_k - \frac{1}{\tau_R}Aike_k . \tag{8.142}$$

Equating (8.141) and (8.142), we derive the desired invariance equation for the function A. This equation coincides with the exact Chapman–Enskog equation (8.138).

As the second step, let us apply the Newton method to the invariance equation (8.138), taking the Euler approximation ($A_0^N \equiv 0$) as the initial condition. Rewriting (8.138) in the form $F(A, k^2) = 0$, we come to the following Newton iterations:

$$\frac{dF(A, k^2)}{dA}\bigg|_{A=A_n} (A_{n+1} - A_n) + F(A_n, k^2) = 0 . \tag{8.143}$$

The first two iterations give:

$$\tau_R^{-1} A_1 = -\frac{1}{3} , \tag{8.144}$$

$$\tau_R^{-1} A_2 = -\frac{1 - \frac{1}{4}y^2}{3(1 - \frac{1}{2}y^2)} . \tag{8.145}$$

The first Newton iteration (8.144) coincides with the first term of the Chapman–Enskog expansion. The second Newton iteration (8.145) is a rational function with the Taylor expansion coinciding with the Chapman–Enskog solution up to the super-Burnett term, and it has a pole at $y_2 = \sqrt{2}$. The further Newton iterations are also rational functions with the relevant poles at points y_n, and the sequence of this points tends very rapidly to the location of the actual singularity $y_c = 1$ ($y_3 \approx 1.17$, $y_4 \approx 1.01$, etc.).

Inclusion of Normal Processes

Accounting for normal processes in the framework of the semi-hydrodynamical models [212] leads to the following generalization of (8.125), (8.126) (written in Fourier variables for the one-dimensional case):

$$\partial_t e_k = -ikc^2 p_k \ , \tag{8.146}$$

$$\partial_t p_k = -\frac{1}{3}ike_k - ikN_k - \frac{1}{\tau_R}p_k \ , \tag{8.147}$$

$$\partial_t N_k = -\frac{4}{15}ikc^2 p_k - \frac{1}{\tau}N_k \ , \tag{8.148}$$

where $\tau = \tau_N \tau_R/(\tau_N + \tau_R)$, τ_N is the characteristic time of normal processes, and N_k is the additional field variable. Following the principle of invariance as explained in the preceding section, we write the closure relation for the non-hydrodynamic variables p_k and N_k as:

$$p_k = ikA_k e_k, \quad N_k = B_k e_k \ , \tag{8.149}$$

where A_k and B_k are two unknown functions of the wave vector k. Further, following the principle of invariance as explained in the previous section, each of the relations (8.149) should be invariant under the dynamics due to (8.146), and due to (8.147) and (8.148). This results in two equations for the functions A_k and B_k:

$$k^2 c^2 A_k^2 = -\frac{1}{\tau_R}A_k - B_k - \frac{1}{3} \ , \tag{8.150}$$

$$k^2 c^2 A_k B_k = -\frac{1}{\tau}B_k + \frac{4}{15}k^2 c^2 A_k \ .$$

When the energy balance equation (8.146) is closed with the relation (8.149), this amounts to a dispersion relation for the extended diffusion mode, $\omega_k = k^2 c^2 A_k$, where A_k is the solution to the invariance equations (8.150), *subject to the condition* $A_k \to 0$ *as* $k \to 0$. Resolving equations (8.150) with respect to A_k, and introducing $\overline{A}_k = k^2 c^2 A_k$, we arrive at the following:

$$\Phi(\overline{A}_k) = \frac{5\overline{A}_k(1 + \tau\overline{A}_k)(\tau_R\overline{A}_k + 1)}{5 + 9\tau\overline{A}_k} = -\frac{1}{3}\tau_R k^2 c^2 \ . \tag{8.151}$$

The invariance equation (8.151) is completely analogous to the (8.138). Written in the form (8.151), it allows for a direct investigation of the critical points. For this purpose, we find zeroes of the derivative, $d\Phi(\overline{A}_k)/d\overline{A}_k = 0$. When the roots of the latter equation, $\overline{A}_k^c$, are found, the critical values of the wave vector are given as $-(1/3)k_c^2 c^2 = \Phi(\overline{A}_k^c)$. The condition $d\Phi(\overline{A}_k)/d\overline{A}_k = 0$ reads:

$$18\tau^2 \tau_R \overline{A}_k^3 + 3\tau(3\tau + 8\tau_R)\overline{A}_k^2 + 10(\tau + \tau_R)\overline{A}_k + 5 = 0 . \qquad (8.152)$$

Let us consider the particularly interesting case, $\epsilon = \tau_N/\tau_R \ll 1$ (the normal events are less frequent than resistive). Then the real-valued root of (8.152), $\overline{A}_k(\epsilon)$, corresponds to the coupling of the extended diffusion mode to the critical non-hydrodynamic mode. The corresponding modification of the critical wave vector k_c (8.139) due to the normal processes amounts to shifts towards shorter waves, and we derive:

$$[k_c(\epsilon)]^2 = k_c^2 + \frac{3\epsilon}{10\tau_R^2 c^2} . \qquad (8.153)$$

Accounting for Anisotropy

The above examples concerned the isotropic Debye model. Let us consider the simplest anisotropic model of a cubic media with a longitudinal (L) and two degenerated transverse (T) phonon modes, taking into account resistive processes only. Introduce the Fourier variables, e_k, e_k^T, $\boldsymbol{p}_k^T$, and $\boldsymbol{p}_k^L$, where $e_k = e_k^L + 2e_k^T$ is the Fourier transform of the total energy of the three phonon modes (the only conserved quantity), while the rest of variables are specific quantities. The isotropic model (8.125), (8.126) generalizes to [212]:

$$\partial_t e_k = -ic_L^2 \boldsymbol{k} \cdot \boldsymbol{p}_k^L - 2ic_T^2 \boldsymbol{k} \cdot \boldsymbol{p}_k^T , \qquad (8.154)$$

$$\partial_t e_k^T = -ic_T^2 \boldsymbol{k} \cdot \boldsymbol{p}_k^T + \frac{1}{\lambda}\left[c_L^3(e_k - 2e_k^T) - c_T^3 e_k^T\right] , \qquad (8.155)$$

$$\partial_t \boldsymbol{p}_k^L = -\frac{1}{3}i\boldsymbol{k}(e_k - 2e_k^T) - \frac{1}{\tau_R^L}\boldsymbol{p}_k^L , \qquad (8.156)$$

$$\partial_t \boldsymbol{p}_k^T = -\frac{1}{3}i\boldsymbol{k}e_k^T - \frac{1}{\tau_R^T}\boldsymbol{p}_k^T , \qquad (8.157)$$

where $\lambda = \tau_R^T c_T^3 + 2\tau_R^L c_L^3$. The term containing the factor λ^{-1} corresponds to the energy exchange between the L and T phonon modes. The invariance constraint for the closure relations,

$$\boldsymbol{p}_k^L = i\boldsymbol{k}A_k e_k , \quad \boldsymbol{p}_k^T = i\boldsymbol{k}B_k e_k , \quad e_k^T = X_k e_k , \qquad (8.158)$$

result in the following invariance equations for the $\boldsymbol{k}$-dependent functions A_k, B_k, and X_k:

$$k^2 c_L^2 A_k^2 + 2k^2 c_T^2 A_k B_k = -\frac{1}{\tau_R^L} A_k - \frac{1}{3}(1 - 2X_k) \ , \tag{8.159}$$

$$2k^2 c_T^2 B_k^2 + k^2 c_L B_k A_k = -\frac{1}{\tau_R^T} B_k - \frac{1}{3} X_k \ , \tag{8.160}$$

$$X_k \left(k^2 c_L^2 A_k + 2k^2 c_T^2 B_k \right) = c_T^2 k^2 B_k + \frac{1}{\lambda} \left[c_L^3 - X_k \left(2c_L^3 + c_T^3 \right) \right] \ . \tag{8.161}$$

When the energy balance equation (8.154) is closed with the relations (8.158), we obtain the dispersion relation for the extended diffusion mode, $\omega_k = \overline{A}_k + 2\overline{B}_k$, where the functions $\overline{A}_k = k^2 c_L^2 A_k$, and $\overline{B}_k = k^2 c_T^2 B_k$, satisfy the conditions: $\overline{A}_k \to 0$, and $\overline{B}_k \to 0$, as $k \to 0$. The resulting dispersion relation is rather complicated in the general case of the four parameters of the problem, c_L, c_T, τ_R^L and τ_R^T. Therefore, introducing the function $\overline{Y}_k = \overline{A}_k + 2\overline{B}_k$, let us consider the following specific situations of closed equations for the $\overline{Y}_k$ on the basis of the invariance equations (8.159):

(i) $c_L = c_T = c$, $\tau_R^L = \tau_R^T = \tau_R$ (complete degeneration of the parameters of the L and T subsystems): The system (8.159) results in two decoupled equations:

$$\overline{Y}_k \left(\tau_R \overline{Y}_k + 1 \right) + \frac{1}{3} k^2 c^2 \tau_R = 0 \ , \tag{8.162}$$

$$\left(\tau_R \overline{Y}_k + 1 \right)^2 + \frac{1}{3} k^2 c^2 \tau_R^2 = 0 \ . \tag{8.163}$$

Equation (8.162) coincides with (8.138) for the isotropic case, and its solution defines the coupling of the extended diffusion to a non-hydrodynamic mode. Equation (8.163) does not have a solution with the required asymptotic behavior $\overline{Y}_k \to 0$ as $k \to 0$, and is therefore irrelevant to the features of the diffusion mode in this completely degenerated case. It describes the two further propagating and damped non-hydrodynamic modes of the (8.154). The nature of these modes, as well of the mode which couples to the diffusion mode well be seen below.

(ii) $c_L = c_T = c$, $\tau_R^L \neq \tau_R^T$ (nondegenerate characteristic time of resistive processes in the L and T subsystems):

$$\left[\overline{Y}_k \left(\tau_R^L \overline{Y}_k + 1 \right) + \frac{1}{3} k^2 c^2 \tau_R^L \right] \times \left[\left(\tau_R' \overline{Y}_k + 3 \right) \left(\tau_R^T \overline{Y}_k + 1 \right) + \frac{1}{3} k^2 c^2 \tau_R^T \tau_R' \right]$$
$$+ \frac{2}{3} k^2 c^2 \left(\tau_R^T - \tau_R^L \right) = 0 \ , \tag{8.164}$$

where $\tau_R' = 2\tau_R^L + \tau_R^T$. As $\tau_R^T - \tau_R^L \to 0$, (8.164) tends to the degenerated case (8.162). At $k = 0$, $\tau_R^L \neq \tau_R^L$, there are four solutions to (8.164). The $\overline{Y}_0 = 0$ is the hydrodynamic solution indicating the beginning of the diffusion mode. The two non-hydrodynamic solutions, $\overline{Y}_0 = -1/\tau_R^L$, and $\overline{Y}_0 = -1/\tau_R^T$, $\overline{Y}_0 = -3/\tau_R'$, are associated with the longitudinal and the transverse phonons, respectively. The difference in relaxational times makes the latter transverse root nondegenerate, and the third non-hydrodynamic mode, $\overline{Y}_0 = -3/\tau'_R$, appears instead.

(iii) $c_L \neq c_T$, $\tau_R^L = \tau_R^T = \tau_R$ (nondegenerate speed of the L and the T sound).

$$\left[\overline{Y}_k \left(\tau_R \overline{Y}_k + 1\right) + \frac{1}{3}k^2 c_L^2 \tau_R\right] \times \left[\left(\tau_R \overline{Y}_k + 1\right)^2 + \frac{1}{3}k^2 c_T^2 \tau_R^2\right]$$
$$+\frac{2}{3}k^2 \tau_R \frac{c_L^3 \left(c_T^2 - c_L^2\right)}{2c_L^3 + c_T^3}\left(\tau_R \overline{Y}_k + 1\right) = 0 . \tag{8.165}$$

As $c_T - c_L \to 0$, (8.165) tends to the degenerate case (8.162). However, this time the non-hydrodynamic mode associated with the transverse phonons degenerates at $k = 0$.

Thus, we are able to identify the modes in (8.162) and (8.163). The non-hydrodynamic mode which couples to the extended diffusion mode is associated with the longitudinal phonons, and is the case (8.162). The case (8.163) is due to the transverse phonons. In the nondegenerate cases, (8.164) and (8.165), both pairs of modes become propagating after a certain critical values of k, and the behavior of the extended diffusion mode is influenced by all three non-hydrodynamic modes just mentioned. It should be stressed, however, that the second sound mode, which is the continuation of the diffusion mode [206, 207], is due to (8.162). The results of the above analysis lead to the following conclusion:

(i) The examples considered above indicate an interesting mechanism of a *kinetic* formation of the second sound regime from the extended diffusion with the participation of the non-hydrodynamic mode. The onset of the propagating mode shows up as the critical point of the extension of the hydrodynamic solution into the domain of finite k, which was found within the Chapman–Enskog and equivalent approaches. These results concern the situation at the high-temperature edge of the Guyer–Krumhansl window, and are complementary to the coupling between the transversal ballistic mode and the second sound at the low-temperature edge [213].

(ii) The crossover from the diffusion-like to the wave-like propagation was previously found in [214–216] in the framework of exact Chapman–Enskog solution to the Boltzmann equation for the Lorentz gas model [202], and for similar models of phonon scattering in anisotropic disordered media [217]. The characteristic common feature of the models studied in [202, 214–217] and the models [212] is the existence of a gap between the hydrodynamic (diffusive) and the non-hydrodynamic components of the spectrum. Therefore, one can expect that the destruction of the extended diffusion is solely due to the *existence* of this gap. In applications to the phonon kinetic theory this amounts to the introduction of the relaxation time approximation. In other words, we may expect that the mechanism of crossover from diffusion to second sound in the simple models [212] is identical to what could be found from the phonon-Boltzmann kinetic equation within the relaxation time approximation. However, it should be noted that the original (i.e., without the relaxation time approximation) phonon kinetic equations are *gapless* (see, for example, [211]). On the other hand, most of the works on heat propagation in solids *do* exploit the idea of the gap, since it is only possible to speak of diffusion if such a gap exists. To conclude this point, the following general hypothesis can be expressed: *the existence of diffusion (and hence of the gap in the relaxational spectrum) leads to its destruction through the coupling with a non-hydrodynamic mode.*

8.4.7 Nonlinear Grad Equations

In the preceding sections, the Chapman–Enskog and other methods were probed explicitly for the linearized Grad equations far beyond the usual Navier–Stokes approximation. This was possible, first of all because the problem of the reduced description was shaped into a rather simple *algebraic* form. Indeed, the algebraic structure of the stress tensor $\boldsymbol{\sigma}_k(\rho_k, \boldsymbol{u}_k, T_k, \boldsymbol{k})$ and of the heat flux $\boldsymbol{q}_k(\rho_k, \boldsymbol{u}_k, T_k, \boldsymbol{k})$ was fairly simple. However, when we attempt to extend the approach onto the nonlinear Grad equations, the algebraic structure of the problem is no longer simple. Indeed, when we proceed along the lines of the Chapman–Enskog method, for example, the number of *types* of terms, $\nabla \boldsymbol{u}$, $\nabla \nabla \boldsymbol{u}$, $(\nabla \boldsymbol{u})^2$, $\nabla T \nabla \rho$, and so on, in the Chapman–Enskog coefficients $\boldsymbol{\sigma}^{(n)}$ and $\boldsymbol{q}^{(n)}$ demonstrates a combinatorial growth with the order n.

Still, progress is possible if we impose some rules for the selection of the relevant terms. As applied to the Chapman–Enskog expansion, these selection rules prescribe that only contributions arising from terms with a definite structure in each order $\boldsymbol{\sigma}^{(n)}$ and $\boldsymbol{q}^{(n)}$ should be retained, and all other terms should be ignored. This approach can be linked again with the partial summation rules for the perturbation series in many-body theories, where usually terms with a definite structure are summed instead of the whole series. Our viewpoint on the problem of the extension of the hydrodynamics in the nonlinear case can be expressed as follows: The exact extension seems to be impossible, and, moreover, quite useless because of the lack of a physical transparency. Instead, certain sub-series of the Chapman–Enskog expansion, selected on clear physical grounds, may lead to less complicated equations, which, at the same time, provide an extension for a certain subclass of hydrodynamic phenomena. This viewpoint is illustrated in this section by considering a sub-series of the Chapman–Enskog expansion which provides the dominating contribution when the flow velocity becomes very large (and thus it is relevant to a high-speed subclass of hydrodynamic phenomena such as strong shock waves).

The approach to the Chapman–Enskog series for the nonlinear Grad equations just mentioned, and which was based on a diagrammatic representation of the Chapman–Enskog method, has been attempted earlier in [237]. In this section, however, we shall take the route of the dynamic invariance equations which leads to the same results more directly.

The Dynamic Viscosity Factor

The starting point is the set of one-dimensional nonlinear Grad equations for
the hydrodynamic variables ρ, u and T, coupled to the non-hydrodynamic
variable σ, where σ is the xx-component of the stress tensor:

$$\partial_t \rho = -\partial_x(\rho u) ; \tag{8.166}$$

$$\partial_t u = -u\partial_x u - \rho^{-1}\partial_x p - \rho^{-1}\partial_x \sigma ; \tag{8.167}$$

$$\partial_t T = -u\partial_x T - (2/3)T\partial_x u - (2/3)\rho^{-1}\sigma\partial_x u ; \tag{8.168}$$

$$\partial_t \sigma = -u\partial_x \sigma - (4/3)p\partial_x u - (7/3)\sigma\partial_x u - \frac{p}{\mu(T)}\sigma . \tag{8.169}$$

Here $\mu(T)$ is the temperature-dependent viscosity coefficient. We shall adopt
the form $\mu(T) = \alpha T^\gamma$, which is characteristic to the point-center models of
particles collisions, where γ varies from $\gamma = 1$ (the Maxwell molecules) to
$\gamma = 1/2$ (hard spheres), and where α is a dimensional factor.

Even in this model, the Chapman–Enskog expansion appears to be ex-
ceedingly complicated in the full setting. Therefore, we address another, simpler problem: *What is the leading correction to the Navier–Stokes approximation when the characteristic value of the average velocity is comparable to
the thermal velocity?*

Our goal is to compute the correction to the Navier–Stokes closure re-
lation, $\sigma_{\mathrm{NS}} = -(4/3)\mu\partial_x u$, for high values of the average velocity. Let us
consider first the Burnett correction from (8.166):

$$\sigma_{\mathrm{B}} = -\frac{4}{3}\mu\partial_x u + \frac{8(2-\gamma)}{9}\mu^2 p^{-1}(\partial_x u)^2 - \frac{4}{3}\mu^2 p^{-1}\partial_x(\rho^{-1}\partial_x p) . \tag{8.170}$$

The correction of the desired type is given by the nonlinear term proportional to $(\partial_x u)^2$. Each further nth term of the Chapman–Enskog expansion
contributes, among others, a nonlinear term proportional to $(\partial_x u)^{n+1}$. Such
terms can be named *high-speed* terms since they dominate the rest of the contributions in each order of the Chapman–Enskog expansion when the characteristic average velocity is comparable to the heat velocity. Indeed, if U is a
characteristic mean velocity, and $u = U\bar{u}$, where $\bar{u}$ is dimensionless, then the
term $(\partial_x u)^{n+1}$ receives the factor U^{n+1} which is the highest possible order
of U among the terms available in the nth order of the Chapman–Enskog expansion. Simple dimensional analysis leads to the conclusion that such terms
are of the form $\mu g^n \partial_x u$, where $g = p^{-1}\mu\partial_x u$ is dimensionless. Therefore, the
Chapman–Enskog expansion for the function σ may be formally rewritten as:

$$\sigma = -\mu\left\{\frac{4}{3} - \frac{8(2-\gamma)}{9}g + r_2 g^2 + \ldots + r_n g^n + \ldots\right\}\partial_x u + \ldots \tag{8.171}$$

The series in the brackets is the collection of the high-speed contributions
of interest, coming from *all* orders of the Chapman–Enskog expansion, while

the dots outside the brackets stand for the terms of other nature. Thus, the high-speed correction to the Navier–Stokes closure relation in the framework of the Grad equations (8.166) takes the form:

$$\sigma_{\mathrm{nl}} = -\mu R(g)\partial_x u \,, \tag{8.172}$$

where $R(g)$ is the function represented by a formal subsequence of Chapman–Enskog terms in the expansion (8.171). The function R can be viewed also as a dynamic modification of the viscosity μ due to the gradient of the average velocity.

We shall now turn to the problem of an explicit derivation of the function R (8.172). Following the principle of dynamic invariance, we first compute the microscopic derivative of the function σ_{nl} by substituting (8.172) into the right hand side of (8.169):

$$\partial_t^{\mathrm{micro}} \sigma_{\mathrm{nl}} = -u\partial_x \sigma_{\mathrm{nl}} - \frac{4}{3}p\partial_x u - \frac{7}{3}\sigma_{\mathrm{nl}}\partial_x u - \frac{p}{\mu(T)}\sigma_{\mathrm{nl}}$$

$$= \left\{ -\frac{4}{3} + \frac{7}{3}gR + R \right\} p\partial_x u + \dots \,, \tag{8.173}$$

where dots denote the terms irrelevant to the closure relation (8.172) (such terms appear, because (8.172) is not the exact closure relation).

Second, computing the macroscopic derivative of the closure relation (8.172) due to (8.166), (8.167), and (8.168), we obtain:

$$\partial_t^{\mathrm{macro}} \sigma_{\mathrm{nl}} = -[\partial_t \mu(T)]R\partial_x u - \mu(T)\frac{dR}{dg}[\partial_t g]\partial_x u - \mu(T)R\partial_x[\partial_t u] \,. \tag{8.174}$$

In the latter expression, the time derivatives of the hydrodynamic variables should be replaced with the right hand sides of (8.166), (8.167), and (8.168), where, in turn, the function σ should be replaced by the function σ_{nl} (8.172). After some computation, we derive the following:

$$\partial_t^{\mathrm{macro}} \sigma_{\mathrm{nl}} = \left\{ gR + \frac{2}{3}(1 - gR) \times \left(\gamma gR + (\gamma - 1)g^2\frac{dR}{dg} \right) \right\} p\partial_x u + \dots \tag{8.175}$$

Again, the dots stand for the terms irrelevant to the present analysis.

Equating the relevant terms in (8.173) and (8.175), we obtain the invariance equation for the function R:

$$(1 - \gamma)g^2(1 - gR)\frac{dR}{dg} + \gamma g^2 R^2 + \left[\frac{3}{2} + g(2 - \gamma) \right] R - 2 = 0 \,. \tag{8.176}$$

For Maxwell molecules ($\gamma = 1$), (8.176) simplifies considerably, and becomes algebraic:

$$g^2 R^2 + \left(\frac{3}{2} + g \right) R - 2 = 0 \,. \tag{8.177}$$

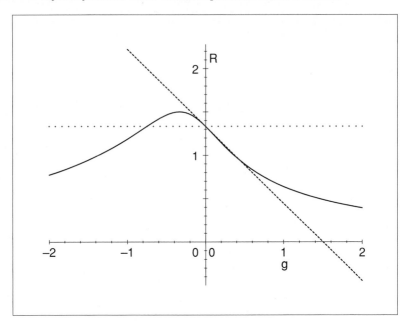

Fig. 8.10. Viscosity factor $R(g)$ for Maxwell molecules. *Solid*: exact solution. *Dash*: the Burnett approximation. *Dots*: the Navier–Stokes approximation

The solution which recovers the Navier–Stokes closure relation in the limit of small g then reads:

$$R_{\mathrm{MM}} = \frac{-3 - 2g + 3\sqrt{1 + (4/3)g + 4g^2}}{4g^2} . \tag{8.178}$$

Function R_{MM} (8.178) is plotted in Fig. 8.10. Note that R_{MM} is positive for all values of its argument g, as is appropriate for the viscosity factor, while the Burnett approximation to the function R_{MM} violates positivity.

For other models ($\gamma \neq 1$), the invariance equation (8.176) is a rather complicated nonlinear ODE with the initial condition $R(0) = 4/3$ (the Navier–Stokes condition). Several ways to derive analytic results are possible. One possibility is to expand the function R into powers of g, in the point $g = 0$. This will bring us back to the original sub-series of the Chapman–Enskog expansion (see (8.171)). Instead, we take advantage of the opportunity offered by the parameter γ. Introduce another parameter $\beta = 1 - \gamma$, and consider the expansion:

$$R(\beta, g) = R_0(g) + \beta R_1(g) + \beta^2 R_2(g) + \dots .$$

Substituting this expansion into the invariance equation (8.176), we derive $R_0(g) = R_{\mathrm{MM}}(g)$,

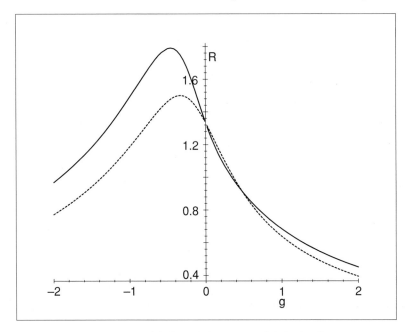

Fig. 8.11. Viscosity factor $R(g)$ for hard spheres. *Solid*: the first approximation. *Dash*: exact solution for the Maxwell molecules

$$R_1(g) = -g(1 - gR_0)\frac{R_0 + g(\mathrm{d}R_0/\mathrm{d}g)}{2g^2 R_0 + g + (3/2)} , \qquad (8.179)$$

etc. That is, the solution for other models is constructed in a form of a series with the exact solution for the Maxwell molecules as the leading term. For hard spheres ($\beta = 1/2$), the result to the first-order term reads: $R_{\mathrm{HS}} \approx R_{\mathrm{MM}} + (1/2)R_1$. The resulting approximate viscosity factor is shown in Fig. 8.11. The features of the approximation obtained are qualitatively the same as in the case of Maxwell molecules.

It is interesting that precisely the same result for the nonlinear elongational viscosity obtained in [17] was derived later by A. Santos [18] from the solution to the BGK kinetic equation in the regime of so-called homoenergetic extension flow, see also [19], where a misprint in [17] in formula (8.179) was detected. For further discussion, see [20]

Attraction to the Invariant Set

Above, we have derived a correction to the Navier–Stokes expression for the stress σ, in the one-dimensional case, for large values of the average velocity u. This correction has the form $\sigma = -\mu R(g)\partial_x u$, where $g \propto \partial_x u$ is the longitudinal rate. The viscosity factor $R(g)$ is a solution to the differential equation (8.176), subject to a certain initial condition. Uribe and Piña [218] have indicated some interesting features of the invariance equation (8.176) for the model of hard spheres. In particular, they have found that a numerical integration from the initial point into the domain of negative longitudinal rates is very difficult. What happens to the relevant solution for negative values of g?

Let us denote as $P = (g, R)$ the points in the (g, R) plane. The relevant solution $R(g)$ emerges from the point $P_0 = (0, 4/3)$, and can be uniquely continued to arbitrary values of g, positive and negative. This solution can be constructed, for example, with the Taylor expansion, and which is identical with the relevant sub-series of the Chapman–Enskog expansion. However, the difficulty in constructing this solution numerically for $g < 0$ originates from the fact that the same point P_0 is the point of *essential singularity* of other (irrelevant) solutions to (8.176). Indeed, for $|g| \ll 1$, let us consider $\tilde{R}(g) = R(g) + \Delta$, where $R(g) = (4/3) + (8/9)(\gamma - 2)g$ is the relevant solution for small $|g|$, and $\Delta(g)$ is a deviation. Neglecting in (8.176) all regular terms (of the order g^2), and also neglecting $g\Delta$ in comparison to Δ, we derive the following equation: $(1 - \gamma)g^2(d\Delta/dg) = -(3/2)\Delta$. The solution is $\Delta(g) = \Delta(g_0)\exp[a(g^{-1} - g_0^{-1})]$, where $a = (3/2)(1 - \gamma)^{-1}$. The essential singularity at $g = 0$ is apparent from this solution, unless $\Delta(g_0) \neq 0$ (that is, no singularity exists only for the relevant solution, $\tilde{R} = R$). Let $\Delta(g_0) \neq 0$. If $g < 0$, then $\Delta \to 0$, together with all its derivatives, as $g \to 0$. If $g > 0$, the solution blows up, as $g \to 0$.

The complete picture for $\gamma \neq 1$ is as follows: The lines $g = 0$ and $P = (g, g^{-1})$ define the boundaries of the basin of attraction $A = A_- \bigcup A_+$, where $A_- = \{P| -\infty < g < 0, R > g^{-1}\}$, and $A_+ = \{P|\infty > g > 0, R < g^{-1}\}$. The graph $G = (g, R(g))$ of the relevant solution belongs to the closure of A, and goes through the points $P_0 = (0, 4/3)$, $P_- = (-\infty, 0)$, and $P_+ = (\infty, 0)$. These points at the boundaries of A are the points of essential singularity of any other (irrelevant) solution with the initial condition $P \in A$, $P \notin A \bigcap G$. Namely, if $P \in A_+$, $P \notin A_+ \bigcap G$, the solution blows up at P_0, and attracts to P_+. If $P \in A_-$, $P \notin A_- \bigcap G$, the solution blows up at P_-, and attracts to P_0.

The above consideration is supported by a numerical study of (8.176). In Fig. 8.12, it is demonstrated how the dynamic viscosity factor $R(g)$ emerges as the attractor of various solutions to the invariance equation (8.176) [the case considered corresponds to hard spheres, $\gamma = 1/2$]. The analytical approximation (8.179) is also shown in Fig. 8.12. It provides a reasonable global approximation to the attractor for both positive and negative g. We conclude with a discussion.

(i) The main feature of the above example of extending the hydrodynamic description into a highly non-equilibrium and nonlinear domain can be expressed as follows: this is an *exact partial summation* of the Chapman–Enskog expansion. "Partial" means that the relevant high-speed terms, dominating the other contributions in the limit of the high average velocity, were

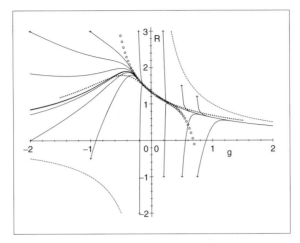

Fig. 8.12. Viscosity factor as an attractor. *Solid* lines: numerical integration with various initial points (crosses). Two poorly resolved lines correspond to the initial conditions $(-100, 0)$ and $(-100, 3)$. *Circles*: Taylor expansion to the 5th order. *Dots*: the analytical approximation of (8.179). *Dash*: boundaries of the basin of attraction

accounted to all orders of the original Chapman–Enskog expansion. "Exact" means that, though we have used the formally different route, the result is indeed the sum of the relevant sub-series of the original Chapman–Enskog expansion. In other words, if we now expand the function $R_{\mathrm{MM}}(g)$ (8.178) in powers of g, around the point $g = 0$, we obtain again to the corresponding series inside the brackets in (8.171). That this is indeed true can be checked up to the few lower orders straightforwardly, although the complete proof requires a more involved analysis. As the final comment to this point, we would like to stress a certain similarity between the problem considered above and the frequent situations in many-body problems: there is no single leading *term*; instead, there is the leading *sub-series* of the perturbation expansions, under certain conditions.

(ii) Let us discuss briefly the features of the resulting hydrodynamics. The hydrodynamic equations are now given by (8.166), (8.167), and (8.168), where σ is replaced by σ_{nl} (8.172). First, the correction concerns the non-linear regime, and, thus, the linearized form of the new equations coincides with the linearized Navier–Stokes equations. Second, the solution (8.178) for Maxwell molecules and the result of the approximation (8.179) for other models suggest that the modified viscosity μR gives a vanishing contribution in the limit of very high values of the average velocity. This feature seems to be of no surprise: if the average velocity is very high in comparison to other characteristic velocities (in our case, to the heat velocity), no mechanisms of momentum transfer are relevant except for the transfer with the stream.

However, a cautious remark is in order since the original "kinetic" description are Grad's equations (8.166) and not the Boltzmann equation.

(iii) The invariance equation (8.176) defines the relevant physical solution to the viscosity factor for all values of g, and demonstrates an interesting phase-space behavior similar to those of finite-dimensional dynamical systems.

8.5 The Main Lesson

Up to now, the problem of the exact relationship between kinetics and hydrodynamics remains unsolved. All the methods used to establish this relationship are not rigorous, and involve approximations. In this work, we have considered situations where hydrodynamics is the exact consequence of kinetics, and in that respect, a new class of exactly solvable models of statistical physics has been established.

The main lesson we can learn from the exact solution is the following: The Chapman–Enskog method is the Taylor series expansion approach to solving the appropriate invariance equation. Alternative iteration methods are much more robust for solving this equation. Therefore, it seems quite important to develop approaches to the problem of reduced description based on the principle of dynamic invariance rather than on particular methods of solving the invariance equations. The exact solutions where these questions can be answered in all the quantitative details provide a sound motivation for such developments.

9 Relaxation Methods

The "large stepping" relaxation method for solution of the invariance equation is developed.

9.1 "Large Stepping" for the Equation of the Film Motion

Relaxation method is an alternative to the Newton iteration method described in Chap. 6: The initial approximation to the invariant manifold F_0 is moved with the film extension (4.5),

$$\frac{\mathrm{d}F_t(y)}{\mathrm{d}t} = (1 - P_{t,y})J(F_t(y)) = \Delta_{F(y)} \,,$$

until a fixed point is reached. The advantage of this method is a relative freedom in its implementation: equation (4.5) needs not to be solved exactly, one is interested only in finding fixed points. Therefore, a "large stepping" in the direction of the defect, $\Delta_{F(y)}$ is possible, while the termination point is defined by the condition that the vector field becomes orthogonal to $\Delta_{F(y)}$. For simplicity, let us consider the procedure of termination in the linear approximation. Let $F_0(y)$ be the initial approximation to the invariant manifold, and we seek the first correction,

$$F_1(y) = F_0(y) + \tau_1(y)\Delta_{F_0(y)} \,,$$

where function $\tau_1(y)$ has the dimension of time, and is found from the condition that the linearized vector field attached to the points of the new manifold is orthogonal to the initial defect,

$$\langle \Delta_{F_0(y)} | (1 - P_y)[J(F_0(y)) + \tau_1(y)(D_x J)_{F_0(y)} \Delta_{F_0(y)}] \rangle_{F_0(y)} = 0 \,. \qquad (9.1)$$

Explicitly,

$$\tau_1(y) = -\frac{\langle \Delta_{F_0(y)} | \Delta_{F_0(y)} \rangle_{F_0(y)}}{\langle \Delta_{F_0(y)} | (D_x J)_{F_0(y)} | \Delta_{F_0(y)} \rangle_{F_0(y)}} \,. \qquad (9.2)$$

Further steps $\tau_k(y)$ are found in the same way. It is clear from the above that the step of the relaxation method for the film extension is equivalent to

Alexander N. Gorban and Iliya V. Karlin: *Invariant Manifolds for Physical and Chemical Kinetics*, Lect. Notes Phys. **660**, 247–277 (2005)
www.springerlink.com

the Galerkin approximation for solving the iteration of the Newton method with incomplete linearization. Actually, the relaxation method was first introduced in these terms in [24]. It was implemented in the method of invariant grids [105] for the grid-based numerical representations of manifolds (see Chap. 10). An idea of using the explicit Euler method to approximate the finite-dimensional inertial manifold was proposed earlier in [25]. In our approach the special choice of the projector field is important. For recent development of the numerical Euler-type methods for the solution of the invariance equation see [28].

The advantage of the equation (9.2) is the explicit form of the size of the steps $\tau_k(y)$. This method was successfully applied to the closure problem in the context of the Fokker-Planck equation [24].

9.2 Example: Relaxation Method for the Fokker-Planck Equation

We address here the problem of closure for the Fokker-Planck equation (FPE) (2.60) in a general setting. First, we review the maximum entropy principle as a source of suitable quasiequilibrium initial approximations for closures. We also discuss a version of the maximum entropy principle, valid for a near-equilibrium dynamics, and which results in explicit formulas for arbitrary potential U and diffusion matrix D.

In this Example we consider the FPE of the form (2.60):

$$\partial_t W(\boldsymbol{x}, t) = \partial_{\boldsymbol{x}} \cdot \{D \cdot [W \partial_{\boldsymbol{x}} U + \partial_{\boldsymbol{x}} W]\} \ . \tag{9.3}$$

Here $W(\boldsymbol{x}, t)$ is the probability density over the configuration space $\boldsymbol{x}$, at the time t, while $U(\boldsymbol{x})$ and $D(\boldsymbol{x})$ are the potential and the positively semi-definite $(y \cdot D \cdot y \geq 0)$ diffusion matrix.

9.2.1 Quasi-Equilibrium Approximations for the Fokker-Planck Equation

The quasiequilibrium closures are almost never invariants of the true moment dynamics. For corrections to the quasiequilibrium closures, we apply the method of invariant manifold, which is carried out (subject to certain approximations explained below) to explicit recurrence formulas for one-moment near-equilibrium closures for arbitrary U and D. As a by-product, these formulas provide also a method for computing the lowest eigenvalue of the problem, which dominates the near-equilibrium FPE dynamics. Results are tested with model potentials, including the FENE-like potentials [151–153]. A generalization of the present approach to many-moment closures is also straightforward.

Let us denote as M the set of linearly independent moments,

$$\{M_0, M_1, \ldots, M_k\}, \text{ where } M_i[W] = \int m_i(x)W(x)\,\mathrm{d}x, \; m_0 = 1 .$$

We assume that there exists a function $W^*(M, x)$ which extremizes the entropy S (2.61) under the constraints of fixed M. This quasiequilibrium distribution function may be written as

$$W^* = W_{\text{eq}} \exp\left[\sum_{i=0}^{k} \Lambda_i m_i(x) - 1\right] , \tag{9.4}$$

where $\Lambda = \{\Lambda_0, \Lambda_1, \ldots, \Lambda_k\}$ are Lagrange multipliers. Closed-form equations for moments M are derived in two steps. First, the quasiequilibrium distribution (9.4) is substituted into the FPE (9.3) or (2.62) to give a formal expression:

$$\partial_t W^* = \hat{M}_{W^*}(\delta S / \delta W)\big|_{W=W^*}$$

, where $\hat{M}_{W^*}$ is given by (2.62). Second, introducing the quasiequilibrium projector Π^*,

$$\Pi^* \bullet = \sum_{i=0}^{k} (\partial W^* / \partial M_i) \int m(x) \bullet \mathrm{d}x ,$$

and applying Π^* on both sides of the formal expression, we derive closed for M in the quasiequilibrium approximation. Further processing requires explicit solution to the constraints, $\int W^*(\Lambda, x)m_i(x)\,\mathrm{d}x = M_i$, to get the dependence of Lagrange multipliers Λ on the moments M. Though typically the functions $\Lambda(M)$ are not known explicitly, one general remark about the moment equations is readily available. Specifically, the moment equations in the quasiequilibrium approximation have the form:

$$\dot{M}_i = \sum_{j=0}^{k} M_{ij}^*(M)(\partial S^*(M)/\partial M_j) , \tag{9.5}$$

where $S^*(M) = S[W^*(M)]$ is the quasiequilibrium entropy, and where M_{ij}^* is an M-dependent $(k+1) \times (k+1)$ matrix:

$$M_{ij}^* = \int W^*(M, x)[\partial_x m_i(x)] \cdot D(x) \cdot [\partial_x m_i(x)]\,\mathrm{d}x .$$

The matrix M_{ij}^* is symmetric, positive semi-definite, and its kernel is the vector δ_{0i}. Thus, *the quasiequilibrium closure reproduces the gradient structure on the macroscopic level* (2.62), the vector field of quasiequilibrium equations (9.5) is a transform of the gradient of the quasiequilibrium entropy given by the symmmetric positive operator.

The following version of the quasiequilibrium closures makes it possible to derive more explicit results in the general case [233, 246–248]: In many cases, one can split the set of moments M in two parts, $M_I = \{M_0, M_1, \ldots, M_l\}$ and $M_{II} = \{M_{l+1}, \ldots, M_k\}$, in such a way that the quasiequilibrium distribution can be constructed explicitly for M_I as $W_I^*(M_I, x)$. The full quasiequilibrium problem for $M = \{M_I, M_{II}\}$ in the "shifted" formulation reads (see the "triangle entropy method" in Chap. 5): extremize the functional $S[W_I^* + \Delta W]$ with respect to ΔW, under the constraints $M_I[W_I^* + \Delta W] = M_I$ and $M_{II}[W_I^* + \Delta W] = M_{II}$. Let us denote as $\Delta M_{II} = M_{II} - M_{II}(M_I)$ deviations of the moments M_{II} from their values in the quasiequilibrium state W_I^*. For small deviations, the entropy is well approximated with its quadratic part

$$\Delta S = - \int \Delta W \left[1 + \ln \frac{W_I^*}{W_{\mathrm{eq}}} \right] \mathrm{d}x - \frac{1}{2} \int \frac{\Delta W^2}{W_I^*} \mathrm{d}x \ .$$

Taking into account the fact that $M_I[W_I^*] = M_I$, we come to the following maximizaton problem:

$$\Delta S[\Delta W] \to max, \ \ M_I[\Delta W] = 0, \ \ M_{II}[\Delta W] = \Delta M_{II} \ . \tag{9.6}$$

The solution to the problem (9.6) is always explicitly found from a $(k+1) \times (k+1)$ system of linear algebraic equations for Lagrange multipliers. This triangle entropy method for Boltzmann equations was discussed in details in Sect. 5.6.

In the remainder of this section we deal solely with one-moment near-equilibrium closures: $M_I = M_0$, (i. e. $W_I^* = W_{\mathrm{eq}}$), and the set M_{II} contains a single moment $M = \int mW \, \mathrm{d}x$, $m(x) \neq 1$. We shall specify notations for the near-equilibrium FPE, writing the distribution function as $W = W_{\mathrm{eq}}(1 + \Psi)$, where the function Ψ satisfies an equation:

$$\partial_t \Psi = W_{\mathrm{eq}}^{-1} \hat{J} \Psi \ , \tag{9.7}$$

where $\hat{J} = \partial_x \cdot [W_{\mathrm{eq}} D \cdot \partial_x]$. The triangle one-moment quasiequilibrium function reads:

$$W^{(0)} = W_{\mathrm{eq}} \left[1 + \Delta M m^{(0)} \right] \tag{9.8}$$

where

$$m^{(0)} = [\langle mm \rangle - \langle m \rangle^2]^{-1}[m - \langle m \rangle] \ . \tag{9.9}$$

Here brackets $\langle \ldots \rangle = \int W_{\mathrm{eq}} \ldots \mathrm{d}x$ denote equilibrium averaging. The super-script (0) indicates that the triangle quasiequilibrium function (9.8) will be considered as the initial approximation to the procedure which we address below. Projector for the approximation (9.8) has the form

$$\Pi^{(0)} \bullet = W_{\mathrm{eq}} \frac{m^{(0)}}{\langle m^{(0)} m^{(0)} \rangle} \int m^{(0)}(x) \bullet \mathrm{d}x \ . \tag{9.10}$$

Substituting the function (9.8) into the FPE (9.7), and applying the projector (9.10) on both the sides of the resulting formal expression, we derive the equation for M:

$$\dot{M} = -\lambda_0 \Delta M \ , \tag{9.11}$$

where $1/\lambda_0$ is an effective relaxation time of the moment M to its equilibrium value, in the quasiequilibrium approximation (9.8):

$$\lambda_0 = \langle m^{(0)} m^{(0)} \rangle^{-1} \langle \partial_x m^{(0)} \cdot D \cdot \partial_x m^{(0)} \rangle \ . \tag{9.12}$$

9.2.2 The Invariance Equation for the Fokker-Planck Equation

Both the quasiequilibrium and the triangle quasiequilibrium closures are almost never invariants of the FPE dynamics. That is, the moments M of solutions of the FPE (9.3) and the solutions of the closed moment equations like (9.5), are different functions of time, even if the initial values coincide. These variations are generally significant even for the near-equilibrium dynamics. Therefore, we ask for corrections to the quasiequilibrium closures to end up with the invariant closures. This problem falls precisely into the framework of the method of invariant manifold [11] (Chap. 6), and we shall apply this method to the one-moment triangle quasiequilibrium closure approximations, as a simple example.

First, the invariant one-moment closure is given by an unknown distribution function $W^{(\infty)} = W_{eq}[1 + \Delta M m^{(\infty)}(x)]$ which satisfies the invariance equation

$$[1 - \Pi^{(\infty)}] \hat{J} m^{(\infty)} = 0 \ . \tag{9.13}$$

Here $\Pi^{(\infty)}$ is the projector, associated with function $m^{(\infty)}$, and which is also yet unknown. Equation (9.13) is a formal expression of the invariance principle for a one-moment near-equilibrium closure: considering $W^{(\infty)}$ as a manifold in the space of distribution functions, parameterized with the values of the moment M, we require that the microscopic vector field $\hat{J} m^{(\infty)}$ be equal to its projection, $\Pi^{(\infty)} \hat{J} m^{(\infty)}$, onto the tangent space of the manifold $W^{(\infty)}$.

Now we turn our attention to solving the invariance equation (9.13) iteratively, beginning with the triangle one-moment quasiequilibrium approximation $W^{(0)}$ (9.8). We apply the following iteration process to (9.13):

$$[1 - \Pi^{(k)}] \hat{J} m^{(k+1)} = 0 \ , \tag{9.14}$$

where $k = 0, 1, \ldots$, and where $m^{(k+1)} = m^{(k)} + \mu^{(k+1)}$, and the correction satisfies the condition $\langle \mu^{(k+1)} m^{(k)} \rangle = 0$. The projector is updated after each iteration, and it has the form

$$\Pi^{(k+1)} \bullet = W_{eq} \frac{m^{(k+1)}}{\langle m^{(k+1)} m^{(k+1)} \rangle} \int m^{(k+1)}(x) \bullet \, dx \ . \tag{9.15}$$

Applying $\Pi^{(k+1)}$ to the formal expression,

$$W_{eq} m^{(k+1)} \dot{M} = \Delta M [1 - \Pi^{(k+1)}] m^{(k+1)} \ ,$$

we derive the $(k+1)$th update of the effective time (9.12):

$$\lambda_{k+1} = \frac{\langle \partial_x m^{(k+1)} \cdot D \cdot \partial_x m^{(k+1)} \rangle}{\langle m^{(k+1)} m^{(k+1)} \rangle} . \qquad (9.16)$$

Specializing to the one-moment near-equilibrium closures, and following the general argument of Chap. 6, solutions to the invariance equation (9.13) are eigenfunctions of the operator $\hat{J}$, while the limit of the iteration process (9.14) is the eigenfunction which corresponds to the eigenvalue with the minimal nonzero absolute value.

9.2.3 Diagonal Approximation

In order to obtain more explicit results, we shall now proceed with to an approximate solution to the problem (9.14) *at each iteration*. The correction $\mu^{(k+1)}$ satisfyes the condition $\langle m^{(k)} \mu^{(k+1)} \rangle = 0$, and can be decomposed as follows: $\mu^{(k+1)} = \alpha_k e^{(k)} + e_{\perp}^{(k)}$. Here $e^{(k)}$ is the defect of the kth approximation: $e^{(k)} = W_{eq}^{-1}[1 - \Pi^{(k)}]\hat{J}m^{(k)} = \lambda_k m^{(k)} + R^{(k)}$, where

$$R^{(k)} = W_{eq}^{-1} \hat{J} m^{(k)} . \qquad (9.17)$$

The function $e_{\perp}^{(k)}$ is orthogonal to both $e^{(k)}$ and $m^{(k)}$ ($\langle e^{(k)} e_{\perp}^{(k)} \rangle = 0$, and $\langle m^{(k)} e_{\perp}^{(k)} \rangle = 0$).

Our *diagonal approximation* (DA) consists in neglecting the part $e_{\perp}^{(k)}$. In other words, we seek an improvement of the non-invariance of the kth approximation *along its defect*, $\Delta = e^{(k)}$. Specifically, we consider the following ansatz at the kth iteration:

$$m^{(k+1)} = m^{(k)} + \alpha_k e^{(k)} . \qquad (9.18)$$

Substituting the ansatz (9.18) into (9.14), we integrate the latter expression with the functon $e^{(k)}$ to evaluate the coefficient α_k:

$$\alpha_k = \frac{A_k - \lambda_k^2}{\lambda_k^3 - 2\lambda_k A_k + B_k} , \qquad (9.19)$$

where functions A_k and B_k are represented by the following equilibrium averages:

$$A_k = \langle m^{(k)} m^{(k)} \rangle^{-1} \langle R^{(k)} R^{(k)} \rangle \qquad (9.20)$$
$$B_k = \langle m^{(k)} m^{(k)} \rangle^{-1} \langle \partial_x R^{(k)} \cdot D \cdot \partial_x R^{(k)} \rangle .$$

Finally, putting together (9.16), (9.17), (9.18), (9.19), and (9.20), we arrive at the following DA recurrence solution:

$$m^{(k+1)} = m^{(k)} + \alpha_k[\lambda_k m^{(k)} + R^{(k)}] , \tag{9.21}$$

$$\lambda_{k+1} = \frac{\lambda_k - (A_k - \lambda_k^2)\alpha_k}{1 + (A_k - \lambda_k^2)\alpha_k^2} . \tag{9.22}$$

Notice that the stationary points of the DA process (9.22) are the true solutions to the invariance equation (9.13). What *may be* lost within the DA is the convergence to the true limit of the procedure (9.14), i.e. to the *minimal* eigenvalue. In a general situation this is highly improbable, though.

In order to test the convergence of the DA process (9.22) we considered two potentials U in the FPE (9.3) with a constant diffusion matrix D. The first test was with the square potential $U = x^2/2$, in the three-dimensional configuration space, since for this potential the spectrum is well-known. We have considered two examples of the initial one-moment quasiequilibrium closures with $m^{(0)} = x_1 + 100(x^2 - 3)$ (example 1), and $m^{(0)} = x_1 + 100x^6 x_2$ (example 2), in (9.9). The result of performance of the DA for λ_k is presented in Table 9.1, together with the error δ_k which was estimated as the norm of the variance at each iteration: $\delta_k = \langle e^{(k)} e^{(k)} \rangle / \langle m^{(k)} m^{(k)} \rangle$. In both examples, we see a good monotonic convergency to the minimal eigenvalue $\lambda_\infty = 1$, corresponding to the eigenfunction x_1. This convergence is even striking in the example 1, where the initial choice was very close to a different eigenfunction $x^2 - 3$, and which can be seen in the non-monotonic behavior of the variance. Thus, we have an example to trust the DA approximation as converging to the proper object.

Table 9.1. Iterations λ_k and the error δ_k for $U = x^2/2$

		0	1	4	8	12	16	20
Ex. 1	λ	1.99998	1.99993	1.99575	1.47795	1.00356	1.00001	1.00000
	δ	$0.16 \cdot 10^{-4}$	$0.66 \cdot 10^{-4}$	$0.42 \cdot 10^{-2}$	0.24	$0.35 \cdot 10^{-2}$	$0.13 \cdot 10^{-4}$	$0.54 \cdot 10^{-7}$

		0	1	2	3	4	5	6
Ex. 2	λ	3.399	2.437	1.586	1.088	1.010	1.001	1.0002
	δ	1.99	1.42	0.83	0.16	$0.29 \cdot 10^{-1}$	$0.27 \cdot 10^{-2}$	$0.57 \cdot 10^{-3}$

For the second test, we have taken a one-dimensional potential $U = -50\ln(1 - x^2)$, the configuration space is the segment $|x| \le 1$. Potentials of this type (a so-called FENE potential) are used in applications of the FPE to models of polymer solutions [151–153]. Results are given in Table 9.2 for the two initial functions, $m^{(0)} = x^2 + 10x^4 - \langle x^2 + 10x^4 \rangle$ (example 3), and $m^{(0)} = x^2 + 10x^8 - \langle x^2 + 10x^8 \rangle$ (example 4). Both examples demonstrate a stabilization of the λ_k at the same value after some ten iterations.

In conclusion, we have developed the principle of invariance to obtain moment closures for the Fokker-Planck equation (9.3), and have derived explicit results for the one-moment near-equilibrium closures, particularly important to get information about the spectrum of the FP operator.

Table 9.2. Iterations λ_k for $U = -50\ln(1 - x^2)$

	0	1	2	3	4	5	6	7	8
Ex. 3 λ	213.17	212.186	211.914	211.861	211.849	211.845	211.843	211.842	211.841
Ex. 4 λ	216.586	213.135	212.212	211.998	211.929	211.899	211.884	211.876	211.871

9.3 Example: Relaxational Trajectories: Global Approximations

Here we describe semi-analytical approximate methods for nonlinear space-independent dissipative systems equipped with the entropy functional. The key point of the analysis is an upper limiting state in the beginning of the relaxation. Extremal properties of this state are described, and explicit estimations are derived. This limiting state is used to construct explicit approximations of the trajectories. Special effort is paid to accomplish positivity, smoothness and the entropy growth along the approximate trajectories. The method is tested for the space-independent Boltzmann equation with various collision mechanisms.

9.3.1 Initial Layer and Large Stepping

For relaxing systems, it is a common place to distinguish three subsequent regimes on a way from an initial non-equilibrium state $f_0(\Gamma)$ to the final equilibrium state $f^0(\Gamma)$, where Γ is the phase variable: the early-time relaxation immediately after the system leaves the initial state f_0, the intermediate regime, and the final regression to the equilibrium state f^0. This model picture is only approximate. For gases, the early-time relaxation occurs in a few first collisions of the molecules, and can be singled out from the whole relaxational process and investigated separately.

Considering the beginning of the relaxation, we may expect that it is dominated by a rate of processes in the initial state. In the case of a dilute gas, in particular, this rate is given by the Boltzmann collision integral, $Q(f)$, evaluated in the state f_0, and equal to $Q_0 = Q(f_0)$. The latter expression is the known function of the phase variable, $Q_0(\Gamma)$. Put differently, our expectation is that states which the system passes through in the beginning are close to those on a ray, $f(\Gamma, a)$:

$$f(\Gamma, a) = f_0(\Gamma) + aQ_0(\Gamma) , \tag{9.23}$$

where $a \geq 0$ is a scalar variable (we use dimensionless variables). It is clear that such an approximation can be valid if only a "is not too large". On the other hand, nothing tells us ultimatively that a must be "strictly infinitesimal" if we want to obtain at least a moderate by accuracy approximation. In general, this consideration can be relevant if the parameter a in (9.23) does not exceed some certain upper value a^*.

In this Example we give an answer to the following question: what is the upper limiting state, f^*, the system *cannot* overcome when driven with the initial rate Q_0? As long as we can consider Q_0 as the dominant direction in the early-time relaxation, the answer amounts to an upper estimate of the parameter a in (9.23), and thus the limiting state f^* is:

$$f^*(\Gamma) = f(\Gamma, a^*) = f_0(\Gamma) + a^* Q_0(\Gamma) , \qquad (9.24)$$

where the value a^* is the subject of the analysis to be performed.

Our approach will be based on the following consideration. Denote as $S(f)$ the entropy of the state $f(\Gamma)$, and as $S(a)$ its value in the state $f(\Gamma, a)$ (9.23). A state $f(\Gamma, a')$ can be regarded *accessible* from the initial state $f(\Gamma, 0) = f_0(\Gamma)$ in the course of the Q_0-dominated dynamics, if and only if the function $S(a)$ increases with an increase of the variable a from 0 to a'. The upper limiting value, a^*, is thus characterized by the following two properties:

1. $S(a)$ increases, as a increases from 0 to a^*.
2. $S(a)$ decreases, as a exceeds a^*.

Assuming the usual convexity properties of the entropy, we conclude that the state $f(\Gamma, a^*)$ with these properties is unique.

In the next subsection, "Extremal properties of the limiting state," we derive an equation for the limiting state $f(\Gamma, a^*)$ in two ways: firstly, as a direct consequence of the two properties just mentioned, and, secondly, as an equilibrium state of an appropriately chosen kinetic model of the Q_0-dominated relaxation. Next we introduce a method to obtain the explicit estimate of the function $f(\Gamma, a^*)$ (details are given in special Subsect. 9.3.5 "Estimations"). With this, we get the answer to the question posed above.

The derivation of the state $f(\Gamma, a^*)$ plays the key role in the section "Approximate phase trajectories". There we aim at constructing explicit approximations to trajectories of a given space-independent kinetic equation. Namely, we construct an explicit function $f(\Gamma, a)$, where parameter a spans a segment $[0, 1]$, and which satisfies the following conditions:

1. $f(\Gamma, 0) = f_0(\Gamma)$.
2. $f(\Gamma, 1) = f^0(\Gamma)$.
3. $f(\Gamma, a)$ is a non-negative function of Γ for each a.
4. $C(a) \equiv C(f(a)) = $ const, where $C(f)$ are linear conserved quantities.
5. $S(a) \equiv S(f(a))$ is a monotonically increasing function of a.
6. $\partial f(\Gamma, a)/\partial a|_{a=0} = k Q_0(\Gamma)$, where $k > 0$.

Function $f(\Gamma, a)$ is a path from the initial state f_0 to the equilibrium state f^0 (conditions 1 and 2). All states of the path make physical sense (condition 3), conserved quantities remain fixed, and the entropy monotonically increases along the path (conditions 4 and 5). Finally, condition 6 requires that the path is tangent to the exact trajectory in their common initial state f_0. A function $f(\Gamma, a)$ with the properties 1-6 is, of course, not unique but a construction of a *definite* example is a rather non-trivial task. Indeed, the major difficulty is to take into account the tangency condition 6 together with the rest of the requirements.

The simplest function with the properties 1-6, and which depends *smoothly* on a, is constructed explicitly in the Subsect. 9.3.3 "Approximate trajectories" (details of the procedure are given in Subsect. 9.3.5). We also discuss the question of the time dependence $f(\Gamma, a(t))$. In the section "Relaxation of the Boltzmann gas", the method is applied to the space-independent nonlinear Boltzmann equation for several collisional mechanisms. In particular, we compare our approximations with the celebrated BKW-mode [255, 256, 262] for the Maxwell molecules, and with solutions to the two-dimensional very hard particles model (VHP) [257, 258].

Before to proceeding any further, it is worthwhile to give here a brief comment on the status of the approximate trajectories considered below. It is well known that the space-independent problem for dissipative kinetic equations is one of the most developed branches of kinetic theory with respect to existence and uniqueness theorems [259–261]. The exact treatment of specific models is also avaiable [263, 286]. On the other hand, there exists a gap of *approximate semi-analytical* methods in this problem. This is not surprising because most of the techniques of the kinetic theory [239] are based on a small parameter expansions, and this is simply not the case of the initial layer problem. The present study fills out this gap. Indeed, as the examples demonstrate, the smooth approximations $f(a, \Gamma)$ constructed below provide a reasonable (and simple) approximation to the exact trajectories.

Moreover, these functions serve for the initial approximation in an iterative method of constructing the exact trajectories for the dissipative systems [26]. This method, in turn, is based on a more general consideration of the paper [11] (Chap. 6). We give additional comments on this iterative method below, as well as we provide an illustration of the correction.

9.3.2 Extremal Properties of the Limiting State

Let us first come to the equation for the limiting state $f(\Gamma, a^*)$ (9.24) in an informal way. The two features of the function $f(\Gamma, a^*)$ indicated above tell us that this is the state of the entropy maximum on the ray $f(\Gamma, a)$ (9.23)[1]. The extremum condition in this state reads:

$$D_f S|_{f=f(\Gamma,a^*)} \left(\frac{\partial f(\Gamma, a)}{\partial a} \right) = \int \frac{\partial f(\Gamma, a)}{\partial a} \frac{\delta S(f)}{\delta f} \Bigg|_{f=f(\Gamma,a^*)} d\Gamma = 0 , \quad (9.25)$$

where $\delta S/\delta f$ denotes the (functional) derivative of the entropy evaluated at the state $f(\Gamma, a^*)$. For a particularly interesting case of

---

[1] The entropy $S(a)$ increases when a runs from zero to a^*, and $S(a)$ starts to decrease when a exceeds a^*.

$$S_B(f) = -\int f(\Gamma) \ln f(\Gamma) \, d\Gamma$$

(the Boltzmann entropy), and $\int Q(f) \, d\Gamma = 0$ (conservation of the number of particles), (9.25) gives:

$$\int Q_0(\Gamma) \ln \{f_0(\Gamma) + a^* Q_0(\Gamma)\} \, d\Gamma = 0 . \tag{9.26}$$

In order to avoid a duplication of formulas, and in a view of the examples considered below, we shall restrict our consideration to the Boltzmann entropy case. The (unique) positive solution to (9.26) is the value a^* which gives the desired upper estimate.

In order to derive (9.26) more formally, an explicit presentation is required for a model dynamics dominated by Q_0. Let us introduce a partition of the phase space into two domains, Γ_+ and Γ_-, in such a way that the function $Q_0(\Gamma)$ is positive on Γ_+, and is negative on Γ_-, and thus $Q_0(\Gamma) = Q_0^+(\Gamma) - Q_0^-(\Gamma)$, where both the functions $Q_0^+(\Gamma)$ and $Q_0^-(\Gamma)$ are positive and concentrated on Γ_+ and Γ_-, respectively[2]. Let us consider the following kinetic equation:

$$\partial_t f = k_1 (Q_0^+(\Gamma) - Q_0^-(\Gamma))(w^-(f) - w^+(f)) , \tag{9.27}$$

where

$$w^-(f) = \exp\left(\int_{\Gamma_-} Q_0^-(\Gamma) \ln f(\Gamma, t) \, d\Gamma\right) , \tag{9.28}$$

$$w^+(f) = \exp\left(\int_{\Gamma_+} Q_0^+(\Gamma) \ln f(\Gamma, t) \, d\Gamma\right) ,$$

and $k_1 > 0$ is an arbitrary positive constant. When supplied with the initial condition $f(\Gamma, 0) = f_0(\Gamma)$, equation (9.27) has a formal solution of the form:

$$f(\Gamma, t) = f_0(\Gamma) + a(t) Q_0(\Gamma), \tag{9.29}$$

provided that $a(t)$ is the solution of the ordinary differential equation

$$\frac{da}{dt} = k_1 (w^-(a) - w^+(a)) ,$$

with the initial condition $a(0) = 0$. Here $w^\pm(a) = w^\pm(f(a))$.

The solution (9.29) describes a relaxation from the initial state f_0 to the equilibrium state f^*, as t tends to infinity[3] . The entropy S_B monotonically

---

[2] For the Boltzmann collision integral, this partition should not be confused with the natural representation in the "gain−loss form" as $\int w(\mathbf{v}_1' \mathbf{v}' | \mathbf{v}_1 \mathbf{v})(f' f_1' - f f_1) \, d\mathbf{v}_1' \, d\mathbf{v}' \, d\mathbf{v}_1$.

[3] The equilibrium state f^* of the model kinetic equation (9.27) is not the global equilibrium f^0, exept for the BGK model of the collision integral.

increases along this solution up to the value $S_B^* = S_B(f^*)$ in the state f^*. Substituting $f^* = f_0 + a^* Q_0$ into the right-hand side of (9.27), we derive the equation for the equilibrium state f^* in the form of the detailed balance:

$$w^-(a^*) = w^+(a^*) . \tag{9.30}$$

The latter equation is precisely (9.26). Note that the parameter k_1 in (9.27) does not appears in the final result (9.26) since it is responsible only for the rate of the approach to the equilibrium state f^* due to the dynamics (9.27) but not for the location of this state on the ray (9.23).

Let us discuss the idea behind the model dynamics presented by (9.27). As long as we disregard any change of Q in the beginning of the relaxation, the function $Q_0(\Gamma)$ represents a distinguished direction of relaxation in the space of states. The partition of the phase space $\Gamma_+ \cup \Gamma_-$ corresponds then to specification of the *gain* (Γ_+) and of the *loss* (Γ_-) of the phase density, while the factors w^+ and w^- (9.28),

$$w^\pm(f) \sim \exp\left\{ -\int_{\Gamma_\pm} Q_0^\pm(\Gamma) \frac{\delta S(f)}{\delta f(\Gamma)} \, d\Gamma \right\} , \tag{9.31}$$

are the rates of the gain and and of the loss in the current state f, respectively. Equation (9.27) implements these processes in the familiar "gain minus loss" form, while the state f^* corresponds to the balance of the gain and of the loss (9.30). One can also observe a formal analogy of the structure of (9.27) with that of the so-called Marcelin-De Donder equations of chemical kinetics [81, 245] (see Chap. 2).

Thus, the limiting state $f^* = f_0 + a^* Q_0$ is described as the equilibrium state of the kinetic equation (9.27), and solves (9.26). Note that the parameter a^* is correctly defined by (9.26), independently of the partition introduced in the (9.27). The existence of the model relaxational equation (9.27) guarantees that f^* is a physical state (f^* is a non-negative function).

In order to complete the analysis, we have to learn to solve the one-dimensional nonlinear equation (9.26). In general, a method of successive approximation is required to find the solution a^* as a limit of a sequence $a_1^*, a_2^*, \dots$. Some care should be taken in order to get all the approximations a_i^* not greater than the unknown exact value a^*, since only the states $f(a, \Gamma)$ with $a \le a^*$ are relevant. Moreover, what one actually needs in computations is *some* definite approximation $a_1^* \le a^*$. In Subsect. 9.3.5, a corresponding method is developed, which is based on the partition of Q_0 introduced above. In particular, the first approximation a_1^* reads:

$$a_1^* = \frac{1 - \exp\{-\sigma_0/q\}}{\alpha + \beta \exp\{-\sigma_0/q\}} , \tag{9.32}$$

where q, σ_0, α, and β are numerical coefficients:

$$\sigma_0 = -\int Q_0(\Gamma) \ln f_0(\Gamma) \, d\Gamma \,, \tag{9.33}$$

$$q = \int_{\Gamma_+} Q_0^+(\Gamma) \, d\Gamma = \int_{\Gamma_-} Q_0^-(\Gamma) \, d\Gamma \,,$$

$$\alpha = \sup_{\Gamma \in \Gamma_-} \frac{Q_0^-(\Gamma)}{f_0(\Gamma)} \,,$$

$$\beta = \int_{\Gamma_+} \frac{(Q_0^+(\Gamma))^2}{q f_0(\Gamma)} \, d\Gamma \,.$$

In the latter expressions, σ_0 is the entropy production in the initial state, q is the normalization factor, α and β reflect the maximal loss and the total gain of the phase density in the initial state, respectively. Finiteness of the parameters collected in (9.33) gives a restriction on the initial state f_0 for wich the estimate (9.32) is valid.

9.3.3 Approximate Trajectories

In this subsection we shall demonstrate how to use the states f^* (9.24) in the problem of constructing the approximate trajectories of the space-independent relaxational equations

$$\partial_t f = Q(f) \,. \tag{9.34}$$

Here $Q(f)$ is a kinetic operator (the collision integral in the case of the Boltzmann equation). We assume that (9.34) describes a relaxation to the global equilibrium state $f^0(\Gamma)$, and the entropy $S_B(f)$ increases monotonically along the solutions. Let $c_1(\Gamma), \ldots c_k(\Gamma)$ be the conserved densities, i.e.

$$\int c_i Q(f) \, d\Gamma = 0 \,.$$

Then the quantities $C_i(f) = \int c_i f \, d\Gamma$ are conserved along the solution. Assume that the set of conserved densities $c_1(\Gamma), \ldots c_k(\Gamma)$ is full. In this case

$$\ln f^0(\Gamma) = \sum_1^k a_i c_i(\Gamma) \,,$$

where a_i are some numbers. A standard example of (9.34) is the space-independent Boltzmann equation which we consider below.

Let $f(\Gamma, t)$ be the solution to (9.34) with the initial condition $f(\Gamma, 0) = f_0(\Gamma)$. The trajectory of this solution can be represented as a function $f(\Gamma, a)$, where a varies from 0 to 1. For each a, the function $f(\Gamma, a)$ is a non-negative function of Γ, and

$$f(\Gamma, 0) = f_0(\Gamma), \quad f(\Gamma, 1) = f^0(\Gamma),$$

$$\int c_i(\Gamma) f(\Gamma, a) \, d\Gamma = \text{const}, \quad \partial_a f(\Gamma, a)|_{a=0} \propto Q_0(\Gamma). \quad (9.35)$$

In other words, as a varies from zero to one, the states $f(\Gamma, a)$ follow the solution $f(\Gamma, t)$ as t varies from zero to infinity. Since the entropy increases with time on the solution $f(\Gamma, t)$, the function $S_B(a) = S_B(f(a))$ is a monotonically increasing function of the variable a. This condition, as well as the conditions (9.35), must be met by any method of constructing an approximation to the trajectory $f(\Gamma, a)$ (see the conditions 1-6 in the Introduction).

The simplest approximation based on the function f^* of the preceding section can be constructed as follows:

$$f(\Gamma, a) = \begin{cases} (1 - 2a) f_0(\Gamma) + 2a f^*(\Gamma) & \text{for } 0 \le a \le \frac{1}{2} \\ 2(1 - a) f^*(\Gamma) + (2a - 1) f^0(\Gamma) & \text{for } \frac{1}{2} \le a \le 1 \end{cases}. \quad (9.36)$$

This approximation amounts to the *two-step* relaxation from the initial state f_0 to the equilibrium state f^0 through the intermediate state f^* (9.24). The first step (parameter a increases from 0 to 1/2) is the relaxation directed along Q_0 up to the state f^* (9.24). The second step (parameter a increases from 1/2 to 1) is the linear relaxation from f^* towards the equilibrium state. The last step can be viewed as the trajectory of a solution to the equation,

$$\partial_t f = -k_2(f - f^0), \quad (9.37)$$

with the initial condition f^* (9.24). In kinetic theory, equation of the form (9.37) is known as the BGK-model (2.17). The entropy increase along the second step is due to the well known properties of the equation (9.37).

Expression (9.36) demonstrates the advantage of using the state f^* for the purpose of approximating the trajectory: all the conditions (9.35) are obviously satisfied, and also we do not worry about the entropy increase. Thus, all the conditions 1-6 mentioned in the Introduction are satisfied by the approximation (9.36) due to the features of the state f^*. For explicit expressions the estimate (9.32) can be used.

A disadvantage of the two-step approximation (9.36) is its non-smoothness at $a = 1/2$. This can be improved as follows: Let us consider a *triangle* T formed by the three states, f_0, f^*, and f^0, i.e. a closed set of convex linear combinations of these functions[4]. This object allows to use a geometrical language. A simple consequence of the properties of the state f^* is that all the elements of the triangle T are non-negative functions, and if f belongs to T then $C_i(f) = C_i(f_0)$, where $i = 1, \ldots, k$ (all the conservation laws are fixed in the triangle). Therefore, a better approximation to the trajectory can be constructed as a smooth curve inscribed into the triangle T in such a way that:

---

[4] The state f belongs to T if $f = a_1 f_0 + a_2 f^* + a_3 f^0$, where $a_i \ge 0$, and $a_1 + a_2 + a_3 = 1$.

1. It begins in the state f_0 at $a = 0$;
2. It is tangent to the side $L_{f_0 f^*} = \{f | f = a_1 f_0 + a_2 f^*, a_1 \geq 0, a_2 \geq 0, a_1 + a_2 = 1\}$ in the state f_0;
3. It ends in the equilibrium state f^0 at $a = 1$.

Notice that the approximation (9.36) corresponds to the path from f_0 to f^0 over the two sides of the triangle T: firstly, over the segment between f_0 and f^*, and, secondly, over the segment between f^* and f^0.

The simplest form of such a smooth curve reads (the MDD spline):

$$f_g(\Gamma, a) = f^0 + (1 - a^2)\{ag(f^* - f_0) + f_0 - f^0\}, \qquad (9.38)$$

where $g, 0 < g \leq 1$, is a parameter which has to be determined in a way that the entropy $S_B(a)$, calculated in the states (9.38), is monotonically increasing function of a. The explicit *sufficient* method to estimate the value of parameter g in (9.38) is rather non-trivial, and it is developed in Subsect. 9.3.5.

Finally, let us consider briefly a question of the time dependence for the approximation $f(\Gamma, a)$. Clearly, this question is relevant as soon as one looks for the approximate trajectories directly, rather than integrating (9.34) in time[5]. The answer assumes a dependence $a(t)$, and requires an ordinary differential equation for a. Such an equation should be obtained upon substitution of the expression $f(\Gamma, a)$ into the originating kinetic equation (9.34), and by a further projecting. Specifically, the equation for $a(t)$ has a form:

$$\frac{da}{dt} \int \varphi(\Gamma, a) \frac{\partial f(\Gamma, a)}{\partial a} d\Gamma = \int \varphi(\Gamma, a) Q(f(\Gamma, a)) d\Gamma, \qquad (9.39)$$

where integration with the function $\varphi(\Gamma, a)$ establishes the projection operation. Usually, this is achieved by some moment projecting (φ is independent of a), but this choice is arbitrary. Another possibility is to use the *thermodynamic projector* (Chap. 5). Then (9.39) becomes the entropy rate equation along the path (9.38):

$$\frac{da}{dt} \frac{dS_B(a)}{da} = \sigma_B(a), \qquad (9.40)$$

where $S_B(a) = -\int f(\Gamma, a) \ln f(\Gamma, a) d\Gamma$
and $\sigma_B(a) = -\int Q(f(\Gamma, a)) \ln f(\Gamma, a) d\Gamma$
are the entropy and the entropy production in the states $f(\Gamma, a)$ (9.38), respectively.

A further consideration of (9.40) is beyond the scope of this Example. Nevertheless, let us consider the asymptotics of (9.40) for the motion from f_0 towards f^*. As above, we take $f(\Gamma, a) = (1 - a)f_0 + af^*$. Equation (9.40) for this function gives:

---

[5] This question is typical to various approximations used in the kinetic theory [9, 11].

$$a(t) \sim \frac{1}{a^*}t, \quad a \ll 1$$

$$a(t) \sim \frac{\sigma_B^*}{(a^*)^2 K_0}\sqrt{t}, \quad 1 - a \ll 1 \,,$$

where σ_B^* is the entropy production in the state f^*, and $K_0 = \int \frac{Q_0^2}{f_0}\, d\Gamma$. The slowing down at the final stage is due to the fact that $dS_B(a)/da \to 0$, as $a \to 1$, and $\sigma_B^* > 0$.

9.3.4 Relaxation of the Boltzmann Gas

The direct and the simplest application of the approach is the space-independent Boltzmann equation. In what follows, Γ is the velocity $\mathbf{v}$, and $f(\Gamma)$ is the one-body distribution function, $f(\mathbf{v})$, which obeys the the equation:

$$\partial_t f(\mathbf{v}, t) = Q(f) \,, \tag{9.41}$$

with $Q(f)$ the Boltzmann collision integral.

In the first example we consider the following form of the collision integral

$$Q(f) = \int d\mathbf{w} \int d\hat{\mathbf{n}} \gamma(\hat{\mathbf{g}} \cdot \hat{\mathbf{n}}) \left\{ f(\mathbf{v}', t)f(\mathbf{w}', t) - f(\mathbf{v}, t)f(\mathbf{w}, t) \right\} \,, \tag{9.42}$$

where the function γ depends only on the scalar product of unit vectors $\hat{\mathbf{g}} = \frac{\mathbf{v} - \mathbf{w}}{|\mathbf{v} - \mathbf{w}|}$ and $\hat{\mathbf{n}} = \frac{\mathbf{v}' - \mathbf{w}'}{|\mathbf{v} - \mathbf{w}|}$, while $\mathbf{v}' = \frac{1}{2}(\mathbf{v} + \mathbf{w} + \hat{\mathbf{n}}|\mathbf{v} - \mathbf{w}|)$, and $\mathbf{w}' = \frac{1}{2}(\mathbf{v} + \mathbf{w} - \hat{\mathbf{n}}|\mathbf{v} - \mathbf{w}|)$. The Boltzmann equation (9.41) with the collision integral (9.42) corresponds to the power-law repelling potential inversly proportional to the fourth degree of the distance (the 3D Maxwell molecules, see e.g. [261]). The reason to consider this model is that it admits the exact solution, the famous BKW-mode discovered by Bobylev [262], and by Krook and Wu [255, 256]. The BKW-mode is the following one-parametric set of the distribution functions $f(c, \mathbf{v})$:

$$f(c, \mathbf{v}) = \frac{1}{2}\left(\frac{2\pi}{c}\right)^{-3/2} \exp\left\{ -\frac{cv^2}{2} \right\} \left((5 - 3c) + c(c - 1)v^2 \right) \,, \tag{9.43}$$

where the parameter c spans the segment $[1, \frac{5}{3}[$, the value $c = 1$ corresponds to the equilibrium Maxwell distribution

$$f^0(\mathbf{v}) = f(1, \mathbf{v}) = (2\pi)^{-3/2} \exp\{-v^2/2\} \,.$$

As c decays from a given value c_0, where $1 < c_0 < 5/3$, to the value $c = 1$, the functions $f(c, \mathbf{v})$ (9.43) describe the trajectory of the BKW-mode (the time dependence of c is unimportant in the present context, see e.g. [286]).

Considering the states (9.43) as the initial states in the procedure described above, we can construct the upper limiting states, $f^*(c, \mathbf{v}) = f(c, \mathbf{v})+$

$a^*(c)Q(c, \mathbf{v})$, for each value of c. First, we compute the collision integral (9.42) in the states (9.43) and obtain the functions $Q(c, \mathbf{v})$:

$$Q(c, \mathbf{v}) = \frac{\lambda}{2}(c-1)^2 \left(\frac{2\pi}{c}\right)^{-3/2} \exp\left\{-\frac{cv^2}{2}\right\} (15 - 10cv^2 + c^2(v^2)^2) , \quad (9.44)$$

where λ is a constant: $\lambda = \frac{1}{8}\int d\hat{\mathbf{n}}\gamma(\hat{\mathbf{k}} \cdot \hat{\mathbf{n}})(1 - (\hat{\mathbf{k}} \cdot \hat{\mathbf{n}})^2)$.

Expression (9.44) suggests a simple structure of the velocity space partition into the domains $V_+(c)$ and $V_-(c)$ (corresponding to the domains $\Gamma_\pm$ (9.31)). Namely, for a given c, the function (9.44) is positive inside a sphere of radius $v_-(c) = \sqrt{c^{-1}(5 - \sqrt{10})}$, and outside a larger sphere of radius $v_+(c) = \sqrt{c^{-1}(5 + \sqrt{10})}$ (both the spheres are centered in $\mathbf{v} = 0$), while it is negative inside the spheric layer between these spheres:

$$V_-(c) = \{\mathbf{v} \mid v_-(c) < |\mathbf{v}| < v_+(c)\} , \quad (9.45)$$
$$V_+(c) = \{\mathbf{v} \mid v_-(c) > |\mathbf{v}|\} \cup \{\mathbf{v} \mid |\mathbf{v}| > v_+(c)\} .$$

The limiting states $f^*(c, \mathbf{v})$ are given by the following expression:

$$f^*(c, \mathbf{v}) = \frac{1}{2}\left(\frac{2\pi}{c}\right)^{-3/2} \exp\left\{-\frac{cv^2}{2}\right\} \quad (9.46)$$
$$\times \left(5 - 3c + 15a^*(c) + (c - 1 - 10a^*(c))cv^2 + a^*(c)c^2(v^2)^2\right) ,$$

where $a^*(c)$ is a solution to (9.26):

$$\int Q(c, \mathbf{v}) \ln\left(f(c, \mathbf{v}) + a^*(c)\frac{Q(c, \mathbf{v})}{\lambda(c - 1)^2}\right) d\mathbf{v} = 0 . \quad (9.47)$$

Taking into account the partition (9.45), all the parameters (9.33) are expressed by definite one-dimensional integrals. Thus, we obtain the first approximate $a_1^*(c)$. Numerical results are presented in Table 9.3 (second column) for the three values of the parameter c taken on the BKW mode. It is interesting to compare $a_1^*(c)$ with $a_{\max}(c)$, for which the function, $f(c, \mathbf{v}) + a\frac{Q(c, \mathbf{v})}{\lambda(c-1)^2}$, looses positivity (i.e., this function becomes negative for some $\mathbf{v}$, as $a > a_{\max}(c)$). The ratio $a_1^*(c)/a_{\max}(c)$ is given in the third column of Table 9.3. The step in the direction $Q(c, \mathbf{v})$ which is allowed due to the entropy estimate reasons is never negligible in comparison to that determined by the positivity reasons, as seen in Table 9.3.

We now use (9.46) to get the approximations of trajectories (9.36) and (9.38). The estimation of the parameter g in the expression (9.38) according to Subsect. 9.3.5 gives the value $g = 1$ for all the initial states (9.43).

In order to make a comparison with the exact result (9.43), we have considered the dependencies of the normalized moments $m_k(m_l)$, where

Table 9.3. The limiting states for the Maxwell molecules

c	a_1^*	$a_1^*/a_{\max}$
1.12	$3.1779 \cdot 10^{-3}$	0.2221
1.24	$1.1660 \cdot 10^{-2}$	0.4291
1.48	$3.8277 \cdot 10^{-2}$	0.7087

$$m_s(f) = \frac{\int (v^2)^s f \, d\mathbf{v}}{\int (v^2)^s f^0 \, d\mathbf{v}}, \quad s = 0, 1, 2, \dots . \tag{9.48}$$

Typical dependencies of the higher-order moments ($k \geq 3$) on the lowest-order non-trivial moment ($l = 2$) are presented in the Fig. 9.1 for a considerably nonequilibrium initial state (9.43) with $c = 1.42$.

The error of the approximation (9.38) was estimated as follows: In each moment plane (m_k, m_l), the approximation (9.38) and the BKW-mode (9.43) generate two sets (two curvilinear segments), X_{kl} and Y_{kl}, respectively. First, in order to eliminate the contribution from the difference in the total variation of the moments, we rescale the variables:

$$\hat{m}_i = m_i/\Delta_i , \quad i = k, l ,$$

where

$$\Delta_i = \max_{\mathbf{x}, \mathbf{x}' \in X_{kl} \bigcup Y_{kl}} |x_i - x_i'| .$$

Second, in the plane $(\hat{m}_k, \hat{m}_l)$, we compute the Hausdorff distance, d_{kl}, between the two corresponding sets, $\hat{X}_{kl}$ and $\hat{Y}_{kl}$:

$$d_{kl} = \max \left\{ \max_{\mathbf{x} \in \hat{X}_{kl}} \min_{\mathbf{y} \in \hat{Y}_{kl}} d(\mathbf{x}, \mathbf{y}), \quad \max_{\mathbf{y} \in \hat{Y}_{kl}} \min_{\mathbf{x} \in \hat{X}_{kl}} d(\mathbf{x}, \mathbf{y}) \right\} , \tag{9.49}$$

where $d(\mathbf{x}, \mathbf{y})$ is the standard Euclidian distance between two points. Finally, the error δ_{kl} was estimated as the normalized distance d_{kl}:

$$\delta_{kl} = \frac{d_{kl}}{D_{kl}} \cdot 100\% , \tag{9.50}$$

where

$$D_{kl} = \max_{\mathbf{x}, \mathbf{y} \in \hat{Y}_{kl} \bigcup \hat{X}_{kl}} d(\mathbf{x}, \mathbf{y}) .$$

The error δ_{k2} of the plots like in Fig. 9.1 is presented in the Table 9.4 for several values of the parameter c.

The quality of the smooth approximation (9.38) is either good or reasonable up to the order of the moment $k \sim 10$, depending on the closeness of the initial state to the equilibrium. when either k increases, or the initial state is taken very far from the equilibria (i.e., when c is close to $5/3$) the comparison

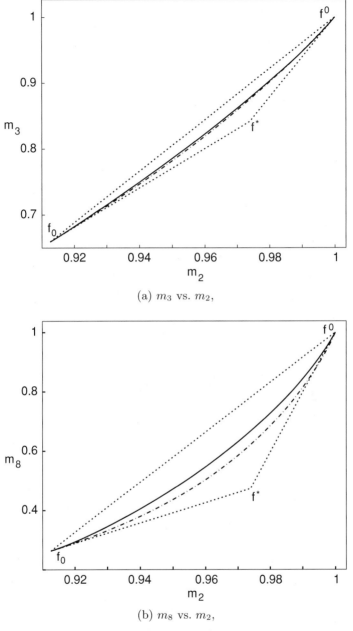

(a) m_3 vs. m_2,

(b) m_8 vs. m_2,

Fig. 9.1. Moment dependencies for the Maxwell molecules: The initial state f_0 is the function (9.43) with $c = 1.42$. Punctuated contour is the image of the triangle T. Punctuated dash line is the BKW-mode. *Solid* line is the smooth approximation (9.38). Punctuated path $f_0 \rightarrow f^* \rightarrow f^0$ is the non-smooth approximation (9.36)

Table 9.4. The error δ_{k2} (9.50) of the approximation (9.38) for the Maxwell molecules with the initial data (9.43)

k	$c = 1.12$	$c = 1.24$	$c = 1.36$	$c = 1.48$	$c = 1.59$
3	0.31	0.33	0.30	0.41	0.70
4	0.44	0.47	0.44	0.81	1.57
5	0.58	0.55	0.71	1.41	2.67
6	0.71	0.57	1.10	2.20	3.97
7	0.81	0.62	1.58	3.14	5.40
8	0.89	0.84	2.19	4.19	6.93
9	0.95	1.11	2.87	5.34	8.52
10	0.99	1.41	3.64	6.55	10.11
20	1.46	6.77	12.91	18.15	22.87
50	10.38	28.47	27.36	30.81	33.94
100	21.76	29.26	32.49	34.78	37.22

becomes worser. For the moments of a very high order, the approximation with the smooth function (9.38) is only qualitative. On the other hand, the two-step (non-smooth) approximation (9.36) provides a much better approximation for higher-order moments ($k \sim 40$ and higher). The explanation is as follows: the BKW-mode (9.43) demonstrates a very rapid relaxation of higher moments to their equilibrium values. Therefore, as expected, the relaxation in the direction Q_0 leads to the state where the higher-order moments are practically the same as in the equilibrium.

The second example is the very hard particles (VHP) model [257, 258]. The distribution function $F(x)$ depends on the phase variable x, where $0 \leq x \leq \infty$, and is governed by the following kinetic equation:

$$\partial_t F(x,t) = \int_x^\infty du \int_0^u dy \left[F(y,t)F(u-y,t) - F(x,t)F(u-x,t) \right] . \quad (9.51)$$

This model has the two conservation laws:

$$N = \int_0^\infty F(x,t)\, dx = 1 \ ,$$

$$E = \int_0^\infty xF(x,t)\, dx = 1 \ ,$$

and has the entropy $S_B(F) = -\int_0^\infty F(x)\ln F(x)\, dx$. The equilibrium distribution reads: $F^0(x) = \exp(-x)$. The general solution to this model is known [257, 258, 286].

The first set of initial conditions which was tested was as follows:

$$F_0(x,\beta) = \beta((2-\beta) + \beta(\beta-1)x)\exp(-\beta x) , \quad (9.52)$$

where $1 \leq \beta < 2$, the value $\beta = 1$ corresponds to the equilibrium state $F_0(x,1) = F^0(x)$.

In accordance with [257, 258, 286], the exact solution solution to (9.51) with the initial data (9.52) reads:

$$F_{ex}(x, \beta, t) = \frac{Az_+ + C}{z_+ - z_-} e^{xz_+} + \frac{Az_- + C}{z_- - z_+} e^{xz_-} , \qquad (9.53)$$

$$z_{\pm} = -\frac{t + 2\beta}{2} \pm \sqrt{\left(\frac{t + 2\beta}{2}\right)^2 - C} ,$$

$$A = 1 - (\beta - 1)^2 e^{-t}; \quad C = t + 2\beta - 1 + e^{-t}(\beta - 1)^2 .$$

Comparison of the smooth approximation (9.38) with the exact solution (9.53) demonstrates the same quality as in the case of the Maxwell molecules. As above, the normalized moments m_k were compared, where

$$m_k = \frac{\int_0^\infty x^k F(x)\, \mathrm{d}x}{\int_0^\infty x^k F^0(x)\, \mathrm{d}x} .$$

In Table 9.5, the error δ_{k2} (9.50) is represented for several values of the parameter β, while Fig. 9.2 illustrate the typical moment behavior. We also represent in this figure the result of the *correction* to the approximation (9.38) due to the first iteration of the Newton method with incomplete linearization (Chap. 6).

Table 9.5. The error δ_{k2} (9.50) of the approximation (9.38) for the VHP model with the initial data (9.52)

k	4	6	8	10	20	100
$\beta = 1.2$	0.95	1.81	2.26	2.23	2.24	9.64
$\beta = 1.6$	0.88	1.89	2.59	2.77	6.14	24.29
$\beta = 1.9$	1.16	1.65	1.45	3.16	12.7	28.22

The second set of the initial conditions for the VHP model (9.51) was considered as follows:

$$F_0(x, \lambda) = \exp(-2x)\left\{1 + \frac{1}{2}\lambda + 2x^2(1 - \lambda) + \frac{1}{3}\lambda x^4\right\} , \qquad (9.54)$$

where $0 < \lambda < 1/5(7 + \sqrt{19})$. The exact solution to (9.51) with the initial condition (9.54) was found in [258]. This solution demonstrates so-called Tjon's overshoot effect [264]. We remind that Tjon's effect takes place when the distribution function becomes overpopulated for some velocities in comparison to both the initial and the equilibrium states. This effect was intensively studied for solvable Boltzmann-like kinetic equations, such as the Maxwell molecules (9.42), the VHP model (9.51), and others (see [286], [265] and

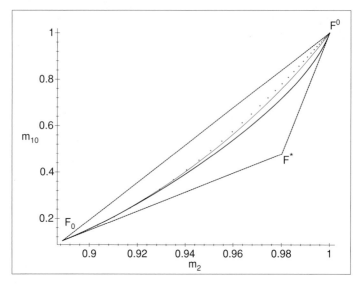

Fig. 9.2. Moment dependency m_{10} vs. m_2 for the VHP model with the initial condition (9.52), $\beta = 1.5$. *Dots* – the exact solution (9.53); *Bold* line is the smooth approximation (9.38); *Solid* line is the first correction to the approximation (9.38)

references therein; it is worthwhile to mention here extensive studies of the Tjon-like effects in chemical kinetics [81, 115]).

The approximation (9.38) for the VHP model (9.51) with the initial condition (9.54) also demonstrates the overshoot just mentioned. In the moment representation, the overshoot of the moments is clearly seen in Fig. 9.3. The quality of the approximation is the same as in the examples above.

9.3.5 Estimations

This is the technical subsection which contains estimations for the limiting state and for the smooth approximation of the trajectory.

Evaluation of the Limiting State. Double-Space Newton Method

Let us introduce a normalization of the partition $Q_0^{\pm}(\Gamma)$:

$$q_0^{\pm}(\Gamma) = q^{-1} Q_0^{\pm}(\Gamma) \,, \quad q = \int_{\Gamma_{\pm}} Q_0^{\pm}(\Gamma) \, d\Gamma \,. \tag{9.55}$$

Switching to the variable $b = qa$, so that $f^* = f_0(\Gamma) + b^* q_0(\Gamma)$, where $q_0(\Gamma) = q^{-1} Q_0(\Gamma)$, equation (9.26) can be rewritten as follows:

$$A_+(b^*) = A_-(b^*) \,, \tag{9.56}$$

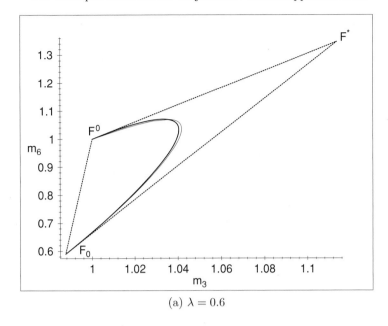

(a) $\lambda = 0.6$

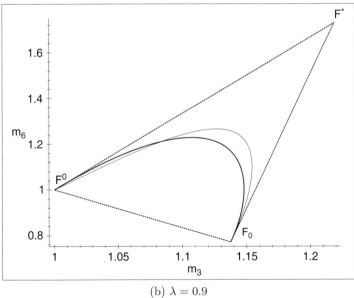

(b) $\lambda = 0.9$

Fig. 9.3. Moment dependencies m_6 vs. m_3 for the VHP model with the initial condition (9.54). *Solid* line is the exact solution [258], *Bold* line is the smooth approximation (9.38)

where

$$A_\pm(b) = \int_{\Gamma_\pm} q_0^\pm(\Gamma) \ln(f_0(\Gamma) \pm b q_0^\pm(\Gamma)) \, d\Gamma \ . \qquad (9.57)$$

It is easy to check the following properties of the functions $A_\pm$ (9.57):

1. The domain of A_+ is the open semi-axis $]b_+, +\infty[$, where $b_+ < 0$, and domain of A_- is the open semi-axis $]-\infty, b_-[$, where $b_- > 0$. The functions $A_\pm$ have logarithmic singularities at points $b_\pm$, respectively.
2. The functions $A_\pm$ are monotonic and concave inside their domains.
3. An inequality holds as: $A_+(0) - A_-(0) = -q^{-1}\sigma_0 < 0$, where $\sigma_0 = -\int Q_0(\Gamma) \ln f_0(\Gamma) \, d\Gamma$ is the entropy production in the state f_0.

One has to solve (9.56) in order to obtain approximations $b_1^*, b_2^*, \ldots$ not greater than the unknown exact value b^*. To obtain a relevant lower estimate of b^*, it is convenient to use the concavity properties of the functions (9.57). Indeed, for positive b, the function A_- is estimated from below as:

$$A_-(b) \geq A_-(0) + \ln(1 - \alpha_1 b) \ . \qquad (9.58)$$

Here α_1 is the inverse of b_-:

$$\alpha_1 = \sup_{\Gamma \in \Gamma_-} \frac{q_0^-(\Gamma)}{f_0(\Gamma)} = \sup_{\Gamma \in \Gamma_-} \frac{Q_0^-(\Gamma)}{q f_0(\Gamma)} = q^{-1}\alpha \ ,$$

while α was introduced in (9.33).

On the contrary, the function A_+ should be estimated from above. Note that a function $\exp A_+$ is also monotonic and concave. We can write for positive b:

$$A_+(b) \leq A_+(0) + \ln\left(1 + b\frac{dA_+(0)}{db}\right) \ , \qquad (9.59)$$

where

$$\frac{dA_+(0)}{db} = \int_{\Gamma_+} \frac{(q_0^+(\Gamma))^2}{f_0(\Gamma)} \, d\Gamma = q^{-1}\beta \ ,$$

and β was introduced in (9.33).

Equating the right-hand side of (9.59) to the right-hand side of (9.58), and solving the linear equation obtained, we get the estimate $b_1^* \leq b^*$. Next, switching back to the variable a, we get the estimate a_1^* (9.32) and (9.33). One can readily recognize that the procedure just described is the first iterate of the Newton method for (9.56) (modified by making use of the concavity to guarantee positivity of the approximate solution, $a_1^* \leq a^*$). We call it the double-space Newton method. Next iterations are performed in the same manner.

Smooth Approximations of the Trajectories

The Triangle of Model Motions

Notation $\overline{\mathrm{conv}}\{f_1, \ldots, f_k\}$ stands for a closed convex linear hull of the functions $f_1, \ldots, f_k$, and we drop the variable Γ. In particular, the triangle T introduced in the section "Approximate trajectories" reads:

$$T = \overline{\mathrm{conv}}\{f_0, f^*, f^0\} \, . \tag{9.60}$$

A function from the triangle T (9.60) can be specified with two parameters, ξ and η, as $f(\xi, \eta)$:

$$f(\xi, \eta) = f^0 + \xi\{\eta(f^* - f_0) + f_0 - f^0\}, \quad 0 \le \xi, \eta \le 1 \, . \tag{9.61}$$

A shift of the function $f(\xi, \eta)$ under a variation of ξ and of η reads:

$$\begin{aligned}
\Delta f(\xi, \eta) &= \partial_\xi f(\xi, \eta)\Delta\xi + \partial_\eta f(\xi, \eta)\Delta\eta + o(\Delta\xi, \Delta\eta) \\
&= (f(\xi, \eta) - f^0)\xi^{-1}\Delta\xi + a^* Q_0 \xi \Delta\eta + o(\Delta\xi, \Delta\eta) \, .
\end{aligned}$$

This shift is a combination of the two: a shift towards f^0, and a shift in the direction Q_0. We further refer to these as to the BGK-motion and the Q_0-motion, respectively. The differential of the entropy $S_B(\xi, \eta) = S_B(f(\xi, \eta))$ is:

$$\mathrm{d}S_B(\xi, \eta) = -\sigma_1(\xi, \eta)\xi^{-1}\,\mathrm{d}\xi + \sigma_2(\xi, \eta)\xi\,\mathrm{d}\eta \, , \tag{9.62}$$

where

$$\sigma_1(\xi, \eta) = \int (f(\xi, \eta) - f^0) \ln \frac{f(\xi, \eta)}{f^0}\,\mathrm{d}\Gamma \, , \tag{9.63}$$

$$\sigma_2(\xi, \eta) = \int (f_0 - f^*) \ln f(\xi, \eta)\,\mathrm{d}\Gamma = -a^* \int Q_0 \ln f(\xi, \eta)\,\mathrm{d}\Gamma \, ,$$

are the entropy productions in the BGK-motion and in the Q_0-motion, respectively.

Introducing smooth dependencies, $\xi(a)$ and $\eta(a)$, where $0 \le a \le 1$, and requiring

$$0 \le \xi(a), \eta(a) \le 1, \ \xi(0) = 1, \ \xi(1) = 0, \ \eta(0) = 0, \ \eta(1) < \infty \, , \tag{9.64}$$

$$\left.\frac{\mathrm{d}\xi(a)}{\mathrm{d}a}\right|_{a=0} = 0, \quad \left.\frac{\mathrm{d}\eta(a)}{\mathrm{d}a}\right|_{a=0} = \gamma, \ 0 < \gamma \le 1 \, ,$$

we obtain a one-parametric set, $f(a) = f(\xi(a), \eta(a))$. Geometrically, $f(a)$ is a smooth curve located in T. This curve begins in f_0 at $a = 0$, ends up in f^0 at $a = 1$, and is tangent to the side of T, $L_{f_0 f^*} = \overline{\mathrm{conv}}\{f_0, f^*\}$, at $a = 0$. Further, only monotonic functions $\xi(a)$ and $\eta(a)$ will be considered:

$$\frac{\mathrm{d}\xi(a)}{\mathrm{d}a} \le 0, \quad \frac{\mathrm{d}\eta(a)}{\mathrm{d}a} \ge 0 \, . \tag{9.65}$$

The crucial point is that the function $f(a)$ should have a correct entropy behavior. Specifically, we require that the entropy $S_B(a) = S_B(f(a)) = S_B(f(\xi(a), \eta(a)))$ is a monotonic function:

$$\frac{\mathrm{d}S_B(a)}{\mathrm{d}a} = -\sigma_1(\xi(a), \eta(a))\xi^{-1}(a)\frac{\mathrm{d}\xi(a)}{\mathrm{d}a} + \sigma_2(\xi(a), \eta(a))\xi(a)\frac{\mathrm{d}\eta(a)}{\mathrm{d}a} \geq 0 .$$
$$(9.66)$$

Since $\sigma_1(\xi, \eta)$ is non-negative everywhere in T, a sufficient condition for inequality (9.66) to be valid for any pair of functions $\xi(a)$ and $\eta(a)$ with the properties (9.64) and (9.65) is that $\sigma_2(\xi, \eta)$ is non-negative everywhere in T. However, this situation might not always be realized for arbitrary f_0 and Q_0. In order to take into account a general situation, we execute the following procedure:

1. We derive a subset of T, inside which σ_2 is non-negative. This subset includes f_0, and will be constructed as a triangle $T' \subseteq T$.
2. We tune the functions $\xi(a)$ and $\eta(a)$ in such a way that $\sigma_1(a)$ dominates $\sigma_2(a)$ outside T'.

The Triangle T'

Let us introduce a different specification of the functions in the triangle T. Denote

$$f_1(y) = (1-y)f_0 + yf^*, \quad f_2(y) = (1-y)f_0 + yf^0, \quad 0 \leq y \leq 1 . \quad (9.67)$$

The functions in T are labeled with two parameters, x and y:

$$f(x,y) = (1-x)f_1(y) + xf_2(y), \quad 0 \leq x, y \leq 1 . \quad (9.68)$$

Let us derive y', where $0 < y' \leq 1$, in such a way that σ_2 is non-negative everywhere in the triangle $T' = \overline{\mathrm{conv}}\{f_0, f_1(y'), f_2(y')\}$.

Introducing a representation $\sigma_2(x,y) = \sigma_2^+(x,y) - \sigma_2^-(x,y)$, where

$$\sigma_2^+(x,y) = \int f_0 \ln f(x,y)\,\mathrm{d}\Gamma, \quad \sigma_2^-(x,y) = \int f^* \ln f(x,y)\,\mathrm{d}\Gamma , \quad (9.69)$$

we notice that the functions $\sigma_2^\pm(x,y)$ are concave in the variable y on the segment $[0,1]$, for any fixed x. Now we apply the standard estimations of a smooth concave function on $[0,1]$ (if $\mathrm{d}^2\psi(t)/\mathrm{d}t^2 \leq 0$ on $[0,1]$, then $\psi(t) \geq (1-t)\psi(0) + t\psi(1)$, and $\psi(t) \leq (\mathrm{d}\psi(t)/\mathrm{d}t|_{t=0})t + \psi(0)$) to the functions (9.69):

$$\sigma_2^+(x,y) \geq (1-y)\sigma_2^+(x,0) + y\sigma_2^+(x,1) ,$$
$$\sigma_2^-(x,y) \leq (\partial_y\sigma_2^-(x,y)|_{y=0})y + \sigma_2^-(x,0) .$$

Furthermore, the function $\sigma_2^+(x,1)$ is concave, hence

$$\sigma_2^+(x,1) \geq (1-x)\sigma_2^+(0,1) + x\sigma_2^+(1,1) \ .$$

Making use of the three inequalities just derived, and taking into account the explicit form of the function $f(x,y)$, we are led to the following estimate of σ_2 in T:

$$\sigma_2(x,y) \geq a^*\sigma_0 - y(xK_1 + K_2) \ , \tag{9.70}$$

where σ_0 is the entropy production in the initial state (9.33), and parameters K_1 and K_2 are:

$$K_1 = \int \frac{f^*}{f_0}(f^0 - f^*)\,d\Gamma + S_B(f^0) - S_B(f^*) \ , \tag{9.71}$$

$$K_2 = \int \frac{f^*}{f_0}(f^* - f_0)\,d\Gamma + S_B(f^*) - S_B(f_0) \ .$$

Here $S_B(f_0)$, $S_B(f^*)$, and $S_B(f^0)$ are values of the entropy in the states f_0, f^*, and f^0, respectively.

Since σ_0 is positive, there always exists such y', where $0 < y' \leq 1$, that the right-hand side of (9.70) is non-negative for all x on the segment $[0,1]$. Specifically, let us introduce a function $\varphi(x) = a^*\sigma_0 - (xK_1 + K_2)$, and denote

$$z = a^*\sigma_0 \min\{K_2^{-1}, (K_1 + K_2)^{-1}\} \ , \tag{9.72}$$

where $\min\{K_2^{-1}, (K_1 + K_2)^{-1}\}$ stands for the minimal of the two numbers, K_2^{-1} and $(K_1 + K_2)^{-1}$. Then y' is defined as:

$$y' = \begin{cases} 1 & \text{if } \varphi(x) \geq 0 \text{ on } [0,1], \text{ or } z \geq 1 \\ z & \text{otherwise} \end{cases} \ . \tag{9.73}$$

Thus, σ_2 is non-negative inside the triangle $T' = \overline{\text{conv}}\{f_0, f_1(y'), f_2(y')\}$, where $f_{1,2}(y')$ are given by (9.67), and y' is given by (9.73). If it happens that $y' = 1$, then $T' = T$, and σ_2 is non-negative everywhere in T. In this case any pair of the functions $\xi(a)$ and $\eta(a)$ with the properties (9.64) and (9.65) give the approximation $f(a)$ consistent with the inequality (9.66). Otherwise, we continue the procedure.

Near-Equilibrium Estimations of the Functions σ_1 and σ_2

Let us come back to the specification (9.61) in order to establish the following inequalities for the functions $\sigma_{1,2}(\xi,\eta)$ (9.63):

$$\sigma_1(\xi,\eta) \geq M_1\xi^2 \ , \tag{9.74}$$
$$\sigma_2(\xi,\eta) \geq M_2\xi \ .$$

Inequalities (9.74) are motivated by the following consideration. Since

$$f(\xi,\eta) \to f^0, \text{ as } \xi \to 0 \ ,$$

parameter ξ controls a deviation of $f(\xi, \eta)$ from f^0 in T. Near the equilibrium state f^0, the function $\sigma_1(\xi, \eta)$ is quadratic in ξ, while the function $\sigma_2(\xi, \eta)$ is linear. Inequalities (9.74) extend these near-equilibrium estimations to other points of T, and they are intended to control dominance of σ_1 over σ_2 outside T' in the case $T' \neq T$.

Writing $\sigma_1(\xi, \eta) = \xi \lambda(\xi, \eta)$, and representing $\lambda(\xi, \eta)$ as a combination of the concave functions, and after making the estimations as above, we come to the following expression for M_1 in the first of the inequalities (9.74):

$$M_1 = S_B(f^0) - S_B(f^*) . \tag{9.75}$$

Since $S_B(f^0) > S_B(f^*)$, expression (9.75) is always positive. The estimate of M_2 is much the same. First, representing $\sigma_2(\xi, \eta)$ in the manner of (9.69), and again estimating the concave functions obtained, we come to the following inequality:

$$\sigma_2(\xi, \eta) \geq \xi(\eta N_1 + N_2) , \tag{9.76}$$

where constants N_1 and N_2 are:

$$N_1 = \int \frac{f^*}{f^0}(f_0 - f^*) \, d\Gamma + S_B(f_0) - S_B(f^*) , \tag{9.77}$$

$$N_2 = \int \frac{f^*}{f^0}(f^0 - f_0) \, d\Gamma + S_B(f_0) - S_B(f^0) .$$

Second, denoting $\min\{N_2, N_1 + N_2\}$ as the minimal of the two numbers, N_2 and $N_1 + N_2$, we derive the constant in the second of the inequalities (9.74):

$$M_2 = \min\{N_2, N_1 + N_2\} \tag{9.78}$$

As above, there are two possibilities:

1. If $M_2 \geq 0$, then σ_2 is non-negative everywhere in T, and any pair of functions $\xi(a)$ and $\eta(a)$ with the properties (9.64) and (9.65) gives $f(a)$ with the correct entropy behavior.
2. If $M_2 < 0$, then we continue the procedure.

Adjustment of the Functions $\xi(a)$ and $\eta(a)$

Let $y' < 1$ and $M_2 < 0$. A further analysis requires an explicit form of the functions $\xi(a)$ and $\eta(a)$ with the properties (9.64) and (9.65), and can be done in any particular case. Consider the simplest choice:

$$\xi(a) = 1 - a^2, \quad \eta(a) = ga , \tag{9.79}$$

where g, $0 < g \leq 1$, is a parameter to be determined. The function (9.61) with the dependencies (9.79) has the form (9.38):

$$f_g(a) = f^0 + (1 - a^2)\{ga(f^* - f_0) + f_0 - f^0\} . \tag{9.80}$$

We should derive the parameter g in (9.79) in such a way that the states $f(a)$ (9.38) belong to T', when a varies from 0 to some a_1, and also that $\sigma_1(a)$ dominates $\sigma_2(a)$ when a varies from a_1 to 1. Under these conditions, the entropy inequality (9.66) is valid for all a on the segment $[0, 1]$.

Substitute now (9.79) into (9.66) and apply the inequalities (9.74) to get:

$$\frac{dS_B(a)}{da} \geq 2a(1 - a^2)M_1 - g(1 - a^2)^2|M_2| . \tag{9.81}$$

We require that $f(a_1) \in \overline{\text{conv}}\{f_1(y'), f_2(y')\}$, and that the right-hand side of the inequality (9.81) is non-negative at a_1:

$$\begin{cases} f_g(a_1) = f(x_1, y') \\ 2a(1 - a^2)M_1 - g(1 - a^2)^2|M_2| \geq 0 \end{cases} . \tag{9.82}$$

Here $f(x_1, y')$ is the specification (9.68) of the function $f_g(a_1)$. Explicitly, condition (9.82) reads:

$$\begin{cases} a_1^2 = x_1 y' \\ a_1 g(1 - a_1^2) = (1 - x_1)y' \\ g(1 - a_1^2) \leq \frac{2M_1}{|M_2|} a_1 \end{cases} . \tag{9.83}$$

Eliminating a_1 and x_1 in (9.83), we are left with the following estimate of the parameter g:

$$g \leq \lambda \frac{\sqrt{y'(1 + \lambda)}}{1 - y' + \lambda} , \tag{9.84}$$

where

$$\lambda = \frac{2M_1}{|M_2|} . \tag{9.85}$$

It may happen that the right-hand side of the inequality (9.84) is greater than 1. In this case we take $g = 1$ in (9.80). Thus, if $y' < 1$, and $M_2 < 0$, the parameter g in (9.80) and (9.38) is estimated as:

$$g = \min\left\{1, \lambda \frac{\sqrt{y'(1 + \lambda)}}{1 - y' + \lambda}\right\} . \tag{9.86}$$

Summary of the Algorithm

The choice of the parameter g in the smooth approximate to the trajectory (9.38) is done in the following four steps:

1. Evaluate K_1 and K_2 (9.71).
2. If $a^*\sigma_0 - (K_1 x + K_2) \geq 0$ on $[0, 1]$, take $g = 1$. Otherwise, evaluate

$$y' = a^*\sigma_0 \min\{K_2^{-1}, (K_1 + K_2)^{-1}\} .$$

3. If $y' \geq 1$, take $g = 1$. Otherwise evaluate N_1 and N_2 (9.77).

4. If $\min\{N_2, N_1 + N_2\} \geq 0$, take $g = 1$. Otherwise evaluate M_1 (9.75) and take

$$g = \min\left\{1, \lambda\frac{\sqrt{y'(1+\lambda)}}{1 - y' + \lambda}\right\}, \qquad \lambda = \frac{2M_1}{|M_2|}.$$

The function $f_g(a)$ (9.38) with g thus derived has the following properties:

1. It begins in f_0 at $a = 0$ and ends in f^0 at $a = 1$.
2. It is a non-negative function of Γ for each a.
3. It satisfies the conservation laws.
4. The entropy $S_B(f_g(a))$ is a monotonic function of a.
5. It is tangent to the exact trajectory at $a = 0$.

In practical computations, the approximation $f_1^* = f_0 + a_1^* Q_0$ with a_1^* (9.32) can be used in this algorithm instead of the exact f^*.

9.3.6 Discussion

Main results of this Example are:

1. The description of the Q_0-dominated kinetics, and of its equilibrium state f^*. The state f^* is explicitly evaluated.
2. The explicit construction of the approximate trajectory $f(\Gamma, a)$ for nonlinear space-independent kinetic equations equipped with the entropy (Lyapunov) function.

The approach used can be termed "geometric" since it avoids integration of kinetic equations in time. In the point 1, it stays at variance with many alternative approaches to the early-time evolution, which usually involve the time integration over the first few collisions. These methods encounter two general difficulties: the time of integration cannot be defined precisely, and approximations involved can violate the entropy increase and the positivity of distribution function. These difficulties are avoided in the present approach. On the other hand, the presentation of the Q_0-dominated relaxation is itself an ansatz, whose relevance to the actual process can be judged only a posteriori. As the examples show, we can indeed speak about such a dynamics. It is remarkable that the limiting state f^* differs significantly from both the initial and equilibrium states. In other words, irrespectively of how short in time the initial stage of the relaxation might be, the change of the state can be large.

Concerning the point 2, it is worthwhile to notice that, though the space-independent problem is too "refined", it nevertheless gives a good example of a problem without small parameters. It is rather remarkable that the global requirements to the trajectory (e.g., the entropy increase) are accomplished with the direct local analysis (Subsect. 9.3.5). Estimations in this part are sufficient, and can be enhanced.

Final comments concern a further treatment of the space-independent relaxation. The goal now is to develop a procedure of *corrections* to the approximate trajectory. In other words, what we need is a sequence of the functions $f_0(\Gamma, a), f_1(\Gamma, a), \ldots$, which converges to the exact trajectory, and where $f_0(\Gamma, a)$ is the initial (global) approximation to the trajectory. Again, a general obstacle is the absence of a small parameter in the problem. However, the method of invariant manifold (Chap. 6) appears to be appropriate (at least formally) since it is based on the Newton method and not on the small parameter expansions. It turns out that smoothness and all the requirements listed in the Introduction should be met by any initial approximation $f_0(\Gamma, a)$ chosen for this procedure. Thus, the approximation (9.38) can be used for this purpose. We have already annonsed this method with a result of the first Newton correction to the approximation (9.38) for the VHP model (see Fig. 9.2).

Finally, the present method recently became a part of the so-called Entropic lattice Boltzman method [136, 137, 140, 141] (see Sect. 2.7) because it enables to implement collision in a numerically stable fashion.

10 Method of Invariant Grids

The method of invariant grids is developed for a grid-based computation of invariant manifolds.

10.1 Invariant Grids

Elsewhere above in this book, we considered the immersions $F(y)$, and the methods for their construction, without addressing the question of how to implement F numerically. In most of the works (of us and of other people on similar problems), analytic forms were required to represent manifolds (see, however, the method of Legendre integrators [254, 266, 369]). However, in order to construct manifolds of a relatively low dimension, grid-based representations of manifolds become a relevant option. The *method of invariant grids* (MIG) was suggested recently in [22].

The main idea of MIG is to find a mapping of the finite-dimensional grids into the phase space of a dynamic system. That is, we construct not just a point approximation of the invariant manifold $F^*(y)$, but an *invariant grid*. When refined, it is expected to converge, of course, to $F^*(y)$, but in any case it is a separate, independently defined object.

Let's denote $L = R^n$, G is a discrete subset of R^n. It is natural to think of a regular grid, but this is not so crucial. For every point $y \in G$, a neighborhood of y is defined: $V_y \subset G$, where V_y is a finite set, and, in particular, $y \in V_y$. On regular grids, V_y includes, as a rule, the nearest neighbors of y. It may also include the points next to the nearest neighbors.

For our purpose, we should define a grid differential operator. For every function, defined on the grid, also all derivatives are defined:

$$\left. \frac{\partial f}{\partial y_i} \right|_{y \in G} = \sum_{z \in V_y} q_i(z, y) f(z), i = 1, \dots n \, . \tag{10.1}$$

where $q_i(z, y)$ are some coefficients.

Here we do not specify the choice of the functions $q_i(z, y)$. We just mention in passing that, as a rule, (10.1) is established using some approximation of f in the neighborhood of y in R^n by some differentiable functions (for example, polynomials). This approximation is based on the values of f at the points of

Alexander N. Gorban and Iliya V. Karlin: *Invariant Manifolds for Physical and Chemical Kinetics*, Lect. Notes Phys. **660**, 279–298 (2005)
www.springerlink.com

V_y. For regular grids, $q_i(z, y)$ are functions of the difference $z - y$. For some of the nodes y which are close to the edges of the grid, functions are defined only on the part of V_y. In this case, the coefficients in (10.1) should be modified appropriately in order to provide an approximation using available values of f. Below we assume this modification is always done. We also assume that the number of points in the neighborhood V_y is always sufficient to make the approximation possible. This assumption restricts the choice of the grids G. Let's call *admissible* all such subsets G, on which one can define differentiation operator in every point.

Let F be a given mapping of some admissible subset $G \subset R^n$ into U. For every $y \in V$ we define tangent vectors:

$$T_y = Lin\{g_i\}_1^n \ , \tag{10.2}$$

where vectors $g_i(i = 1, \ldots n)$ are partial derivatives (10.1) of the vector-function F:

$$g_i = \frac{\partial F}{\partial y_i} = \sum_{z \in V_y} q_i(z, y) F(z) \ , \tag{10.3}$$

or in the coordinate form:

$$(g_i)_j = \frac{\partial F_j}{\partial y_i} = \sum_{z \in V_y} q_i(z, y) F_j(z) \ . \tag{10.4}$$

Here $(g_i)_j$ is the jth coordinate of the vector (g_i), and $F_j(z)$ is the jth coordinate of the point $F(z)$.

The grid G is *invariant*, if for every node $y \in G$ the vector field $J(F(y))$ belongs to the tangent space T_y (here J is the right hand side of the kinetic equations (3.1)).

So, the definition of the invariant grid includes:

1. The finite admissible subset $G \subset R^n$;
2. A mapping F of this admissible subset G into U (where U is the phase space of kinetic equation (3.1));
3. The differentiation formulas (10.1) with given coefficients $q_i(z, y)$;

The *grid invariance equation* has a form of an inclusion:

$$J(F(y)) \in T_y \text{ for every } y \in G \ ,$$

or a form of an equation:

$$(1 - P_y) J(F(y)) = 0 \text{ for every } y \in G \ ,$$

where P_y is the thermodynamic projector (5.25).

The grid differentiation formulas (10.1) are needed, in the first place, to establish the tangent space T_y, and the null space of the thermodynamic projector P_y in each node. It is important to realize that the locality of the construction of the thermodynamic projector enables this without a global parametrization.

Basically, in our approach, the grid specifics is in: (a) differentiation formulas, (b) grid construction strategy (the grid can be extended, contracted, refined, etc.) The invariance equations (3.3), equations of the film dynamics extension (4.5), the iteration Newton method (6.2), and the formulae of the relaxation approximation (9.2) do not change at all. For convenience, let us rewrite all these formulas in the grid context.

Let $x = F(y)$ be the location of the grid's node y immersed into U. We have the set of tangent vectors $g_i(x)$, defined in x (10.3), (10.4). Thus, the tangent space T_y is defined by (10.2). Also, one has the entropy function $S(x)$, the linear functional $D_x S|_x$, and the subspace $T_{0y} = T_y \cap \ker D_x S|_x$ in T_y. Let $T_{0y} \neq T_y$. In this case we have a vector $e_y \in T_y$, orthogonal to T_{0y}, $D_x S|_x(e_y) = 1$. Then the thermodynamic projector is defined as:

$$P_y \bullet = P_{0y} \bullet + e_y D_x S|_x \bullet , \tag{10.5}$$

where P_{0y} is the orthogonal projector on T_{0y} with respect to the entropic scalar product $\langle | \rangle_x$.

If $T_{0y} = T_y$, then the thermodynamic projector is the orthogonal projector on T_y with respect to the entropic scalar product $\langle | \rangle_x$.

For the Newton method with incomplete linearization, the equations for calculation the new node location $x' = x + \delta x$ are:

$$\begin{cases} P_y \delta x = 0 \\ (1 - P_y)(J(x) + DJ(x)\delta x) = 0 . \end{cases} \tag{10.6}$$

Here $DJ(x)$ is a matrix of derivatives of J evaluated at x. The self-adjoint linearization can be used too (see Chap. 7).

Equation (10.6) is a system of linear algebraic equations. In practice, it proves convenient to choose some orthonormal (with respect to the entropic scalar product) basis b_i in $\ker P_y$. Let $r = \dim(\ker P_y)$. Then $\delta x = \sum_{i=1}^r \delta_i b_i$, and system (10.6) takes the form

$$\sum_{k=1}^r \delta_k \langle b_i \mid DJ(x)b_k \rangle_x = -\langle J(x) \mid b_i \rangle_x, i = 1 \ldots r . \tag{10.7}$$

This is the system of linear equations for adjusting the node location according to the Newton method with incomplete linearization.

For the relaxation method, one needs to calculate the defect $\Delta_x = (1 - P_y)J(x)$, and the relaxation step

$$\tau(x) = -\frac{\langle \Delta_x | \Delta_x \rangle_x}{\langle \Delta_x | DJ(x)\Delta_x \rangle_x} . \tag{10.8}$$

Then, the new node location x' is computed as

$$x' = x + \tau(x)\Delta_x .$$ (10.9)

This is the equation for adjusting the node location according to the relaxation method.

10.2 Grid Construction Strategy

From all the reasonable strategies of the invariant grid construction we consider here the following two: the *growing lump* and the *invariant flag*.

10.2.1 Growing Lump

The construction is initialized from the equilibrium point y^*. The first approximation is constructed as $F(y^*) = x^*$, and for some initial V_0 ($V_{y^*} \subset V_0$) one has $F(y) = x^* + A(y - y^*)$, where A is an isometric embedding (in the standard Euclidean metrics) of R^n in E.

For this initial grid one makes a fixed number of iterations of one of the methods chosen (Newton's method with incomplete linearization or the relaxation method), and, after that, puts $V_1 = \bigcup_{y \in V_0} V_y$ and extends F from V_0 onto V_1 using the linear extrapolation, and the process continues. One of the possible variants of this procedure is to extend the grid from V_i to V_{i+1} not after a fixed number of iterations, but only after the invariance defect Δ_y becomes less than a given ϵ (in a given norm, which is entropic, as a rule), for all nodes $y \in V_i$. The lump stops growing after it reaches the boundary and is within a given accuracy $\|\Delta\| < \epsilon$.

10.2.2 Invariant Flag

In order to construct the invariant flag one uses sufficiently regular grids G, in which many points are located on the coordinate lines, planes, etc. One considers the standard flag $R^0 \subset R^1 \subset R^2 \subset \ldots \subset R^n$ (every next space is constructed by adding one more coordinate). It corresponds to a sequence of grids $\{y^*\} \subset G^1 \subset G^2 \ldots \subset G^n$, where $\{y^*\} = R^0$, and G^i is a grid in R^i.

First, y^* is mapped on x^* and further $F(y^*) = x^*$. Then the invariant grid is constructed on $V^1 \subset G^1$ (up to the boundaries and within a given accuracy $\|\Delta\| < \epsilon$). After that, the neighborhoods in G^2 are added to the points V^1, and the grid $V^2 \subset G^2$ is constructed (up to the boundaries and within a given accuracy) and so on, until $V^n \subset G^n$ is constructed.

While constructing the kth-order grid $V^k \subset G^k$, the important role of the grids of lower dimension $V^0 \subset \ldots \subset V^{k-1} \subset V^k$ embedded in it, is preserved. The point $F(y^*) = x^*$ (equilibrium) remains fixed. For every $y \in V^q$ ($q < k$) the tangent vectors $g_1, \ldots, g_q$ are constructed, using the differentiation operators (10.1) on the whole V^k. Using the tangent space $T_y = Lin\{g_1, \ldots, g_q\}$, the projector P_y is constructed, the iterations are applied and so on. All this is done in order to obtain a sequence of embedded invariant grids, given by the same map F.

10.2.3 Boundaries Check and the Entropy

We construct grid mapping of F onto a finite set $V \in G$. The technique of checking whether the grid still belongs to the phase space U of the kinetic system $(F(V) \subset U)$ is quite straightforward: all the points $y \in V$ are checked whether they belong to U. If at the next iteration a point $F(y)$ leaves U, then it is pulled inside by a homothety transform with the center in x^*. Since the entropy is a concave function, the homothety contraction with the center in x^* increases the entropy monotonically. Another variant to cut off the points which leave U.

By construction (5.25), the kernel of the entropic projector is annulled by the entropy differential. Thus, in the first order, the steps in the Newton method with incomplete linearization (6.2) as well as in the relaxation method (9.1), (9.2) do not change the entropy. But if the steps are quite large, then the increase of the entropy may become essential, and the points are returned on their entropy levels by the homothety contraction with the center in the equilibrium point.

10.3 Instability of Fine Grids

When one reduces the grid spacing in order to refine the grid, then, once the grid spacing becomes small enough, one can face the problem of the *Courant instability* [269–271]. Instead of converging, at every iteration the grid becomes more and more entangled (see Fig. 10.1).

A way to avoid such instability is well-known. This is decreasing the time step. In our problem, instead of a true time step, we have a shift in the Newtonian direction. Formally, we can assign the value $h = 1$ for one complete step in the Newtonian direction. Let us extend now the Newton method to arbitrary h. For this, let us find $\delta x = \delta F(y)$ from (10.6), but update δx proportionally to h; the new value of $x_{n+1} = F_{n+1}(y)$ is equal to

$$F_{n+1}(y) = F_n(y) + h_n \delta F_n(y) \tag{10.10}$$

where n denotes the number of iteration.

One way to choose the step value h is to make it adaptive, by controlling the average value of the invariance defect $\|\Delta_y\|$ at every step. Another way is the convergence control: then $\sum h_n$ plays a role of time.

Elimination of the Courant instability for the relaxation method can be done quite analogously. Everywhere the step h is maintained as large as it is possible without running into convergence problems.

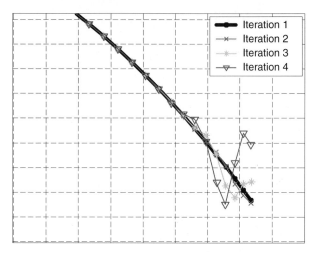

Fig. 10.1. Grid instability. For small grid steps approximations in the calculation of grid derivatives lead to the grid instability effect. Several successive iterations of the algorithm without adaptation of the time step are shown that lead to undesirable "oscillations", which eventually destroy the grid starting from one of its ends

10.4 Which Space is Most Appropriate for the Grid Construction?

For kinetic systems, there are two distinguished representations of the phase space:

- The density space (concentrations, energy or probability densities, etc.)
- The space of conjugated intensive variables, (temperature, chemical potentials, etc.)

The density space is convenient for the construction of the quasi-chemical representations. Here the balance relations are linear and the constraints are in the form of linear inequalities (the densities themselves or some of their linear combinations must be positive).

The conjugated variables space is convenient in the sense that the equilibrium conditions are linear in terms of the conjugate variables. In these spaces the quasiequilibrium manifolds exist in the form of linear subspaces and, vice versa, linear balance equations turn out to be equations of the conditional entropy maximum.

The duality we have just mentioned is well-known and studied in detail in many works on thermodynamics and Legendre transformation [274,275]. This viewpoint of nonequilibrium thermodynamics unifies many well-established mesoscopic dynamical theories, as for example the Boltzmann kinetic theory and the Navier–Stokes–Fourier hydrodynamics [189]. To this end, preceding the grids in the density space were discussed. However, the use of the space of conjugated variables seems to be even more appealing for the grid construction. The main argument is the specific role of quasiequilibrium, which a linear manifold in the conjugated space. Therefore, a linear extrapolation gives a thermodynamically justified quasiequilibrium approximation. A linear approximation of the slow invariant manifold in the neighborhood of the equilibrium in terms of the conjugate variables space already gives the readily global quasiequilibrium manifold which corresponds to the motion separation in the neighborhood of the equilibrium point.

For the mass action law, transition to the conjugate variables is simply the logarithmic transformation of the coordinates.

10.5 Carleman's Formula
in the Analytical Invariant Manifolds Approximations.
First Benefit of Analyticity: Superresolution

When constructing invariant grids, one must define the differential operators (10.1) for every grid's node. For calculating the differential operators in some point y, an interpolation procedure in the neighborhood of y is used. As a rule, it is an interpolation by a low-order polynomial, which is constructed using the function values in the nodes belonging to the neighbourhood of y in G. This approximation (using values in the nearest neighborhood nodes) is natural for smooth functions. But we are looking for the *analytical* invariant manifold (see discussion in Chap. 4). Analytical functions have a much more "rigid" structure than the smooth ones. One can change a smooth function in the neighborhood of any point in such a way, that outside this neighborhood the function will not change. In general, this is not possible for analytical functions: a kind of a "long-range" effect takes place (as is well known) .

The idea is to make use of this effect and to reconstruct some analytical function f_G using a function given on G. There is one important requirement: if the values given on G are values of some function f which is analytical in a neighborhood U, then, if the G is refined "correctly", one must have $f_G \to f$ in U. The sequence of reconstructed function f_G should converge to the "right" function f.

What is the "correct refinement"? For smooth functions for the convergence $f_G \to f$ it is necessary and sufficient that, in the course of refinement, G would approximate the whole U with arbitrary accuracy. For analytical functions it is necessary only that, under the refinement, G would approximate some uniqueness set[1] $A \subset U$. Suppose we have a sequence of grids G, each next is finer than the previous, which approximate a set A. For smooth functions using function values defined on the grids one can reconstruct the function in A. For analytical functions, if the analyticity domain U is known, and

---

[1] Let's remind to the reader that $A \subset U$ is called *uniqueness set* in U if for analytical in U functions ψ and φ from $\psi|_A \equiv \varphi|_A$ it follows $\psi \equiv \varphi$.

A is a uniqueness set in U, then one can reconstruct the function in U. The set U can be essentially bigger than A; because of this such extension was named as *superresolution effect* [276]. There exist formulas for construction of analytical functions f_G for different domains U, uniqueness sets $A \subset U$ and for different ways of discrete approximation of A by a sequence of refined grids G [276]. Here we provide only one Carleman's formula which is the most appropriate for our purposes.

Let domain $U = Q_\sigma^n \subset C^n$ be a product of strips $Q_\sigma \subset C$, $Q_\sigma = \{z | \mathrm{Im} z < \sigma\}$. We shall construct functions holomorphic in Q_σ^n. This is effectively equivalent to the construction of real analytical functions f in the whole R^n with a condition on the convergence radius $r(x)$ of the Taylor series for f as a function of each coordinate: $r(x) \geq \sigma$ in every point $x \in R^n$.

The sequence of refined grids is constructed as follows: let for every $l = 1, \ldots, n$ a finite sequence of distinct points $N_l \subset Q_\sigma$ be defined:

$$N_l = \{x_{lj} | j = 1, 2, 3 \ldots\}, x_{lj} \neq x_{li} \ \ for \ \ i \neq j \tag{10.11}$$

The countable uniqueness set A, which is approximated by a sequence of refined grids, has the form:

$$A = N_1 \times N_2 \times \ldots \times N_n = \{(x_{1i_1}, x_{2i_2}, \ldots, x_{ni_n}) | i_{1,\ldots,n} = 1, 2, 3, \ldots\} \tag{10.12}$$

The grid G_m is defined as the product of initial fragments N_l of length m:

$$G_m = \{(x_{1i_1}, x_{2i_2} \ldots x_{ni_n}) | 1 \leq i_{1,\ldots,n} \leq m\} \tag{10.13}$$

Let us denote $\lambda = 2\sigma/\pi$ (σ is a half-width of the strip Q_σ). The key role in the construction of the Carleman's formula is played by the functional $\omega_m^\lambda(u, p, l)$ of 3 variables: $u \in U = Q_\sigma^n$, p is an integer, $1 \leq p \leq m$, l is an integer, $1 \leq p \leq n$. Further u will be the coordinate value at the point where the extrapolation is calculated, l will be the coordinate number, and p will be an element of multi-index $\{i_1, \ldots, i_n\}$ for the point $(x_{1i_1}, x_{2i_2}, \ldots, x_{ni_n}) \in G$:

$$\omega_m^\lambda(u, p, l) = \frac{(e^{\lambda x_{lp}} + e^{\lambda \bar{x}_{lp}})(e^{\lambda u} - e^{\lambda x_{lp}})}{\lambda(e^{\lambda u} + e^{\lambda \bar{x}_{lp}})(u - x_{lp})e^{\lambda x_{lp}}}$$
$$\times \prod_{j=1 j \neq p}^{m} \frac{(e^{\lambda x_{lp}} + e^{\lambda \bar{x}_{lj}})(e^{\lambda u} - e^{\lambda x_{lj}})}{(e^{\lambda x_{lp}} - e^{\lambda x_{lj}})(e^{\lambda u} + e^{\lambda \bar{x}_{lj}})} \tag{10.14}$$

For real-valued x_{pk} formula (10.14) simplifyes:

$$\omega_m^\lambda(u, p, l) = 2\frac{e^{\lambda u} - e^{\lambda x_{lp}}}{\lambda(e^{\lambda u} + e^{\lambda x_{lp}})(u - x_{lp})} \times \prod_{j=1 j \neq p}^{m} \frac{(e^{\lambda x_{lp}} + e^{\lambda x_{lj}})(e^{\lambda u} - e^{\lambda x_{lj}})}{(e^{\lambda x_{lp}} - e^{\lambda x_{lj}})(e^{\lambda u} + e^{\lambda x_{lj}})} \tag{10.15}$$

The Carleman formula for extrapolation from G_M on $U = Q_\sigma^n$ ($\sigma = \pi\lambda/2$) has the form ($z = (z_1, \ldots, z_n)$):

$$f_m(z) = \sum_{k_1,\dots,k_n=1}^{m} f(x_k) \prod_{j=1}^{n} \omega_m^\lambda(z_j, k_j, j) \,, \qquad (10.16)$$

where $k = k_1, \dots, k_n$, $x_k = (x_{1k_1}, x_{2k_2}, \dots, x_{nk_n})$.

There exists a theorem [276]:

If $f \in H^2(Q_\sigma^n)$, then $f(z) = \lim_{m\to\infty} f_m(z)$, where $H^2(Q_\sigma^n)$ is the Hardy class of holomorphic in Q_σ^n functions.

It is useful to present the asymptotics of (10.16) for large $|\mathrm{Re}z_j|$. For this purpose, we shall consider the asymptotics of (10.16) for large $|\mathrm{Re}u|$:

$$|\omega_m^\lambda(u, p, l)| = \left| \frac{2}{\lambda u} \prod_{j=1 j\neq p}^{m} \frac{e^{\lambda x_{lp}} + e^{\lambda x_{lj}}}{e^{\lambda x_{lp}} - e^{\lambda x_{lj}}} \right| + o(|\mathrm{Re}u|^{-1}) \,. \qquad (10.17)$$

From the formula (10.16) one can see that for the finite m and $|\mathrm{Re}z_j| \to \infty$ function $|f_m(z)|$ behaves like $const \cdot \prod_j |z_j|^{-1}$.

This property (zero asymptotics) must be taken into account when using the formula (10.16). When constructing invariant manifolds $F(W)$, it is natural to use (10.16) not for the immersion $F(y)$, but for the deviation of $F(y)$ from some analytical ansatz $F_0(y)$ [277–280].

The analytical ansatz $F_0(y)$ can be obtained using Taylor series, just as in the Lyapunov auxiliary theorem [3] (see also Chap. 4). Another variant is to use Taylor series for the construction of Pade-approximations.

It is natural to use approximations (10.16) in terms of dual variables as well, since there exists for them (as the examples demonstrate) a simple and effective linear ansatz for the invariant manifold. This is the slow invariant subspace E_{slow} of the operator of linearized system (3.1) in dual variables at the equilibrium point. This invariant subspace corresponds to the set of "slow" eigenvalues (with small $|\mathrm{Re}\lambda|$, $\mathrm{Re}\lambda < 0$). In the space of concentrations this invariant subspace is the quasiequilibrium manifold. It consists of the maximum entropy points on the affine manifolds of the form $x + E_{\mathrm{fast}}$, where E_{fast} is the "fast" invariant subspace of the operator of the linearized system (3.1) at the equilibrium point. It corresponds to the "fast" eigenvalues (large $|\mathrm{Re}\lambda|$, $\mathrm{Re}\lambda < 0$).

Carleman's formulas can be useful for the invariant grids construction in two places: first, for the definition of the grid differential operators (10.1), and second, for the analytical continuation of the manifold from the grid.

10.6 Example: Two-Step Catalytic Reaction

Let us consider a two-step four-component reaction with one catalyst A_2 (the Michaelis-Menten mechanism):

$$A_1 + A_2 \leftrightarrow A_3 \leftrightarrow A_2 + A_4 \,. \qquad (10.18)$$

We assume the Lyapunov function of the form

$$S = -G = -\sum_{i=1}^{4} c_i [\ln(c_i/c_i^{\text{eq}}) - 1] \,.$$

The kinetic equation for the four-component vector of concentrations, $c = (c_1, c_2, c_3, c_4)$, has the form

$$\dot{c} = \gamma_1 W_1 + \gamma_2 W_2 \,. \tag{10.19}$$

Here $\gamma_{1,2}$ are stoichiometric vectors,

$$\gamma_1 = (-1, -1, 1, 0) \,, \quad \gamma_2 = (0, 1, -1, 1) \,, \tag{10.20}$$

while functions $W_{1,2}$ are reaction rates:

$$W_1 = k_1^+ c_1 c_2 - k_1^- c_3 \,, \quad W_2 = k_2^+ c_3 - k_2^- c_2 c_4 \,. \tag{10.21}$$

Here $k_{1,2}^{\pm}$ are reaction rate constants. The system under consideration has two conservation laws,

$$c_1 + c_3 + c_4 = B_1 \,, \quad c_2 + c_3 = B_2 \,, \tag{10.22}$$

or $\langle \boldsymbol{b}_{1,2}, \boldsymbol{c} \rangle = B_{1,2}$, where $\boldsymbol{b}_1 = (1, 0, 1, 1)$ and $\boldsymbol{b}_1 = (0, 1, 1, 0)$. The non-linear system (10.18) is effectively two-dimensional, and we consider a one-dimensional reduced description. For our example, we chosed the following set of parameters:

$$\begin{aligned} &k_1^+ = 0.3 \,, \ k_1^- = 0.15 \,, \ k_2^+ = 0.8, \ k_2^- = 2.0 \,; \\ &c_1^{\text{eq}} = 0.5 \,, \ c_2^{\text{eq}} = 0.1, c_3^{\text{eq}} = 0.1 \,, \ c_4^{\text{eq}} = 0.4 \,; \\ &B_1 = 1.0, \ B_2 = 0.2 \end{aligned} \tag{10.23}$$

The one-dimensional invariant grid is shown in Fig. 10.2 in the (c_1, c_4, c_3) coordinates. The grid was constructed by the growing lump method, as described above. We used Newton iterations to adjust the nodes. The grid was grown up to the boundaries of the phase space.

The grid in this example is a one-dimensional ordered sequence $\{x_1, \ldots, x_n\}$. The grid derivatives for calculating the tangent vectors g were taken as $g(x_i) = (x_{i+1} - x_{i-1})/||x_{i+1} - x_{i-1}||$ for the internal nodes, and $g(x_1) = (x_1 - x_2)/||x_1 - x_2||$, $g(x_n) = (x_n - x_{n-1})/||x_n - x_{n-1}||$ for the grid's boundaries.

Close to the phase space boundaries we had to apply an adaptive algorithm for choosing the time step h: if, after the next growing step (adding new nodes to the grid and after completing $N = 20$ Newtonian steps, the grid did not converged, then we choose a new step size $h_{n+1} = h_n/2$ and recalculate the grid. The final (minimal) value for h was $h \approx 0.001$.

The location of the nodes was parametrized with the entropic distance to the equilibrium point measured in the quadratic metrics given by the matrix

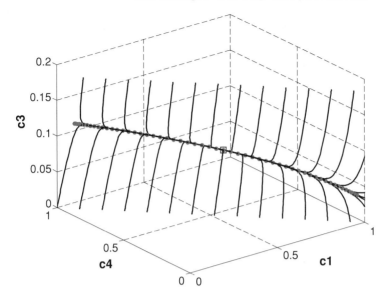

Fig. 10.2. One-dimensional invariant grid (*circles*) for the two-dimensional chemical system. Projection into the 3d-space of c_1, c_4, c_3 concentrations. The trajectories of the system in the phase space are shown by lines. The equilibrium point is marked by the square. The system quickly reaches the grid and further moves along it

$\boldsymbol{H}_c = -||\partial^2 S(\boldsymbol{c})/\partial c_i \partial c_j||$ in the equilibrium c^{eq}. It means that every node is located on a sphere in this metrics with a given radius, which increases linearly with number of the node. In this figure the step of the increase is chosen to be 0.05. Thus, the first node is at the distance 0.05 from the equilibrium, the second is at the distance 0.10 and so on. Figure 10.3 shows several important quantities which facilitate understanding of the object (invariant grid) extracted. The sign on the x-axis of the graphs at Fig. 10.3 is meaningless since the distance is always positive, but in this situation it indicates two possible directions from the equilibrium point.

Figure 10.3a,b represents the slow one-dimensional component of the dynamics of the system. Given any initial condition, the system quickly finds the corresponding point on the manifold and starting from this point the dynamics is given by a part of the graph on the Fig. 10.3a,b.

One of the useful quantities is shown on the Fig. 10.3c. It is the relation between the relaxation times "toward" and "along" the grid (λ_2/λ_1, where λ_1, λ_2 are the smallest and the next smallest by absolute value non-zero eigenvalue of the system, symmetrically linearized at the point of the grid node). The figure demonstrates that the system is very stiff close to the equilibrium point (λ_1 and λ_2 are well separated from each other), and becomes less stiff (by order of magnitude) near the boundary. This leads to the conclusion that the one-dimensional reduced model is more adequate in the neighborhood

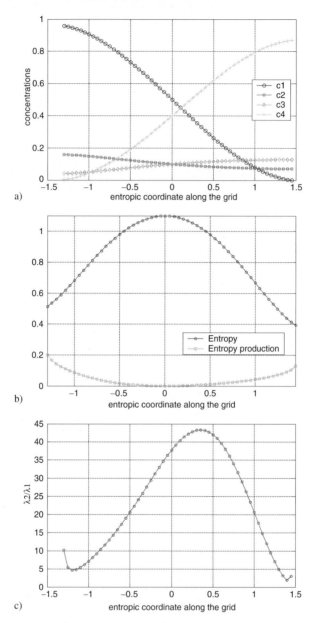

a)

b)

c)

Fig. 10.3. One-dimensional invariant grid for the two-dimensional chemical system. (**a**) Values of the concentrations along the grid. (**b**) Values of the entropy and the entropy production $(-\mathrm{d}G/\mathrm{d}t)$ along the grid. (**c**) Ratio of the relaxation times "towards" and "along" the manifold. The nodes positions are parametrized with entropic distance measured in the quadratic metrics given by $\boldsymbol{H}_c = -||\partial^2 S(\boldsymbol{c})/\partial c_i \partial c_j||$ in the equilibrium c^{eq}. Entropic coordinate equal to zero corresponds to the equilibrium

of the equilibrium where fast and slow motions are separated by two orders of magnitude. On the end-points of the grid the one-dimensional reduction ceases to be well-defined.

10.7 Example: Model Hydrogen Burning Reaction

In this section we consider a more complicated example, where the concentration space is 6-dimensional, while the system is 4-dimensional. We construct an invariant flag which consists of 1- and 2-dimensional invariant manifolds.

We consider a chemical system with six species called H_2 (hydrogen), O_2 (oxygen), H_2O (water), H, O, OH (radicals). We assume the Lyapunov function of the form $S = -G = -\sum_{i=1}^{6} c_i[\ln(c_i/c_i^{\text{eq}}) - 1]$. The subset of the hydrogen burning reaction and corresponding (direct) rate constants have were taken as:

$$
\begin{array}{lll}
1. & H_2 \leftrightarrow 2H & k_1^+ = 2 \\
2. & O_2 \leftrightarrow 2O & k_2^+ = 1 \\
3. & H_2O \leftrightarrow H + OH & k_3^+ = 1 \\
4. & H_2 + O \leftrightarrow H + OH & k_4^+ = 10^3 \\
5. & O_2 + H \leftrightarrow O + OH & k_5^+ = 10^3 \\
6. & H_2 + O \leftrightarrow H_2O & k_6^+ = 10^2
\end{array}
\tag{10.24}
$$

The conservation laws are:

$$
\begin{aligned}
2c_{H_2} + 2c_{H_2O} + c_H + c_{OH} &= b_H \\
2c_{O_2} + c_{H2O} + c_O + c_{OH} &= b_O
\end{aligned}
\tag{10.25}
$$

For parameter values we took $b_H = 2$, $b_O = 1$, and the equilibrium point:

$$
c_{H_2}^{\text{eq}} = 0.27 \; c_{O_2}^{\text{eq}} = 0.135 \; c_{H_2O}^{\text{eq}} = 0.7 \; c_H^{\text{eq}} = 0.05 \; c_O^{\text{eq}} = 0.02 \; c_{OH}^{\text{eq}} = 0.01
\tag{10.26}
$$

Other rate constants k_i^-, $i = 1 \ldots 6$ were calculated from c^{eq} value and k_i^+. For this system the stoichiometric vectors are:

$$
\begin{aligned}
\gamma_1 &= (-1, 0, 0, 2, 0, 0) & \gamma_2 &= (0, -1, 0, 0, 2, 0) \\
\gamma_3 &= (0, 0, -1, 1, 0, 1) & \gamma_4 &= (-1, 0, 0, 1, -1, 1) \\
\gamma_5 &= (0, -1, 0, -1, 1, 1) & \gamma_6 &= (-1, 0, 1, 0, -1, 0)
\end{aligned}
\tag{10.27}
$$

The system under consideration is fictitious in the sense that the subset of equations corresponds to the simplified picture of this chemical process and the rate constants do not correspond to any experimentally measured quantities, rather they reflect only orders of magnitudes relevant real-world systems. In that sense we consider here a qualitative model system, which allows us to illustrate the invariant grids method. Nevertheless, modeling of more realistic systems differs only in the number of species and equations. This leads, of course, to computationally harder problems, but difficulties are not crucial.

Figure 10.4a presents a one-dimensional invariant grid constructed for the system. Figure 10.4b demonstrates the reduced dynamics along the manifold (for the explanation of the meaning of the x-coordinate, see the previous subsection). In Fig. 10.4c the three smallest by the absolute value non-zero eigenvalues of the symmetrically linearized Jacobian matrix of the system are shown. One can see that the two smallest eigenvalues almost interchange on one of the grid ends. This means that the one-dimensional "slow" manifold faces definite problems in this region, it is just not well defined there. In practice, it means that one has to use at least a two-dimensional grids there.

Figure 10.5a gives a view of the two-dimensional invariant grid, constructed for the system, using the "invariant flag" strategy. The grid was raised starting from the 1D-grid constructed at the previous step. At the first iteration for every node of the initial grid, two nodes (and two edges) were added. The direction of the step was chosen as the direction of the eigenvector of the matrix A^{sym} (at the point of the node), corresponding to the second "slowest" direction. The value of the step was chosen to be $\epsilon = 0.05$ in terms of entropic distance. After several Newton's iterations done until convergence was reached, new nodes were added in the direction "ortogonal" to the 1D-grid. This time it was done by linear extrapolation of the grid on the same step $\epsilon = 0.05$. Once some new nodes become one or several negative coordinates (the grid reaches the boundaries) they were cut off. If a new node has only one edge, connecting it to the grid, it was excluded (since it was impossible to calculate 2D-tangent space for this node). The process was continued until the expansion was possible (the ultimate state is when every new node had to be cut off).

The method for calculating tangent vectors for this regular rectangular 2D-grid was chosen to be quite simple. The grid consists of *rows*, which are co-oriented by construction to the initial 1D-grid, and *columns* that consist of the adjacent nodes in the neighboring rows. The direction of the columns corresponds to the second slowest direction along the grid. Then, every row and column is considered as a 1D-grid, and the corresponding tangent vectors are calculated as it was described before:

$$g_{row}(x_{k,i}) = (x_{k,i+1} - x_{k,i-1})/\|x_{k,i+1} - x_{k,i-1}\|$$

for the internal nodes and

$$g_{row}(x_{k,1}) = (x_{k,1} - x_{k,2})/\|x_{k,1} - x_{k,2}\|, g_{row}(x_{k,n_k})$$

$$= (x_{k,n_k} - x_{k,n_k-1})/\|x_{k,n_k} - x_{k,n_k-1}\|$$

for the nodes which are close to the grid's edges. Here $x_{k,i}$ denotes the vector of the node in the kth row, ith column; n_k is the number of nodes in the kth row. Second tangent vector $g_{col}(x_{k,i})$ is calculated analogously. In practice, it proves convenient to orthogonalize $g_{row}(x_{k,i})$ and $g_{col}(x_{k,i})$.

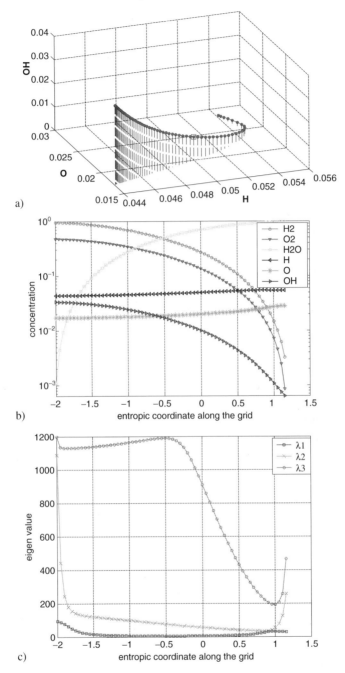

Fig. 10.4. One-dimensional invariant grid for model hydrogen burning reaction. (**a**) Projection into the 3d-space of c_H, c_O, c_{OH} concentrations. (**b**) Concentration values along the grid. (**c**) Three smallest by the absolute value non-zero eigenvalues of the symmetrically linearized system

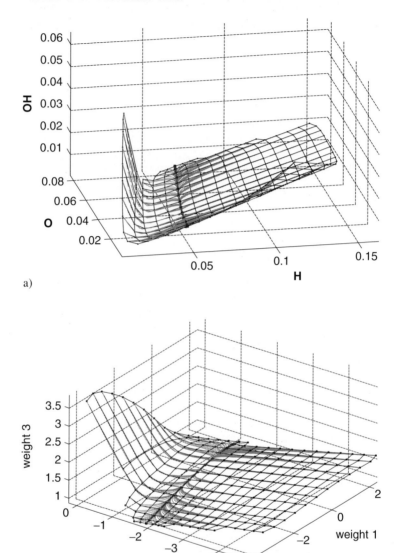

a)

b)

Fig. 10.5. Two-dimensional invariant grid for the model hydrogen burning reaction. (**a**) Projection into the 3d-space of c_H, c_O, c_{OH} concentrations. (**b**) Projection into the principal 3D-subspace. Trajectories of the system are shown coming out from every node. *Bold* line denotes the one-dimensional invariant grid, starting from which the 2D-grid was constructed

10.8 Invariant Grid as a Tool for Data Visualization

Invariant grids provide a possibility of data visualization. In this section we demonstrate this possibility on the model hydrogen burning reaction. Since the phase space is four-dimensional, it is impossible to visualize the grid in one of the coordinate 3D-views, as it was done in the previous subsection. To facilitate visualization one can utilize traditional methods of multi-dimensional data visualization. Here we make use of the principal components analysis (see, for example, [273]), which constructs a three-dimensional linear subspace with maximal dispersion of the othogonally projected data (grid nodes in our case). In other words, the method of principal components constructs in a multi-dimensional space a three-dimensional box such that the grid can be placed maximally tightly inside the box (in the mean square distance meaning). After projection of the grid nodes into this space, we get more or less adequate representation of the two-dimensional grid embedded into the six-dimensional concentrations space (Fig. 10.5b). The disadvantage of the approach is that the axes now do not bear any explicit physical meaning, they are just some linear combinations of the concentrations.

One attractive feature of two-dimensional grids is the possibility to use them as a screen, on which one can display different functions $f(c)$ defined in the concentrations space. This technology was exploited widely in the nonlinear data analysis by the elastic maps method [272]. The idea is to "unfold" the grid on a plane (to present it in the two-dimensional space, where the nodes form a regular lattice). In other words, we are going to work in the internal coordinates of the grid. In our case, the first internal coordinate (let's call it s_1) corresponds to the direction, co-oriented with the one-dimensional invariant grid, the second one (let us call it s_2) corresponds to the second slow direction. By the construction, the coordinate line $s_2 = 0$ line corresponds to the one-dimensional invariant grid. Units of s_1 and s_2 is the entropic distance.

Every grid node has two internal coordinates (s_1, s_2) and, simultaneously, corresponds to a vector in the concentration space. This allows us to map any function $f(c)$ from the multi-dimensional concentration space to the two-dimensional space of the grid. This mapping is defined in a finite number of points (grid nodes), and can be interpolated (linearly, in the simplest case) between them. Using *coloring* and *isolines* one can visualize the values of the function in the neighborhood of the invariant manifold. This is meaningful, since, by the definition, the system spends most of the time in the vicinity of the invariant manifold, thus, one can visualize the behavior of the system. As a result of applying this technology, one obtains a set of color illustrations (a stack of information layers), put onto the grid as a map. This enables applying the whole family of the well developed methods of working with the stack of information layers, such as the *geographical information systems* (GIS) methods.

Briefly, this technique of the visualization is a useful tool for understanding of dynamical systems. It allows to see simultaneously many different

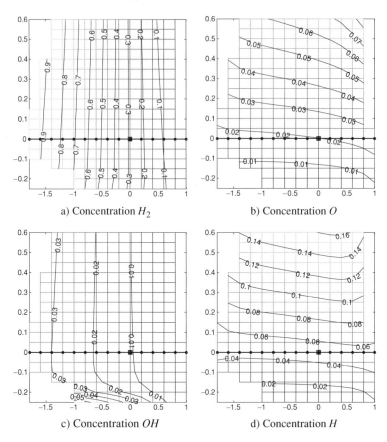

a) Concentration H_2

b) Concentration O

c) Concentration OH

d) Concentration H

Fig. 10.6. Two-dimensional invariant grid as a screen for visualizing different functions defined in the concentrations space. The coordinate axes are entropic distances (see the text for the explanations) along the first and the second slowest directions on the grid. The corresponding 1D invariant grid is denoted by bold line, the equilibrium is denoted by square

scenarios of the system behavior, together with different system's characteristics.

Let us use the invariant grids for the the model hydrogen burning system as a screen for visualisation. The simplest functions to visualize are the coordinates: $c_i(\boldsymbol{c}) = c_i$. In Fig. 10.6 we displayed four colorings, corresponding to the four arbitrarily chosen concentrations functions (of H_2, O, H and OH; Fig. 10.6a-d). The qualitative conclusion that can be made from the graphs is that, for example, the concentration of H_2 practically does not change during the first fast motion (towards the 1D-grid) and then, gradually changes to the equilibrium value (the H_2 coordinate is "slow"). The O coordinate is the opposite case, it is the "fast" coordinate which changes quickly (on the first stage of the motion) to the almost equilibrium value, and it almost does not change after that. Basically, the slopes of the coordinate isolines give some impression of how "slow" a given concentration is Fig. 10.6c shows an interesting behavior of the OH concentration. Close to the 1D grid it behaves like a "slow coordinate", but there is a region on the map where it has a clear "fast" behavior (middle bottom of the graph).

The next two functions which one could wish to visualize are the entropy S and the entropy production $\sigma(\boldsymbol{c}) = -dG/dt(\boldsymbol{c}) = \sum_i \ln(c_i/c_i^{\text{eq}})\dot{c}_i$. They are shown on Fig. 10.7a,b.

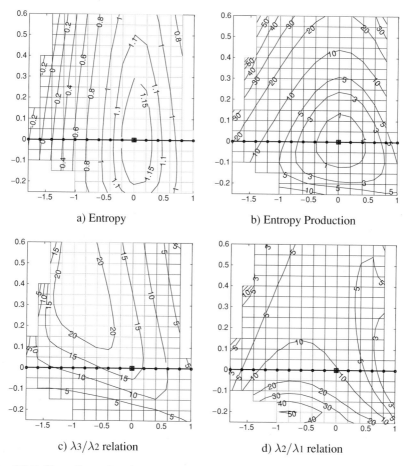

a) Entropy

b) Entropy Production

c) λ_3/λ_2 relation

d) λ_2/λ_1 relation

Fig. 10.7. Two-dimensional invariant grid as a screen for visualizing different functions defined in the concentrations space. The coordinate axes are entropic distances (see the text for the explanations) along the first and the second slowest directions on the grid. The corresponding 1D invariant grid is denoted by bold line, the equilibrium is denoted by square

Finally, we visualize the relation between the relaxation times of the fast motion towards the 2D-grid and the slow motion along it. This is given on the Fig. 10.7c. This picture allows to make a conclusion that two-dimensional consideration can be appropriate for the system (especially in the "high H_2, high O" region), since the relaxation times "towards" and "along" the grid are well separated. One can compare this to the Fig. 10.7d, where the relation between relaxation times towards and along the 1D-grid is shown.

11 Method of Natural Projector

P. and T. Ehrenfest introdused in 1911 a model of dynamics with a coarse-graining of the original conservative system in order to introduce irreversibility [15]. Ehrenfests considered a partition of the phase space into small cells, and they have suggested to combine the motions of the phase space ensemble due to the reversible dynamics with the coarse-graining ("shaking") steps – averaging of the density of the ensemble over the phase cells. This generalizes to the following: alternations of the motion of the phase ensemble due to the microscopic equations with returns to the quasiequilibrium manifold while preserving the values of the macroscopic variables. We here develop a formalism of nonequilibrium thermodynamics based on this generalization. The Ehrenfests' coarse-graining can be treated as a a result of interaction of the system with a generalized thermostat. There are many ways for introduction of thermostat in computational statistical physics [283], but the Ehrenfests' approach remains the basic for understanding the irreversibility phenomenon.

11.1 Ehrenfests' Coarse-Graining Extended to a Formalism of Nonequilibrium Thermodynamics

The idea of the Ehrenfests is the following: One partitions the phase space of the Hamiltonian system into cells. The density distribution of the ensemble over the phase space evolves in time according to the Liouville equation within the time segments $n\tau < t < (n+1)\tau$, where τ is the fixed coarse-graining time step. Coarse-graining is executed at discrete times $n\tau$, densities are averaged over each cell. This alternation of the regular flow with the averaging describes the irreversible behavior of the system.

The most general construction extending the Ehrenfests' idea is given below. Let us stay with notation of Chap. 3, and let a submanifold $F(W)$ be defined in the phase space U. Furthermore, we assume a map (a projection) is defined, $\Pi : U \to W$, with the properties:

$$\Pi \circ F = 1, \quad \Pi(F(y)) = y . \tag{11.1}$$

In addition, one requires some mild properties of regularity, in particular, surjectivity of the differential, $D_x \Pi : E \to L$, in each point $x \in U$.

Alexander N. Gorban and Iliya V. Karlin: *Invariant Manifolds for Physical and Chemical Kinetics*, Lect. Notes Phys. **660**, 299–323 (2005)
www.springerlink.com

Let us fix the coarse-graining time $\tau > 0$, and consider the following problem: Find a vector field Ψ in W,

$$\frac{dy}{dt} = \Psi(y) , \qquad (11.2)$$

such that, for every $y \in W$,

$$\Pi(T_\tau F(y)) = \Theta_\tau y , \qquad (11.3)$$

where T_τ is the shift operator for the system (3.1), and Θ_τ is the (yet unknown!) shift operator for the system in question (11.2).

Equation (11.3) means that one projects not the vector fields but segments of trajectories. The resulting vector field $\Psi(y)$ is called *the natural projection* of the vector field $J(x)$.

Let us assume that there is a very stiff hierarchy of relaxation times in the system (3.1): The motions of the system tend very rapidly to a slow manifold, and next proceed slowly along it. Then there is a smallness parameter, the ratio of these times. Let us take F for the initial condition to the film equation (4.5). If the solution F_t relaxes to the positively invariant manifold F_∞, then in the limit of a very stiff decomposition of motions, the natural projection of the vector field $J(x)$ tends to the usual infinitesimal projection of the restriction of J on F_∞, as $\tau \to \infty$:

$$\Psi_\infty(y) = D_x\Pi|_{x=F_\infty(y)}J(F_\infty(y)) . \qquad (11.4)$$

For stiff dynamic systems, the limit (11.4) is qualitatively almost obvious: After some relaxation time τ_0 (for $t > \tau_0$), the motion $T_\tau(x)$ is located in an ϵ-neighborhood of $F_\infty(W)$. Thus, for $\tau \gg \tau_0$, the natural projection Ψ (equations (11.2) and (11.3)) is defined by the vector field attached to F_∞ with any predefined accuracy. Rigorous proofs of (11.4) requires existence and uniqueness theorems, as well as uniform continuous dependence of solutions on the initial conditions and right hand sides of equations.

The method of natural projector is applied not only to dissipative systems but also (and even mostly) to conservative systems. One of the methods to study the natural projector is based on series expansion[1] in powers of τ. Various other approximation schemes like the Padé approximation are possible too.

The construction of the natural projector was rediscovered in a rather different context by Chorin, Hald and Kupferman [282]. They constructed the optimal prediction methods for an estimation of the solution of nonlinear time-dependent problems when that solution is too complex to be fully

[1] In the well-known work of Lewis [281], this expansion was executed incorrectly (terms of different orders were matched on the left and on the right hand sides of equation (11.3)). This created an obstacle in the development of the method. See a more detailed discussion in the example below.

resolved or when data are missing. The initial conditions for the unresolved components of the solution are drawn from a probability distribution, and their effect on a small set of variables that are actually computed is evaluated via statistical projection. The formalism resembles the projection methods of irreversible statistical mechanics, supplemented by the systematic use of conditional expectations and methods of solution for the fast dynamics equation, needed to evaluate a non-Markovian memory term. The authors claim [282] that result of the computations is close to the best possible estimate that can be obtained given the partial data.

The majority of the methods of invariant manifold can be discussed as development of the Chapman–Enskog method. The central idea is to construct the manifold of distribution functions, where the slow dynamics occurs. The (implicit) change-over from solving the Boltzmann equation to construction of invariant manifold was the crucial idea of Enskog and Chapman. On the other hand, the method of natural projector gives development to the ideas of the Hilbert method. The Hilbert method was historically the first in the solution of the Boltzmann equation. This method is not very popular nowadays, nevertheless, for some purposes it may be more convenient than the Chapman–Enskog method, for example, for a study of stationary solutions [284]. In the method of natural projector we are looking for solutions of kinetic equations with the quasiequilibrium initial state (and in the Hilbert method we start from the local equilibrium too). The main new element in the method of natural projector with respect to the Hilbert method is the construction of the macroscopic equation (11.3). In the next Example the solution for the matching condition (11.3) will be found in a form of Taylor series expansion.

11.2 Example: From Reversible Dynamics to Navier–Stokes and Post-Navier–Stokes Hydrodynamics by Natural Projector

The starting point of our construction are microscopic equations of motion. A traditional example of the microscopic description is the Liouville equation for classical particles. However, we need to stress that the distinction between "micro" and "macro" is always context dependent. For example, Vlasov's equation describes the dynamics of the one-particle distribution function. In one statement of the problem, this is a microscopic dynamics in comparison to the evolution of hydrodynamic moments of the distribution function. In a different setting, this equation itself is a result of reducing the description from the microscopic Liouville equation.

The problem of reducing the description includes a definition of the microscopic dynamics, and of the macroscopic variables of interest, for which equations of the reduced description must be found. The next step is the construction of the initial approximation. This is the well known quasiequilibrium approximation, which is the solution to the variational problem, $S \to \max$, where S in the entropy, under given constraints. This solution assumes that the microscopic distribution functions depend on time only through their dependence on the macroscopic variables. Direct substitution of the quasiequilibrium distribution function into the microscopic equation of motion gives the initial approximation to the macroscopic dynamics. All further corrections can be obtained from a more precise approximation of the microscopic as well as of the macroscopic trajectories within a given time interval τ which is the parameter of the method of natural projector.

The method described here has several clear advantages:

(i) It allows to derive complicated macroscopic equations, instead of writing them *ad hoc*. This fact is especially significant for the description of complex fluids. The method gives explicit expressions for relevant variables with one unknown parameter (τ). This parameter can be obtained from the experimental data.

(ii) Another advantage of the method is its simplicity. For example, in the case where the microscopic dynamics is given by the Boltzmann equation, the approach avoids evaluation of the Boltzmann collision integral.

(iii) The most significant advantage of this formalizm is that it is applicable to nonlinear systems. Usually, in the classical approaches to reduced description, the microscopic equation of motion is linear. In that case, one can formally write the evolution operator in the exponential form. Obviously, this does not work for nonlinear systems, such as, for example, systems with mean field interactions. The method which we are presenting here is based on mapping the expanded microscopic trajectory into the consistently expanded macroscopic trajectory. This does not require linearity. Moreover, the order-by-order recurrent construction can be, in principle, enhanced by restoring to other types of approximations, like Padé approximation, for example, but we do not consider these options here.

In the present section we discuss in detail applications of the method of natural projector [29, 30, 34] to derivations of macroscopic equations, and demonstrate how computations are performed in the higher orders of the expansion. The structure of the Example is as follows: In the next subsection, we describe the formalization of Ehrenfests approach [29, 30]. We stress the role of the quasiequilibrium approximation as the starting point for the constructions to follow. We derive explicit expressions for the correction to the quasiequilibrium dynamics, and conclude this section with the entropy production formula and its discussion. After that, we use the present formalism in order to derive hydrodynamic equations. Zeroth approximation of the scheme is the Euler equations of the compressible nonviscous fluid. The first approximation leads to the Navier–Stokes equations. Moreover, the approach allows to obtain the next correction, so-called post-Navier–Stokes equations. The latter example is of particular interest. Indeed, it is well known that the post-Navier–Stokes equations as derived from the Boltzmann kinetic equation by the Chapman–Enskog method (the Burnett and the super-Burnett hydro-dynamics) suffer from unphysical instability already in the linear approxima-tion [72]. We demonstrate it by the explicit computation that the linearized higher-order hydrodynamic equations derived within the method of natural projector are free from this drawback.

11.2.1 General Construction

Let us consider a microscopic dynamics given by an equation,

$$\dot{f} = J(f) , \tag{11.5}$$

where $f(x, t)$ is a distribution function over the phase space x at time t, and where operator $J(f)$ may be linear or nonlinear. We consider linear macro-scopic variables $M_k = \mu_k(f)$, where operator μ_k maps f into M_k. The prob-lem is to obtain closed macroscopic equations of motion, $\dot{M}_k = \phi_k(M)$. This is achieved in two steps: First, we construct an initial approximation to the macroscopic dynamics and, second, this approximation is further corrected on the basis of the coarse-gaining.

The initial approximation is the quasiequilibrium approximation, and it is based on the entropy maximum principle under fixed constraints (Chap. 5:

$$S(f) \rightarrow \max, \quad \mu(f) = M , \tag{11.6}$$

where S is the entropy functional, which is assumed to be strictly concave, and M is the set of the macroscopic variables $\{M_k\}$, and μ is the set of the corresponding operators. If the solution to the problem (11.6) exists, it is unique thanks to the concavity of the entropy functional. The solution to equation (11.6) is called the quasiequilibrium state, and it will be denoted as $f^*(M)$. The classical example is the local equilibrium of the ideal gas: f is the one-body distribution function, S is the Boltzmann entropy, μ are five linear operators, $\mu(f) = \int\{1, \boldsymbol{v}, v^2\}f \, d\boldsymbol{v}$, with $\boldsymbol{v}$ the particle's velocity; the corresponding $f^*(M)$ is called the local Maxwell distribution function.

If the microscopic dynamics is given by equation (11.5), then the quasi-equilibrium dynamics of the variables M reads:

$$\dot{M}_k = \mu_k(J(f^*(M)) = \phi_k^* . \tag{11.7}$$

The quasiequilibrium approximation has important property, it conserves the type of the dynamics: If the entropy monotonically increases (or not de-creases) due to equation (11.5), then the same is true for the quasiequilibrium entropy, $S^*(M) = S(f^*(M))$, due to the quasiequilibrium dynamics (11.7). That is, if

$$\dot{S} = \frac{\partial S(f)}{\partial f}\dot{f} = \frac{\partial S(f)}{\partial f}J(f) \geq 0 ,$$

then

$$\dot{S}^* = \sum_k \frac{\partial S^*}{\partial M_k} \dot{M}_k = \sum_k \frac{\partial S^*}{\partial M_k} \mu_k(J(f^*(M))) \geq 0 . \qquad (11.8)$$

Summation in k always implies summation or integration over the set of labels of the macroscopic variables.

Conservation of the type of dynamics by the quasiequilibrium approximation is a simple yet a general and useful fact. If the entropy S is an integral of motion of equation (11.5) then $S^*(M)$ is the integral of motion for the quasiequilibrium equation (11.7). Consequently, if we start with a system which conserves the entropy (for example, with the Liouville equation) then we end up with the quasiequilibrium system which conserves the quasiequilibrium entropy. For instance, if M is the one-body distribution function, and (11.5) is the (reversible) Liouville equation, then (11.7) is the Vlasov equation which is reversible, too. On the other hand, if the entropy was monotonically increasing on the solutions of equation (11.5), then the quasiequilibrium entropy also increases monotonically on the solutions of the quasiequilibrium dynamic equations (11.7). For instance, if equation (11.5) is the Boltzmann equation for the one-body distribution function, and M is a finite set of moments (chosen in such a way that the solution to the problem (11.6) exists), then (11.7) are closed moment equations for M which increase the quasiequilibrium entropy (this is the essence of a well known generalization of Grad's moment method, Chap. 5).

11.2.2 Enhancement of Quasiequilibrium Approximations for Entropy-Conserving Dynamics

The goal of the present subsection is to describe the simplest analytic implementation, the microscopic motion with periodic coarse-graining. The notion of coarse-graining was introduced by P. and T. Ehrenfest in their seminal work [15]: The phase space is partitioned into cells, the coarse-grained variables are the amounts of the phase density inside the cells. Dynamics is described by the two processes, by the Liouville equation for f, and by periodic coarse-graining, replacement of $f(x)$ in each cell by its average value in this cell. The coarse-graining operation means forgetting the microscopic details, or of the history.

From the perspective of the general quasiequilibrium approximations, periodic coarse-graining amounts to the return of the true microscopic trajectory on the quasiequilibrium manifold with the preservation of the macroscopic variables. The motion starts at the quasiequilibrium state f_i^*. Then the true solution $f_i(t)$ of the microscopic equation (11.5) with the initial condition $f_i(0) = f_i^*$ is coarse-grained at a fixed time $t = \tau$, solution $f_i(\tau)$ is replaced by the quasiequilibrium function $f_{i+1}^* = f^*(\mu(f_i(\tau)))$. This process is sketched in Fig. 11.1.

From the features of the quasiequilibrium approximation it follows that for the motion with the periodic coarse-graining, the inequality is valid,

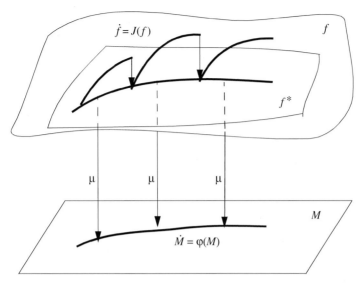

Fig. 11.1. Coarse-graining scheme. f is the space of microscopic variables, M is the space of the macroscopic variables, f^* is the quasiequilibrium manifold, μ is the mapping from the microscopic to the macroscopic space

$$S(f_i^*) \leq S(f_{i+1}^*) \, , \tag{11.9}$$

the equality occurs if and only if the quasiequilibrium is the invariant manifold of the dynamic system (11.5). Whenever the quasiequilibrium is *not* the solution to equation (11.5), the strict inequality in (11.9) demonstrates the entropy increase. Following Ehrenfests, the sequence of the quasiequilibrium states is called the *H-curve*.

In other words, let us assume that the trajectory begins at the quasi-equilibrium manifold, then it takes off from this manifold according to the microscopic evolution equations. Then, after some time τ, the trajectory is coarse-grained, that is the, state is brought back on the quasiequilibrium manifold while keeping the current values of the macroscopic variables. The irreversibility is born in the latter process, and this construction clearly rules out quasiequilibrium manifolds which are invariant with respect to the microscopic dynamics, as candidates for a coarse-graining.

The coarse-graining indicates the way to derive equations for the macroscopic variables from the condition that the macroscopic trajectory, $M(t)$, which governs the motion of the quasiequilibrium states, $f^*(M(t))$, should match precisely the same points on the quasiequilibrium manifold, $f^*(M(t + \tau))$, and this matching should be independent of both the initial time, t, and the initial condition, $M(t)$. The problem is then how to derive the continuous time macroscopic dynamics which would be consistent with this picture. The simplest realization suggested in [29, 30] is based on matching an expansion of both the microscopic and the macroscopic trajectories. Here we present this construction to the third order accuracy [29, 30].

Let us write down the solution to the microscopic equation (11.5), and approximate this solution by the polynomial of the third order in τ. Introducing notation, $J^* = J(f^*(M(t)))$, we write,

$$f(t+\tau) = f^* + \tau J^* + \frac{\tau^2}{2}\frac{\partial J^*}{\partial f}J^* + \frac{\tau^3}{3!}\left(\frac{\partial J^*}{\partial f}\frac{\partial J^*}{\partial f}J^* + \frac{\partial^2 J^*}{\partial f^2}J^*J^*\right) + o(\tau^3).$$

$$(11.10)$$

Evaluation of the macroscopic variables on the function (11.10) gives

$$M_k(t+\tau) = M_k + \tau\phi_k^* + \frac{\tau^2}{2}\mu_k\left(\frac{\partial J^*}{\partial f}J^*\right) \qquad (11.11)$$

$$+ \frac{\tau^3}{3!}\left\{\mu_k\left(\frac{\partial J^*}{\partial f}\frac{\partial J^*}{\partial f}J^*\right) + \mu_k\left(\frac{\partial^2 J^*}{\partial f^2}J^*J^*\right)\right\} + o(\tau^3),$$

where $\phi_k^* = \mu_k(J^*)$ is the quasiequilibrium macroscopic vector field (the right hand side of equation (11.7)), and all the functions and derivatives are taken in the quasiequilibrium state at time t.

We shall now establish the macroscopic dynamic by matching the macroscopic and the microscopic dynamics. Specifically, the macroscopic dynamic equations (11.7) with the right-hand side not yet defined, give the following third-order result:

$$M_k(t+\tau) = M_k + \tau\phi_k + \frac{\tau^2}{2}\sum_j\frac{\partial\phi_k}{\partial M_j}\phi_j \qquad (11.12)$$

$$+ \frac{\tau^3}{3!}\sum_{ij}\left(\frac{\partial^2\phi_k}{\partial M_i M_j}\phi_i\phi_j + \frac{\partial\phi_k}{\partial M_i}\frac{\partial\phi_i}{\partial M_j}\phi_j\right) + o(\tau^3).$$

Expanding functions ϕ_k into a series

$$\phi_k = R_k^{(0)} + \tau R_k^{(1)} + \tau^2 R_k^{(2)} + \dots \quad (R_k^{(0)} = \phi^*),$$

and requiring that the microscopic and the macroscopic dynamics coincide to the order of τ^3, we obtain the sequence of approximations to the right-hand side of the equation for the macroscopic variables. Zeroth order is the quasiequilibrium approximation to the macroscopic dynamics. The first-order correction gives:

$$R_k^{(1)} = \frac{1}{2}\left\{\mu_k\left(\frac{\partial J^*}{\partial f}J^*\right) - \sum_j\frac{\partial\phi_k^*}{\partial M_j}\phi_j^*\right\}. \qquad (11.13)$$

The next, second-order correction has the following explicit form:

$$R_k^{(2)} = \frac{1}{3!} \left\{ \mu_k \left(\frac{\partial J^*}{\partial f} \frac{\partial J^*}{\partial f} J^* \right) + \mu_k \left(\frac{\partial^2 J^*}{\partial f^2} J^* J^* \right) \right\} - \frac{1}{3!} \sum_{ij} \left(\frac{\partial \phi_k^*}{\partial M_i} \frac{\partial \phi_i^*}{\partial M_j} \phi_j^* \right)$$

$$- \frac{1}{3!} \sum_{ij} \left(\frac{\partial^2 \phi_k^*}{\partial M_i \partial M_j} \phi_i^* \phi_j^* \right) - \frac{1}{2} \sum_{j} \left(\frac{\partial \phi_k^*}{\partial M_j} R_j^{(1)} + \frac{\partial R_j^{(1)}}{\partial M_j} \phi_j^* \right) . \qquad (11.14)$$

Further corrections are found by the same token. Equations (11.13)–(11.14) give explicit closed expressions for corrections to the quasiequilibrium dynamics to the order of accuracy specified above.

11.2.3 Entropy Production

The most important consequence of the above construction is that the resulting continuous time macroscopic equations retain the dissipation property of the discrete time coarse-graining (11.9) on each order of approximation $n \geq 1$. Let us first consider the entropy production formula for the first-order approximation. In order to shorten notations, it is convenient to introduce the quasiequilibrium projection operator,

$$P^* g = \sum_{k} \frac{\partial f^*}{\partial M_k} \mu_k(g) . \qquad (11.15)$$

It has been demonstrated in [30] that the entropy production,

$$\dot{S}_{(1)}^* = \sum_{k} \frac{\partial S^*}{\partial M_k} (R_k^{(0)} + \tau R_k^{(1)}) ,$$

equals

$$\dot{S}_{(1)}^* = -\frac{\tau}{2} (1 - P^*) J^* \left. \frac{\partial^2 S^*}{\partial f \partial f} \right|_{f^*} (1 - P^*) J^* . \qquad (11.16)$$

Expression (11.16) is nonnegative definite due to concavity of the entropy. The entropy production (11.16) is equal to zero only if the quasiequilibrium approximation is the true solution to the microscopic dynamics, that is, if $(1 - P^*) J^* \equiv 0$. While quasiequilibrium approximations which solve the Liouville equation are uninteresting objects (except, of course, for the equilibrium itself), vanishing of the entropy production in this case is a simple test of consistency of the theory. Note that the entropy production (11.16) is proportional to τ. Note also that the projection operator does not appear in our consideration a priory, rather, it is the result of exploring the coarse-graining condition in the previous subsection.

Though equation (11.16) looks very natural, its existence is rather subtle. Indeed, equation (11.16) is a difference of the two terms, $\sum_k \mu_k (J^* \partial J^* / \partial f)$

(contribution of the second-order approximation to the microscopic trajectory), and $\sum_{ik} R_i^{(0)} \partial R_k^{(0)} / \partial M_i$ (contribution of the derivative of the quasi-equilibrium vector field). Each of these expressions separately gives a positive contribution to the entropy production, and equation (11.16) is the difference of the two positive definite expressions. In the higher order approximations, these subtractions are more involved, and explicit demonstration of the entropy production formulae becomes a formidable task. Yet, it is possible to demonstrate the increase-in-entropy without explicit computation, though at a price of smallness of τ. Indeed, let us denote $\dot{S}^*_{(n)}$ the time derivative of the entropy on the nth order approximation. Then

$$\int_t^{t+\tau} \dot{S}^*_{(n)}(s)\, ds = S^*(t+\tau) - S^*(t) + O(\tau^{n+1}) \,,$$

where $S^*(t+\tau)$ and $S^*(t)$ are true values of the entropy at the adjacent states of the H-curve. The difference $\delta S = S^*(t+\tau) - S^*(t)$ is strictly positive for any fixed τ, and, by equation (11.16), $\delta S \sim \tau^2$ for small τ. Therefore, if τ is small enough, the right hand side in the above expression is positive, and

$$\tau \dot{S}^*_{(n)}(\theta_{(n)}) > 0 \,,$$

where $t \leq \theta_{(n)} \leq t + \tau$. Finally, since $\dot{S}^*_{(n)}(t) = \dot{S}^*_{(n)}(s) + O(\tau^n)$ for any s on the segment $[t, t + \tau]$, we can replace $\dot{S}^*_{(n)}(\theta_{(n)})$ in the latter inequality by $\dot{S}^*_{(n)}(t)$. The sense of this consideration is as follows: Since the entropy production formula (11.16) is valid in the leading order of the construction, the entropy production will not collapse in the higher orders at least if the coarse-graining time is small enough. More refined estimations can be obtained only from the explicit analysis of the higher-order corrections.

11.2.4 Relation to the Work of Lewis

Among various realizations of the coarse-graining procedures, the work of Lewis [281] appears to be most close to our approach. It is therefore pertinent to discuss the differences. Both methods are based on the coarse-graining condition,

$$M_k(t + \tau) = \mu_k \left(T_\tau f^*(M(t)) \right) \,, \tag{11.17}$$

where T_τ is the formal solution operator of the microscopic dynamics. Above, we applied a consistent expansion of both, the left hand side and the right hand side of the coarse-graining condition (11.17), in terms of the coarse-graining time τ. In the work of Lewis [281], it was suggested, as a general way to exploring the condition (11.17), to write the first-order equation for M in the form of the differential pursuit,

$$M_k(t) + \tau \frac{dM_k(t)}{dt} \approx \mu_k \left(T_\tau f^*(M(t)) \right) \,. \tag{11.18}$$

In other words, in the work of Lewis [281], the expansion to the first order was considered on the left (macroscopic) side of equation (11.17), whereas the right hand side containing the microscopic trajectory $T_\tau f^*(M(t))$ was not treated on the same footing. Clearly, expansion of the right hand side to first order in τ is the only equation which is common in both approaches, and this is the quasiequilibrium dynamics. However, the difference occurs already in the next, second-order term (see [29,30] for details). Namely, the expansion to the second order of the right hand side of Lewis' equation (11.18) results in a dissipative equation (in the case of the Liouville equation, for example) which remains dissipative even if the quasiequilibrium approximation is the exact solution to the microscopic dynamics, that is, when microscopic trajectories once started on the quasiequilibrium manifold belong to it in all the later times, and thus no dissipation can be born by any coarse-graining.

On the other hand, our approach assumes a certain smoothness of trajectories so that the application of the low-order expansion bears physical significance. For example, while using lower-order truncations it is not possible to derive the Boltzmann equation because in that case the relevant quasiequilibrium manifold (N-body distribution function is proportional to the product of one-body distributions, or uncorrelated states) is almost invariant during the long time (of the order of the mean free flight of particles), while the trajectory steeply leaves this manifold during the short-time pair collision. It is clear that in such a case lower-order expansions of the microscopic trajectory do not lead to useful results. It has been clearly stated by Lewis [281], that the exploration of the condition (11.17) depends on the physical situation, and how one makes approximations. In fact, derivation of the Boltzmann equation given by Lewis on the basis of the condition (11.17) does not follow the differential pursuit approximation: As is well known, the expansion in terms of particle's density of the solution to the BBGKY hierarchy is singular, and begins with the *linear* in time term. Assuming the quasiequilibrium approximation for the N-body distribution function under fixed one-body distribution function, and that collisions are well localized in space and time, one gets on the right hand side of equation (11.17),

$$f(t + \tau) = f(t) + n\tau J_B(f(t)) + o(n) ,$$

where n is particle's density, f is the one-particle distribution function, and J_B is the Boltzmanns collision integral. Next, using the mean-value theorem on the left hand side of the equation (11.17), the Boltzmann equation is derived (see also a recent elegant renormalization-group argument for this derivation [55]).

Our approach of matched expansions for exploring the coarse-graining condition (11.17) is, in fact, the exact (formal) statement that the unknown macroscopic dynamics which causes the shift of M_k on the left hand side of equation (11.17) can be reconstructed order-by-order to any degree of accuracy, whereas the low-order truncations may be useful for certain physical situations. A thorough study of the cases beyond the lower-order truncations is of great importance which is left for Chap. 12.

11.2.5 Equations of Hydrodynamics

The method discussed above enables one to establish in a simple way the form of equations of the macroscopic dynamics to various degrees of approximation.

In this subsection, the microscopic dynamics is given by the simplest one-particle Liouville equation (the equation of free flight). For the macroscopic variables we take the density, average velocity, and temperature (average kinetic energy) of the fluid. Under this condition the solution to the quasi-equilibrium problem (11.6) is the local Maxwell distribution. For the hydrodynamic equations, the zeroth (quasiequilibrium) approximation is given by Euler's equations of compressible nonviscous fluid. The next order approximation are the Navier–Stokes equations which have dissipative terms.

Higher-order approximations to the hydrodynamic equations, when they are derived from the Boltzmann kinetic equation (the so-called Burnett approximation), are subject to various difficulties, in particular, they exhibit an instability of acoustic waves at sufficiently short wave length (see, e.g. [42] for a recent review). Here we demonstrate how model hydrodynamic equations, including the post-Navier–Stokes approximations, can be derived on the basis of the coarse-graining idea, and study the linear stability of the obtained equations. We found that the resulting equations are stable.

Two points need a clarification before we proceed further [30]. First, below we consider the simplest Liouville equation for the one-particle distribution, describing freely moving particles without interactions. The procedure of coarse-graining we use is an implementation of collisions leading to dissipation. If we had used the full interacting N-particle Liouville equation, the result would be different, in the first place, in the expression for the local equilibrium pressure. Whereas in the present case we have the ideal gas pressure, in the N-particle case the non-ideal gas pressure would arise.

Second, and more essential is that, to the order of the Navier–Stokes equations, the *result* of our method is identical to the lowest-order Chapman–Enskog method as applied to the Boltzmann equation with a single relaxation time model collision integral (the Bhatnagar–Gross–Krook model [116]). However, this happens only at this particular order of approximation, because already the next, post-Navier–Stokes approximation, is different from the Burnett hydrodynamics as derived from the BGK model (the latter is unstable).

11.2.6 Derivation of the Navier–Stokes Equations

Let us assume that reversible microscopic dynamics is given by the one-particle Liouville equation,

$$\frac{\partial f}{\partial t} = -v_i \frac{\partial f}{\partial r_i} \,, \tag{11.19}$$

where $f = f(\boldsymbol{r}, \boldsymbol{v}, t)$ is the one-particle distribution function, and index i runs over spatial components $\{x,\ y,\ z\}$. Subject to appropriate boundary conditions which we assume, this equation conserves the Boltzmann entropy $S = -k_\mathrm{B} \int f \ln f \, \mathrm{d}\boldsymbol{v} \, \mathrm{d}\boldsymbol{r}$.

We introduce the following hydrodynamic moments as the macroscopic variables: $M_0 = \int f \, \mathrm{d}\boldsymbol{v}$, $M_i = \int v_i f \, \mathrm{d}\boldsymbol{v}$, $M_4 = \int v^2 f \, \mathrm{d}\boldsymbol{v}$. These variables are related to the more conventional density, average velocity and temperature, n, $\boldsymbol{u}$, T as follows:

$$M_0 = n \,, \quad M_i = n u_i \,, \quad M_4 = \frac{3 n k_\mathrm{B} T}{m} + n u^2 \,,$$

$$n = M_0 \,, \quad u_i = M_0^{-1} M_i \,, \quad T = \frac{m}{3 k_\mathrm{B} M_0} (M_4 - M_0^{-1} M_i M_i) \,. \tag{11.20}$$

The quasiequilibrium distribution function (local Maxwellian) reads:

$$f_0 = n \left(\frac{m}{2 \pi k_\mathrm{B} T} \right)^{3/2} \exp \left(\frac{-m(v - u)^2}{2 k_\mathrm{B} T} \right) \,. \tag{11.21}$$

Here and below, n, $\boldsymbol{u}$, and T depend on $\boldsymbol{r}$ and t.

Based on the microscopic dynamics (11.19), the set of macroscopic variables (11.20), and the quasiequilibrium (11.21), we can derive the equations of the macroscopic motion.

A specific feature of the present example is that the quasiequilibrium equation for the density (the continuity equation),

$$\frac{\partial n}{\partial t} = -\frac{\partial n u_i}{\partial r_i} \,, \tag{11.22}$$

should be excluded out of the further corrections. This rule should be applied generally: If a part of the chosen macroscopic variables (momentum flux $n\boldsymbol{u}$ here) correspond to fluxes of other macroscopic variables, then the quasiequilibrium equation for the latter is already exact, and has to be exempted of corrections.

The quasiequilibrium approximation for the rest of the macroscopic variables is derived in the usual way. In order to derive the equation for the velocity, we substitute the local Maxwellian into the one-particle Liouville equation, and act with the operator $\mu_k = \int v_k \cdot \mathrm{d}\boldsymbol{v}$ on both the sides of the equation (11.19). We have:

$$\frac{\partial n u_k}{\partial t} = -\frac{\partial}{\partial r_k} \frac{n k_\mathrm{B} T}{m} - \frac{\partial n u_k u_j}{\partial r_j} \,.$$

Similarly, we derive the equation for the energy density, and the complete system of equations of the quasiequilibrium approximation reads (compressible Euler equations):

$$\frac{\partial n}{\partial t} = -\frac{\partial n u_i}{\partial r_i} , \tag{11.23}$$

$$\frac{\partial n u_k}{\partial t} = -\frac{\partial}{\partial r_k} \frac{n k_{\mathrm B} T}{m} - \frac{\partial n u_k u_j}{\partial r_j} ,$$

$$\frac{\partial \varepsilon}{\partial t} = -\frac{\partial}{\partial r_i} \left(\frac{5 k_{\mathrm B} T}{m} n u_i + u^2 n u_i \right) .$$

where $varepsilon = \frac{3}{2} n k_{\mathrm B} T$ is the energy density.

Now we are going to derive the next order approximation to the macroscopic dynamics (first order in the coarse-graining time τ). For the velocity equation we have:

$$R_{n u_k} = \frac{1}{2} \left(\int v_k v_i v_j \frac{\partial^2 f_0}{\partial r_i \partial r_j} \, \mathrm{d}\boldsymbol{v} - \sum_j \frac{\partial \phi_{n u_k}}{\partial M_j} \phi_j \right) ,$$

where ϕ_j are the corresponding right hand sides of the Euler equations (11.23). In order to take derivatives with respect to macroscopic moments $\{M_0, M_i, M_4\}$, we need to rewrite equations (11.23) in terms of these variables instead of $\{n, u_i, T\}$. After some computation, we obtain:

$$R_{n u_k} = \frac{1}{2} \frac{\partial}{\partial r_j} \left(\frac{n k_{\mathrm B} T}{m} \left[\frac{\partial u_k}{\partial r_j} + \frac{\partial u_j}{\partial r_k} - \frac{2}{3} \frac{\partial u_n}{\partial r_n} \delta_{kj} \right] \right) . \tag{11.24}$$

For the energy we obtain:

$$R_\varepsilon = \frac{1}{2} \left(\int v^2 v_i v_j \frac{\partial^2 f_0}{\partial r_i \partial r_j} \, \mathrm{d}\boldsymbol{v} - \sum_j \frac{\partial \phi_\varepsilon}{\partial M_j} \phi_j \right)$$

$$= \frac{5}{2} \frac{\partial}{\partial r_i} \left(\frac{n k_{\mathrm B}^2 T}{m^2} \frac{\partial T}{\partial r_i} \right) . \tag{11.25}$$

Thus, we get the system of the Navier–Stokes equations in the following form:

$$\frac{\partial n}{\partial t} = -\frac{\partial n u_i}{\partial r_i} ,$$

$$\frac{\partial n u_k}{\partial t} = -\frac{\partial}{\partial r_k} \frac{n k_{\mathrm B} T}{m} - \frac{\partial n u_k u_j}{\partial r_j}$$

$$+ \frac{\tau}{2} \frac{\partial}{\partial r_j} \frac{n k_{\mathrm B} T}{m} \left(\frac{\partial u_k}{\partial r_j} + \frac{\partial u_j}{\partial r_k} - \frac{2}{3} \frac{\partial u_n}{\partial r_n} \delta_{kj} \right) , \tag{11.26}$$

$$\frac{\partial \varepsilon}{\partial t} = -\frac{\partial}{\partial r_i} \left(\frac{5 k_{\mathrm B} T}{m} n u_i + u^2 n u_i \right) + \tau \frac{5}{2} \frac{\partial}{\partial r_i} \left(\frac{n k_{\mathrm B}^2 T}{m^2} \frac{\partial T}{\partial r_i} \right) .$$

We see that the kinetic coefficients (viscosity and heat conductivity) are proportional to the coarse-graining time τ. Note that they are identical with kinetic coefficients as derived from the Bhatnagar–Gross–Krook model [116] in the first approximation of the Chapman–Enskog method [70].

11.2.7 Post-Navier–Stokes Equations

Now we are going to obtain the second-order approximation to the hydrodynamic equations in the framework of the present approach. We shall compare qualitatively the result with the Burnett approximation. The comparison concerns stability of the hydrodynamic modes near the global equilibrium. Stability of the global equilibrium is violated in the Burnett approximation. Though the derivation is straightforward also in the general, nonlinear case, we shall consider only the linearized equations which is appropriate to our purpose here.

Linearizing the local Maxwell distribution function, we obtain:

$$
\begin{aligned}
f &= n_0 \left(\frac{m}{2\pi k_B T_0} \right)^{3/2} \left(\frac{n}{n_0} + \frac{m v_n}{k_B T_0} u_n + \left(\frac{m v^2}{2 k_B T_0} - \frac{3}{2} \right) \frac{T}{T_0} \right) e^{-\frac{m v^2}{2 k_B T_0}} \\
&= \left\{ (M_0 + 2 M_i c_i + \left(\frac{2}{3} M_4 - M_0 \right) \left(c^2 - \frac{3}{2} \right) \right\} e^{-c^2},
\end{aligned}
\tag{11.27}
$$

where we have introduced dimensionless variables:

$$
c_i = \frac{v_i}{v_T}, \; M_0 = \frac{\delta n}{n_0}, \; M_i = \frac{\delta u_i}{v_T}, \; M_4 = \frac{3}{2} \frac{\delta n}{n_0} + \frac{\delta T}{T_0},
$$

$v_T = \sqrt{2 k_B T_0/m}$ is the thermal velocity, Note that δn, and δT determine deviations of these variables from their equilibrium values, n_0, and T_0.

The linearized Navier–Stokes equations read:

$$
\begin{aligned}
\frac{\partial M_0}{\partial t} &= -\frac{\partial M_i}{\partial r_i}, \\
\frac{\partial M_k}{\partial t} &= -\frac{1}{3} \frac{\partial M_4}{\partial r_k} + \frac{\tau}{4} \frac{\partial}{\partial r_j} \left(\frac{\partial M_k}{\partial r_j} + \frac{\partial M_j}{\partial r_k} - \frac{2}{3} \frac{\partial M_n}{\partial r_n} \delta_{kj} \right), \\
\frac{\partial M_4}{\partial t} &= -\frac{5}{2} \frac{\partial M_i}{\partial r_i} + \tau \frac{5}{2} \frac{\partial^2 M_4}{\partial r_i \partial r_i}.
\end{aligned}
\tag{11.28}
$$

Let us first compute the post-Navier–Stokes correction to the velocity equation. In accordance with the equation (11.14), the first part of this term in the linear approximation is:

$$
\frac{1}{3!} \mu_k \left(\frac{\partial J^*}{\partial f} \frac{\partial J^*}{\partial f} J^* \right) - \frac{1}{3!} \sum_{ij} \left(\frac{\partial \phi_k^*}{\partial M_i} \frac{\partial \phi_i^*}{\partial M_j} \phi_j^* \right) = -\frac{1}{6} \int c_k \frac{\partial^3}{\partial r_i \partial r_j \partial r_n} c_i c_j c_n
$$

$$
\times \left\{ (M_0 + 2 M_i c_i + \left(\frac{2}{3} M_4 - M_0 \right) \left(c^2 - \frac{3}{2} \right) \right\} e^{-c^2} d^3 c
$$

$$
+ \frac{5}{108} \frac{\partial}{\partial r_i} \frac{\partial^2 M_4}{\partial r_s \partial r_s} = \frac{1}{6} \frac{\partial}{\partial r_k} \left(\frac{3}{4} \frac{\partial^2 M_0}{\partial r_s \partial r_s} - \frac{\partial^2 M_4}{\partial r_s \partial r_s} \right) + \frac{5}{108} \frac{\partial}{\partial r_k} \frac{\partial^2 M_4}{\partial r_s \partial r_s}
$$

$$
= \frac{1}{8} \frac{\partial}{\partial r_k} \frac{\partial^2 M_0}{\partial r_s \partial r_s} - \frac{13}{108} \frac{\partial}{\partial r_k} \frac{\partial^2 M_4}{\partial r_s \partial r_s}.
\tag{11.29}
$$

The part of equation (11.14) proportional to the first-order correction is:

$$-\frac{1}{2}\sum_j \left(\frac{\partial \phi_k^*}{\partial M_j} R_j^{(1)} + \frac{\partial R_k^{(1)}}{\partial M_j}\phi_j^* \right) = \frac{5}{6}\frac{\partial}{\partial r_k}\frac{\partial^2 M_4}{\partial r_s \partial r_s} + \frac{1}{9}\frac{\partial}{\partial r_k}\frac{\partial^2 M_4}{\partial r_s \partial r_s}. \quad (11.30)$$

Combining together terms (11.29), and (11.30), we obtain:

$$R_{M_k}^{(2)} = \frac{1}{8}\frac{\partial}{\partial r_k}\frac{\partial^2 M_0}{\partial r_s \partial r_s} + \frac{89}{108}\frac{\partial}{\partial r_k}\frac{\partial^2 M_4}{\partial r_s \partial r_s}.$$

Similar calculation for the energy equation leads to the following result:

$$\int c^2 \frac{\partial^3}{\partial r_i \partial r_j \partial r_k} c_i c_j c_k \left\{ \left(M_0 + 2M_i c_i + \left(\frac{2}{3}M_4 - M_0\right)\left(c^2 - \frac{3}{2}\right) \right) \right\} e^{-c^2}\, d^3 c$$

$$-\frac{25}{72}\frac{\partial}{\partial r_i}\frac{\partial^2 M_i}{\partial r_s \partial r_s} = \frac{1}{6}\left(\frac{21}{4}\frac{\partial}{\partial r_i}\frac{\partial^2 M_i}{\partial r_s \partial r_s} + \frac{25}{12}\frac{\partial}{\partial r_i}\frac{\partial^2 M_i}{\partial r_s \partial r_s} \right) = \frac{19}{36}\frac{\partial}{\partial r_i}\frac{\partial^2 M_i}{\partial r_s \partial r_s}.$$

The term proportional to the first-order corrections gives:

$$\frac{5}{6}\left(\frac{\partial^2}{\partial r_s \partial r_s}\frac{\partial M_i}{\partial r_i} \right) + \frac{25}{4}\left(\frac{\partial^2}{\partial r_s \partial r_s}\frac{\partial M_i}{\partial r_i} \right).$$

Thus, we obtain:

$$R_{M_4}^{(2)} = \frac{59}{9}\left(\frac{\partial^2}{\partial r_s \partial r_s}\frac{\partial M_i}{\partial r_i} \right). \quad (11.31)$$

Finally, combining together all the terms, we obtain the following system of linearized hydrodynamic equations:

$$\frac{\partial M_0}{\partial t} = -\frac{\partial M_i}{\partial r_i},$$

$$\frac{\partial M_k}{\partial t} = -\frac{1}{3}\frac{\partial M_4}{\partial r_k} + \frac{\tau}{4}\frac{\partial}{\partial r_j}\left(\frac{\partial M_k}{\partial r_j} + \frac{\partial M_j}{\partial r_k} - \frac{2}{3}\frac{\partial M_n}{\partial r_n}\delta_{kj} \right) +$$

$$\tau^2 \left\{ \frac{1}{8}\frac{\partial}{\partial r_k}\frac{\partial^2 M_0}{\partial r_s \partial r_s} + \frac{89}{108}\frac{\partial}{\partial r_k}\frac{\partial^2 M_4}{\partial r_s \partial r_s} \right\}, \quad (11.32)$$

$$\frac{\partial M_4}{\partial t} = -\frac{5}{2}\frac{\partial M_i}{\partial r_i} + \tau\frac{5}{2}\frac{\partial^2 M_4}{\partial r_i \partial r_i} + \tau^2\frac{59}{9}\left(\frac{\partial^2}{\partial r_s \partial r_s}\frac{\partial M_i}{\partial r_i} \right).$$

Now we are in a position to investigate the dispersion relation of this system. Substituting $M_i = \tilde{M}_i \exp(\omega t + i(\boldsymbol{k}, \boldsymbol{r}))$ $(i = 0, k, 4)$ into equation (11.32), we reduce the problem to finding the spectrum of the matrix:

$$\begin{pmatrix} 0 & -ik_x & -ik_y \\ -ik_x\frac{k^2}{8} & -\frac{1}{4}k^2 - \frac{1}{12}k_x^2 & -\frac{k_x k_y}{12} \\ -ik_y\frac{k^2}{8} & -\frac{k_x k_y}{12} & -\frac{1}{4}k^2 - \frac{1}{12}k_y^2 \\ -ik_z\frac{k^2}{8} & -\frac{k_x k_z}{12} & -\frac{k_y k_z}{12} \\ 0 & -ik_x\left(\frac{5}{2} + \frac{59k^2}{9}\right) & -ik_y\left(\frac{5}{2} + \frac{59k^2}{9}\right) \end{pmatrix}$$

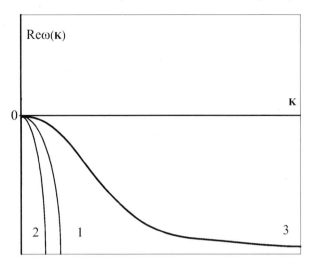

Fig. 11.2. Attenuation rates of various modes of the post-Navier–Stokes equations as functions of the wave vector. Attenuation rate of the twice degenerated shear mode is curve 1. Attenuation rate of the two sound modes is curve 2. Attenuation rate of the diffusion mode is curve 3

$$
\begin{pmatrix}
-ik_z & 0 \\
-\frac{k_x k_z}{12} & -ik_x\left(\frac{1}{3}+\frac{89k^2}{108}\right) \\
-\frac{k_y k_z}{12} & -ik_y\left(\frac{1}{3}+\frac{89k^2}{108}\right) \\
-\frac{1}{4}k^2-\frac{1}{12}k_z^2 & -ik_z\left(\frac{1}{3}+\frac{89k^2}{108}\right) \\
-ik_z\left(\frac{5}{2}+\frac{59k^2}{9}\right) & -\frac{5}{2}k^2
\end{pmatrix}
$$

This matrix has five eigenvalues. The real parts of these eigenvalues responsible for the decay rate of the corresponding modes are shown in Fig.11.2 as functions of the wave vector k. We see that *all real parts of all the eigenvalues are non-positive for any wave vector.* In other words, this means that the present system is linearly stable. For the Burnett hydrodynamics as derived from the Boltzmann or from the single relaxation time Bhatnagar–Gross–Krook model, it is well known that the decay rate of the acoustic branch becomes positive after some value of the wave vector [42, 72], which leads to the instability. While the method suggested here is clearly semi-phenomenological (coarse-graining time τ remains unspecified), the consistency of the expansion with the entropy requirements, and especially the latter result of the linear stable limit of the post-Navier–Stokes correction strongly indicates that it might be more suited to establishing models of highly nonequilibrium hydrodynamics.

11.3 Example: Natural Projector
for the Mc Kean Model

In this section the fluctuation–dissipation formula derived by the method of natural projector [31] is illustrated by the explicit computation for McKean's kinetic model [285]. It is demonstrated that the result is identical, on the one hand, to the sum of the Chapman–Enskog expansion, and, on the other hand, to the solution of the invariance equation. The equality between all the three results holds up to the crossover from the hydrodynamic to the kinetic domain.

11.3.1 General Scheme

Let us consider a microscopic dynamics (3.1) given by an equation for the distribution function $f(x,t)$ over a configuration space x:

$$\partial_t f = J(f) \,, \tag{11.33}$$

where operator $J(f)$ may be linear or nonlinear. Let $\boldsymbol{m}(f)$ be a set of linear functionals whose values, $\boldsymbol{M} = \boldsymbol{m}(f)$, represent the macroscopic variables, and also let $f(\boldsymbol{M}, x)$ be a set of distribution functions satisfying the consistency condition,

$$\boldsymbol{m}(f(\boldsymbol{M})) = \boldsymbol{M} \,. \tag{11.34}$$

The choice of the relevant distribution functions is the point of central importance which we discuss later on but for the time being we need the condition (11.34) only.

Given a finite time interval τ, it is possible to reconstruct uniquely the macroscopic dynamics from a single condition of the coarse-graning. For the sake of completeness, we shall formulate this condition here. Let us denote as $\boldsymbol{M}(t)$ the initial condition at the time t to the *yet unknown* equations of the macroscopic motion, and let us take $f(\boldsymbol{M}(t), x)$ for the initial condition of the microscopic equation (11.33) at the time t. Then the condition for the reconstruction of the macroscopic dynamics reads as follows: For every initial condition $\{\boldsymbol{M}(t), t\}$, solution to the macroscopic dynamic equations at the time $t + \tau$ is equal to the value of the macroscopic variables on the solution to equation (11.33) with the initial condition $\{f(\boldsymbol{M}(t), x), t\}$:

$$\boldsymbol{M}(t + \tau) = \boldsymbol{m}\left(T_\tau f(\boldsymbol{M}(t))\right) \,, \tag{11.35}$$

where T_τ is the formal solution operator of the microscopic equation (11.33). The right hand side of equation (11.35) represents an operation on trajectories of the microscopic equation (11.33), introduced in a particular form by Ehrenfests' [15] (the coarse-graining): The solution at the time $t + \tau$ is replaced by the state on the manifold $f(\boldsymbol{M}, x)$. Notice that the coarse-graining time τ in equation (11.35) is finite, and we stress the importance of the required independence from the initial time t, and from the initial condition at t.

The essence of the reconstruction of the macroscopic equations from the condition just formulated is in the following [29, 30]: Seeking the macroscopic equations in the form,

$$\partial_t \boldsymbol{M} = \boldsymbol{R}(\boldsymbol{M}, \tau) , \qquad (11.36)$$

we proceed with the Taylor expansion of the unknown functions $\boldsymbol{R}$ in terms of powers τ^n, where $n = 0, 1, \ldots$, and require that each approximation, $\boldsymbol{R}^{(n)}$, of the order n, is such that the resulting macroscopic solutions satisfy the condition (11.36) to the order τ^{n+1}. This process of successive approximation is solvable. Thus, the unknown macroscopic equation (11.36) can be reconstructed to any given accuracy.

Coming back to the problem of choosing the distribution function $f(\boldsymbol{M}, x)$, we recall that many physically relevant cases of the microscopic dynamics (11.33) are characterized by existence of a concave functional $S(f)$ (the entropy functional; discussions of S can be found in [115, 191, 192]). Traditionally, two cases are distinguished, the conservative [$\mathrm{d}S/\mathrm{d}t \equiv 0$ due to equation (11.33)], and the dissipative [$\mathrm{d}S/\mathrm{d}t \geq 0$ due to equation (11.33), where equality sign corresponds to the stationary solution]. The approach (11.35) and (11.36) is applicable to both these situations. In both of these cases, among the possible sets of distribution functions $f(\boldsymbol{M}, x)$, the distinguished role is played by the well known quasiequilibrium approximations, $f^*(\boldsymbol{M}, x)$, which are maximizers of the functional $S(f)$ for fixed $\boldsymbol{M}$. We recall that, due to convexity of the functional S, if such maximizer exists then it is unique.

The special role of the quasiequilibrium approximations is due to the fact that they preserve the type of dynamics (Chap. 5): If $\mathrm{d}S/\mathrm{d}t \geq 0$ due to equation (11.33), then $\mathrm{d}S^*/\mathrm{d}t \geq 0$ due to the quasiequilibrium dynamics, where $S^*(\boldsymbol{M}) = S(f^*(\boldsymbol{M}))$ is the quasiequilibrium entropy, and where the quasiequilibrium dynamics coincides with the zeroth order in the above construction, $\boldsymbol{R}^{(0)} = \boldsymbol{m}(J(f^*(\boldsymbol{M})))$.

In particular, the strict increase in the quasiequilibrium entropy has been demonstrated for the first and higher order approximations (see preceding sections of this chapter and [30]). Examples have been provided focusing on the conservative case, and demonstrating that several well known dissipative macroscopic equations, such as the Navier–Stokes equation and the diffusion equation for the one-body distribution function, are derived as the lowest order approximations of this construction.

The advantage of the method of natural projector is the locality of construction, because only Taylor series expansion of the microscopic solution is involved. This is also its natural limitation. From the physical standpoint, finite and fixed coarse-graining time τ remains a phenomenological device which makes it possible to infer the form of the macroscopic equations by a non-complicated computation rather than to derive a full form thereof. For instance, the form of the Navier–Stokes equations can be derived from the simplest model of free motion of particles, in which case the coarse-graining is a substitution for collisions (see previous example.

Going away from the limitations imposed by the finite coarse graining time [29, 30] can be recognized as the major problem of a consistent formulation of the nonequilibrium statistical thermodynamics. Intuitively, this requires taking the limit $\tau \to \infty$, allowing for all the relevant correlations to be developed by the microscopic dynamics, rather than to be cut off at the finite τ (see Chap. 12).

11.3.2 Natural Projector for Linear Systems

However, there is one important exception when the "$\tau \to \infty$ problem" is readily solved [30, 31]. This is the case where equation (11.33) is linear,

$$\partial_t f = L f \ , \tag{11.37}$$

and where the quasiequilibrium is a linear function of $\boldsymbol{M}$. This is, in particular, the classical case of the linear irreversible thermodynamics where one considers the linear macroscopic dynamics near the equilibrium, f^{eq}, $L f^{\mathrm{eq}} = 0$. We assume, for simplicity, that the macroscopic variables $\boldsymbol{M}$ are equal to zero at the equilibrium, and are normalized in such a way that $\boldsymbol{m}(f^{\mathrm{eq}}\boldsymbol{m}^\dagger) = \mathbf{1}$, where $\dagger$ denotes transposition, and $\mathbf{1}$ is an appropriate identity operator. In this case, the linear dynamics of the macroscopic variables $\boldsymbol{M}$ has the form,

$$\partial_t \boldsymbol{M} = \boldsymbol{R}\boldsymbol{M} \ , \tag{11.38}$$

where the linear operator $\boldsymbol{R}$ is determined by the coarse-graining condition (11.35) in the limit $\tau \to \infty$:

$$\boldsymbol{R} = \lim_{\tau \to \infty} \frac{1}{\tau} \ln \left[\boldsymbol{m} \left(e^{\tau L} f^{\mathrm{eq}} \boldsymbol{m}^\dagger \right) \right] \ . \tag{11.39}$$

Formula (11.39) has been already briefly mentioned in [30], and its relation to the Green-Kubo formula has been demonstrated in [31]. The Green-Kubo formula reads:

$$\boldsymbol{R}_{\mathrm{GK}} = \int_0^\infty \langle \dot{\boldsymbol{m}}(0)\dot{\boldsymbol{m}}(t) \rangle \, \mathrm{d}t \ , \tag{11.40}$$

where angular brackets denote equilibrium averaging, and where $\dot{\boldsymbol{m}} = L^\dagger \boldsymbol{m}$. The difference between the formulae (11.39) and (11.40) stems from the fact that condition (11.35) does not use an a priori hypothesis about the separation of the macroscopic and the microscopic time scales. For the classical N-particle dynamics, equation (11.39) is a very complicated expression, involving a logarithm of non-commuting operators. It is therefore desirable to gain its understanding in simple model situations.

11.3.3 Explicit Example of the Fluctuation–Dissipation Formula

In this subsection we want to give an explicit example of the formula (11.39). In order to make our point, we consider here dissipative rather than conservative dynamics in the framework of the well known toy kinetic model introduced by McKean [285] for the purpose of testing various ideas in kinetic theory. In the dissipative case with a clear separation of time scales, existence of the formula (11.39) is underpinned by the entropy growth in both the fast and slow dynamics. This physical idea underlies generically the extraction of the slow (hydrodynamic) component of motion through the concept of normal solutions to kinetic equations, as pioneered by Hilbert [16], and has been discussed by many authors, e.g. . [112, 197, 201]. Case studies for linear kinetic equation help clarifying the concept of this extraction [202, 203, 285].

Therefore, since for the dissipative case there exist well established approaches to the problem of reducing the description, and which are exact in the present setting, it is very instructive to see their relation to the formula (11.39). Specifically, we compare the result with the exact sum of the Chapman–Enskog expansion [70], and with the exact solution in the framework of the method of invariant manifold. We demonstrate that both the three approaches, different in their nature, give the same result as long as the hydrodynamic and the kinetic regimes are separated.

The McKean model is the kinetic equation for the two-component vector function $\boldsymbol{f}(r,t) = (f_+(r,t), f_-(r,t))^\dagger$:

$$\partial_t f_+ = -\partial_r f_+ + \epsilon^{-1}\left(\frac{f_+ + f_-}{2} - f_+\right) , \tag{11.41}$$

$$\partial_t f_- = \partial_r f_- + \epsilon^{-1}\left(\frac{f_+ + f_-}{2} - f_-\right) .$$

Equation (11.41) describes the one-dimensional kinetics of particles with velocities $+1$ and -1 as a combination of the free flight and a relaxation with the rate ϵ^{-1} to the local equilibrium. Using the notation, $(\boldsymbol{x}, \boldsymbol{y})$, for the standard scalar product of the two-dimensional vectors, we introduce the fields, $n(r,t) = (\boldsymbol{n}, \boldsymbol{f})$ [the local particle's density, where $\boldsymbol{n} = (1,1)$], and $j(r,t) = (\boldsymbol{j}, \boldsymbol{f})$ [the local momentum density, where $\boldsymbol{j} = (1,-1)$]. Equation (11.41) can be equivalently written in terms of the moments,

$$\partial_t n = -\partial_r j , \tag{11.42}$$

$$\partial_t j = -\partial_r n - \epsilon^{-1} j .$$

The local equilibrium,

$$\boldsymbol{f}^*(n) = \frac{n}{2}\boldsymbol{n} , \tag{11.43}$$

is the conditional maximum of the entropy,

$$S = -\int (f_+ \ln f_+ + f_- \ln f_-) \, dr ,$$

under the constraint which fixes the density, $(\boldsymbol{n}, \boldsymbol{f}^*) = n$. The quasiequilibrium manifold (11.43) is linear in our example, as well as the kinetic equation.

The problem of reducing the description for the model (11.41) amounts to finding the closed equation for the density field $n(r, t)$. When the relaxation parameter ϵ^{-1} is small enough (the relaxation dominance), then the first Chapman–Enskog approximation to the momentum variable, $j(r, t) \approx -\epsilon \partial_r n(r, t)$, amounts to the standard diffusion approximation. Let us consider now how the formula (11.39), and other methods, extend this result.

Because of the linearity of the equation (11.41) and of the local equilibrium, it is natural to use the Fourier transform, $h_k = \int \exp(ikr) h(r) \, dr$. Equation (11.41) is then written,

$$\partial_t \boldsymbol{f}_k = \boldsymbol{L}_k \boldsymbol{f}_k \ , \tag{11.44}$$

where

$$\boldsymbol{L}_k = \begin{pmatrix} -ik - \frac{1}{2\epsilon} & \frac{1}{2\epsilon} \\ \frac{1}{2\epsilon} & ik - \frac{1}{2\epsilon} \end{pmatrix} \ . \tag{11.45}$$

Derivation of the fluctuation-dissipation formula (11.39) in our example goes as follows: We seek the macroscopic dynamics of the form,

$$\partial_t n_k = R_k n_k \ , \tag{11.46}$$

where the function R_k is yet unknown. In the left-hand side of equation (11.35) we have:

$$n_k(t + \tau) = e^{\tau R_k} n_k(t) \ . \tag{11.47}$$

In the right-hand side of equation (11.35) we have:

$$\left(\boldsymbol{n}, e^{\tau \boldsymbol{L}_k} \boldsymbol{f}^*(n_k(t)) \right) = \frac{1}{2} \left(\boldsymbol{n}, e^{\tau \boldsymbol{L}_k} \boldsymbol{n} \right) n_k(t) \ . \tag{11.48}$$

After equating the expressions (11.47) and (11.48), we require that the resulting equality holds in the limit $\tau \to \infty$ independently of the initial data $n_k(t)$. Thus, we arrive at the formula (11.39):

$$R_k = \lim_{\tau \to \infty} \frac{1}{\tau} \ln \left[\left(\boldsymbol{n}, e^{\tau \boldsymbol{L}_k} \boldsymbol{n} \right) \right] \ . \tag{11.49}$$

Equation (11.49) defines the macroscopic dynamics (11.46) within the present approach. Explicit evaluation of the expression (11.49) is straightforward in the present model. Indeed, operator $\boldsymbol{L}_k$ has two eigenvalues, $\Lambda_k^\pm$, where

$$\Lambda_k^\pm = -\frac{1}{2\epsilon} \pm \sqrt{\frac{1}{4\epsilon^2} - k^2} \tag{11.50}$$

Let us denote as $\boldsymbol{e}_k^\pm$ two (arbitrary) eigenvectors of the matrix $\boldsymbol{L}_k$, corresponding to the eigenvalues $\Lambda_k^\pm$. Vector $\boldsymbol{n}$ has a representation, $\boldsymbol{n} = \alpha_k^+ \boldsymbol{e}_k^+ +$

$\alpha_k^- e_k^-$, where $\alpha_k^\pm$ are complex-valued coefficients. With this, we obtain in equation (11.49),

$$R_k = \lim_{\tau \to \infty} \frac{1}{\tau} \ln \left[\alpha_k^+(\boldsymbol{n}, \boldsymbol{e}_k^+) e^{\tau \Lambda_k^+} + \alpha_k^-(\boldsymbol{n}, \boldsymbol{e}_k^-) e^{\tau \Lambda_k^-} \right] . \tag{11.51}$$

For $k \le k_c$, where $k_c^2 = 4\epsilon$, we have $\Lambda_k^+ > \Lambda_k^-$. Therefore,

$$R_k = \Lambda_k^+, \text{ for } k < k_c . \tag{11.52}$$

As was expected, formula (11.39) in our case results in the exact hydrodynamic branch of the spectrum of the kinetic equation (11.41). The standard diffusion approximation is recovered from equation (11.52) as the first non-vanishing approximation in terms of the $(k/k_c)^2$.

At $k = k_c$, the crossover from the extended hydrodynamic to the kinetic regime takes place, and $\mathrm{Re}\Lambda_k^+ = \mathrm{Re}\Lambda_k^-$. However, we may still extend the function R_k for $k \ge k_c$ on the basis of the formula (11.49):

$$R_k = \mathrm{Re}\,\Lambda_k^+ \text{ for } k \ge k_c \tag{11.53}$$

Notice that the function R_k as given by equations (11.52) and (11.53) is continuous but non-analytic at the crossover.

11.3.4 Comparison with the Chapman–Enskog Method and Solution of the Invariance Equation

Let us now compare this result with the Chapman–Enskog method. Since the exact Chapman–Enskog solution for the systems like equation (11.43) has been recently discussed in detail elsewhere [40, 42, 205, 219–221], we shall be brief here. Following the Chapman–Enskog method, we seek the momentum variable j in terms of an expansion,

$$j^{\mathrm{CE}} = \sum_{n=0}^{\infty} \epsilon^{n+1} j^{(n)} \tag{11.54}$$

The Chapman–Enskog coefficients, $j^{(n)}$, are found from the recurrence equations,

$$j^{(n)} = -\sum_{m=0}^{n-1} \partial_t^{(m)} j^{(n-1-m)} , \tag{11.55}$$

where the Chapman–Enskog operators $\partial_t^{(m)}$ are defined by their action on the density n:

$$\partial_t^{(m)} n = -\partial_r j^{(m)} . \tag{11.56}$$

The recurrence equations (11.54), (11.55), and (11.56), become well defined as soon as the aforementioned zero-order approximation $j^{(0)}$ is specified,

$$j^{(0)} = -\partial_r n \ . \tag{11.57}$$

From equations (11.55), (11.56), and (11.57), it follows that the Chapman–Enskog coefficients $j^{(n)}$ have the following structure:

$$j^{(n)} = b_n \partial_r^{2n+1} n \ , \tag{11.58}$$

where coefficients b_n are found from the recurrence equation,

$$b_n = \sum_{m=0}^{n-1} b_{n-1-m} b_m, \quad b_0 = -1 \ . \tag{11.59}$$

Notice that coefficients (11.59) are real-valued, by the sense of the Chapman–Enskog procedure. The Fourier image of the Chapman–Enskog solution for the momentum variable has the form,

$$j_k^{\mathrm{CE}} = ik B_k^{\mathrm{CE}} n_k \ , \tag{11.60}$$

where

$$B_k^{\mathrm{CE}} = \sum_{n=0}^{\infty} b_n (-\epsilon k^2)^n \ . \tag{11.61}$$

Equation for the function B (11.61) is easily found upon multiplying equation (11.59) by $(-k^2)^n$, and summing in n from zero to infinity:

$$\epsilon k^2 B_k^2 + B_k + 1 = 0 \ . \tag{11.62}$$

Solution to the latter equation which respects condition (11.57), and which constitutes the exact Chapman–Enskog solution (11.61) is:

$$B_k^{\mathrm{CE}} = \begin{cases} k^{-2} \Lambda_k^+, & k < k_c \\ \mathrm{none}, & k \geq k_c \end{cases} \tag{11.63}$$

Thus, the exact Chapman–Enskog solution derives the macroscopic equation for the density as follows:

$$\partial_t n_k = -ik j_k^{\mathrm{CE}} = R_k^{\mathrm{CE}} n_k \ , \tag{11.64}$$

where

$$R_k^{\mathrm{CE}} = \begin{cases} \Lambda_k^+, & k < k_c \\ \mathrm{none}, & k \geq k_c \end{cases} \tag{11.65}$$

The Chapman–Enskog solution does not extend beyond the crossover at k_c. This happens because the full Chapman–Enskog solution appears as a continuation the diffusion approximation, whereas formula (11.49) is not based on such an extension.

Finally, let us discuss briefly the comparison with the solution within the method of invariant manifold [9, 11, 14]. Specifically, the momentum variable

$j_k^{inv} = ikB_k^{inv}n_k$ is required to be invariant of both the microscopic and the macroscopic dynamics, that is, the time derivative of j_k^{inv} due to the macroscopic subsystem,

$$\frac{\partial j_k^{inv}}{\partial n_k}\partial_t n_k = ikB_k^{inv}(-ik)[ikB_k^{inv}] , \qquad (11.66)$$

should be equal to the derivative of j_k^{inv} due to the microscopic subsystem,

$$\partial_t j_k^{inv} = -ikn_k - \epsilon^{-1}ikB_k^{inv}n_k , \qquad (11.67)$$

and that the equality of the derivatives (11.66) and (11.67) should hold independently of the specific value of the macroscopic variable n_k. This amounts to a condition for the unknown function B_k^{inv}, which is essentially the same as equation (11.62), and it is straightforward to show that the same selection procedure of the hydrodynamic root as above in the Chapman–Enskog case results in equation (11.65).

In conclusion, in this Example we have given the explicit illustration for the formula (11.39). The example demonstrates that the fluctuation-dissipation formula (11.39) gives the exact macroscopic evolution equation, which is identical to the sum of the Chapman–Enskog expansion, as well as to the invariance principle. This identity holds up to the point where the hydrodynamics and the kinetics cease to be separated. Whereas the Chapman–Enskog solution does not extend beyond the crossover point, the formula (11.39) demonstrates a non-analytic extension.

12 Geometry of Irreversibility:
The Film of Nonequilibrium States

A geometrical framework of nonequilibrium thermodynamics is developed in this chapter. The notion of *macroscopically definable ensembles* is introduced. A thesis about macroscopically definable ensembles is suggested. This thesis should play the same role in the nonequilibrium thermodynamics, as the well-known Church-Turing thesis in the theory of computability. The *primitive macroscopically definable ensembles* are described. These are ensembles with macroscopically prepared initial states. A method for computing trajectories of primitive macroscopically definable nonequilibrium ensembles is elaborated. These trajectories are represented as sequences of deformed quasiequilibrium ensembles and simple quadratic models between them. The primitive macroscopically definable ensembles form a manifold in the space of ensembles. We call this manifold the *film of nonequilibrium states*. The equation for the film and the equation for the ensemble motion on the film are written down. The notion of the invariant film of non-equilibrium states, and the method of its approximate construction transform the problem of nonequilibrium kinetics into a series of problems of equilibrium statistical physics. The developed methods allow us to solve the problem of macro-kinetics even when there are no autonomous equations of macro-kinetics.

12.1 The Thesis About Macroscopically Definable Ensembles
and the Hypothesis About
Primitive Macroscopically Definable Ensembles

The goal of this chapter is to discuss the nonlinear problem of irreversibility, and to revise previous attempts to solve it. The interest to the problem of irreversibility persists during decades. It has been intensively discussed in the past, and nice accounts of these discussions can be found in the literature (see, for example, [194, 195, 286, 287]). We here intend to develop a more geometrical viewpoint on the subject. First, in Sect. 12.2, we discuss in an informal way the origin of the problem, and demonstrate how the basic constructions arise. Second, in Sect. 12.3, we give a consistent geometric formalization of these constructions. Our presentation is based on the notion

Alexander N. Gorban and Iliya V. Karlin: *Invariant Manifolds for Physical and Chemical Kinetics*, Lect. Notes Phys. **660**, 325–366 (2005)
www.springerlink.com

of the natural projection introduced in section 12.4. We discuss in detail the method of natural projector as the consistent formalization of Ehrenfests' ideas of coarse-graining.

In Sect. 12.4.2 we introduce a one-dimensional model of nonequilibrium states. In the background of many derivations of nonequilibrium kinetic equations one can imagine the following picture: Above each point of the quasiequilibrium manifold there is located a huge subspace of nonequilibrium distributions with the same values of the macroscopic variables, as in the quasiequilibrium state. It seems that the motion of the nonequilibrium ensemble decomposes into two projections, transversal to the quasiequilibrium manifold, and in the projection on this manifold. The motion in each layer above the quasiequilibrium points is highly complicated, but fast, and everything quickly settles in this fast motion.

However, upon a more careful looking into the motions of the ensembles which start from the quasiequilibrium points, we recognize that above each point of the quasiequilibrium manifold it is located *just a single and in some sense monotonic curve,* and all the relevant nonequilibrium (not-quasiequilibrium) states form just a one-dimensional manifold.

The one-dimensional models of nonequilibrium states form a *film of nonequilibrium states.* In Sect. 12.5 we present a collection of methods for the film construction. One of the benefits from this new technique is the possibility to solve the problem of macro-kinetics even when there are no autonomous equations of macro-kinetic for moment variables. The notion of the invariant film of non-equilibrium states, and the method of its approximate construction transform the problem of nonequilibrium kinetics into a series of problems of equilibrium statistical physics.

The most important results of this chapter are:

1. The notion of *macroscopically definable ensembles* is developed.
2. The *primitive macroscopically definable ensembles* are described.
3. The method for computing trajectories of primitive macroscopically definable nonequilibrium ensembles is elaborated. These trajectories are represented a sequence of deformed quasiequilibrium ensembles connected by quadratic models.

Let us give here an introductory description of these results.

The notion of macroscopically definable ensembles consists of three ingredients:

1. The macroscopic variables, the variables which values *can be controlled by us*;
2. The quasiequilibrium state, the conditional equilibrium state for fixed values of the macroscopic variables;
3. The natural dynamics of the system, or the microscopic dynamics.

We use the simplest representation of the control: At certain moments of time we fix some values of the macroscopic variables (one can fix all of all these macroscopic variables, or only a part of them; for the whole system, or for macroscopically defined part of it; the current values, or some arbitrary values of these variables), and the system settles in the corresponding conditional equilibrium state. We can also keep fixed values of some macroscopic variables during a time interval.

These control operations are discrete in time. The continuous control can be obtained by a closure: the limit of a sequence of macroscopically definable ensembles is macroscopically definable too.

The role of the macroscopic variables for the irreversibility problem was clarified by M. Leontovich and J. Lebowitz several decades ago [288–292]. This was the first step. Now we do need the elaborate notion of ensembles which can be obtained by the macroscopic tools. The Maxwell Demon gives the early clear picture of a difference between the macroscopic and microscopic tools for the ensembles control (books are devoted to the studies of this Demon [293, 294]). Nevertheless, a further step towards the notion of the macroscopic definability in the context of constructive transition from the microdynamics to macrokinetics was not done before the paper [33]. Our analysis is an analog of the Church-Turing thesis in the theory of computability [295, 296]. This thesis concerns the notion of an effective (or mechanical) method in mathematics. As a "working hypothesis", Church proposed: A function of positive integers is effectively calculable only if recursive.

We introduce a class of "macroscopically definable ensembles" and formulate the thesis: An ensemble can be macroscopically obtained only if macroscopically definable according to the introduced notion. This is the thesis about the success of the formalization, as the Church-Turing thesis, and nobody can prove or disprove it in a rigorous sense, as well as this famous thesis.

Another important new notion is the *"macroscopically definable transformation"* of the ensemble: If one got an ensemble, how can he transform it? First, it is possible just to let them evolve due to the natural dynamics, second, it can be controlled by the macroscopic tools in the prescribed way (it is necessary just to keep values of some macroscopic variables during some time).

The *primitive macroscopically definable ensembles* are ensembles with quasiequilibrium initial states and without further macroscopic control. These ensembles are prepared macroscopically, and evolve due to the natural dynamics. The significance of this class of ensembles is determined by the *hypothesis about the primitive macroscopically definable ensembles*: Any macroscopically definable ensemble can be approximated by primitive macroscopically definable ensembles with appropriate accuracy. After that there remains no other effective way to decribe the nonequilibrium state.

The primitive macroscopically definable ensembles form the manifold in the space of ensembles. We call this manifold the "film of nonequilibrium states". The equation for the film and the equation for the ensemble motion on the film are written down.

The film of nonequilibrium states is the trajectory of the manifold of initial quasiequilibrium states due to the natural (microscopic) dynamics. For every value of macroscopic variables this film gives us a curve. The curvature of this curve defines kinetic coefficients and the entropy production.

The main technical problem is the computation of this curve for arbitrary values of the macroscopic variables. We represent it as a sequence of distinguished states and second-order polynomial (Kepler) models for the trajectory between these points. This can be viewed as a further development of the method for initial layer problem in the Boltzmann kinetics (see Sect. 9.3 and [26,27]). For the dissipative (Boltzmann) microkinetics it was sufficient to use the first-order models (with or without smoothing). For conservative microkinetics it is necessary to use the higher-order models. Applications of this method to the lattice kinetic equations (Sect. 2.7) allowed

- To create the lattice Boltzmann method with the H-theorem [137];
- To transform the lattice Boltzmann method into the numerically stable computational tool for fluid flows and other dissipative systems out of equilibrium [136];
- To develop the entropic lattice Boltzmann method as a basis for the formulation of a new class of turbulence models based on genuinely kinetic principles [66].

In this chapter we extend the method elaborated for dissipative systems [26,27] to the higher-order models for conservative systems. The constructing of the method of physically consistent computation is the central part of this chapter.

The main results of this chapter were presented in the talk given at the First Mexican Meeting on Mathematical and Experimental Physics, Mexico City, September 10–14, 2001 [33], and in the lectures given on the V Russian National Seminar "Modeling of Nonequilibrium systems", Krasnoyarsk, October 18–20, 2002 [298].

12.2 The Problem of Irreversibility

12.2.1 The Phenomenon of the Macroscopic Irreversibility

The best way to get a feeling about the problem of irreversibility is the following thought experiment (*Gedankenexperiment*): Let us watch a movie: It's raining, people are running, cars rolling. Let us now wind this movie in the opposite direction, and we shall see a strange and funny picture: Drops of the rain are raising up to the clouds, people run with their backs forward, cars also behave quite strange, and so forth. This cannot be true, and we "know" this for sure, we have never seen anything like this in our life. Let us now imagine that we watch the same movie with a magnitude of 10^8–10^9 so that we can resolve individual particles. And all of the sudden we discover that we cannot notice any substantial difference between the direct and the reverse demonstration: Everywhere the particles move, collide, react according to the laws of physics, and nowhere there is a violation of anything. We cannot tell the direct progressing of the time from the reversed. So, we have the irreversibility of the macroscopic picture under the reversibility of the microscopic one.

Rain, people, cars – all this is too complicated. One of the simplest examples of the irreversible macroscopic picture under the apparent reversibility of the microscopic picture is given by R. Feynman in his lectures on the character of physical law [297]. We easily label it as self-evident the fact that particles of different colors mix together, and we would deem it wonderful the reverse picture of a spontaneous decomposition of their mixture. However, by itself, an appreciation of one picture as usual, and of the other as unusual and wonderful – this is not yet physics. It is desirable to measure somehow this transition from order to disorder.

12.2.2 Phase Volume and Dynamics of Ensembles

Let there be n blue and n white particles in a box, and let the box is separated in two halves, the left and the right. Location of all the particles in the box is described by the assembly of $2n$ vectors of locations of the individual particles. The set of all the assemblies is a "box" in the $6n$-dimensional space. A point in this $6n$-dimensional box describes a configuration. The motion of this point is defined by equations of mechanics.

"Order" is the configuration in which the blue particles are all in the right half, and all the white particles are in the left half. The set of all such configurations has a rather small volume. It makes only $(1/2)^{2n}$ of the total volume of the $6n$-dimensional box. If $n = 10$, this is of the order of one per million of the total volume. It is practically unthinkable to land into such a configuration by a chance. It is also highly improbable that, by forming more or less voluntary the initial conditions, we can observe that the system becomes ordered by itself. From this standpoint, the motion goes from the states of "order" to the state of "disorder", just because there are many more states of "disorder".

However, we have defined it in this way. The well known question of what has more order, a fine castle or a pile of stones, has a profound answer: It depends on which pile you mean. If "piles" are thought as all configurations of stones which are not castles, then there are many more such piles, and so there is less order in such a pile. However, if these are specially and uniquely placed stones (for example, a garden of stones), then there is the same amount of order in such a pile as in a fine castle. *Not a specific configuration is important but an assembly of configurations embraced by one notion.*

This transition from single configurations to their assemblies (ensembles) play the pivotal role in the understanding of irreversibility: The irreversible transition from the ordered configuration (blue particles are on the right, white particles are on the left) to the disordered one occurs simply because there are many more of the disordered (in the sense of the volume). Here, strictly speaking, we have to add also a reference to the Liouville theorem: The volume in the phase space which is occupied by the ensemble does not change in time as the mechanical system evolves. Because of this fact, the phase volume V is a good measure to compare the assemblies of configurations. However, more often the quantity $\ln V$ is used, this is called the entropy.

The point representing the configuration, very rapidly leaves a small neighborhood and for a long time (in practice, never) does not return into it. In this, seemingly idyllic picture, there are still two rather dark clouds left. First, the arrow of time has not appeared. If we move from the ordered initial state (separated particles) backwards in time, then everything will stay the same as when we move forward in time, that is, the order will be changing into the disorder. Second, let us wind the film backwards, let us shoot the movie about mixing of colored particles, and then let us watch in the reverse order their demixing. Then the initial configurations for the reverse motion will only seem to be disordered. Their "order" is in the fact that they were obtained from the separated mixture by letting the system to evolve for the time t. There are also very few such configurations, just the same number as of the ordered (separated particles) states. If we start with these configurations, then we obtain the ordered system after the time t. Then why this most obvious consequence of the laws of mechanics looks so improbable on the screen? Perhaps, it should be accepted that states which are obtained from the ordered state by a time shift, and by inversion of particle's velocities (in order to initialize the reverse motion in time), *cannot be prepared using macroscopic means* of preparation. In order to prepare such states, one would have to employ an army of Maxwell's Demons which would invert individual velocities with sufficient accuracy (here, it is much more into the phrase "sufficient accuracy" but this has to be discussed separately and next time).

For that reason, we lump the distinguished initial conditions, for which the mixture decomposes spontaneously ("piles" of special form, or "gardens of stones") together with other configurations into *macroscopically definable ensembles*. And already for these ensembles the spontaneous demixing becomes improbable. This way we come to a new viewpoint: (i). We cannot prepare individual systems but only representatives of ensembles. (ii) We cannot prepare ensembles at our will but only "macroscopically definable ensembles". What are these macroscopically definable ensembles? It seems that one has to give some constructions, the universality of which can only be proven by time and experience.

There is one property that distinguishes an arbitrary ensemble with phase volume V and ensembles (with the same volume) that we usually associate with the order. This property is *observability*. Usually we can fix a configuration within some error only, this means that we cannot distinguish points, if the distance between them is less then some $\varepsilon > 0$. Hence, the *observable ensemble* should not change its volume significantly, if we replace all points by the ε-small balls (i.e. if we just add a small ball to the set of states, or, if the ensemble is presented by the distribution density, just average the density over such balls). This operation, averaging over small balls or cells, is called *coarse graining*. The observable state should not significantly change its volume after the coarse-graining. The ordered state (the blue particles are all in the right half, and all the white particles are in the left half, for example) is observable, but dynamics makes it unobservable after some time. Of course, the notion of macroscopically definable ensembles should meet the expectation concerning observability as well as implementability and controlability of these ensembles.

12.2.3 Macroscopically Definable Ensembles and Quasiequilibria

The main tool in the study of the macroscopically definable ensembles is the notion of the macroscopic variables, and of the quasiequilibria. In the dynamics of the ensembles, the macroscopic variables are defined as linear functionals (moments) of the density distribution of the ensemble. Macroscopic variables M usually include the hydrodynamic fields: density of particles, density of momentum, and density of energy. This list may also include the stress tensor, the reaction rates and other quantities. In the present context, it is solely important that the list the macroscopic variables is identified for the system under consideration.

A single system is characterized by a single point x in the phase space. The ensemble of the systems is defined by the probability density F over the phase space. The density F must satisfy a set of restrictions, the most important of which are: Nonnegativity, $F(x) \geq 0$, normalization,

$$\int_X F(x)\, dV(x) = 1 \, , \tag{12.1}$$

and that the entropy is defined, that is, there exists the integral,

$$S(F) = - \int_X F(x) \ln F(x)\, dV(x) \, . \tag{12.2}$$

The function $F \ln F$ is continuously extended to zero values of F: $0 \ln 0 = 0$). Here, $dV(x)$ is the invariant measure (phase volume.

The quasiequilibrium ensemble describes the "equilibrium under restrictions". It is assumed that some external forcing keeps the given values of the macroscopic variables M, with this, "all the rest" comes to the equilibrium.

The corresponding (generalized) canonical ensemble F which is the solution to the problem:

$$S(F) \to \max, \; M(F) = M \; . \tag{12.3}$$

where $S(F)$ is the entropy, $M(F)$ is the set of macroscopic variables.

The thesis about the macroscopically definable ensembles. Macroscopically definable ensembles are obtained as the result of two operations:

1. Bringing the system into the quasiequilibrium state corresponding to either the whole set of the macroscopic variables M, or to its subset;
2. Evolution of the ensemble according to the microscopic dynamics (due to the Liouville equation) during some time t.

These operations can be applied in the interchanging order any number of times, and for arbitrary time segments t. The limit of macroscopically definable ensembles will also be termed macroscopically definable. One always begins with the first operation.

In order to work out the notion of macroscopic definability, one has to pay more attention to partitioning the system into subsystems. This involves a partition of the phase space X with the measure dV into a direct product of spaces, $X = X_1 \times X_2$ with the measure $dV_1 dV_2$. To each admissible ("macroscopic") partition into sub-systems, it corresponds the operation of taking a "partial quasiequilibrium", applied to some density $F_0(x_1, x_2)$:

$$S(F) \to \max \, , \tag{12.4}$$

$$M(F) = M, \; \int_{X_2} F(x_1, x_2) \, dV_2(x_2) = \int_{X_2} F_0(x_1, x_2) \, dV_2(x_2) \; .$$

where M is some subset of macroscopic variables (not necessarily the whole list of the macroscopic variables). In (12.4), the state of the first subsystem is not changing, whereas the second subsystem is brought into the quasiequilibrium. In fact, the problem (12.4) is a version of the problem (12.3) with additional "macroscopic variables",

$$\int_{X_2} F(x_1, x_2) \, dV_2(x_2) \; . \tag{12.5}$$

The *extended thesis* about the macroscopically definable ensembles allows to use also operations (12.4) with only one restriction: The initial state should be the "true quasiequilibrium", that is, macroscopic variables related to all possible partitions into subsystems should appear only after the sequence of operations has started with the solution to the problem (12.3) for some initial M. This does not exclude a possibility of including operators (12.5) into the list of the basic macroscopic variables M. The standard example of such an inclusion are few-body distribution functions treated as macroscopic variables in derivations of kinetic equations from the Liouville equation.

Irreversibility is related to the choice of the initial conditions. The extended set of macroscopically definable ensembles is thus given by three objects:

1. The set of macroscopic variables M which are linear (and, in an appropriate topology, continuous) mappings of the space of distributions onto the space of values of the macroscopic variables;
2. Macroscopically admissible partitions of the system into sub-systems;
3. Equations of the microscopic dynamics (the Liouville equation, for example).

The choice of the macroscopic variables and of the macroscopically admissible partitions is a distinguished topic. The main question is: which variables are under the macroscopic control? Here the macroscopic variables are represented as formal elements of the construction, and the arbitrariness is removed only at solving specific problems. Usually we can postulate some properties of macroscopic variables, for example, symmetry with respect to any permutation of equivalent particles.

We have discussed the *prepared* ensembles. But there is another statement of the problem: Let an ensemble be just given. The way it emerged it may be irrelevant or unknown, for example, some demon or *oracle*[1] can prepare the ensemble for us. How can we transform this ensemble by the macroscopic tools? First, it is possible just to let it evolve, second, it can be controlled by the macroscopic tools in the prescribed way (it is necessary just to keep values of some macroscopic variables during some time).

The thesis about the macroscopically definable transformation of ensembles. Macroscopically definable transformation of ensembles are obtained as the result of two operations:

1. Bringing the system into the quasiequilibrium state corresponding to either the whole set of the macroscopic variables M, or to its subset.
2. Changing the ensemble according to the microscopic dynamics (due to the Liouville equation, for example) during some time t.

These operations can be applied in the interchanging order any number of times, and for arbitrary time segments t. The limit of macroscopically definable transformations will also be termed macroscopically definable. The main difference of this definition (macroscopically definable transformation) from the definition of the macroscopically definable ensembles is the absence of the restriction on the initial state, one can start from an arbitrary ensemble.

The class of macroscopically definable ensembles includes a simpler, but important subclass. Let us reduce the macroscopic control to preparation of the initial quasiequilibrium ensemble: we just prepare the ensemble by macroscopic tools and then let it evolve due to the natural dynamics (Liouville

---

[1] In the theory of computation, if there is a device which could answer questions beyond those that a Turing machine can answer, then it is called the oracle.

equation, for example). Let us call this class *the primitive macroscopically definable ensembles*. These ensembles appear as the results (for $t > 0$) of motions which start from the quasiequilibrium state (at $t = 0$). The main technical focus of our work concerns the computation of the manifold of primitive macroscopically definable ensembles for a given system.

The importance of this subclass of ensembles is determined by the following hypothesis. **The hypothesis about the primitive macroscopically definable ensembles**. Any macroscopically definable ensemble can be approximated by primitive macroscopically definable ensembles with an appropriate accuracy. In certain limits we can attempt to say: "with any accuracy". Moreover, this hypothesis with "arbitrary accuracy" can be found as the basic but implicit foundation of all nonequilibrium kinetics theories which claim derivation the macrokinetics from microdymamics, for example Zubarev's nonequilibrium statistical operator theory [195]. This hypothesis allows us to describe nonequilibrium state as a result of evolution of quasiequilibrium state in time.

The hypothesis about the primitive macroscopically definable ensembles is a hypothesis indeed, it can hold for different systems with different accuracy, it can be valid or invalid. In some limits the set of primitive macroscopically definable ensembles can be dense in the set of all macroscopically definable ensembles, or, in some cases it can be not dense. There is a significant difference between this hypothesis and the *thesis* about macroscopically definable ensembles. The thesis can be accepted, or not, the reasons for its acceptance can be discussed, but nobody can prove or disprove the definition, even the definition of the macroscopically definable ensembles.

12.2.4 Irreversibility and Initial Conditions

The choice of the initial state of the ensemble plays the crucial role in the thesis about the macroscopically definable ensembles. The initial state is always taken as the quasiequilibrium distribution which realizes the maximum of the entropy for given values of the macroscopic variables. The choice of the initial state splits the time axis into two semi-axes: moving forward in time, and moving backward in time. In both cases the observed disorder increases (the simplest example is the mixing of the particles of different colors).

In some works, in order to achieve the "true nonequilibrium", that is, the irreversible motion along the whole time axis, the quasiequilibrium initial condition is shifted to $-\infty$ in time. This trick, however, casts some doubts, the major being this: Most of the known equations of the macroscopic dynamics describing irreversible processes have solutions which can be extended backwards in time only for finite times (or cannot be extended at all). Such equations as the Boltzmann kinetic equation, diffusion equation, equations of chemical kinetics and like do not allow for almost all their solutions to be extended backward in time for indefinitely long. All motions have a "beginning" beyond which some physical properties of a solution will be lost (often, positivity of distributions), although formally solutions may even exist, as in the case of ordinary differential equations of chemical kinetics.

12.2.5 Weak and Strong Tendency to Equilibrium, Shaking and Short Memory

One aspect of irreversibility is the special choice of the initial conditions. Roughly speaking, the arrow of time is defined by the fact that the quasi-equilibrium initial condition was in the past.

This remarkably simple observation does not, however, exhaust the problem of transition from the reversible equations to the irreversible macroscopic equations. One more aspect deserves a serious consideration. Indeed, distribution functions tend to the equilibrium state according to the macroscopic equations in a strong sense: deviations from the equilibrium tends to zero in the sense of most relevant norms (in the L^1 sense, for example, or even uniformly). On the contrast, for the Liouville equation, the tendency to equilibrium occurs (if at all) only in the weak sense: the average values of sufficiently "regular" functions on the phase space do tend to their equilibrium values but the distribution function itself does not tend to the equilibrium with respect to any norm, not even point-wise. This is especially easy to appreciate if the initial state was the equipartition over some small bounded subset of the phase space (the "phase drop" with small, but non-zero volume). This phase drop can mix over the phase space, but for all the times it will remain "the drop of oil in the water", the density will be always taking only two values, 0 and $p > 0$, and the volume of the set where the density is larger than zero will not be changing in time, of course. So, how to arrive from the weak convergence (in the sense of the convergence of the mean values), to the strong convergence (to the L^1 or to the uniform convergence, for example)? In order to do this, there are two basic constructions: The coarse-graining (shaking) in the sense of Ehrenfests', and the short memory approximation.

The idea of coarse-graining dates back to P. and T. Ehrenfests, and it has been most clearly expressed in their famous paper of 1911 [15]. Ehrenfests considered a partition of the phase space into small cells, and they have suggested to supplement the motions of the phase space ensemble due to the Liouville equation with "shaking" – averaging of the density of the ensemble over the phase cells. In the result of this process, the convergence to the equilibrium becomes strong out of the weak. It is not difficult to recognize that ensembles with constant densities over the phase cells are quasiequilibria; corresponding macroscopic variables are integrals of the density over the phase cells ("occupation numbers" of the cells). This generalizes to the following: alternations of the motion of the phase ensemble due to microscopic equations with returns to the quasiequilibrium manifold, preserving the values of the macroscopic variables. The formalization of this idea was given in the previous chapter.

12.2.6 Subjective Time and Irreversibility

In our discussion, the source of the arrow of time is, after all, the asymmetry of the subjective time of the experimentalist. *We prepare* initial conditions, and *after* that *we watch* what will happen in the future but not what happened in the past. Thus, we obtain kinetic equations for specifically prepared systems. How is this related to the dynamics of the real world? These equations are applicable to real systems to the extent that the reality can be modeled with systems with specifically prepared quasiequilibrium initial conditions. This is anyway less demanding than the condition of quasi-staticity of processes in classical thermodynamics. For this reason, versions of nonequilibrium thermodynamics and kinetics based on this understanding of irreversibility allowed to include such a variety of situations, and moreover, they include all classical equations of nonequilibrium thermodynamics and kinetics.

12.3 Geometrization of Irreversibility

12.3.1 Quasiequilibrium Manifold

We remind here some of the constructions from Chap. 5. Let E be a linear space, and $U \subset E$ be a convex subset, with a nonempty interior int$U$. Let a twice differentiable concave functional S be defined in int$U$, and S be continuous on U. According to the familiar interpretation, S is the entropy, E is an appropriate space of distributions, U is the cone of nonnegative distributions from E. Space E is chosen in such a way that the entropy is well defined on U.

Let K be a closed linear subspace of space E, and $m : E \to E/K$ be the natural projection on the factor-space. The factor-space $L = E/K$ will further play the role of the space of macroscopic variables (in examples, the space of moments of the distribution).

For each $M \in \text{int}m(U)$ we define the quasiequilibrium, $f_M^* \in \text{int}U$, as the solution to the problem,

$$S(f) \to \max, \ m(f) = M \ . \tag{12.6}$$

We assume that, for each $M \in m(U)$, there exists the (unique) solution to the problem (12.6). This solution, f_M^*, is called the quasiequilibrium, corresponding to the value M of the macroscopic variables. The set of quasiequilibria f_M^* forms a manifold in int$U$, parameterized by the values of the macroscopic variables $M \in \text{int}U/L$ (Fig. 12.1).

Let us specify some notations: E^T is the adjoint to the E space. Adjoint spaces and operators will be indicated by ^{T}, whereas notation * is earmarked for equilibria and quasiequilibria.

Furthermore, $[l, x]$ is the result of application of the functional $l \in E^T$ to the vector $x \in E$. We recall that, for an operator $A : E_1 \to E_2$, the adjoint

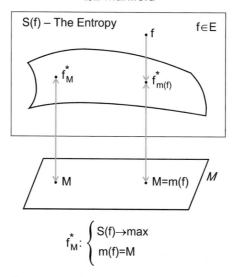

Fig. 12.1. Relations between a microscopic state f, the corresponding macroscopic state $M = m(f)$, and quasiequilibria f^*_M

operator, $A^T : E^T_1 \rightarrow E^T_2$ is defined by the following relation: For any $l \in E^T_2$ and $x \in E_1$,

$$[l, Ax] = [A^T l, x] \ .$$

Next, $D_f S(f) \in E^T$ is the differential of the functional $S(f)$, $D^2_f S(f)$ is the second differential of the functional $S(f)$. Corresponding quadratic functional $D^2_f S(f)(x, x)$ on E is defined by the Taylor formula,

$$S(f + x) = S(f) + \left[D_f S(f), x \right] + \frac{1}{2} D^2_f S(f)(x, x) + o(\|x\|^2) \ . \qquad (12.7)$$

We keep the same notation for the corresponding symmetric bilinear form, $D^2_f S(f)(x, y)$, and also for the linear operator, $D^2_f S(f) : E \rightarrow E^T$, defined by the formula,

$$[D^2_f S(f)x, y] = D^2_f S(f)(x, y) \ .$$

In this formula, on the left hand side there is the operator, on the right hand side there is the bilinear form. Operator $D^2_f S(f)$ is symmetric on E, $D^2_f S(f)^T = D^2_f S(f)$.

Concavity of S means that for any $x \in E$ the inequality holds,

$$D^2_f S(f)(x, x) \leq 0 \ ;$$

in the restriction onto the affine subspace parallel to $K = \ker m$ we assume the strict concavity,

$$D_f^2 S(f)(x, x) < 0 \text{ if } x \in K \text{ and } x \neq 0 .$$

A comment on the degree of rigor is in order: the statements which will be made below become theorems or plausible hypotheses in specific situations. Moreover, specialization is always done with an account for these statements in such a way as to simplify the proofs.

Let us compute the derivative $D_M f_M^*$. For this purpose, let us apply the method of Lagrange multipliers: There exists such a linear functional $\Lambda(M) \in L^T$, that

$$D_f S(f)\big|_{f_M^*} = \Lambda(M) \cdot m, \quad m(f_M^*) = M , \tag{12.8}$$

or

$$D_f S(f)\big|_{f_M^*} = m^T \cdot \Lambda(M), \quad m(f_M^*) = M . \tag{12.9}$$

From equation (12.9) we get,

$$m(D_M f_M^*) = 1_{(L)} , \tag{12.10}$$

where we have indicated the space in which the unit operator is acting. Next, using the latter expression, we transform the differential of the equation (12.8),

$$D_M \Lambda = (m(D_f^2 S)_{f_M^*}^{-1} m^T)^{-1} , \tag{12.11}$$

and, consequently, from (12.9)

$$D_M f_M^* = (D_f^2 S)_{f_M^*}^{-1} m^T (m(D_f^2 S)_{f_M^*}^{-1} m^T)^{-1} . \tag{12.12}$$

Notice that, elsewhere in equation (12.12), operator $(D_f^2 S)^{-1}$ acts on the linear functionals from $\text{im}(m^T)$. These functionals are precisely those which become zero on K (that is, on $\ker(m)$), or, which is the same, those which can be represented as functionals of macroscopic variables.

The tangent space to the quasiequilibrium manifold in the point f_M^* is the image of the operator $D_M f_M^*$:

$$\text{im}(D_M f_M^*) = (D_f^2 S)_{f_M^*}^{-1} \text{im}(m^T) = (D_f^2 S)_{f_M^*}^{-1} \text{Ann} K \tag{12.13}$$

where $\text{Ann} K$ (the annulator of K) is the set of linear functionals which become zero on K. Another way to write equation (12.13) is the following:

$$x \in \text{im}(D_M f_M^*) \Leftrightarrow (D_f^2 S)_{f_M^*}(x, y) = 0, \ y \in K \tag{12.14}$$

This means that $\text{im}(D_M f_M^*)$ is the orthogonal completement of K in E with respect to the scalar product,

QE Projector

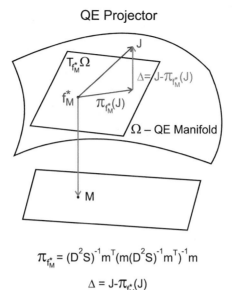

$$\pi_{f_M^*} = (D^2 S)^{-1} m^T (m (D^2 S)^{-1} m^T)^{-1} m$$

$$\Delta = J - \pi_{f_M^*}(J)$$

Fig. 12.2. Quasiequilibrium manifold Ω, tangent space $T_{f_M^*}\Omega$, quasiequilibrium projector $\pi_{f_M^*}$, and defect of invariance, $\Delta = \Delta_{f_M^*} = J - \pi_{f_M^*}(J)$

$$\langle x | y \rangle_{f_M^*} = -(D_f^2 S)_{f_M^*}(x, y) \, . \tag{12.15}$$

The entropic scalar product (12.15) appears often in the constructions below. (Usually, this becomes the scalar product indeed after the conservation laws are excluded). Let us denote as $T_{f_M^*} = \mathrm{im}(D_M f_M^*)$ the tangent space to the quasiequilibrium manifold in the point f_M^*. An important role in the construction of quasiequilibrium dynamics and its generalizations is played by the *quasiequilibrium projector*, an operator which projects E on $T_{f_M^*}$ parallel to K. This is the orthogonal projector with respect to the entropic scalar product, $\pi_{f_M^*} : E \to T_{f_M^*}$:

$$\pi_{f_M^*} = (D_M f_M^*)_M \, m = \left(D_f^2 S\right)_{f_M^*}^{-1} m^T \left(m \left(D_f^2 S\right)_{f_M^*}^{-1} m^T \right)^{-1} m \, . \tag{12.16}$$

It is straightforward to check the equality $\pi_{f_M^*}^2 = \pi_{f_M^*}$, and the self-adjointness of $\pi_{f_M^*}$ with respect to entropic scalar product (12.15). Thus, we have introduced the basic constructions: quasiequilibrium manifold, entropic scalar product, and quasiequilibrium projector (Fig. 12.2).

12.3.2 Quasiequilibrium Approximation

Let a kinetic equation be defined in U:

$$\frac{\mathrm{d}f}{\mathrm{d}t} = J(f) \, . \tag{12.17}$$

(This can be the Liouville equation, the Boltzmann equation, and so on, dependent on which level of precision is taken for the microscopic description.) One seeks the dynamics of the macroscopic variables M. If we adopt the thesis that the solutions of the equation (12.17) of interest for us begin on the quasiequilibrium manifold, and stay close to it for all the later times, then, as the first approximation, we can take the quasiequilibrium approximation. It is constructed this way: We regard f as the quasiequilibrium, and write,

$$\frac{\mathrm{d}M}{\mathrm{d}t} = m\left(J\left(f_M^*\right)\right) . \tag{12.18}$$

With this, the corresponding to M point on the quasiequilibrium manifold moves according to the following equation:

$$\frac{\mathrm{d}f_{M(t)}^*}{\mathrm{d}t} = (D_M f_M^*)m(J(f_M^*)) = \pi_{f_M^*} J(f_M^*) , \tag{12.19}$$

where $\pi_{f_M^*}$ is the quasiequilibrium projector (12.16).

Let us term function $S(M) = S(f_M^*)$ *the quasiequilibrium entropy*. Let us denote as $\mathrm{d}S(M)/\mathrm{d}t$ the derivative of the quasiequilibrium entropy due to the quasiequilibrium approximation (12.18). Then,

$$\frac{\mathrm{d}S(M)}{\mathrm{d}t} = \left.\frac{\mathrm{d}S(f)}{\mathrm{d}t}\right|_{f=f_M^*} . \tag{12.20}$$

From the identity (12.20), it follows **the theorem about preservation of the type of dynamics:**

(i) If for the original kinetic equation (12.17) $\mathrm{d}S(f)/\mathrm{d}t = 0$ at $f = f_M^*$, then the entropy is conserved due to the quasiequilibrium system (12.19).

(ii) If for the original kinetic equation (12.17) $\mathrm{d}S(f)/\mathrm{d}t \geq 0$ at $f = f_M^*$, then, at the same points f_M^*, $\mathrm{d}S(M)/\mathrm{d}t \geq 0$ due to the quasiequilibrium system (12.18).

The theorem about the preservation of the type of dynamics[2] demonstrates that if there was no dissipation in the original system (12.17) (if the entropy was conserved) then there is also no dissipation in the quasiequilibrium approximation. The passage to the quasiequilibrium does not introduce irreversibility. The reverse may happen, for example, there is no dissipation in the quasiequilibrium approximation for hydrodynamic variables as obtained from the Boltzmann kinetic equation (the compressible Euler equations). Though dissipation is present in the Boltzmann equation, it occurs in different points but on the quasiequilibrium manifold of local Maxwellians the entropy production is equal to zero. The same statement also holds for

---

[2] This is a rather old theorem, one of us had published this theorem in 1984 already as a textbook material ([115], chapter 3 "Quasiequilibrium and entropy maximum", p. 37, see also the paper [29]), but from time to time different particular cases of this theorem are continued to be published as new results.

the thermodynamic projectors described in Sect. 5.3. On the other hand, the entropy production in the quasiequilibrium state is the same as for the quasiequilibrium system in the corresponding point, hence, if the initial system is dissipative, then quasiequilibrium entropy production is nonnegative.

Usually, the original dynamics (12.17) does not leave the quasiequilibrium manifold invariant, that is, the vector field $J(f)$ is not tangent to the quasiequilibrium manifold in all its points f_M^*. In other words, the *condition of invariance* (see Chap. 3),

$$(1 - \pi_{f_M^*})J(f_M^*) = 0 , \qquad (12.21)$$

is not satisfied on the quasiequilibrium manifold. The left hand side of the invariance condition (12.21) is the *defect of invariance*, and we denote it as $\Delta_{f_M^*}$ (Chap. 3). It is possible to consider the invariance condition as an equation, and to compute corrections to the quasiequilibrium approximation f_M^* in such a way as to make it "more invariant". If the original equation (12.17) is already dissipative, this route of corrections, supplemented by the construction of the projector as in Sect. 5.3, leads to an appropriate macroscopic kinetics [11].

However, here, we are mainly interested in the route "from the very beginning", from conservative systems to dissipative. And here solving the invariance equation does not help since it will lead us to "more invariant" but still conservative dynamics. In all the approaches to this problem (passage from the conservative to the dissipative systems), dissipation is introduced in a more or less explicit fashion by various assumptions about the "short memory". The originating point of our constructions is the absolutely transparent and explicit approach of Ehrenfests.

12.4 Natural Projector
and Models of Nonequilibrium Dynamics

12.4.1 Natural Projector

So, let the original system (12.17) be conservative, and thus, $dS(f)/dt = 0$. The idea of Ehrenfests is to supplement the dynamics (12.17) by coarse-graining ("shakings"). The coarse-graining steps are external perturbations which are applied periodically with a fixed time interval τ, and which lead to "forgetting" of the small scale (nonequilibrium) details of the dynamics. For us here the coarse-graining is the replacement of f by the quasiequilibrium distribution $f_{m(f)}^*$. In the particular case which was originally considered in by Ehrenfests, the macroscopic variables $m(f)$ were the averages of f over cells in the phase space, while $f_{m(f)}^*$ was the cell-homogeneous distribution with the constant density within each cell equal to the corresponding cell-average of f. In the limit $\tau \to 0$, one gets back the quasiequilibrium approximation –

and the type of the dynamics is preserved. In this limit we obtain just the usual projection of the vector field $J(f)$ (12.17) on the tangent bundle to the quasiequilibrium manifold.

So, the natural question appears: What will happen, if we shall not just send τ to zero but will consider finite, and even large, τ? In such an approach, not just the vector fields are projected but segments of trajectories. We shall term this way of projecting the *natural*. Let us now pose the problem of the *natural projector* formally. Let $T_t(f)$ be the phase flow of the system (12.17). We must derive a phase flow of the macroscopic system, $\Theta_t(M)$ (that is, the phase flow of the macroscopic system, $dM/dt = F(M)$, which we are looking for), such that, for any M,

$$m(T_\tau(f_M^*)) = \Theta_\tau(M) . \tag{12.22}$$

That is, when moving along the macroscopic trajectory, after the time τ we must obtain the same values of the macroscopic variables as if we were moving along the true microscopic trajectory for the same time τ, starting with the quasiequilibrium initial condition (Fig. 12.3).

The final form of the equation for the macroscopic variables M (see Chap. 11) may be written:

$$\frac{dM}{dt} = F(M) = m(J(f_M^*)) + (\tau/2)m(D_f J(f)|_{f_M^*} \Delta f_M^*) + o(\tau^2) . \tag{12.23}$$

Natural projector

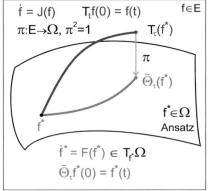

$$\forall f^* \in \Omega \quad \pi(T_\tau f^*) = \tilde{\Theta}_\tau(f^*)$$

Fig. 12.3. Projection of segments of trajectories: The microscopic motion above the manifold Ω and the macroscopic motion on this manifold. If these motions began in the same point on Ω, then, after time τ, projection of the microscopic state onto Ω should coincide with the result of the macroscopic motion on Ω. For quasiequilibrium Ω, projector $\pi : E \to \Omega$ acts as $\pi(f) = f_{m(f)}^*$

It is remarkable the appearance of the defect of invariance in the second term (proportional to τ): If the quasiequilibrium manifold is invariant with respect to the microscopic dynamics, then $F(M)$ is the quasiequilibrium state.

The formula for the entropy production follows from (12.23):

$$\frac{\mathrm{d}S(f_M^*)}{\mathrm{d}t} = (\tau/2)\langle \Delta_{f_M^*} | \Delta_{f_M^*} \rangle_{f_M^*} . \qquad (12.24)$$

The quasiequilibrium entropy increases due to the equation of the macroscopic dynamics (12.23) in those points of the quasiequilibrium manifold where the defect of invariance is not equal to zero. This way we see how the problem of the natural projector (projected are not vector fields but segments of trajectories) results in the dissipative equations. For specific examples see [30] and Chap. 11. The second term in equation (12.23) results in viscosity and heat conductivity terms in the Navier–Stokes equations, diffusion and other dissipative contributions. However, it remains the undetermined parameter τ. Formula (12.24) gives the entropy production just proportional to the time interval between subsequent coarse-grainings. Of course, this could be true only for small enough τ, whereas we are mostly interested in the limit $\tau \to \infty$. It is only in this limit where one can eliminate the arbitrariness in the choice of τ present in equations (12.23) and (12.24). In order to do this, we need to study more carefully the structure of the trajectories which begin on the quasiequilibrium manifold.

12.4.2 One-Dimensional Model of Nonequilibrium States

In the background of many derivations of nonequilibrium kinetic equations one can recognize the following picture: Above each point of the quasiequilibrium manifold there is located a huge subspace of nonequilibrium distributions with the same values of the macroscopic variables, as in the quasiequilibrium. It is as if the motion decomposes into two projections, above the point on the quasiequilibrium manifold, and in the projection on this manifold. The motion in each layer above the quasiequilibria is extremely complicated, but fast, and everything quickly settles in this fast motion.

However, upon a more careful looking into the motions which begin in the quasiequilibrium points, we shall observe that, above each point of the quasiequilibrium manifold it is located just a single and in certain sense monotonic curve. All the nonequilibrium (not-quasiequilibrium) states which come into the game form just a one-dimensional manifold. This is the curve of *the primitive macroscopically definable ensembles*. These ensembles appear as the result (for $t > 0$) of motions which start from the quasiequilibrium state (at $t = 0$). It is namely this curve the construction of which we shall be dealing with in this chapter.

For each value of the macroscopic variables M, and for every time $\tau \geq 0$, we define $M_{-\tau}$ by the following equality:

$$m(T_\tau(f^*_{M-\tau})) = M . \tag{12.25}$$

In other words, $M_{-\tau}$ are those values of macroscopic variables which satisfy $\Theta_\tau(M_{-\tau}) = M$ for the natural projector (12.22). Of course, it may well happen that such $M_{-\tau}$ exists not for every pair (M, τ) but we shall assume here that for every M there exists $\tau_M > 0$, so that there exists $M_{-\tau}$ for $0 < \tau < \tau_M$.

A set of distributions, $q_{M,\tau} = T_\tau(f^*_{M-\tau})$, forms precisely the desired curve of nonequilibrium states with the given values of M. Notice that, for each τ, it holds, $m(q_{M,\tau}) = M$. The set $\{q_{M,\tau}\}$ for all possible M and τ is positive invariant: If the motion of the system starts on it at some time t_0, it stays on it also at $t > t_0$. If the dependence $q_{M,\tau}$ is known, equations of motion in the coordinate system (M, τ) have a simple form:

$$\frac{d\tau}{dt} = 1 , \tag{12.26}$$

$$\frac{dM}{dt} = m(J(q_{M,\tau})) .$$

The simplest way to study $q_{M,\tau}$ is through a consideration of a sequence of its derivatives with respect to τ at fixed M. The first derivative is readily written as,

$$\left.\frac{dq_{M,\tau}}{d\tau}\right|_{\tau=0} = J(f^*_M) - \pi_{f^*_M} J(f^*_M) = \Delta_{f^*_M} . \tag{12.27}$$

By the construction of the quasiequilibrium manifold (we remind that $K = \ker m$), for any $x \in K$,

$$S(f^*_M + \tau x) = S(f^*_M) - (\tau^2/2)\langle x|x\rangle_{f^*_M} + o(\tau^2) .$$

Therefore,

$$S(q_{M,\tau}) = S(f^*_M) - (\tau^2/2)\langle \Delta_{f^*_M}|\Delta_{f^*_M}\rangle_{f^*_M} + o(\tau^2) .$$

Thus, to first order in τ, we have, as expected,

$$q_{M,\tau} = f^*_M + \tau\Delta_{f^*_M} + o(\tau) .$$

Let us find $q_{M,\tau}$ to the accuracy of the order $o(\tau^2)$. To this end, we expand all the functions in equation (12.25) to the order of $o(\tau^2)$. With

$$M_{-\tau} = M - \tau m(J(f^*_M)) + \tau^2 B(M) + o(\tau^2) ,$$

where function B is yet unknown, we write:

$$f^*_{M-\tau} = f^*_M - \tau D_M f^*_M m(J(f^*_M)) + \tau^2 D_M f^*_M B(M) + (\tau^2/2)A_2(M) + o(\tau^2) ,$$

where

$$A_2(M) = \left.\frac{\mathrm{d}^2 f^*_{M+tm(J(f^*_M))}}{\mathrm{d}t^2}\right|_{t=0} , \qquad (12.28)$$

and

$$T_\tau(x+\tau\alpha) = x + \tau\alpha + \tau J(x) + \tau^2 D_x J(x)\big|_x \alpha$$
$$+ (\tau^2/2) D_x J(x)\big|_x J(x) + o(\tau^2) ,$$
$$T_\tau(f^*_{M-\tau}) = f^*_M - \tau D_M f^*_M m(J(f^*_M)) + \tau^2 D_M f^*_M B(M) + (\tau^2/2) A_2(M)$$
$$+ \tau J(f^*_M) - \tau^2 D_f J(f)\big|_{f^*_M} D_M f^*_M m(J(f^*_M))$$
$$+ (\tau^2/2) D_f J(f)\big|_{f^*_M} J(f^*_M) + o(\tau^2)$$
$$= f^*_M + \tau \Delta_{f^*_M} + (\tau^2/2) A_2(M)$$
$$+ (\tau^2/2) D_f J(f)\big|_{f^*_M} (1 - 2\pi_{f^*_M}) J(f^*_M)$$
$$+ \tau^2 D_M f^*_M B(M) + o(\tau^2) .$$

The latter somewhat lengthy expression simplifies significantly under the action of m. Indeed,

$$m(A_2(M)) = \mathrm{d}^2[M + tm(J(f^*_M))]/\mathrm{d}t^2 = 0 ,$$
$$m(1 - \pi_{f^*_M}) = 0 ,$$
$$m(D_M f^*_M) = 1 .$$

Thus,

$$m(T_\tau(f^*_{M-\tau})) = M + (\tau^2/2) m(D_f J(f)\big|_{f^*_M} (1 - 2\pi_{f^*_M}) J(f^*_M)) + \tau^2 B(M) + o(\tau^2) ,$$

$$B(M) = (1/2) m(D_f J(f)\big|_{f^*_M} (2\pi_{f^*_M} - 1) J(f^*_M)) .$$

Accordingly, to second order in τ,

$$q_{M,\tau} = T_\tau(f^*_{M-\tau}) \qquad (12.29)$$
$$= f^*_M + \tau \Delta_{f^*_M} + (\tau^2/2) A_2(M)$$
$$+ (\tau^2/2)(1 - \pi_{f^*_M}) D_f J(f)\big|_{f^*_M} (1 - 2\pi_{f^*_M}) J(f^*_M) + o(\tau^2) .$$

Notice that, besides the dynamic contribution of the order of τ^2 (the last term), there appears also the term A_2 (12.28) which is related to the curvature of the quasiequilibrium manifold along the quasiequilibrium trajectory.

Let us address the behavior of the entropy production in the neighborhood of f^*_M. Let $x \in K$ (that is, $m(x) = 0$). The production of the quasiequilibrium entropy, $\sigma^*_M(x)$, equals, by definition,

$$\sigma^*_M(x) = D_M S(f^*_M) \cdot m(J(f^*_M + x)) . \qquad (12.30)$$

Equation (12.30) gives the rate of the entropy change under the motion of the projection of the state onto the quasiequilibrium manifold if the true trajectory passes the point $f_M^* + x$. In order to compute the right hand side of equation (12.30), we use essentially the same argument, as in the proof of the entropy production formula (12.24). Namely, in the point f_M^*, we have $K \subset \ker D_f S(f)\big|_{f_M^*}$, and thus $D_f S(f)\big|_{f_M^*} \pi_{f_M^*} = D_f S(f)\big|_{f_M^*}$. Using this, and the fact that the entropy production in the quasiequilibrium approximation is equal to zero, equation (12.30) may be written,

$$\sigma_M^*(x) = D_f S(f)\big|_{f_M^*} \left(J(f_M^* + x) - J(f_M^*) \right) . \qquad (12.31)$$

To the linear order in x, the latter expression reads:

$$\sigma_M^*(x) = D_f S(f)\big|_{f_M^*} D_f J(f)\big|_{f_M^*} x . \qquad (12.32)$$

Using the identity

$$D_f^2 S(f)\big|_f J(f) + D_f S(f)\big|_f D_f J(f)\big|_f = 0 , \qquad (12.33)$$

we obtain in equation (12.32),

$$\sigma_M^*(x) = -D_f^2 S(f)\big|_{f_M^*} \left(J(f_M^*), x \right) = \langle J(f_M^*) | x \rangle_{f_M^*} . \qquad (12.34)$$

Because $x \in K$, we have $(1 - \pi_{f_M^*})x = x$, and

$$\begin{aligned}
\langle J(f_M^*) | x \rangle_{f_M^*} &= \langle J(f_M^*) | (1 - \pi_{f_M^*})x \rangle_{f_M^*} \\
&= \langle (1 - \pi_{f_M^*}) J(f_M^*) | x \rangle_{f_M^*} = \langle \Delta_{f_M^*} | x \rangle_{f_M^*} .
\end{aligned}$$

Thus, finally, the entropy production in the formalism developed here, to the linear order reads,

$$\sigma_M^*(x) = \langle \Delta_{f_M^*} | x \rangle_{f_M^*} . \qquad (12.35)$$

The above consideration gives us the simplest way to study the primitive macroscopically definable ensembles using Taylor expansion in τ. This way has obvious limitations because τ remains a parameter of the theory.

12.4.3 Curvature and Entropy Production: Entropic Circle and First Kinetic Equations

In a consequent geometric approach to the problem of constructing the one-dimensional model of nonequilibrium states it is more relevant to consider the entropic parameter, $\delta S = S^*(M) - S$ instead of τ. Within this parameterization of the one-dimensional curve of the nonequilibrium states one has to address functions $\sigma_M(\Delta S)$, rather than $\sigma_M(\tau)$.

In order to give an example here, we notice that the simplest geometric estimate amounts to approximating the trajectory $q_{M,\tau}$ with a second order curve[3]. Given $\dot{q}_{M,\tau}$ and $\ddot{q}_{M,\tau}$ (12.29), we construct a tangent circle (in the entropic metrics, $\langle|\rangle_{f_M^*}$, since the entropy is the integral of motion of the original equations). For the radius of this circle we compute

$$R = \frac{\langle \dot{q}_{M,0}|\dot{q}_{M,0}\rangle_{f_M^*}}{\sqrt{\langle \ddot{q}_{\perp\ M,0}|\ddot{q}_{\perp\ M,0}\rangle_{f_M^*}}} \ , \tag{12.36}$$

where

$$\dot{q}_{M,0} = \Delta_{f_M^*} \ ,$$

$$\ddot{q}_{\perp\ M,0} = \ddot{q}_{M,0} - \frac{\langle \ddot{q}_{M,0}|\Delta_{f_M^*}\rangle_{f_M^*}\Delta_{f_M^*}}{\langle \Delta_{f_M^*}|\Delta_{f_M^*}\rangle_{f_M^*}} \ ,$$

$$\ddot{q}_{M,0} = (1 - \pi_{f_M^*})D_f J(f)\big|_{f_M^*}(1 - 2\pi_{f_M^*})J(f_M^*) + \left(D_M\pi_{f_M^*}\right)m(J(f_M^*)) \ .$$

Let us represent the microscopic motion as a circular motion along this entropic circle with the constant "linear velocity" $\dot{q}_{M,0} = \Delta_{f_M^*}$. After the microscopic motion passed the quarter of the circle, the entropy production begins decreasing and it becomes equal to zero after passing the semicircle. Hence, after passing the quarter of the circle, this model should be changed. The time of the motion along the quarter of the entropic circle is:

$$\tau \approx \frac{\pi}{2}\sqrt{\frac{\langle \Delta_{f_M^*}|\Delta_{f_M^*}\rangle_{f_M^*}}{\langle \ddot{q}_{\perp\ M,0}|\ddot{q}_{\perp\ M,0}\rangle_{f_M^*}}} \ . \tag{12.37}$$

After averaging over the $1/4$ of this circle we obtain the macroscopic equations

$$\frac{dM}{dt} = m\left(J\left(f_M^* + \frac{2}{\pi}R\frac{\Delta_{f_M^*}}{\|\Delta_{f_M^*}\|} + \left(1 - \frac{2}{\pi}\right)R\frac{\ddot{q}_{\perp\ M,0}}{\|\ddot{q}_{\perp\ M,0}\|}\right)\right)$$

$$= m(J(f_M^*)) + \frac{2}{\pi}\frac{R}{\|\Delta_{f_M^*}\|}m\left(D_f J(f)\big|_{f_M^*}(\Delta_{f_M^*})\right)$$

$$+ \left(1 - \frac{2}{\pi}\right)\frac{R}{\|\ddot{q}_{\perp\ M,0}\|}m\left(D_f J(f)\big|_{f_M^*}(\ddot{q}_{\perp\ M,0})\right) + o(R) \ . \tag{12.38}$$

where $\|y\| = \sqrt{\langle y|y\rangle_{f_M^*}}$.

Equations (12.38) contain no undetermined parameters. This is the simplest example of a general macroscopic equations obtained by the natural projector. The coefficients ($2/\pi$, etc.) can be corrected, but the form is more universal. The entropy production for equations (12.38) is proportional both to the defect of invariance and to the radius of curvature:

---

[3] We shall argue below in detail, why the first-order estimates, $q_{M,\tau} = f_M^* + \tau\Delta_{f_M^*}$, are insufficient in the case of the conservative dynamics.

$$\sigma_M = \frac{2}{\pi} R \| \Delta f_M^* \| \,. \tag{12.39}$$

This equation demonstrates the thermodynamical sense of curvature of the curve of the nonequilibrium states. The combination

$$\frac{\textbf{defect of invariance}}{\textbf{curvature}} \tag{12.40}$$

is the dissipation (recall that all the scalar products and norms are *entropic*).

12.5 The Film of Non-Equilibrium States

12.5.1 Equations for the Film

The set $q_{M,\tau}$ in the space E forms a "surface" parameterized by "two variables": A scalar, $\tau \geq 0$, and the value of the macroscopic variables, M, subject to the condition

$$M = m(q_{M,\tau}) \,. \tag{12.41}$$

We call this surface *the film of non-equilibrium states* or simply *the film*. It consists of *the primitive macroscopically definable ensembles*, the result (for $t > 0$) of motions which start from the quasiequilibrium state (at $t = 0$).

For each $\tau \geq 0$ *the section of the film* is defined: the set, $q_{M,\tau}$, for a given τ. It is parameterized by the value of M. For $\tau = 0$ the section of the film coincides with the quasiequilibrium manifold. The film itself can be considered as a trajectory of motion of the section under the variation of $\tau \in [0; +\infty)$. It is not difficult to write down equations of this motion using the definition of $q_{M,\tau}$:

$$q_{M,\tau} = T_\tau f_{M_{-\tau}}^* \,, \tag{12.42}$$

where T_τ is the phase flow of the microscopic dynamical system, $M_{-\tau}$ is defined with equation (12.25).

For small $\Delta\tau$

$$q_{M,\tau+\Delta\tau} = q_{M-\Delta M,\tau} + J(q_{M,\tau})\Delta\tau + o(\Delta\tau) \,, \tag{12.43}$$

where $\Delta M = m J(q_{M,\tau})\Delta\tau$. Hence,

$$\frac{dq_{M,\tau}}{d\tau} = (1 - D_M q_{M,\tau} m) J(q_{M,\tau}) \,. \tag{12.44}$$

The initial condition for equation (12.44) is the quasiequilibrium

$$q_{M,0} = f_M^* \,. \tag{12.45}$$

Equation (12.44), subject to the initial condition (12.45), defines the film of non-equilibrium states in the space E. This film is a minimal positive invariant set (i.e invariant with respect to the shift T_τ for positive times $\tau > 0$), including the quasiequilibrium manifold, f_M^*. All of the macroscopic kinetics take place only on this film.

Thus, the study of the non-equilibrium kinetics can be separated into two problems:

1. Construction of the film of non-equilibrium states: solution of equation (12.44) with the initial condition (12.45).
2. Investigation of the motion of the system on the film.

Of course, one should assume that the film will be constructed only approximately. Therefore, the second problem in turn should be separated in two subproblems:

- Construction of projection of the microscopic vector field J on the approximately found film, and construction of equations for M and τ.
- Investigation and solution of equations for M and τ.

It should be emphasized that the existence of the film is not significantly questionable (though, of course, proving theorems about existence and uniqueness for (12.44), (12.45) can turn into a hard mathematical problem). In a contrast, existence of kinetic coefficients (viscosity etc.), and generally, of the fast convergence of dM/dt to a certain dependence dM/dt of M is essentially a hypothesis which is not expected to always be true.

Below we mostly deal with the problem of construction of equations: the problems ii1) and ii2). And we shall begin with the problem ii2). Thus, let the film be approximately constructed.

12.5.2 Thermodynamic Projector on the Film

We need the projector in order to project the vector field on the tangent space. The method of the thermodynamic projector ([9,10] and Chap. 5) allows to characterize every manifold (subject to certain requirements of transversality) as the quasiequilibrium one. This is achieved by a construction of a projection of a neighborhood of the manifold. The projection of the neighborhood on the manifold should satisfy essentially only one condition: a point of the manifold must be the point of maximum of the entropy on its preimage. If the preimage of the point f^* is a domain in the affine subspace, $K_{f^*} \subset E$, then the required condition is the property **A** (5.37):

$$(D_f S)_{f^*}(K_{f^*} - f^*) \equiv 0 . \tag{12.46}$$

where $K_{f^*} - f^*$ is the linear subspace in E because $f^* \in K_{f^*}$.

For the projections with the property **A** (5.37), a dissipative vector field is projected into a dissipative one, and a conservative vector field (with the entropy conservation) is projected into a conservative one, i.e. the entropy balance is exact. Thus, let the film, $q_{M,\tau}$, be defined, and let us construct for it the projector.

Under small variation of variables M and τ

$$\Delta q_{M,\tau} = D_M q_{M,\tau} \Delta M + D_\tau q_{M,\tau} \Delta \tau + o(\Delta M, \Delta \tau) \,,$$
$$\Delta S = D_f S|_{q_{M,\tau}} \Delta q_{M,\tau} + o(\Delta M, \Delta \tau) \,. \tag{12.47}$$

After simple transformations we obtain:

$$\Delta \tau = \frac{\Delta S - D_f S|_{q_{M,\tau}} D_M q_{M,\tau} \Delta M}{D_f S|_{q_{M,\tau}} D_\tau q_{M,\tau}} + o(\Delta M, \Delta S) \,,$$

$$\Delta q_{M,\tau} = \left[1 - \frac{D_\tau q_{M,\tau} D_f S|_{q_{M,\tau}}}{D_f S|_{q_{M,\tau}} D_\tau q_{M,\tau}} \right] D_M q_{M,\tau} \Delta M$$
$$+ \frac{D_\tau q_{M,\tau} \Delta S}{D_f S|_{q_{M,\tau}} D_\tau q_{M,\tau}} + o(\Delta M, \Delta S) \,. \tag{12.48}$$

From this formulae we obtain the projector with the property **A** for J, π_A:

$$\pi_A|_{q_{M,\tau}} J = \left[1 - \frac{D_\tau q_{M,\tau} D_f S|_{q_{M,\tau}}}{D_f S|_{q_{M,\tau}} D_\tau q_{M,\tau}} \right] D_M q_{M,\tau} m J$$
$$+ \frac{D_\tau q_{M,\tau} D_f S|_{q_{M,\tau}}}{D_f S|_{q_{M,\tau}} D_\tau q_{M,\tau}} J \,. \tag{12.49}$$

It is straightforward to check the equality $\pi_A^2 = \pi_A$. For the conservative vector fields $J(f)$, the second term in (12.49) vanishes because $D_f S|_f(J(f)) = 0$, and

$$\pi_A|_{q_{M,\tau}} J = \left[1 - \frac{D_\tau q_{M,\tau} D_f S|_{q_{M,\tau}}}{D_f S|_{q_{M,\tau}} D_\tau q_{M,\tau}} \right] D_M q_{M,\tau} m J \,. \tag{12.50}$$

The equation for M corresponding to (12.50) has the form:

$$\frac{dM}{dt} = m(\pi_A|_{q_{M,\tau}} (J(q_{M,\tau})))$$
$$= m \left[1 - \frac{D_\tau q_{M,\tau} D_f S|_{q_{M,\tau}}}{D_f S|_{q_{M,\tau}} D_\tau q_{M,\tau}} \right] D_M q_{M,\tau} m J(q_{M,\tau})$$
$$= m J(q_{M,\tau}) \,. \tag{12.51}$$

By the definition of the projector with the property **A** the equation for M (12.51) should be supplemented with the equation for S:

$$\frac{dS}{dt} = 0 \,, \tag{12.52}$$

or for τ, in accordance with (12.48),

Dynamic Equation on The Film

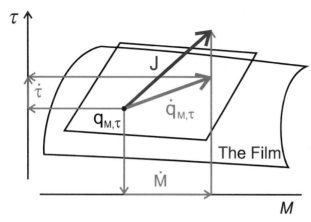

Fig. 12.4. Dynamics on the film: $\dot{M} = mJ(q_{M,\tau})$, $\dot{\tau} = -\dfrac{D_f S|_{q_{M,\tau}} D_M q_{M,\tau} \dot{M}}{D_f S|_{q_{M,\tau}} D_\tau q_{M,\tau}}$

$$\frac{d\tau}{dt} = \frac{\dot{S} - D_f S|_{q_{M,\tau}} D_M q_{M,\tau} \dot{M}}{D_f S|_{q_{M,\tau}} D_\tau q_{M,\tau}} = -\frac{D_f S|_{q_{M,\tau}} D_M q_{M,\tau} \dot{M}}{D_f S|_{q_{M,\tau}} D_\tau q_{M,\tau}} \ , \quad (12.53)$$

where $\dot{M}$ is defined in accordance with (12.51). The numerator in (12.53) has a simple meaning: it is the rate of the entropy production by dynamic equations (12.51) when τ is constant (for frozen τ). Expression (12.53) can be obtained from the condition of the constant entropy for the motion on the film in accordance with (12.51,12.53). Equations (12.51,12.53) describe dynamics on the film (Fig. 12.4).

The system of equations (12.51,12.53) has a very simple sense:

$$\frac{dM}{dt} = mJ(q_{M,\tau}); \quad \frac{dS}{dt} = 0 \ . \quad (12.54)$$

It is just the standard moment equation supplied by the equation of entropy production (in this case by the equation of entropy conservation).

It should be emphasized that the projector with the property **A** is not unique, and here we made the simplest choice.

Let us further assume that condition (12.27) is satisfied:

$$q_{M,\tau} = f_M^* + \tau \Delta_{f_M^*} + o(\tau) \ .$$

In expressions (12.48,12.51,12.53) the denominator, $D_f S|_{q_{M,\tau}} D_\tau q_{M,\tau}$, is present. For $\tau \to 0$ this expression vanishes:

$$D_\tau q_{M,\tau}|_{\tau=0} = \Delta_{f_M^*} \ ,$$
$$D_f S|_{f=f_M^*} x = 0, \text{ for } x \in \ker m \ , \quad (12.55)$$

$m(\Delta_{f_M^*}) = 0$, therefore $D_f S|_{q_{M,\tau}} D_\tau q_{M,\tau} \to 0$ for $\tau \to 0$. For $\tau \to 0$ indeterminate forms $0/0$ appear in expressions (12.48–12.50,12.52,12.53). Let us resolve the indeterminate forms and calculate the corresponding limits.

Two indeterminate forms are present:

$$N_1 = \frac{(D_\tau q_{M,\tau})(D_f S|_{q_{M,\tau}}) D_M q_{M,\tau} m J}{D_f S|_{q_{M,\tau}} D_\tau q_{M,\tau}} \qquad (12.56)$$

and the right hand side of equation (12.53), $N_2(\tau)$. Let us evaluate the form (12.56). We obtain:

$$\lim_{\tau \to 0} N_1(\tau) = \frac{\Delta_{f_M^*} D_f S|_{f_M^*} \pi_{f_M^*} D_f J(f)|_{f_M^*}}{\langle \Delta_{f_M^*} | \Delta_{f_M^*} \rangle_{f_M^*}} \qquad (12.57)$$

using identity (12.33), similar to (12.24), we obtain:

$$\lim_{\tau \to 0} N_2(\tau) = -\frac{\Delta_{f_M^*} \langle \Delta_{f_M^*} | \Delta_{f_M^*} \rangle_{f_M^*}}{\langle \Delta_{f_M^*} | \Delta_{f_M^*} \rangle_{f_M^*}} = -\Delta_{f_M^*} .$$

Therefore, for $\tau \to 0$

$$\pi_A|_{q_{M,\tau}} J(q_{M,\tau}) \to D_M f_M^* m J(f_M^*) + \Delta_{f_M^*}$$
$$= \pi_{f_M^*} J(f_M^*) + (1 - \pi_{f_M^*}) J(f_M^*) = J(f_M^*) . \quad (12.58)$$

Similarly, after simple calculations we obtain that:

$$\frac{d\tau}{dt} \to 1, \text{ for } \tau \to 0 . \qquad (12.59)$$

The fact that for $\tau \to 0$ the action of the projector π_A on J becomes trivial, $\pi_A J = J$, can be obtained (without calculations) from the construction of $q_{M,\tau}$ in the vicinity of zero. We have chosen this dependence in such a way that $J(q_{M,\tau})$ becomes transverse to the film for $\tau \to 0$. This follows from the condition (12.27). Let us emphasize, however, that derivation of the formulas (12.50–12.53) themselves was not based on (12.27), and they are applicable to any ansatz, $q_{M,\tau}$, not necessarily with the right behavior near the quasiequilibrium (if one needs such an ansatz for anything).

12.5.3 Fixed Points of the Film Equation

What features can one expect from the dynamics of the film according to equation (12.44)? A naive expectation that $q_{M,\tau}$ tends to a stable fixed point of equation (12.44) leads to somewhat strange consequences. Fixed point for equation (12.44) is the invariant manifold q_M. On this manifold,

$$J(q_M) = D_M q_M m J(q_M) , \qquad (12.60)$$

i.e. the projection of the vector field, J, onto q_M coincides with J. Were the condition $q_{M,\tau} \to q_M$ satisfied for $\tau \to \infty$, the dynamics would become "more and more conservative". On the limit manifold q_M, the entropy should be conserved. This leads to unusual consequences. The first of them is the limited extendability backwards "in the entropy".

Indeed, let us consider the set of points $M_{-\tau}$ (12.25) for a given M. There exists the limit,

$$\lim_{\tau \to \infty} T_\tau(f^*_{M_{-\tau}}) = q_M ,$$

The flow T_τ conserves the entropy, hence, the difference of the values of the quasiequilibrium entropy, $S(M) - S(M_{-\tau}) = \Delta S_\tau$, is bounded on the half-axis, $\tau \in [0; +\infty)$: $\Delta S_\tau < \Delta S_\infty(M)$. This means that it is impossible to get into the values of macroscopic variables, M, from the quasiequilibrium initial conditions, M_1, for that $S(M) - S(M_1) > \Delta S_\infty(M)$. Thus, possible fixed points of the equation (12.44), regardless of their obvious interest, likely demonstrate some exotic possibilities.

12.5.4 The Failure of the Simplest Galerkin-Type Approximations for Conservative Systems

Usually, the simplest approach to the problem is the projection approximation: one considers a projection of the vector field, $J(f)$, onto the manifold in question and investigates the obtained equations of motion. However, it is not difficult to see sure that such an approach is unfruitful in the present case of conservative systems. If the orthogonal with respect the entropic scalar product projection is taken, then only the quasiequilibrium approximations with increased number of moments could be obtained.

For the dissipative systems, in contrast, such a projection approximations leads to quite satisfactory results. For example, if for the Boltzmann equation and the hydrodynamic moments the approximate invariant manifold is to be searched in the form $f^\#_M = f^*_M + a(M)\Delta_{f^*_M}$, where f^*_M is local Maxwellian, then we obtain the Navier–Stokes equations with the viscosity and heat conductivity calculated within the first Sonine polynomials approximation. Using another scalar product simply leads to unphysical results.

In order to highlight the pitfall in the conservative case, let us give an example with a linear field, $J(f) = Af$, and a quadratic entropy, $S(f) = (1/2)\langle f|f \rangle$. The conservativity of J means that for each f it holds

$$\langle f|Af \rangle = 0 . \tag{12.61}$$

The quasiequilibrium subspace corresponding to the moments $M = mf$ is the orthogonal complement, $\ker M$. The quasiequilibrium projector, π, is an orthogonal projector on this subspace. For the defect of invariance $\Delta_{f^*_M}$ we obtain:

$$\Delta_{f^*_M} = (A - \pi A)f^*_M . \tag{12.62}$$

Under the simplest projection approximation we write

$$q_{M,\tau} = f_M^* + a(M,\tau)\Delta_{f_M^*} . \tag{12.63}$$

Projector on $\Delta_{f_M^*}$ is

$$\frac{|\Delta_{f_M^*}\rangle\langle\Delta_{f_M^*}|}{\langle\Delta_{f_M^*}|\Delta_{f_M^*}\rangle} . \tag{12.64}$$

Thus, we pass from the equation of motion of the film (12.44) to the Galerkin-type approximation for $a(M,\tau)$.

$$
\begin{aligned}
\dot{a} = 1 &+ a\frac{\langle\Delta_{f_M^*}|A\Delta_{f_M^*}\rangle}{\langle\Delta_{f_M^*}|\Delta_{f_M^*}\rangle} - a\frac{\langle\Delta_{f_M^*}|A\pi A\Delta_{f_M^*}\rangle}{\langle\Delta_{f_M^*}|\Delta_{f_M^*}\rangle} \\
&- a^2\frac{\langle\Delta_{f_M^*}|A\pi A\Delta_{f_M^*}\rangle}{\langle\Delta_{f_M^*}|\Delta_{f_M^*}\rangle} - (D_M a)m\frac{Af_M^* + aA\Delta_{f_M^*}}{\langle\Delta_{f_M^*}|\Delta_{f_M^*}\rangle} .
\end{aligned}
\tag{12.65}
$$

One can try to find fixed points (solving $\dot{a} = 0$). This is the projected invariance equation. Due to the properties of the operator A, and the self-adjoint projector, π, we obtain for conservative systems

$$\langle\Delta_{f_M^*}|A\Delta_{f_M^*}\rangle = 0 , \tag{12.66}$$

$$\langle\Delta_{f_M^*}|A\pi A\Delta_{f_M^*}\rangle = -\langle\pi A\Delta_{f_M^*}|(\pi A^2 - (\pi A)^2)\Delta_{f_M^*}\rangle . \tag{12.67}$$

On the other hand, for the dissipative systems the form (12.66) is negatively definite, and it is this form that determines the Navier–Stokes equations (in the first Sonine's polynomials approximation) in the derivation of these equations from the Boltzmann equation. For the conservative equations this main part vanishes, while the second term in equation (12.65), generally speaking, is sign-indefinite.

The failure of the projection approximations becomes even more obvious in the equations of motions on the film. Here everything is very simple:

$$\dot{a} = 1 + a\frac{\langle\Delta_{f_M^*}|A\Delta_{f_M^*}\rangle}{\langle\Delta_{f_M^*}|\Delta_{f_M^*}\rangle} . \tag{12.68}$$

For the dissipative systems under frozen M, a relaxes to the stable point

$$a = -\frac{\langle\Delta_{f_M^*}|\Delta_{f_M^*}\rangle}{\langle\Delta_{f_M^*}|A\Delta_{f_M^*}\rangle} > 0 . \tag{12.69}$$

This fixed point is "the leading order term" in the solution of the invariance equation, $\dot{a} = 0$ (12.65).

However, for the conservative systems $\dot{a} = 1$. This result was expected from the entropy production formula (12.24), and

$$- S(f) = (1/2)\langle f|f\rangle = (1/2)\langle\pi f|\pi f\rangle + (1/2)\langle(1 - \pi)f|(1 - \pi)f\rangle .$$

12.5.5 Second Order Kepler Models of the Film

In the problems of the dissipative kinetics (namely, in the problem of the initial layer for the Boltzmann equation) it was found efficient to approximate the trajectories by segments (with further smoothing and corrections, or without them). These segments were constructed in the following way: the initial direction of motion was taken, and f evolved along this direction for as long as the entropy increases. Further, the procedure was repeated from the obtained point (for details see [26, 27] and Sect. 9.3).

Unfortunately, in the problem of the initial layer for the conservative systems there are no termination points during the motion along a straight line (more precisely, the beginning of the motion itself can be considered as a termination point because under the linear approximation the relation (12.66) is valid). In the initial layer for the dissipative systems the motion of the system along the straight line $x = \tau\Delta$ in any case increases the entropy. For the conservative systems one needs to "rotate the phase", and the models of motion should be arcs of ellipses (in linear space), or the constant entropy lines, rather than straight lines. In the film problem the simplest "good" model is a general conic section. A simple example: $J(f) = Af$, A is generator of rotation around the axis with the direction $r = e_x + \alpha e_y$, $M = x$, the film is the lateral surface of the cone, obtained by rotation of the quasiequilibrium manifold, the axis $\{xe_x\}$, around the axis $\{\varphi r\}$. For $\alpha < 1$ the curve $q_{M,\tau}$ is an ellipse, for $\alpha > 1$ it is a hyperbole, for $\alpha = 1$ it is a parabola.

The curve $q_{M,\tau}$ is an intersection of two manifolds: one of them is the result of the motion of the quasiequilibrium manifold along the vector field $J(f)$, other is the linear manifold $f_M^* + \ker m$.

Already in the finite-dimensional space, and under linear approximation (J is linear, S is quadratic), we have an interesting geometrical picture: quasiequilibrium manifold is an orthogonal complement to $\ker m$, A is the rotation generator. $(\ker m)^\perp$ is rotated under action of $e^{A\tau}$, the unknown curve is the section:

$$(f_M^* + \ker m) \bigcap e^{AR_+}(\ker m)^\perp , \qquad (12.70)$$

where $R_+ = [0; \infty)$, $f_M^* \in (\ker m)^\perp$.

Thus, the simplest model motion is a second order curve. However, it is not sufficient to know the first and the second derivatives. We need information about the third-order derivative. If we consider the curve $q_{M,\tau}$ as a trajectory in the Kepler problem, then the location, r, of the center of attraction (repulsion) is (Fig. 12.5):

$$r = q_0 - \ddot{q}\frac{\langle \dot{q}_\perp | \dot{q}_\perp \rangle}{\langle \dddot{q} | \dot{q}_\perp \rangle} , \qquad (12.71)$$

where q_0 is the initial point where all the derivatives are taken. The force is:

Second order models

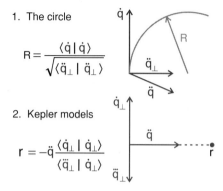

1. The circle

$$R = \frac{\langle \dot{q} | \dot{q} \rangle}{\sqrt{\langle \ddot{q}_\perp | \ddot{q}_\perp \rangle}}$$

2. Kepler models

$$r = -\ddot{q} \frac{\langle \dot{q}_\perp | \dot{q}_\perp \rangle}{\langle \ddot{q}_\perp | \dot{q}_\perp \rangle}$$

Fig. 12.5. The definition of the second-order models

$$F = \alpha \frac{r - q}{\langle r - q | r - q \rangle^{3/2}} \; ;$$

$$\alpha^2 = \langle \ddot{q} | \ddot{q} \rangle \langle r - q | r - q \rangle^2 = \langle \ddot{q} | \ddot{q} \rangle^3 \frac{\langle \ddot{q}_\perp | \ddot{q}_\perp \rangle^4}{\langle \ddot{q} | \dot{q}_\perp \rangle^4} \; ; \qquad (12.72)$$

$$\begin{array}{llll} \alpha > 0 & \text{(attraction)} & \text{if} & \langle \ddot{q} | \dot{q}_\perp \rangle < 0 \; ; \\ \alpha < 0 & \text{(repulsion)} & \text{if} & \langle \ddot{q} | \dot{q}_\perp \rangle > 0 \; . \end{array} \qquad (12.73)$$

It is necessary to point out that the Kepler problem defines an approximation of the trajectory $q_{M,\tau}$, but not the dependence on τ.

An important question is the finiteness of the film. Is the model motion finite? The answer is simple in terms of the Kepler problem [182]:

$$\frac{\|\dot{q}\|^2}{2} < \frac{\alpha}{\|r - q_0\|} \; ,$$

or

$$\frac{\|\dot{q}\|^2 |\langle \dot{q}_\perp | \ddot{q} \rangle|}{2 \|\dot{q}_\perp\|^2 \|\ddot{q}\|^2} < 1 \; . \qquad (12.74)$$

Here $\| \; \| = (\langle | \rangle_{f_M^*})^{1/2}$ is the norm in the entropic scalar product, as usual.

12.5.6 The Finite Models: Termination at the Horizon Points

In order to construct a step-by-step approximation it is necessary to solve two problems: the choice of the direction of the next step, and the choice of the size of this step.

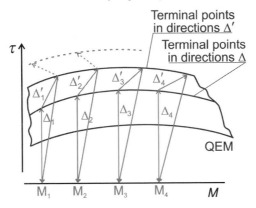

Fig. 12.6. The stepwise construction of the film for dissipative system. First-order models: The motion along the defect of invariance

If the motion $q_{M,\tau}$ is taken along the straight line (dissipative systems), the direction of the step is $\dot{q}_{M,\tau_0}$ (let us remind that $\dot{q}_{M,\tau_0}$ is the defect of the invariance of the manifold $q_M = q_{M,\tau_0}$ at fixed $\tau = \tau_0$), and the size of the step should be adjusted in such a way as to reach a stable point, that is, the point where the direction $\dot{q}_{M,\tau}$ becomes orthogonal to the initial one, $\dot{q}_{M,\tau_0}$ (Fig. 12.6). The current direction of $\dot{q}_{M,\tau}$ is calculated with the help of (12.44), where the projector is frozen ($D_M q_{M,\tau_0} m$ instead of $D_M q_{M,\tau} m$).

For the conservative systems we have chosen the second order models instead of the linear ones. For finiteness of the models we need to define the moments of termination of motion. It is suggested to operate in a manner similar to the case of the dissipative systems: to stop at the moment when the direction of the motion becomes orthogonal to the initial one.

Thus, if q_{M,τ_0} is a starting point of motion, and $\tilde{q}_{M,\tau_0+\theta}$ is a motion on the finite second order model, then the condition for the transition to the next model is

$$\left\langle \dot{q}_{M,\tau_0} \left| \frac{d\tilde{q}_{M,\tau_0+\theta}}{d\theta} \right\rangle = 0 \right. \tag{12.75}$$

(in the entropic scalar product).

Let us call *the horizon points* such points, $q_{M,\tau_0+\theta_0}$, where the scalar product (12.75) for the first time becomes equal to zero (for $0 \leq \theta < \theta_0$ this scalar product is positive). This notion is motivated by the fact that for $\theta > \theta_0$ the motion on the second order model "disappears behind the

The Film for conservative systems:
second order steps

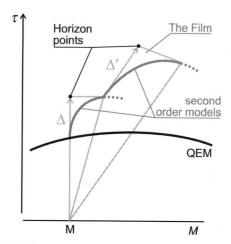

Fig. 12.7. The stepwise construction of the film for conservative system. Finite second-order models: The motion starts in the direction of the defect of invariance, and stops when the direction of motion becomes orthogonal to the defect of invariance

horizon", and its orthogonal projection on the line parallel to $\dot{q}_{M,\tau_0}$ starts to move back passing the same points for the second time.

The convention about the change of the model in the horizon points seems quite natural. The following sequence of calculations becomes self-explaining (Fig. 12.7):

1. we seed the film with the quasiequilibrium manifold, $q_{M,0} = f_M^*$;
2. we calculate $\dot{q}_{M,0}$, $\ddot{q}_{M,0}$, ... in accordance with equation (12.44);
3. we construct the (finite) second order models, $q_{M,\theta}$;
4. we find the horizon points, $q_{M,\theta_0(M)}$, from (12.75);
5. then we take the manifold of the horizon points as a new initial manifold, and dio the next iteration.

At a first glance, this sequence contradicts the original statement of the film problem. The manifold $q_{M,\theta_0(M)}$ does not have the form of $q_{M,\tau}$ for a fixed τ and thus it is not a shift of the quasiequilibrium manifold by the given time along the true microscopic equations of motion.

The second difficulty was already mentioned: the time of motion along the model curve does not coincide with the proper time, τ. More precisely, it coincides only within the second order. However, now global, not local approximation are constructed. Therefore, global corrections to the time, or ways to circumvent these corrections, are required.

The following two subsections are devoted to the elimination of these difficulties.

12.5.7 The Transversal Restart Lemma

Let $q_{M,\tau}$ ($\tau \in [0; +\infty)$) be the solution to (12.44) under initial condition (12.45) (the film). We call *the transverse section* of the film, $q_{M,\tau}$, the manifold, $q_{M,\theta(M)}$, where $\theta(M)$ is a smooth function $0 \leq \theta(M) \leq t < \infty$.

Let *the transversality condition* be satisfied. Namely, for every bounded domain that does not include equilibrium there exists $\varepsilon > 0$ such that in this patch

$$\frac{\|J(q_{M,\theta(M)}) - D_M q_{M,\theta(M)} m J(q_{M,\theta(M)})\|}{\|J(q_{M,\theta(M)})\|} > \varepsilon \qquad (12.76)$$

in an appropriate norm. Let $\tilde{q}_{M,\tau}$ be the solution to (12.44) under the initial condition $\tilde{q}_{M,0} = q_{M,\theta(M)}$. Then the following *transverse restart lemma* is valid:

$$q_{M,[0;+\infty)} = q_{M,[0;\theta(M)]} \bigcup \tilde{q}_{M,[0;+\infty)} . \qquad (12.77)$$

here $q_{M,[a;b]} = \{q_{M,\tau} | \tau \in [a; b]\}$.

The transversality condition (12.76) can be understood as a condition of an "uniform noninvariance". As we already know, fixed points of the film equations are irrelevant.

The transversal restart lemma is the statement about the correctness of the film. One way to derive the film is to seed it at the quasiequilibrium edge and to evolve in τ to $+\infty$ along the film equation (12.44). Another way is to evolve it to some transverse section, not obligatory uniformly in time, and then continue growing the film from this new edge. The result will be the same.

In order to "prove"[4] this lemma, we notice that it is equivalent to the following statement. For every $\widetilde{M}$ the segment of the trajectory, $T_{\tilde{\tau}} f^*_{\widetilde{M}}$ ($\tilde{\tau} \in [0; t]$), crosses the manifold $q_{M,\theta(M)}$, and only once.

In order to demonstrate the unicity of the section, we consider the film in another coordinates, for each point q we set $\widetilde{M}$ and $\tilde{\tau}$: $q = T_{\tilde{\tau}} f^*_{\widetilde{M}}$. In these coordinates the transversality condition excludes folds on $q_{M,\theta(M)}$.

In order to demonstrate the existence of the crossing point, q^*, of the segment $T_{\tilde{\tau}} f^*_{\widetilde{M}}$ ($\tilde{\tau} \in [0; t]$) with the section manifold $q_{M,\theta(M)}$, we define in the neighborhood of the point $f^*_{\widetilde{M}}$ on the quasiequilibrium manifold the mapping into the neighborhood of this section point. Image of the point $f^*_{\widetilde{M}}$ is section of the trajectory $T_{\tilde{\tau}} f^*_{\widetilde{M}}$ ($\tilde{\tau} \in [0; t]$) with the manifold $q_{M,\theta(M)}$ in the neighborhood of q^*. Due to the transversality condition, it performs an isomorphism

---

[4] Let us remind that within the degree of generality used here there are no proofs to the theorems of existence and uniqueness.

of the neighborhoods. Therefore, the set of $\widetilde{M}$ for which the section of the trajectory with $q_{M,\theta(M)}$ exists is open. Furthermore, it is closed, because the limit of section points is a section point (and segment $[0;t]$ is compact). Obviously, it is not empty. Consequently, it is the set of all possible M.

12.5.8 The Time Replacement, and the Invariance of the Projector

Let the film of nonequilibrium states be constructed as $\tilde{q}_{M,\theta}$, where relation between θ and τ is implicit; $\tau = \tau(M,\theta)$, $\theta = \theta(M,\tau)$. In order to determine these functions one needs to solve equation obtained from (12.44) with substitution $q_{M,\tau} = \tilde{q}_{M,\theta(M,\tau)}$ (and projection, because $\tilde{q}$ is only an approximation). The calculation itself presents no difficulties. However, is it possible to avoid the inversion in replacing of time for a derivation of the kinetic equations? In another words, could we use the constructed geometrical object, the film, without an exact reconstruction of the time, τ, on it?

For a positive answer to this question it is sufficient to demonstrate that the equations of motion, constructed with the projector (12.51–12.53), describe the same motion on the film after the time replacement.

This property of the π_A is evident: while deriving equations (12.51–12.53), we did not use that τ is the "true time" from the equation (12.44), and made the local replacement of variables, passing from ΔM, $\Delta \tau$ to ΔM, ΔS.

Thus, the projector π_A is invariant with respect to the time replacement, and, when constructing equations of motion, it is not necessary to restore the "true time".

Results of this and previous subsections allow to apply the sequence of operations suggested in Subsect. 12.5.6.

12.5.9 Correction to the Infinite Models

Let an infinite model $q_{M,\theta}$, $(\theta \in [0;+\infty))$, $q_{M,0} = f_M^*$ be constructed for the film. Actually, it means that an approximation is constructed for the whole film $q_{M,\tau}$ (not just for its initial segment, as it was for the finite models). Naturally, there arises a problem of correction to this approximation, and, in general, construction of a step-by-step computational procedure.

The projector π_A on the film is defined (12.50). Correspondingly, the invariance defect of the film is determined too

$$\Delta q_{M,\theta} = (1 - \pi_A|_{q_{M,\theta}}) J(q_{M,\theta})$$
$$= \left[1 - \left(1 - \frac{D_\theta q_{M,\theta} D_f S|_{q_{M,\theta}}}{D_f S|_{q_{M,\theta}} D_\theta q_{M,\theta}}\right) D_M q_{M,\theta} m\right] J(q_{M,\theta}) \quad (12.78)$$

It is easy to verify, that if $q_{M,\theta}$ is a solution to (12.44), then $\Delta q_{M,\theta} \equiv 0$.

Subsequently we calculate the corrections to $q_{M,\tau}$ using an iterative method for the manifold correction (see Chaps. 6 and 9).

Generally speaking, one could (and should) calculate these corrections also for the finite models. However, the infinite models are distinguished, because they require such corrections.

12.5.10 The Film, and the Macroscopic Equations

Let the film of nonequilibrium states be constructed. What next? There are two routes.

1. Investigation of the conservative dynamics of "$N + 1$" variables, where "N" is moments for the moments M, and "$+1$" is for the coordinate τ on the film;
2. Derivation of the macroscopic equations for M.

Actually, the second route is more desirable, it leads to familiar classes of kinetic equations. The first one, however, is always available, because the film exists always (at least formally) but the existence of equations for M is not guaranteed.

The route of obtaining equations for M is essentially the same as suggested by us [29], [30–33] following Ehrenfests [15], and Zubarev [195]. That is,

– One chooses a time T.
– For arbitrary M_0 one solves the problem of the motion on the film (12.51), (12.53) under initial conditions $M(0) = M_0$, $\tau(0) = \tau_0$ on the segment $t \in [0; T]$. The solution is $M(t, M_0)$.
– For the mapping $M_0 \to M(T)$ the system $\mathrm{d}M/\mathrm{d}t = F(M)$ is constructed. It has the property that for its phase flow, $\theta_t(M)$, the identity

$$\theta_T(M_0) \equiv M(T, M_0) \tag{12.79}$$

is satisfied. This is the method of natural projector once again (see (12.22) and Chap. 11).

In this sequence of actions there are two nontrivial problems: solution to the equations on the film, and reconstruction of the vector field by transformation of the phase flow, θ_T, under fixed T.

The natural method for solving the first problem is the averaging method. The equations of motion on the film read

$$\dot{M} = \varepsilon P(M, \tau) ; \quad \dot{\tau} = Q(M, \tau) \tag{12.80}$$

where ε is (formally) small parameter.

Assuming that the motion of M is slow, one can write down the series of the Bogoliubov-Krylov averaging method [183]. The first term of this series is a simple averaging over the period $T : \tau_1(T, M)$ is solution to the equation $\dot{\tau} = Q(M, \tau)$ under fixed M,

$$M_1(t, M_0) = M_0 + \varepsilon t \left(\frac{1}{T} \int_0^T P(M_0, \tau_1(\theta, M_0)) \, \mathrm{d}\theta \right) \qquad (12.81)$$

for $t \in [0; T]$, and

$$M_1(T, M_0) = M_0 + \varepsilon \int_0^T P(M_0, \tau(\theta, M_0)) \, \mathrm{d}\theta , \qquad (12.82)$$

correspondingly.

The first correction to the reconstruction of the vector field, $F(M)$, by the transformation of the phase flow, $\theta_T(M)$, is very simple too:

$$F_1(M) = \frac{1}{T}(\theta_T(M) - M) . \qquad (12.83)$$

Hence, we obtain the first correction to the macroscopic equations:

$$\dot{M} = F_1(M) = \frac{1}{T} \int_0^T m(J(q_{M,\tau(t,M)})) \, \mathrm{d}t , \qquad (12.84)$$

where $\tau(t, M)$ is a solution to the equation (12.53) under fixed M (actually, $mJ(q_{M,\tau})$ should be substituted into (12.53) instead of $\dot{M}$).

The second and higher approximations are much more cumbersome, but their construction is not a principal problem.

In general, the sequence of the horizon points of the second order finite Kepler models and corresponding $\dot{q}_i$, $\ddot{q}_i$ determines the macroscopic kinetic equations. Only the values of the coefficients remain unknown. Let us start from linearized in layers system (12.17)

$$\dot{f} = J(f^*_{m(f)}) + L_{m(f)}(f - f^*_{m(f)}) , \qquad (12.85)$$

where linear operator L_M parameterized by macroscopic variables $M = m(f)$. For the system (12.85) the second order finite Kepler models give the macroscopic equation

$$\dot{M} = m(J(f^*_M)) + \sum_i (\alpha_i m(L_M(\dot{q}_i)) + \beta_i m(L_M(\ddot{q}_i))) , \qquad (12.86)$$

with α_i, $\beta_i > 0$.

The final comment on the positivity of the "kinetic coefficients" α_i and β_i is important, and cannot be easily verified every time. However, in the case under consideration it is so by the following theorem.

The theorem about the positivity of kinetic constants. The motion on the Kepler ellipse from start to the horizon point always satisfies the property

$$q - q_0 = \alpha \dot{q} + \beta \ddot{q} ; \quad \alpha, \beta > 0 , \qquad (12.87)$$

where q_0 is a starting point, $\dot{q}$, and $\ddot{q}$ are the velocity, and the acceleration, correspondingly.

This theorem follows from elementary theorems about analytical geometry of second-order curves: Let a chord in an ellipse is passing through a focus, and $l_{1,2}$ are the tangents to the ellipse at the ends of this chord. Then the angle between $l_{1,2}$ that is based on the chord is acute. The starting point q_0 is one of the ends of the chord, the vector of acceleration $\ddot{q}$ is the direction of the chord (from q_0 to the focus), the velocity vector $\dot{q}$ is the tangent direction at the point q_0. Following these elementary facts, the horizon point belongs to the arc on which the angle between $l_{1,2}$ is based, hence the positivity condition (12.87) holds.

For the model motion on the entropic circle, strictly speaking, this is not always the case. Positivity of the coefficients is guaranteed only for $m(L(\dot{q}))$, and $m(L(\ddot{q}_\perp))$.

Two phenomena can be related to the increase of the number of terms in (12.86) as compared to the short-memory approximation: (i) alteration of the kinetic constants (terms are not orthogonal to each other, therefore, new terms contribute to the previous processes), (ii) birth of new processes.

Motion on an infinite film can lead to stabilization of kinetic coefficients as the functions of M, but it can also lead to their permanent transformation. In the second case one has to introduce into macroscopic equations an additional variable, the coordinate τ on the film.

From the applications point of view, another form of equations of motion on the film could be more natural. In these equations kinetic coefficients are used as dynamic variables. Essentially, this is just another representation of equations (12.51), (12.53). For every kinetic coefficient, k, expression $\mathrm{d}k/\mathrm{d}t = \psi_k(\tau, M) = \varphi_k(k, M)$ is calculated in accordance with (12.51), (12.53). Substitution of variables $(\tau, M) \to (k, M)$ in this equation is possible (at least locally) if value k does not stabilize during the motion on the film. Finally, we have the system in the form:

$$\dot{M} = m(J(f_M^*)) + \sum_j k_j F_j(M) \ ; \quad \dot{k}_j = \varphi_j(k_j, M) \ . \tag{12.88}$$

For the motion starting from the quasiequilibrium state the initial conditions are $k_j = 0$.

12.5.11 New in the Separation of the Relaxation Times

Originally, there are no dissipative possesses in the quasiequilibrium state (the theorem of preservation of the type of dynamics for the quasiequilibrium approximation).

The first thing that occurs during the motion out of the quasiequilibrium initial conditions is the emergence of the dissipation. It can be described (in the first non-vanishing approximation) by equation (12.23). It is of special importance that there is yet no separation into dissipation processes with various relaxation times and kinetic coefficients on that stage. This separation occurs at further stages: Various processes appear, their kinetic coefficients are determined (see, for example, (12.86)) (or, in certain cases, the dynamics of the kinetic coefficients is determined).

Generalizing, we can distinguish three stages:

1. birth of dissipation;
2. branching of dissipation: appearance of various processes;
3. macroscopic relaxation.

It is important to notice in this scheme that the determination of the kinetic coefficients can occur at both stages: at the second stage when macroscopic (hydrodynamic) relaxation can be described in the usual form with kinetic coefficient as functions of the macroscopic parameters, as well as in the third phase (motion on the film), when the hydrodynamic description includes dynamics of the kinetic coefficients also.

12.6 The Main Results

In order to solve the problem of irreversibility we have introduced the notion of the *macroscopically definable ensembles*. They result from the evolution of ensembles out of the quasiequilibrium initial conditions *under macroscopic control*.

Technically, the solution to the problem of irreversibility looks as follows: we can operate only with the macroscopically definable ensembles; *the class of these ensembles is not invariant with respect to the time inversion*. The notion of the macroscopically definable ensembles casts the problem of irreversibility into a new setting. It could be called a *control theory point of view*. The key question is: Which parameters can we control? These those parameters are fixed until "all the rest" come into equilibrium. The quasiequilibrium states are obtained in such a way.

A further development of this direction should lead to investigation of the macro-dynamics under controlled macro-parameters. This will be a supplement of the postulated quasiequilibrium initial conditions with an investigation of a general case of an evolution of the controlled ensembles.

The method of the natural projector allows us to construct an approximate dynamics of macro-variables. When the time of projection, τ, tends to infinity, these equations should tend to the actual equations of macro-dynamics, if the latter exist. This hypothesis about their existence in the thermodynamic limit (first, the number of particles $N \to \infty$, and after that, the time of projection $\tau \to \infty$) is the basis of Zubarev's nonequilibrium statistical operator approach [195].

Here, we need to make a remark. Frequently, physicists use mathematical objects whose existence and uniqueness are not proven: solutions to the equations of hydro- and gaso-dynamics, kinetic equations etc. Often, the failure to prove theorems of existence and uniqueness is viewed as a lack of an adequate mathematical statement of the problem (definition of spaces, etc.). For all this, it is assumed that essential obstacles either are absent, or can be sorted out separately, independently of the theorem proof in physically trivial situations. Existence (or non-existence) of the macroscopic dynamics is a problem of a different kind. The cases of non-existence can be found as frequently as the physically expected existence.

The notion of the invariant film of non-equilibrium states, and the method of its approximate construction allows us to solve the problem of macro-kinetics even when there are no autonomous equations of macro-kinetics. The existence of the film seems to be one of the physically trivial problems of existence and uniqueness of solutions. Further computations will show how productive the methods of film construction are.

The formula for *entropy production*,

$$\sigma \sim \frac{\text{defect of invariance}}{\text{curvature}}$$

clarifies the geometrical sense of the dissipation. Here, "defect of invariance" is the defect of invariance of the quasiequilibrium manifold, and "curvature" is the curvature of the film of nonequilibrium states in the direction of the defect of invariance of the quasiequilibrium manifold.

At least one essential problem remains unsolved. This is the *problem of indivisible events*: For a macroscopically small time, a small microscopic subsystems can go through "its whole life", from the beginning to the limit state (or, more accurate, to the limit behaviour which may be not only a state, but a type of motion, etc.). The microscopic evolution of the system in a small interval of the macroscopic time *cannot* be written in the form

$$\Delta f = \dot{f} \Delta t .$$

The evolution of the microscopic subsystems in a macroscopically small time Δt should be described as an *"ensemble of indivisible events"*. An excellent hint is given by the Boltzmann equation with its indivisible collisions, another good hint gives the chemical kinetics with indivisible events of elementary reactions. The useful formalism for a description such ensembles of indivisible events is well developed. It is the "quasi-chemical" representation (see Chap. 7). But the way from general system to such ensembles remains unclear and presents *the challenge* to the future works (see, however, section "Neurons and particles" in the paper [10]).

There is an important link between the theory of invariant film and the Hilbert method in the theory of the Boltzmann equation (see Chap. 2). The Hilbert method constructs the invariant film for the Boltzmann equation, and the initial manifold for this film is the local Maxwellian manifold (the local equilibrium manifold). The significant novelty of the theory of the invariant film of non-equilibrium states is the splitting of the problem in two parts: the geometrical part (construction of the film) and the dynamical part (dynamics on the film). The first (geometrical) part is solved here by the method of "large stepping" instead of a Taylor series expansion as in the original Hilbert method.

13 Slow Invariant Manifolds for Open Systems

13.1 Slow Invariant Manifold for a Closed System Has Been Found. What Next?

Suppose that the slow invariant manifold is found for a dissipative system. What have we constructed it for? First of all, for solving the Cauchy problem, in order to separate motions. This means that the Cauchy problem is divided in the following two subproblems:

- Reconstruct the "fast" motion from the initial conditions to the slow invariant manifold (*the initial layer problem*).
- Solve the Cauchy problem for the "slow" motions on the manifold.

Thus, solving the Cauchy problem becomes easier (and in some complicated cases it just becomes possible).

Let us stress here that for any sufficiently reliable solution of the Cauchy problem one must solve not only the reduced Cauchy problem for the slow motion, but also the initial layer problem for fast motions.

While solving the latter problem it was found to be surprisingly effective to use piece-wise linear approximations with smoothing or even without it [26, 27]. This method was used for the Boltzman equation.

There exists a different way to model the initial layer in kinetics problems: it is the route of model equations. For example, the Bhatnagar–Gross–Krook (BGK) equation [116] is the simplest model for the Boltzmann equation. It describes relaxation into a small neighborhood of the local Maxwell distribution. There are many types and hierarchies of the model equations [22, 112, 116, 117, 166]. The principal idea of any model equation is to replace the fast processes by a simple relaxation term. As a rule, it has a form $\mathrm{d}x/\mathrm{d}t = \ldots - (x - x_{sl}(x))/\tau$, where $x_{sl}(x)$ is a point of the approximate slow manifold. Such form is used in the BGK-equation, or in the quasi-equilibrium models [117]. It also can take a gradient form, like in the gradient models [22, 166]. These simplifications not only allows to study the fast motions separately but it also allows to zoom in the details of the interaction of fast and slow motions in the vicinity of the slow manifold.

What concerns solving the Cauchy problem for the "slow" motions, this is the basic problem of the hydrodynamics, of the gas dynamics (if the initial "big" systems describes kinetics of a gas or a fluid), etc. Here invariant

Alexander N. Gorban and Iliya V. Karlin: *Invariant Manifolds for Physical and Chemical Kinetics*, Lect. Notes Phys. **660**, 367–417 (2005)
www.springerlink.com

manifold methods provide equations for a further study. However, even a preliminary consideration of the practical aspects of these studies shows a definite shortcoming. In practice, obtained equations are exploited not only for "closed" systems. The initial equations (3.1) describe a dissipative system that approaches the equilibrium. The equations of slow motion describe dissipative system too. Then these equations are supplied with various forces and flows, and after that they describe systems with more or less complex dynamics.

Because of this, there is a different answer to our question, what have we constructed the invariant manifold for? *First of all, in order to construct models of open system dynamics in the neighborhood of the slow manifold.* Various approaches to this modeling are described in the following subsections.

13.2 Slow Dynamics in Open Systems. Zero-Order Approximation and the Thermodynamic Projector

Let the initial dissipative system (3.1),

$$\frac{\mathrm{d}x}{\mathrm{d}t} = J(x),\ x \in U\ ,$$

be "spoiled" by an additional term ("external vector field" $J_{ex}(x, t)$):

$$\frac{\mathrm{d}x}{\mathrm{d}t} = J(x) + J_{ex}(x, t), x \subset U\ . \tag{13.1}$$

For this driven system the entropy does not increase everywhere. In the system (13.1) various nontrivial dynamic effects become possible, such as a non-uniqueness of stationary states, auto-oscillations, etc. The "inertial manifold" effect is well-known: solutions of (13.1) approach some relatively low-dimensional manifold on which all the non-trivial dynamics takes place [173, 317, 318]. This "inertial manifold" can have a finite dimension even for infinite-dimensional systems, for example, for the "reaction+diffusion" systems [334].

In the theory of nonlinear control of partial differential equations systems a strategy based on the approximate inertial manifolds [342] is suggested to facilitate the construction of finite-dimensional systems of ordinary differential equations (ODE), whose solutions can be arbitrarily close to the solutions of the infinite-dimensional system [344].

It is natural to expect that the inertial manifold of the system (13.1) is located somewhere close to the slow manifold of the initial dissipative system (3.1). This hypothesis has the following motivation. Suppose that the vector field $J_{ex}(x, t)$ is sufficiently small. Let us introduce, for example, a small parameter $\varepsilon > 0$, and consider $\varepsilon J_{ex}(x,t)$ instead of $J_{ex}(x,t)$. Let us assume that for the system (3.1) a separation of motions into "slow" and "fast" takes place. In that case, there exists such an interval of positive ε that $\varepsilon J_{ex}(x,t)$ is comparable to J only in a small neighborhood of the given slow manifold of the system (3.1). Outside this neighborhood, $\varepsilon J_{ex}(x,t)$ is negligibly small in comparison with J and its influence on the motion is negligible. For this statement to be true, it is important that the system (3.1) is dissipative and every solution comes in finite time to a small neighborhood of the given slow manifold.

Precisely this perspective on the system (13.1) allows to exploit slow invariant manifolds constructed for the dissipative system (3.1) as the ansatz and the zero-order approximation in a construction of the inertial manifold of the open system (13.1). In the zero-order approximation, the right part of the equation (13.1) is simply projected onto the tangent space of the slow manifold.

The choice of the projector is determined by the motion separation which was described above, because the fast component of the vector field (13.1) is taken from the dissipative system (3.1). A projector which is suitable for all dissipative systems with the given entropy function is unique. It is constructed in the following way (detailed consideration was given above in Chap. 5 and in [10]). Let a point $x \in U$ be defined and some vector space T, on which one needs to construct a projection (T is the tangent space to the slow manifold at the point x). We introduce the entropic scalar product $\langle | \rangle_x$:

$$\langle a \mid b \rangle_x = -(a, D_x^2 S(b)) . \tag{13.2}$$

Let us consider T_0, a subspace of T, which is annulled by the differential of S at the point x

$$T_0 = \{a \in T | D_x S(a) = 0\} . \tag{13.3}$$

Suppose[1] that $T_0 \neq T$. Let $e_g \in T$, $e_g \perp T_0$ with respect to the entropic scalar product $\langle | \rangle_x$, and $D_x S(e_g) = 1$. These conditions uniquely define vector the e_g.

The projector onto T is defined by the formula

$$P(J) = P_0(J) + e_g D_x S(J) \tag{13.4}$$

where P_0 is the orthogonal projector onto T_0 with respect to the entropic scalar product $\langle | \rangle_x$. For example, if T is a finite-dimensional space, then the projector (13.4) is constructed in the following way. Let $e_1, \ldots, e_n$ be a basis in T, and for definiteness, $D_x S(e_1) \neq 0$.

(1) Let us construct a system of vectors

$$b_i = e_{i+1} - \lambda_i e_1, (i = 1, \ldots, n-1) , \tag{13.5}$$

---

[1] If $T_0 = T$, then the thermodynamic projector is the orthogonal projector on T with respect to the entropic scalar product $\langle | \rangle_x$.

where $\lambda_i = D_x S(e_{i+1})/D_x S(e_1)$, and hence $D_x S(b_i) = 0$. Thus, $\{b_i\}_1^{n-1}$ is a basis in T_0.

(2) Let us orthogonalize $\{b_i\}_1^{n-1}$ with respect to the entropic scalar product $\langle | \rangle_x$ (3.1). We thus derived an orthonormal with respect to $\langle | \rangle_x$ basis $\{g_i\}_1^{n-1}$ in T_0.

(3) We find $e_g \in T$ from the conditions:

$$\langle e_g \mid g_i \rangle_x = 0, (i = 1, \dots, n - 1), D_x S(e_g) = 1 . \tag{13.6}$$

and, finally we get

$$P(J) = \sum_{i=1}^{n-1} g_i \langle g_i \mid J \rangle_x + e_g D_x S(J) . \tag{13.7}$$

If $D_x S(T) = 0$, then the projector P is simply the orthogonal projector with respect to the $\langle | \rangle_x$ scalar product. This happens if x is the point of the global maximum of entropy (equilibrium). Then

$$P(J) = \sum_{i=1}^{n} g_i \langle g_i | J \rangle_x, \langle g_i | g_j \rangle = \delta_{ij} . \tag{13.8}$$

Remark. In applications, equation (3.1) often has additional linear balance constraints (conservation laws) such as numbers of particles, momentum, energy, etc. When solving the closed dissipative system (3.1), we simply fix the balance values and consider the dynamics of (3.1) on the corresponding affine balance subspace.

For driven system (13.1) the conservation laws can be violated by external flows and fields. Because of this, for the open system (13.1) the natural balance subspace includes the balance subspace of (3.1) with different balance values. For every set of balance values there is a corresponding equilibrium. Slow invariant manifold of the dissipative systems that is applied to the description of the driven systems (13.1) is usually the *union* of slow manifolds for all possible balance values. The equilibrium of the dissipative closed system corresponds to the entropy maximum *given the balance values are fixed*. In the phase space of the driven system (13.1) the entropy gradient in the equilibrium points of the system (3.1) is not necessarily equal to zero.

In particular, for the Boltzmann entropy in the local finite-dimensional case one gets the thermodynamic projector in the following form.

$$S = - \int f(v)(\ln(f(v)) - 1) \, dv ,$$

$$D_f S(J) = - \int J(v) \ln f(v) \, dv ,$$

$$\langle \psi \mid \varphi \rangle_f = -(\psi, D_f^2 S(\varphi)) = \int \frac{\psi(v)\varphi(v)}{f(v)} \, dv$$

$$P(J) = \sum_{i=1}^{n-1} g_i(v) \int \frac{g_i(v)J(v)}{f(v)} \, dv - e_g(v) \int J(v) \ln f(v) \, dv, \tag{13.9}$$

where $g_i(v)$ and $e_g(v)$ are constructed according to the scheme described above,

$$\int \frac{g_i(v)g_j(v)}{f(v)}\,\mathrm{d}v = \delta_{ij}\,, \tag{13.10}$$

$$\int g_i(v)\ln f(v)\,\mathrm{d}v = 0\,, \tag{13.11}$$

$$\int g_i(v)e_g(v)\,\mathrm{d}v = 0\,, \tag{13.12}$$

$$\int e_g(v)\ln f(v)\,\mathrm{d}v = 1\,. \tag{13.13}$$

If for all $g \in T$ we have $\int g(v)\ln f(v)\,\mathrm{d}v = 0$, then the projector P is defined as the orthogonal projector with respect to the $\langle\,|\,\rangle_f$ scalar product.

13.3 Slow Dynamics in Open Systems. First-Order Approximation

The thermodynamic projector (13.4) defines the duality of slow and fast motions: if T is the tangent space of the slow manifold, then $T = \mathrm{im}P$, and $\ker P$ is the plane of fast motions. Let us denote by P_x the projector at a point x of a given slow manifold.

The vector field $J_{ex}(x,t)$ can be decomposed in two components:

$$J_{ex}(x,t) = P_x J_{ex}(x,t) + (1 - P_x)J_{ex}(x,t)\,. \tag{13.14}$$

Let us denote $J_{ex\,s} = P_x J_{ex}$, $J_{ex\,f} = (1 - P_x)J_{ex}$. The slow component $J_{ex\,s}$ gives a correction to the motion along the slow manifold. This is a zero-order approximation. The "fast" component shifts the slow manifold in the fast motions plane. This shift changes $P_x J_{ex}$ accordingly. Consideration of this effect gives a first-order approximation. In order to find it, let us rewrite the invariance equation taking J_{ex} into account:

$$\begin{cases} (1 - P_x)(J(x + \delta x) + \varepsilon J_{ex}(x,t)) = 0\,; \\ P_x\delta x = 0\,. \end{cases} \tag{13.15}$$

The first iteration of the Newton method subject to incomplete linearization gives:

$$\begin{cases} (1 - P_x)(D_x J(\delta x) + \varepsilon J_{ex}(x,t)) = 0\,; \\ P_x\delta x = 0\,. \end{cases} \tag{13.16}$$

$$(1 - P_x)D_x J(1 - P_x)J(\delta x) = -\varepsilon J_{ex}(x,t)\,. \tag{13.17}$$

Thus, we have derived a linear equation in the space $\ker P_x$. The operator $(1 - P_x)D_x J(1 - P_x)$ is defined in this space.

Taking into account of the self-adjoint linearization instead of the operator $D_x J$ (see Chap. 7) considerably simplifies solving and studying equation (13.17). It is necessary to take into account here that the projector P_x is a sum of the orthogonal projector with respect to the entropic scalar product $\langle | \rangle_x$ and a projector of rank one.

Assume that the first-order approximation equation (13.17) has been solved and the following function is found:

$$\delta_1 x(x, \varepsilon J_{ex\,f}) = -[(1 - P_x)D_x J(1 - P_x)]^{-1} \varepsilon J_{ex\,f} , \qquad (13.18)$$

where $D_x J$ is either the differential of J or symmetrized differential of J (7.17).

Let x be a point on the initial slow manifold. At the point $x + \delta x(x, \varepsilon J_{ex\,f})$ the right-hand side of equation (13.1) in the first-order approximation is given by

$$J(x) + \varepsilon J_{ex}(x, t) + D_x J(\delta x(x, \varepsilon J_{ex\,f})) . \qquad (13.19)$$

Due to the first-order approximation (13.19), the motion projected onto the manifold is given by the following equation

$$\frac{dx}{dt} = P_x(J(x) + \varepsilon J_{ex}(x, t) + D_x J(\delta x(x, \varepsilon J_{ex\,f}(x, t)))) . \qquad (13.20)$$

Note that in equation (13.20), the vector field $J(x)$ enters only in the form of projection, $P_x J(x)$. For the invariant slow manifold it holds $P_x J(x) = J(x)$, but actually we always deal with approximately invariant manifolds, hence, it is necessarily to use the projection $P_x J$ instead of J in (13.20).

Remark. The notion "projection of a point onto the manifold" needs to be specified. For every point x of the slow invariant manifold Ω there are defined both the thermodynamic projector P_x (13.4) and the fast motions plane $\ker P_x$. Let us define a projector Π of some neighborhood of *motion* onto *motion* in the following way:

$$\Pi(z) = x, \quad \text{if } P_x(z - x) = 0 . \qquad (13.21)$$

Qualitatively, it means that z, after all fast motions were completed, comes into a small neighborhood of x. The operation (13.4) is defined uniquely in some small neighborhood of the manifold *motion*.

A derivation of slow motion equations requires not only an assumption that εJ_{ex} is small but it must be slow as well: $\frac{d}{dt}(\varepsilon J_{ex})$ must be small too.

One can get further approximations for slow motions of the system (13.1), taking into account the time derivatives of J_{ex}. This approach is considered in a more detail in the following Example for a particularly interesting driven system of dilute polymeric solutions. A short description of the scheme is given in the next section. That is an alternative to the using the projection operators methods [194].

13.4 Beyond the First-Order Approximation: Higher-Order Dynamic Corrections, Stability Loss and Invariant Manifold Explosion

Let us pose formally the invariance problem for the driven system (13.1) in the neighborhood of the slow manifold Ω of the initial (dissipative) system.

Let for a given neighborhood of Ω an operator Π (13.21) be defined. One needs to define the function $\delta x(x, \ldots) = \delta x(x, J_{ex}, \dot{J}_{ex}, \ddot{J}_{ex}, \ldots)$, $x \in \Omega$, with the following properties:

$$P_x(\delta x(x, \ldots)) = 0 \,,$$
$$J(x + \delta x(x, \ldots)) + J_{ex}(x + \delta x(x, \ldots), t)$$
$$= \dot{x}_{sl} + D_x \delta x(x, \ldots)\dot{x}_{sl} + \sum_{n=0}^{\infty} D_{J_{ex}^{(n)}} \delta x(x, \ldots) J_{ex}^{(n+1)} \,, \qquad (13.22)$$

where

$$\dot{x}_{sl} = P_x(J(x + \delta x(x, \ldots)) + J_{ex}(x + \delta x(x, \ldots), t)), \quad J_{ex}^{(n)} = \frac{d^n J_{ex}}{dt^n} \,,$$

$D_{J_{ex}^{(n)}} \delta x(x, \ldots)$ is a partial differential of the function

$$\delta x(x, J_{ex}, \dot{J}_{ex}, \ddot{J}_{ex}, \ldots, J_{ex}^{(n)}, \ldots)$$

with respect to the variable $J_{ex}^{(n)}$. One can rewrite equations (13.22) in the following form:

$$(1 - P_x - D_x \delta x(x, \ldots))(J(x + \delta x(x, \ldots)) + J_{ex}(x + \delta x(x, \ldots), t))$$
$$= \sum_{n=0}^{\infty} D_{J_{ex}^{(n)}} \delta x(x, \ldots) J_{ex}^{(n+1)}. \qquad (13.23)$$

For solving (13.23) one can use iterations method and also take into account smallness considerations. The series in the right hand side of equation (13.23) can be rewritten as

$$\text{R.H.S.} = \sum_{n=0}^{k-1} \varepsilon^{n+1} D_{J_{ex}^{(n)}} \delta x(x, \ldots) J_{ex}^{(n+1)} \qquad (13.24)$$

at the kth iteration, considering the terms only to order less than k. The first iteration equation was solved in the previous section. On the second iteration one gets the following equation:

$$(1 - P_x - D_x \delta_1 x(x, J_{ex}))(J(x + \delta_1 x(x, J_{ex})))$$
$$+ D_z J(z)|_{z=x+\delta_1 x(x, J_{ex})} \cdot (\delta_2 x - \delta_1 x(x, J_{ex})) + J_{ex})$$
$$= D_{J_{ex}} \delta_1 x(x, J_{ex}) \dot{J}_{ex} \,. \qquad (13.25)$$

This is a linear equation with respect to $\delta_2 x$. The solution $\delta_2 x(x, J_{ex}, \dot{J}_{ex})$ depends linearly on $\dot{J}_{ex}$, but non-linearly on J_{ex}. Let us remind that the first iteration equation solution depends linearly on J_{ex}.

In all these iteration equations the field J_{ex} and its derivatives are included in the formulas as if they were functions of time t only. Indeed, for any solution $x(t)$ of the equations (13.1) $J_{ex}(x, t)$ can be substituted for $J_{ex}(x(t), t)$. The function $x(t)$ will be a solution of the system (13.1) in which $J_{ex}(x, t)$ is substituted for $J_{ex}(t)$ in this way.

However, in order to obtain the macroscopic equations (13.20) one must return to $J_{ex}(x, t)$. For the first iteration such return is quite simple as one can see from (13.19). There $J_{ex}(x, t)$ is calculated in points of the initial slow manifold. In the general case, suppose that $\delta x = \delta x(x, J_{ex}, \dot{J}_{ex}, \ldots, J_{ex}^{(k)})$ has been found. The equations for x (13.20) have the following form:

$$\frac{\mathrm{d}x}{\mathrm{d}t} = P_x(J(x + \delta x) + J_{ex}(x + \delta x, t)) . \tag{13.26}$$

In these equations the shift δx must be a function of x and t (or a function of x, t, α, where α are external fields, see example below. One calculates the shift $\delta x(x, t)$ using the following equation:

$$J_{ex} = J_{ex}(x + \delta x(x, J_{ex}, \dot{J}_{ex}, \ldots, J_{ex}^{(k)}), t) . \tag{13.27}$$

It can be solved, for example, by the iterative method, taking $J_{ex0} = J_{ex}(x, t)$:

$$J_{ex(n+1)} = J_{ex}(x + \delta x(x, J_{ex(n)}, \dot{J}_{ex(n)}, \ldots, J_{ex(n)}^{(k)}), t) . \tag{13.28}$$

We hope that using J_{ex} in the equations (13.27) and (13.28) both as a variable and as a symbol of an unknown function $J_{ex}(x, t)$ will not lead to a confusion.

In all the constructions introduced above it was assumed that δx is sufficiently small and the driven system (13.1) will not deviate too far from the slow invariant manifold of the initial system. However, a stability loss is possible: solutions of the equation (13.1) can deviate arbitrarily far if the strength of the perturbations exceeds a certain level. The invariant manifold can loose it's stability. Qualitatively, this effect of *invariant manifold explosion* can be represented as follows.

Suppose that J_{ex} includes the parameter ε: one has εJ_{ex} in the equation (13.1). When ε is small, the system's motions are located in a small neighborhood of the initial manifold. This neighborhood grows monotonically with increase of ε, but after some ε_0 a sudden change happens ("explosion") and the neighborhood, in which the motion takes place, becomes significantly wider at $\varepsilon > \varepsilon_0$ than at $\varepsilon < \varepsilon_0$. The stability loss is not necessarily associated with the invariance loss. In the last example to this chapter it is shown how the invariant manifold (which is at the same time the quasiequilibrium manifold in this example) can loose its stability. This "explosion" of the invariant manifold leads to essential physical consequences.

13.5 Example: The Universal Limit in Dynamics of Dilute Polymeric Solutions

The method of invariant manifold is developed for a derivation of reduced description in kinetic equations of dilute polymeric solutions. It is demonstrated that this reduced description becomes universal in the limit of small Deborah and Weissenberg numbers, and it is represented by the (revised) Oldroyd 8 constants constitutive equation for the polymeric stress tensor. Coefficients of this constitutive equation are expressed in terms of the microscopic parameters. A systematic procedure of corrections to the revised Oldroyd 8 constants equations is developed. Results are tested with simple flow situations.

Kinetic equations arising in the theory of polymer dynamics constitute a wide class of microscopic models of complex fluids. Same as in any branch of kinetic theory, the problem of reduced description becomes actual as soon as the kinetic equation is established. However, in spite of an enormous amount of work in the field of polymer dynamics [151–153, 354, 364], this problem remains less studied as compared to other classical kinetic equations.

It is the purpose of this section to suggest a systematic approach to the problem of reduced description for kinetic models of polymeric fluids. First, we would like to specify our motivation by comparing the problem of the reduced description for that case with a similar problem in the familiar case of the rarefied gas obeying the classical Boltzmann kinetic equation [70, 112].

The problem of reduced description begins with establishing a set of slow variables. For the Boltzmann equation, this set is represented by five hydrodynamic fields (density, momentum and energy) which are low-order moments of the distribution function, and which are conserved quantities of the dissipation process due to particle's collisions. The reduced description is a closed system of equations for these fields. One starts with the manifold of local equilibrium distribution functions (local Maxwellians), and finds a correction by the Chapman–Enskog method [70]. The resulting reduced description (the compressible Navier–Stokes hydrodynamic equations) is universal in the sense that the form of equations does not depend on details of particle's interaction whereas the latter shows up explicitly only in the transport coefficients (viscosity, temperature conductivity, etc.).

Coming back to the complex fluids, we shall consider the case of dilute polymer solutions represented by dumbbell models studied below. Two obstacles preclude an application of the traditional techniques. First, the question which variables should be regarded as slow is at least less evident because the dissipative dynamics in the dumbbell models has no nontrivial conservation laws as compared to the Boltzmann case. Consequently, a priori, there are no distinguished manifolds of distribution functions like the local equilibria which can be regarded as a starting point. Second, while the Boltzmann kinetic equation provides a self-consistend closed description, the dumbbell kinetic equations are coupled to the hydrodynamic equations. This coupling manifests itself as an external flux in the kinetic equation.

The distinguished macroscopic variable associated with the polymer kinetic equations is the polymeric stress tensor [151, 364]. This variable is not a conserved quantity but nevertheless it should be treated as a relevant slow variable because it actually contributes to the macroscopic (hydrodynamic) equations. Equations for the stress tensor are known as the constitutive equations, and the problem of reduced description for the dumbbell models consists in deriving such equations from the kinetic equation.

Our approach is based on the method of invariant manifold [11], modified for systems coupled with external fields. This method suggests constructing invariant sets (or manifolds) of distribution functions that represent the asymptotic states of the slow evolution of the kinetic system. In the case of dumbbell models, the reduced description is produced by equations which constitute stress-strain relations, and two physical requirements are met by our approach: The first is the principle of *frame-indifference* with respect to any time-dependent reference frame. This principle requires that the resulting equations for the stresses contain only frame-indifferent quantities. For example, the frame-dependent vorticity tensor should not show up in these equations unless being presented in the frame-indifferent combinations with another tensors. The second principle is the *thermodynamic stability*. In the absence of the flow, the constitutive model should be purely dissipative, in other words, it should describe the relaxation of the stresses to their equilibrium values.

The physical picture addressed below takes into account two assumptions: (i) In the absence of the flow, deviations from the equilibrium are small. Then the invariant manifold is represented by eigenvectors corresponding to the slowest relaxation modes. (ii). When the external flow is taken into account, it is assumed to cause a small deformation of the invariant manifolds of the purely dissipative dynamics. Two characteristic parameters are necessary to describe this deformation. The first is the characteristic time variation of the external field. The second is the characteristic intensity of the external field. For dumbbell models, the first parameter is associated with the conventional Deborah number while the second one is usually called the Weissenberg number. An iteration approach which involves these parameters is developed.

The two main results of the study are as follows: First, the lowest-order constitutive equations with respect to the characteristic parameters mentioned above has the form of the revised phenomenological *Oldroyd 8 constants model*. This result is interpreted as the macroscopic limit of the microscopic dumbbell dynamics whenever the rate of the strain is low, and the Deborah number is small. This limit is valid generically, in the absence or in the presence of the hydrodynamic interaction, and for the arbitrary nonlinear elastic force. The phenomenological constants of the Oldroyd model are expressed in a closed form in terms of the microscopic parameters of the model.

The universality of this limit is similar to that of the Navier–Stokes equations which are the macroscopic limit of the Boltzmann equation at small Knudsen numbers for arbitrary hard-core molecular interactions. The test calculation for the nonlinear FENE force demonstrates a good quantitative agreement of the constitutive equations with solutions to the microscopic kinetic equation within the domain of their validity.

The second result is a regular procedure of finding corrections to the zero-order model. These corrections extend the model into the domain of higher strain rates, and to flows which alternate faster in time. Same as in the zero-order approximation, the higher-order corrections are linear in the stresses, while their dependence on the gradients of the flow velocity and its time derivatives becomes highly nonlinear.

The section is organized as follows: For the sake of completeness, we present the nonlinear dumbbell kinetic models in the next subsection, "The problem of reduced description in polymer dynamics". In the section, "The method of invariant manifold for weakly driven systems", we describe in details our approach to the derivation of macroscopic equations for an abstract kinetic equation coupled to external fields. This derivation is applied to the dumbbell models in the section, "Constitutive equations". The zero-order constitutive equation is derived and discussed in detail in this section, as well as the structure of the first correction. Tests of the zero-order constitutive equation for simple flow problems are given in the section, "Tests on the FENE dumbbell model".

13.5.1 The Problem of Reduced Description in Polymer Dynamics

Elastic Dumbbell Models

The elastic dumbbell model is the simplest microscopic model of polymer solutions [151]. It dumbbell reflects the two basic features of the real-world macromolecules to be orientable and stretchable by a flowing solvent. The polymeric solution is represented by a set of identical elastic dumbbells placed in an isothermal incompressible fluid. In this example we adopt notations used in kinetic theory of polymer dynamics [151]. Let $\boldsymbol{Q}$ be the connector vector between the beads of a dumbbell, and $\Psi(\boldsymbol{x}, \boldsymbol{Q}, t)$ be the configuration distribution function which depends on the location in the space $\boldsymbol{x}$ at time t. We assume that dumbbells are distributed uniformly, and consider the normalization, $\int \Psi(\boldsymbol{x}, \boldsymbol{Q}, t) \, \mathrm{d}\boldsymbol{Q} = 1$. The Brownian motion of beads in the physical space causes a diffusion in the phase space described by the Fokker–Planck equation (FPE) [151]:

$$\frac{\mathrm{D}\Psi}{\mathrm{D}t} = -\frac{\partial}{\partial \boldsymbol{Q}} \cdot \boldsymbol{k} \cdot \boldsymbol{Q}\Psi + \frac{2k_\mathrm{B}T}{\xi} \frac{\partial}{\partial \boldsymbol{Q}} \cdot \boldsymbol{D} \cdot \left(\frac{\partial}{\partial \boldsymbol{Q}}\Psi + \frac{\boldsymbol{F}}{k_\mathrm{B}T}\Psi \right) . \tag{13.29}$$

Here, $\mathrm{D}/\mathrm{D}t = \partial/\partial t + \boldsymbol{v} \cdot \nabla$ is the substantional derivative, ∇ is the spatial gradient, $\boldsymbol{k}(\boldsymbol{x}, t) = (\nabla \boldsymbol{v})^\dagger$ is the gradient of the velocity of the solvent $\boldsymbol{v}$, † denotes transposition of tensors, $\boldsymbol{D}$ is the dimensionless diffusion matrix, k_{B} is the Boltzmann constant, T is the temperature, ξ is the dimensional coefficient characterizing a friction exerted on beads moving through solvent media (the friction coefficient [151, 152]), and $\boldsymbol{F} = \partial\phi/\partial\boldsymbol{Q}$ is the elastic spring force defined by the potential ϕ. We consider forces of the form $\boldsymbol{F} = Hf(Q^2)\boldsymbol{Q}$, where $f(Q^2)$ is a dimensionless function of the variable $Q^2 = \boldsymbol{Q} \cdot \boldsymbol{Q}$, and H is the dimensional constant. Incompressibility of solvent implies $\sum_i k_{ii} = 0$.

Let us introduce a time dimensional constant

$$\lambda_{\mathrm{r}} = \frac{\xi}{4H} \, ,$$

which coincides with a characteristic relaxation time of dumbbell configuration in the case when the force $\boldsymbol{F}$ is linear: $f(Q^2) = 1$. It proves convenient to rewrite the FPE (13.29) in the dimensionless form:

$$\frac{D\Psi}{D\widehat{t}} = -\frac{\partial}{\partial\widehat{\boldsymbol{Q}}} \cdot \widehat{\mathbf{k}} \cdot \widehat{\boldsymbol{Q}}\Psi + \frac{\partial}{\partial\widehat{\boldsymbol{Q}}} \cdot \boldsymbol{D} \cdot \left(\frac{\partial}{\partial\widehat{\boldsymbol{Q}}}\Psi + \widehat{\boldsymbol{F}}\Psi\right) . \tag{13.30}$$

Various dimensionless quantities used are: $\widehat{\boldsymbol{Q}} = (H/k_{\mathrm{B}}T)^{1/2}\boldsymbol{Q}$, $D/D\widehat{t} = \partial/\partial\widehat{t} + \boldsymbol{v} \cdot \overline{\nabla}$, $\widehat{t} = t/\lambda_{\mathrm{r}}$ is the dimensionless time, $\overline{\nabla} = \lambda_{\mathrm{r}}\nabla$ is the reduced space gradient, and $\widehat{\mathbf{k}} = \boldsymbol{k}\lambda_{\mathrm{r}} = (\overline{\nabla}\boldsymbol{v})^{\dagger}$ is the dimensionless tensor of the gradients of the velocity. In the sequel, only dimensionless quantities $\widehat{\boldsymbol{Q}}$ and $\widehat{\boldsymbol{F}}$ are used, and we keep notations $\boldsymbol{Q}$ and $\boldsymbol{F}$ for them for the sake of simplicity.

The quantity of interest is the stress tensor introduced by Kramers [151]:

$$\boldsymbol{\tau} = -\nu_{\mathrm{s}}\dot{\boldsymbol{\gamma}} + nk_{\mathrm{B}}T(\mathbf{1} - \langle\boldsymbol{FQ}\rangle) , \tag{13.31}$$

where ν_{s} is the viscosity of the solvent, $\dot{\boldsymbol{\gamma}} = \boldsymbol{k} + \boldsymbol{k}^{\dagger}$ is the rate-of-strain tensor, n is the concentration of polymer molecules, and the angle brackets stand for the averaging with the distribution function Ψ: $\langle\bullet\rangle \equiv \int \bullet\Psi(\boldsymbol{Q})\,\mathrm{d}\boldsymbol{Q}$. The tensor

$$\boldsymbol{\tau}_{\mathrm{p}} = nk_{\mathrm{B}}T(\mathbf{1} - \langle\boldsymbol{FQ}\rangle) \tag{13.32}$$

gives a contribution to the stresses caused by the presence of polymer molecules.

The stress tensor is required in order to write down a closed system of hydrodynamic equations:

$$\frac{D\boldsymbol{v}}{Dt} = -\rho^{-1}\nabla p - \nabla \cdot \boldsymbol{\tau}[\Psi] . \tag{13.33}$$

Here p is the pressure, and $\rho = \rho_{\mathrm{s}} + \rho_{\mathrm{p}}$ is the mass density of the solution where ρ_{s} is the solvent, and ρ_{p} is the polymeric contributions.

Several models of the elastic force are known in the literature. The Hookean law is relevant to small perturbations of the equilibrium configuration of the macromolecule:

$$F = Q \; . \tag{13.34}$$

In that case, the differential equation for τ is easily derived from the kinetic equation, and is the well known *Oldroyd–B* constitutive model [151].

Another model, the Finitely Extendible Nonlinear Elastic (FENE) force law [355], was derived as an approximation to the inverse Langevin force law [151] for a more realistic description of the elongation of a polymeric molecule in a solvent:

$$F = \frac{Q}{1 - Q^2/Q_0^2} \; . \tag{13.35}$$

This force law takes into account the nonlinear stiffness and the finite extendibility of dumbbells, where Q_0 is the maximal extendibility.

The properties of the diffusion matrix are important for both the microscopic and the macroscopic behavior. The isotropic diffusion is represented by the simplest diffusion matrix

$$D_{\mathrm{I}} = \frac{1}{2}\mathbf{1} \; . \tag{13.36}$$

Here $\mathbf{1}$ is the unit matrix. When the hydrodynamic interaction between the beads is taken into account, this results in an anisotropic contribution to the diffusion matrix (13.36). The original form of this contribution is the Oseen-Burgers tensor D_{H} [356, 357]:

$$D = D_{\mathrm{I}} - \kappa D_{\mathrm{H}} \; , \qquad D_{\mathrm{H}} = \frac{1}{Q}\left(1 + \frac{QQ}{Q^2}\right) \; , \tag{13.37}$$

where

$$\kappa = \left(\frac{H}{k_{\mathrm{B}}T}\right)^{1/2}\frac{\xi}{16\pi\nu_{\mathrm{s}}} \; .$$

Several modifications of the Oseen-Burgers tensor can be found in the literature (the *Rotne-Prager-Yamakawa* tensor [358, 359]), but here we consider only the classical version.

Properties of the Fokker–Planck Operator

Let us review some of the properties of the Fokker–Planck operator J in the right hand side of (13.30) relevant to what will follow. This operator can be written as $J = J_{\mathrm{d}} + J_{\mathrm{h}}$, and it represents two processes.

The first term, J_{d}, is the dissipative part,

$$J_{\mathrm{d}} = \frac{\partial}{\partial Q}\cdot D\cdot\left(\frac{\partial}{\partial Q} + F\right) \; . \tag{13.38}$$

This part is responsible for the diffusion and friction which affect internal configurations of dumbbells, and it drives the system to the unique equilibrium state,

$$\Psi_{\mathrm{eq}} = c^{-1} \exp(-\phi(Q^2)) ,$$

where $c = \int \exp(-\phi) \, \mathrm{d}Q$ is the normalization constant.

The second part, J_h, describes the hydrodynamic drag of the beads in the flowing solvent:

$$J_\mathrm{h} = -\frac{\partial}{\partial Q} \cdot \hat{\mathbf{k}} \cdot Q . \tag{13.39}$$

The dissipative nature of the operator J_d is reflected by its spectrum. We assume that this spectrum consists of real-valued nonpositive eigenvalues, and that the zero eigenvalue is not degenerated. In the sequel, the following scalar product will be useful:

$$\langle g, h \rangle_\mathrm{s} = \int \Psi_{\mathrm{eq}}^{-1} gh \, \mathrm{d}Q .$$

The operator J_d is symmetric and nonpositive definite in this scalar product:

$$\langle J_\mathrm{d}g, h \rangle_\mathrm{s} = \langle g, J_\mathrm{d}h \rangle_\mathrm{s}, \quad \text{and} \quad \langle J_\mathrm{d}g, g \rangle_\mathrm{s} \leq 0 . \tag{13.40}$$

Since

$$\langle J_\mathrm{d}g, g \rangle_\mathrm{s} = -\int \Psi_{\mathrm{eq}}^{-1} (\partial g/\partial Q) \cdot \Psi_{\mathrm{eq}} D \cdot (\partial g/\partial Q) \, \mathrm{d}Q ,$$

the above inequality is valid if the diffusion matrix D is positive semidefinite. This happens if $D = D_\mathrm{I}$ (13.36) but is not generally valid in the presence of the hydrodynamic interaction (13.37). Let us split the operator J_d according to the splitting of the diffusion matrix D:

$$J_\mathrm{d} = J_\mathrm{d}^\mathrm{I} - \kappa J_\mathrm{d}^\mathrm{H}, \quad \text{where } J_\mathrm{d}^\mathrm{I,H} = \partial/\partial Q \cdot D_\mathrm{I,H} \cdot (\partial/\partial Q + F) .$$

Both the operators J_d^I and J_d^H have nondegenerated eigenvalue 0 which corresponds to their common eigenfunction Ψ_{eq}: $J_\mathrm{d}^\mathrm{I,H} \Psi_{\mathrm{eq}} = 0$, while the rest of the spectrum of both operators belongs to the nonpositive real semi-axis. Then the spectrum of the operator $J_\mathrm{d} = J_\mathrm{d}^\mathrm{I} - \kappa J_\mathrm{d}^\mathrm{H}$ remains nonpositive for sufficiently small values of the parameter κ. The spectral properties of both operators $J_\mathrm{d}^\mathrm{I,H}$ depend only on the choice of the spring force F. Thus, in the sequel we assume that the hydrodynamic interaction parameter κ is sufficiently small so that the *thermodynamic stability* property (13.40) holds.

We note that the scalar product $\langle \bullet, \bullet \rangle_\mathrm{s}$ coincides with the second differential $D_\Psi^2 S|_{\Psi_{\mathrm{eq}}}$ of an entropy functional $S[\Psi]$:

$$\langle \bullet, \bullet \rangle_\mathrm{s} = -D_\Psi^2 S|_{\Psi_{\mathrm{eq}}} [\bullet, \bullet] ,$$

where the entropy has the form:

$$S[\Psi] = -\int \Psi \ln \left(\frac{\Psi}{\Psi_{\mathrm{eq}}} \right) \mathrm{d}Q = -\left\langle \ln \left(\frac{\Psi}{\Psi_{\mathrm{eq}}} \right) \right\rangle . \tag{13.41}$$

The entropy S grows in the course of dissipation:

$$D_\Psi S[J_\mathrm{d}\Psi] \geq 0 .$$

This inequality, similar to inequality (13.40), is satisfied for sufficiently small κ. Symmetry and nonpositiveness of operator J_d in the scalar product defined by the second differential of the entropy is a common property of linear dissipative systems.

Statement of the Problem

Given the kinetic equation (13.29), we aim at deriving differential equations for the stress tensor $\boldsymbol{\tau}$ (13.31). The latter includes the moments $\langle \boldsymbol{FQ} \rangle = \int \boldsymbol{FQ}\Psi \, \mathrm{d}\boldsymbol{Q}$.

In general, when the diffusion matrix is non-isotropic and/or the spring force is nonlinear, closed equations for these moments are not available, and approximations are required. With this, any derivation should be consistent with the three requirements:

(i)   *Dissipativity or thermodynamic stability:* The macroscopic dynamics should be dissipative in the absence of the flow.
(ii) *Slowness*: The macroscopic equations should represent the slow degrees of freedom of the kinetic equation.
(iii) *Material frame indifference*: The form of equations for the stresses should be invariant with respect to the Eucluidian, time dependent transformations of the reference frame [151, 360].

While these three requirements should be met by any approximate derivation, the validity of our approach will be restricted by two additional assumptions:

(a) Let us denote θ_1 the inertial time of the flow, which we define via the characteristic value of the gradient of the flow velocity: $\theta_1 = |\nabla \boldsymbol{v}|^{-1}$, and θ_2 the characteristic time of the variation of the flow velocity. We assume that the characteristic relaxation time of the molecular configuration θ_r is small as compared to both the characteristic times θ_1 and θ_2:

$$\theta_\mathrm{r} \ll \theta_1 \text{ and } \theta_\mathrm{r} \ll \theta_2 . \tag{13.42}$$

(b) In the absence of the flow, the initial deviation of the distribution function from the equilibrium is small so that the linear approximation is valid.

While the assumption (b) is merely of a technical nature, and it is intended to simplify the treatment of the dissipative part of the Fokker–Planck operator (13.38) for elastic forces of a complicated form, the assumption (a) is crucial for taking into account the flow in an adequate way. We have assumed that the two parameters characterizing the composed system 'relaxing polymer configuration + flowing solvent' should be small: These two parameters are:

$$\varepsilon_1 = \theta_r/\theta_1 \ll 1 , \quad \varepsilon_2 = \theta_r/\theta_2 \ll 1 . \tag{13.43}$$

The characteristic relaxation time of the polymeric configuration is defined via the coefficient λ_r: $\theta_r = c\lambda_r$, where c is some positive dimensionless constant which is estimated by the absolute value of the lowest nonzero eigenvalue of the operator J_d. The first parameter ε_1 is usually termed the *Weissenberg number* while the second one ε_2 is the *Deborah number* ([361], Sect. 7.2).

13.5.2 The Method of Invariant Manifold for Weakly Driven Systems

The Newton Iteration Scheme

In this section we introduce an extension of the method of invariant manifold [11] onto systems coupled with external fields. We consider a class of dynamic systems of the form

$$\frac{d\Psi}{dt} = J_d\Psi + J_{ex}(\alpha)\Psi , \tag{13.44}$$

where J_d is a linear operator representing the dissipative part of the dynamic vector field, while $J_{ex}(\alpha)$ is a linear operator which represents an external flux and depends on a set of external fields $\alpha = \{\alpha_1, \ldots, \alpha_k\}$. Parameters α are either known functions of the time, $\alpha = \alpha(t)$, or they obey a set of equations,

$$\frac{d\alpha}{dt} = \Phi(\Psi, \alpha) . \tag{13.45}$$

Without any restriction, parameters α are adjusted in such a way that $J_{ex}(\alpha = 0) \equiv 0$. Kinetic equation (13.30) has the form (13.44), and general results of this section will be applied to the dumbbell models below in a straightforward way.

We assume that the vector field $J_d\Psi$ has the same dissipative properties as the Fokker–Planck operator (13.38). Namely there exists the globally convex entropy function S which obeys: $D_\Psi S[J_d\Psi] \geq 0$, and the operator J_d is symmetric and nonpositive in the scalar product $\langle \bullet, \bullet \rangle_s$ defined by the second differential of the entropy: $\langle g, h \rangle_s = -D_\Psi^2 S[g, h]$. Thus, the vector field $J_d\Psi$ drives the system irreversibly to the unique equilibrium state Ψ_{eq}.

We consider a set of n real-valued functionals, $M_i^*[\Psi]$ (macroscopic variables), in the phase space $\mathcal{F}$ of the system (13.44). A macroscopic description is obtained once we have derived a closed set of equations for the variables M_i^*.

Our approach is based on constructing a relevant invariant manifold in the phase space $\mathcal{F}$. This manifold is thought as a finite-parametric set of solutions $\Psi(M)$ to equations (13.44) which depends on time implicitly via the n variables $M_i[\Psi]$. The latter may differ from the macroscopic variables M_i^*. For systems with external fluxes (13.44), we assume that the invariant manifold depends also on the parameters α, and on their time derivatives taken to arbitrary order: $\Psi(M, \mathcal{A})$, where $\mathcal{A} = \{\alpha, \alpha^{(1)}, \dots\}$ is the set of time derivatives $\alpha^{(k)} = \mathrm{d}^k \alpha / \mathrm{d}t^k$. It is convenient to consider time derivatives of α as independent parameters. This assumption is important because then we do not need an explicit form of (13.45) in the course of construction of the invariant manifold.

By the definition, the dynamic invariance postulates the equality of the "macroscopic" and the "microscopic" time derivatives:

$$J\Psi(M, \mathcal{A}) = \sum_{i=1}^{n} \frac{\partial \Psi(M, \mathcal{A})}{\partial M_i} \frac{\mathrm{d}M_i}{\mathrm{d}t} + \sum_{n=0}^{\infty} \sum_{j=1}^{k} \frac{\partial \Psi(M, \mathcal{A})}{\partial \alpha_j^{(n)}} \alpha_j^{(n+1)}, \qquad (13.46)$$

where $J = J_{\mathrm{d}} + J_{\mathrm{ex}}(\alpha)$. The time derivatives of the macroscopic variables, $\mathrm{d}M_i/\mathrm{d}t$, are calculated as follows:

$$\frac{\mathrm{d}M_i}{\mathrm{d}t} = D_\Psi M_i [J\Psi(M, \mathcal{A})], \qquad (13.47)$$

where $D_\Psi M_i$ stands for differentials of the functionals M_i.

Let us introduce the projector operator associated with the parameterization of the manifold $\Psi(M, \mathcal{A})$ by the values of the functionals $M_i[\Psi]$.:

$$P_M = \sum_{i=1}^{n} \frac{\partial \Psi(M, \mathcal{A})}{\partial M_i} D_\Psi M_i[\bullet] \qquad (13.48)$$

It projects vector fields from the phase space $\mathcal{F}$ onto the tangent space $T\Psi(M, \mathcal{A})$ of the manifold $\Psi(M, \mathcal{A})$. Then (13.46) is rewritten as the *invariance equation*:

$$(1 - P_M)J\Psi(M, \mathcal{A}) = \sum_{n=0}^{\infty} \sum_{j=1}^{k} \frac{\partial \Psi}{\partial \alpha_j^{(n)}} \alpha_j^{(n+1)}, \qquad (13.49)$$

which has the invariant manifolds as its solutions.

Furthermore, we assume the following: (i). The external flux $J_{\mathrm{ex}}(\alpha)\Psi$ is small in comparison to the dissipative part $J_{\mathrm{d}}\Psi$, i.e. with respect to some norm we require:

$$\|J_{\mathrm{ex}}(\alpha)\Psi\| \ll \|J_{\mathrm{d}}\Psi\|.$$

This allows us to introduce a small parameter ε_1, and to replace the operator J_{ex} with $\varepsilon_1 J_{\mathrm{ex}}$ in (13.44). Parameter ε_1 is proportional to the characteristic value of the external variables α. (ii). The characteristic time θ_α of the variation of the external fields α is large in comparison to the characteristic relaxation time θ_{r}, and the second small parameter is $\varepsilon_2 = \theta_{\mathrm{r}}/\theta_\alpha \ll 1$. The parameter ε_2 does not enter the vector field J explicitly but it shows up in the invariance equation. Indeed, with a substitution, $\alpha^{(i)} \rightarrow \varepsilon_2^i \alpha^{(i)}$, the invariance equation (13.46) is rewritten in a form which incorporates both the parameters ε_1 and ε_2:

$$(1 - P_M)\{J_d + \varepsilon_1 J_{ex}\}\Psi = \varepsilon_2 \sum_i \sum_{j=1}^k \frac{\partial \Psi}{\partial \alpha_j^{(i)}} \alpha_j^{(i+1)} \qquad (13.50)$$

We develop a modified Newton scheme for solution of this equation. Let us assume that we have some initial approximation to desired manifold $\Psi_{(0)}$. We seek the correction of the form $\Psi_{(1)} = \Psi_{(0)} + \Psi_1$. Substituting this expression into (13.50), we derive:

$$(1 - P_M^{(0)})\{J_d + \varepsilon_1 J_{ex}\}\Psi_1 - \varepsilon_2 \sum_i \sum_{j=1}^k \frac{\partial \Psi_1}{\partial \alpha_j^{(i)}} \alpha_j^{(i+1)} =$$

$$-(1 - P_M^{(0)})J\Psi_{(0)} + \varepsilon_2 \sum_i \sum_{j=1}^k \frac{\partial \Psi_{(0)}}{\partial \alpha_j^{(i)}} \alpha_j^{(i+1)}. \qquad (13.51)$$

Here $P_M^{(0)}$ is a projector onto tangent bundle of the manifold $\Psi_{(0)}$. Further, we neglect two terms in the left hand side of this equation, which are multiplied by parameters ε_1 and ε_2, regarding them small in comparison to the first term. In the result we arrive at the equation,

$$(1 - P_M^{(0)})J_d \Psi_1 = -(1 - P_M^{(0)})J\Psi_{(0)} + \varepsilon_2 \sum_i \sum_{j=1}^k \frac{\partial \Psi_{(0)}}{\partial \alpha_j^{(i)}} \alpha_j^{(i+1)}. \qquad (13.52)$$

For $(n + 1)$-th iteration we obtain:

$$(1 - P_M^{(n)})J_d \Psi_{n+1} = -(1 - P_M^{(0)})J\Psi_{(n)} + \varepsilon_2 \sum_i \sum_{j=1}^k \frac{\partial \Psi_{(n)}}{\partial \alpha_j^{(i)}} \alpha_j^{(i+1)}, \qquad (13.53)$$

where $\Psi_{(n)} = \sum_{i=0}^n \Psi_i$ is the approximation of n-th order and $P_M^{(n)}$ is the projector onto its tangent bundle.

It should be noted that deriving equations (13.52) and (13.53) we have not varied the projector P_M with respect to yet unknown term Ψ_{n+1}, i.e. we have kept $P_M = P_M^{(n)}$ and have neglected the contribution from the term Ψ_{n+1}. The motivation for this action comes from the original paper [11], where it was shown that such modification generates iteration schemes properly converging to slow invariant manifold.

In order to gain the solvability of (13.53) an additional condition is required:

$$P_M^{(n)} \Psi_{n+1} = 0. \qquad (13.54)$$

This condition is sufficient to provide the existence of the solution to linear system (13.53), while the additional restriction onto the choice of the projector is required in order to guarantee the uniqueness of the solution. This condition is

$$\ker[(1 - P_M^{(n)})J_d] \cap \ker P_M^{(n)} = \mathbf{0} \ . \tag{13.55}$$

Here ker denotes a null space of the corresponding operator. How this condition can be met is discussed in the next subsection.

It is natural to begin the iteration procedure (13.53) starting from the invariant manifold of the non-driven system. In other words, we choose the initial approximation $\Psi_{(0)}$ as the solution of the invariance equation (13.50) corresponding to $\varepsilon_1 = 0$ and $\varepsilon_2 = 0$:

$$(1 - P_M^{(0)})J_d\Psi_{(0)} = 0 \ . \tag{13.56}$$

We shall return to the question how to construct solutions to this equation in the subsection "Linear zero-order equations".

The above recurrent equations (13.53), (13.54) present the Newton method for the solution of invariance equation (13.50), which involves the small parameters. A similar procedure for the Grad equations of the Boltzmann kinetic theory was used recently in [21]. When these parameters are not small, one should proceed directly with equations (13.51).

Above, we have focused our attention on how to organize the iterations to construct invariant manifolds of weakly driven systems. The only question we have not yet answered is how to choose the projectors in iterative equations in a consistent way. In the next subsection we discuss the problem of derivation of the reduced dynamics and its relation to the problem of the choice of the projector.

Projector and Reduced Dynamics

Below we suggest the projector which is equally applicable for constructing invariant manifolds by the iteration method (13.53), (13.54) and for generating macroscopic equations on a given manifold.

Let us discuss the problem of constructing closed equations for macroparameters. Having some approximation to the invariant manifold, we nevertheless deal with a non-invariant manifold and we face the problem how to construct the dynamics on it. If the n-dimensional manifold $\tilde{\Psi}$ is found then the macroscopic dynamics is induced by a projector P onto the tangent bundle of $\tilde{\Psi}$ as follows [11]:

$$\frac{\mathrm{d}M_i^*}{\mathrm{d}t} = D_\Psi M_i^*\big|_{\tilde{\Psi}} \left[PJ\tilde{\Psi}\right] \ . \tag{13.57}$$

In order to specify the projector we apply the two above mentioned principles: dissipativity and slowness. The dissipativity is required to have the unique and stable equilibrium solution for macroscopic equations when the external fields are absent ($\alpha = 0$). The slowness condition requires the *induced* vector field $PJ\Psi$ to match the slow modes of the original vector field $J\Psi$.

Let us consider the parameterization of the manifold $\widetilde{\Psi}(M)$ by the parameters $M_i[\Psi]$. This parameterization generates associated projector $P = P_M$ by (13.48). This leads us to look for the admissible parameterization of this manifold, where by admissibility we understand the concordance with the dissipativity and the slowness requirements. We solve the problem of the admissible parameterization in the following way. Let us define the functionals M_i $i = 1, \ldots, n$ by the set of the eigenvectors φ_i of the operator J_{d}:

$$M_i[\widetilde{\Psi}] = \langle \varphi_i, \widetilde{\Psi} \rangle_{\mathrm{s}} \ ,$$

where $J_{\mathrm{d}}\varphi_i = \lambda_i \varphi_i$. The eigenvectors $\varphi_1, \ldots, \varphi_n$ are taken as a union of orthonormal basises in the eigenspaces corresponding to the eigenvalues with smallest absolute values: $0 < |\lambda_1| \leq |\lambda_2| \leq \ldots \leq |\lambda_n|$, $\langle \varphi_i, \varphi_j \rangle_{\mathrm{s}} = \delta_{ij}$. Since the function Ψ_{eq} is the eigenvector corresponding to the eigenvalue zero, we have: $M_i[\Psi_{\mathrm{eq}}] = \langle \varphi_i, \Psi_{\mathrm{eq}} \rangle_{\mathrm{s}} = 0$.

The associated projector P_M,

$$P_M = \sum_{i=1}^{n} \frac{\partial \widetilde{\Psi}}{\partial M_i} \langle \varphi_i, \bullet \rangle_{\mathrm{s}} \ , \tag{13.58}$$

generates the equations of the macroscopic dynamics in terms of the parameters M_i:

$$\mathrm{d}M_i/\mathrm{d}t = \langle \varphi_i P_M J\widetilde{\Psi} \rangle_{\mathrm{s}} = \langle \varphi_i J\widetilde{\Psi} \rangle_{\mathrm{s}} \ .$$

Their explicit form is

$$\frac{\mathrm{d}M_i}{\mathrm{d}t} = \lambda_i M_i + \langle J_{\mathrm{ex}}^+(\alpha) g_i, \widetilde{\Psi}(M) \rangle_{\mathrm{s}} \ , \tag{13.59}$$

where the J_{ex}^+ is the adjoint to operator J_{ex} with respect to the scalar product $\langle \bullet, \bullet \rangle_{\mathrm{s}}$.

Apparently, in the absence of forcing ($\alpha \equiv 0$) the macroscopic equations $\mathrm{d}M_i/\mathrm{d}t = \lambda_i M_i$ are thermodynamically stable. They represent the dynamics of the slowest eigenmodes the of the equation $\mathrm{d}\Psi/\mathrm{d}t = J_{\mathrm{d}}\Psi$. Thus, the projector (13.58) complies with the requirements of dissipativity and slowness in the absence the external flow.

In order to rewrite the macroscopic equations (13.59) in terms of the required set of macroparameters, $M_i^*[\Psi] = \langle m_i^*, \Psi \rangle_{\mathrm{s}}$, we use the formula (13.57) which is equivalent to the change of variables $\{M\} \to \{M^*(M)\}$, $M_i^* = \langle m_i^*, \widetilde{\Psi}(M) \rangle_{\mathrm{s}}$ in the equations (13.59). Indeed, this is seen from the relation:

$$D_\Psi M_i^* \big|_{\tilde{\Psi}} \left[P_M J\widetilde{\Psi} \right] = \sum_j \frac{\partial M_i^*}{\partial M_j} D_\Psi M_j \big|_{\tilde{\Psi}} [J\widetilde{\Psi}] \ .$$

We have constructed the dynamics with the help of the projector P_M associated with the lowest eigenvectors of the operator J_d. It is directly verified that such projector (13.58) fulfills the condition (13.54) for arbitrary manifold $\Psi_{(n)} = \tilde{\Psi}$. For this reason it is natural to use the projector (13.58) for both procedures, constructing the invariant manifold, and deriving the macroscopic equations.

We note that the above approach to defining the dynamics via the spectral projector is a specification of the concept of "thermodynamic parameterization" proposed in [9, 11].

13.5.3 Linear Zero-Order Equations

In this section we focus our attention on the solution of the zero-order invariance equation (13.56). We seek the linear invariant manifold of the form

$$\Psi_{(0)}(a) = \Psi_{\text{eq}} + \sum_{i=1}^{n} a_i m_i \, , \tag{13.60}$$

where a_i are coordinates on this manifold. This manifold can be considered as an expansion of the relevant slow manifold near the equilibrium state. This limits the domain of validity of the manifolds (13.60) because they may be not positively definite. This remark indicates that nonlinear invariant manifolds should be considered for large deviations from the equilibrium but this goes beyond the scope of this Example.

The linear n-dimensional manifold representing the slow motion for the linear dissipative system (13.44) is associated with the n slowest eigenmodes. This manifold should be built up as the linear hull of the eigenvectors φ_i of the operator J_d, corresponding to the lower part of its spectrum. Thus we choose $m_i = \varphi_i$.

Dynamic equations for the macroscopic variables M^* are derived in two steps. First, following the subsection, "Projector and reduced dynamics", we parameterize the linear manifold $\Psi_{(0)}$ with the values of the moments $M_i[\Psi] = \langle \varphi_i, \Psi \rangle_s$. We obtain the parameterization of the manifold (13.60) in terms of $a_i = M_i$,

$$\Psi_{(0)}(M) = \Psi_{\text{eq}} + \sum_{i=1}^{n} M_i \varphi_i \, ,$$

The reduced dynamics in terms of variables M_i reads:

$$\frac{dM_i}{dt} = \lambda_i M_i + \sum_j \langle J_{\text{ex}}^+ \varphi_i, \varphi_j \rangle_s M_j + \langle J_{\text{ex}}^+ \varphi_i, \Psi_{\text{eq}} \rangle_s \, , \tag{13.61}$$

where $\lambda_i = \langle \varphi_i, J_d \varphi_i \rangle_s$ are eigenvalues which correspond to eigenfunctions φ_i.

Second, we switch from the variables M_i to the variables $M_i^*(M) = \langle m_i^*, \Psi_{(0)}(M) \rangle_s$ in (13.61). Resulting equations for the variables M^* are also linear:

$$\frac{dM_i^*}{dt} = \sum_{jkl} (B^{-1})_{ij} \Lambda_{jk} B_{kl} \Delta M_l^* + \sum_{jk} (B^{-1})_{ij} \langle J_{ex}^+ \varphi_j, \varphi_k \rangle_s \Delta M_k^*$$
$$+ \sum_j (B^{-1})_{ij} \langle J_{ex}^+ \varphi_j, \Psi_{eq} \rangle_s . \tag{13.62}$$

Here, $\Delta M_i^* = M_i^* - M_{eq|i}^*$ is the deviation of the variable M_i^* from its equilibrium value $M_{eq|i}^*$, and $B_{ij} = \langle m_i^*, \varphi_j \rangle$, and $\Lambda_{ij} = \lambda_i \delta_{ij}$.

13.5.4 Auxiliary Formulas. 1. Approximations to Eigenfunctions of the Fokker–Planck Operator

In this subsection we discuss the question how to find the lowest eigenvectors $\Psi_{eq} m_0(Q^2)$ and $\Psi_{eq} m_1(Q^2) \overset{\circ}{\mathbf{Q}\mathbf{Q}}$ of the operator J_d (13.38) in the classes of functions of the form: $w_0(Q)$ and $w_1(Q) \overset{\circ}{\mathbf{Q}\mathbf{Q}}$. The results presented in this subsection will be used below in the subsections: "Constitutive equations" and "Tests on the FENE dumbbell model".

It is directly verified that:

$$J_d w_0 = G_0^h w_0 ,$$
$$J_d w_1 \overset{\circ}{\mathbf{Q}\mathbf{Q}} = (G_1^h w_1) \overset{\circ}{\mathbf{Q}\mathbf{Q}} ,$$

where the operators G_0^h and G_1^h are given by:

$$G_0^h = G_0 - \kappa H_0 , \qquad G_1^h = G_1 - \kappa H_1 . \tag{13.63}$$

The operators $G_{0,1}$ and $H_{0,1}$ act in the space of isotropic functions (i.e. dependent only on $Q = (\mathbf{Q} \cdot \mathbf{Q})^{1/2}$ as follows:

$$G_0 = \frac{1}{2} \left(\frac{\partial^2}{\partial Q^2} - fQ \frac{\partial}{\partial Q} + \frac{2}{Q} \frac{\partial}{\partial Q} \right) , \tag{13.64}$$

$$G_1 = \frac{1}{2} \left(\frac{\partial^2}{\partial Q^2} - fQ \frac{\partial}{\partial Q} + \frac{6}{Q} \frac{\partial}{\partial Q} - 2f \right) , \tag{13.65}$$

$$H_0 = \frac{2}{Q} \left(\frac{\partial^2}{\partial Q^2} - fQ \frac{\partial}{\partial Q} + \frac{2}{Q} \frac{\partial}{\partial Q} \right) , \tag{13.66}$$

$$H_1 = \frac{2}{Q} \left(\frac{\partial^2}{\partial Q^2} - fQ \frac{\partial}{\partial Q} + \frac{5}{Q} \frac{\partial}{\partial Q} - 2f + \frac{1}{Q^2} \right) . \tag{13.67}$$

The following two properties of the operators $G_{0,1}^h$ are important for our analysis: Let us define two scalar products $\langle \bullet, \bullet \rangle_0$ and $\langle \bullet, \bullet \rangle_1$:

$$\langle y, x \rangle_0 = \langle xy \rangle_e ,$$

$$\langle y, x \rangle_1 = \langle xy Q^4 \rangle_e .$$

Here, $\langle \bullet \rangle_e$ is the equilibrium average as defined in (13.80). For sufficiently small κ the operators G_0^h and G_1^h are symmetric and nonpositive in the scalar products $\langle \bullet, \bullet \rangle_0$ and $\langle \bullet, \bullet \rangle_1$ respectively. Thus, for obtaining the desired eigenvectors of the operator J_d we need to find the eigenfunctions m_0 and m_1 related to the lowest nonzero eigenvalues of the operators $G_{0,1}^h$.

Since we regard the parameter κ small it is convenient, first, to find the lowest eigenfunctions $g_{0,1}$ of the operators $G_{0,1}$ and, second, to use the standard perturbation technique in order to obtain $m_{0,1}$. For the first-order perturbation one finds [367]:

$$m_0 = g_0 + \kappa h_0 , \quad h_0 = -g_0 \frac{\langle g_0 H_0 G_0 g_0 \rangle_0}{\langle g_0, g_0 \rangle_0} - G_0 H_0 g_0 ;$$

$$m_1 = g_1 + \kappa h_1 , \quad h_1 = -g_1 \frac{\langle g_1 H_1 G_1 g_1 \rangle_1}{\langle g_1, g_1 \rangle_1} - G_1 H_1 g_1 . \qquad (13.68)$$

For the rest of this subsection we describe one recurrent procedure for obtaining the functions m_0 and m_1 in a constructive way. Let us solve this problem by minimizing the functionals $Lambda_{0,1}$:

$$\Lambda_{0,1}[m_{0,1}] = -\frac{\langle m_{0,1}, G_{0,1}^h m_{0,1} \rangle_{0,1}}{\langle m_{0,1}, m_{0,1} \rangle_{0,1}} \longrightarrow \min , \qquad (13.69)$$

by means of the *gradient descent method*.

Let us denote $e_{0,1}$ the eigenfunctions of the zero eigenvalues of the operators $G_{0,1}^h$, $e_0 = 1$ and $e_1 = 0$. Let the initial approximations $m_{0,1}^{(0)}$ to the lowest eigenfunctions $m_{0,1}$ be so chosen that $\langle m_{0,1}^{(0)}, e_{0,1} \rangle_{0,1} = 0$. We define the variational derivative $\delta \Lambda_{0,1} / \delta m_{0,1}$ and look for the correction in the form:

$$m_{0,1}^{(1)} = m_{0,1}^{(0)} + \delta m_{0,1}^{(0)} , \quad \delta m_{0,1}^{(0)} = \alpha \frac{\delta \Lambda_{0,1}}{\delta m_{0,1}} , \qquad (13.70)$$

where scalar parameter $\alpha < 0$ is found from the condition:

$$\frac{\partial \Lambda_{0,1}[m_{0,1}^{(1)}(\alpha)]}{\partial \alpha} = 0 .$$

In the explicit form the result reads:

$$\delta m_{0,1}^{(0)} = \alpha_{0,1}^{(0)} \Delta_{0,1}^{(0)} ,$$

where

$$\Delta_{0,1}^{(0)} = \frac{2}{\langle m_{0,1}^{(0)}, m_{0,1}^{(0)}\rangle_{0,1}} \left(m_{0,1}^{(0)}\lambda_{0,1}^{(0)} - G_{0,1}^{\mathrm{h}}m_{0,1}^{(0)}\right) , \tag{13.71}$$

$$\lambda_{0,1}^{(0)} = \frac{\langle m_{0,1}^{(0)}, G_{0,1}^{\mathrm{h}}m_{0,1}^{(0)}\rangle_{0,1}}{\langle m_{0,1}^{(0)}, m_{0,1}^{(0)}\rangle_{0,1}} ,$$

$$\alpha_{0,1}^{(0)} = q_{0,1} - \sqrt{q_{0,1}^2 + \frac{\langle m_{0,1}^{(0)}, m_{0,1}^{(0)}\rangle_{0,1}}{\langle \Delta_{0,1}^{(0)}, \Delta_{0,1}^{(0)}\rangle_{0,1}}} ,$$

$$q_{0,1} = \frac{1}{\langle \Delta_{0,1}^{(0)}, \Delta_{0,1}^{(0)}\rangle_{0,1}} \left(\frac{\langle m_{0,1}^{(0)}, G_{0,1}^{\mathrm{h}}m_{0,1}^{(0)}\rangle_{0,1}}{\langle m_{0,1}^{(0)}, m_{0,1}^{(0)}\rangle_{0,1}} - \frac{\langle \Delta_{0,1}^{(0)}, G_{0,1}^{\mathrm{h}}\Delta_{0,1}^{(0)}\rangle_{0,1}}{\langle \Delta_{0,1}^{(0)}, \Delta_{0,1}^{(0)}\rangle_{0,1}} \right) .$$

With the new correction $m_{0,1}^{(1)}$, we can repeat the procedure and eventually generate recurrence scheme. Since by the construction all iterative approximations $m_{0,1}^{(n)}$ remain orthogonal to the zero eigenfunctions $e_{0,1}$, $\langle m_{0,1}^{(n)}, e_{0,1}\rangle_{0,1} = 0$ we avoid the convergence of this recurrence procedure to the eigenfunctions $e_{0,1}$. (Note that this method resembles the relaxation method, Chap. 9.)

The quantities $\delta_{0,1}^{(n)}$:

$$\delta_{0,1}^{(n)} = \frac{\langle \Delta_{0,1}^{(n)}, \Delta_{0,1}^{(n)}\rangle_{0,1}}{\langle m_{0,1}^{(n)}, m_{0,1}^{(n)}\rangle_{0,1}}$$

can serve as a relative error for controlling the convergence of the iteration procedure (13.70).

13.5.5 Auxiliary Formulas. 2. Integral Relations

Let Ω be a sphere in $\mathbf{R}^3$ centered at the origin, or the entire space $\mathbf{R}^3$. For any function $s(x^2)$, where $x^2 = \boldsymbol{x} \cdot \boldsymbol{x}$, $\boldsymbol{x} \in \mathbf{R}^3$, and any 3×3 matrices $\boldsymbol{A}$, $\boldsymbol{B}$, $\boldsymbol{C}$ independent of $\boldsymbol{x}$, the following integral relations are valid:

$$\int_\Omega s(x^2)\, \overset{\circ}{\boldsymbol{xx}}\, (\overset{\circ}{\boldsymbol{xx}} : \boldsymbol{A})\, \mathrm{d}\boldsymbol{x} = \frac{2}{15}\, \overset{\circ}{\boldsymbol{A}} \int_\Omega sx^4\, \mathrm{d}\boldsymbol{x} ;$$

$$\int_\Omega s(x^2)\, \overset{\circ}{\boldsymbol{xx}}\, (\overset{\circ}{\boldsymbol{xx}} : \boldsymbol{A})(\overset{\circ}{\boldsymbol{xx}} : \boldsymbol{B})\, \mathrm{d}\boldsymbol{x} = \frac{4}{105}\, (\boldsymbol{A} \cdot \boldsymbol{B} \overset{\circ}{+ \boldsymbol{B}} \cdot \boldsymbol{A}) \int_\Omega sx^6\, \mathrm{d}\boldsymbol{x} ;$$

$$\int_\Omega s(x^2)\, \overset{\circ}{\boldsymbol{xx}}\, (\overset{\circ}{\boldsymbol{xx}} : \boldsymbol{A})(\overset{\circ}{\boldsymbol{xx}} : \boldsymbol{B})(\overset{\circ}{\boldsymbol{xx}} : \boldsymbol{C})\, \mathrm{d}\boldsymbol{x} =$$

$$\frac{4}{315} \left\{ \overset{\circ}{\boldsymbol{A}} (\boldsymbol{B} : \boldsymbol{C}) \overset{\circ}{+ \boldsymbol{B}} (\boldsymbol{A} : \boldsymbol{C}) \overset{\circ}{+ \boldsymbol{C}} (\boldsymbol{A} : \boldsymbol{B}) \right\} \int_\Omega sx^8\, \mathrm{d}\boldsymbol{x} .$$

13.5.6 Microscopic Derivation of Constitutive Equations

Iteration Scheme

In this section we apply the above developed formalism to the elastic dumb-bell model (13.30). External field variables α are the components of the tensor $\widehat{\mathbf{k}}$.

Since we aim at constructing a closed description for the stress tensor τ (13.31) with the six independent components, the relevant manifold in the problem should be six-dimensional. Moreover, we allow a dependence of the manifold on the material derivatives of the tensor $\widehat{\mathbf{k}}$: $\widehat{\mathbf{k}}^{(i)} = D^i\mathbf{k}/Dt^i$. Let $\Psi^*(M, \mathcal{K})$ $\mathcal{K} = \{\widehat{\mathbf{k}}, \widehat{\mathbf{k}}^{(1)}, \ldots\}$ be the desired manifold parameterized by the six variables M_i $i = 1, \ldots, 6$ and the independent components (maximum eight for each $\widehat{\mathbf{k}}^{(l)}$) of the tensors $\widehat{\mathbf{k}}^{(l)}$. Small parameters ε_1 and ε_2, introduced in the section: "The problem of reduced description in polymer dynamics", are established by (13.43). We then write the invariance equation:

$$(1 - P_M)(J_d + \varepsilon_1 J_h)\Psi = \varepsilon_2 \sum_{i=0}^{\infty} \sum_{lm} \frac{\partial \Psi}{\partial \widehat{k}_{lm}^{(i)}} \widehat{k}_{lm}^{(i+1)} , \qquad (13.72)$$

where $P_M = (\partial\Psi/\partial M_i)D_\Psi M_i[\bullet]$ is the projector associated with chosen parameterization and summation indexes l, m run only eight independent components of tensor $\widehat{\mathbf{k}}$.

Following the further steps of the procedure we obtain the recurrent equations:

$$(1 - P_M^{(n)})J_d\Psi_{n+1} = -(1 - P_M^{(n)})[J_d + \varepsilon_1 J_h]\Psi_{(n)} + \varepsilon_2 \sum_i \sum_{lm} \frac{\partial\Psi_{(n)}}{\partial \widehat{k}_{lm}^{(i)}} \widehat{k}_{lm}^{(i+1)} ,$$

$$\qquad (13.73)$$

$$P_M^{(n)}\Psi_{n+1} = 0 , \qquad (13.74)$$

where Ψ_{n+1} is the correction to the manifold $\Psi_{(n)} = \sum_{i=0}^{n} \Psi_i$.

The zero-order manifold is found as the relevant solution to the equation:

$$(1 - P_M^{(0)})J_d\Psi_{(0)} = 0 \qquad (13.75)$$

We construct zero-order manifold $\Psi_{(0)}$ in the subsection, "Zero-order constitutive equation".

The Dynamics in the General Form

Let us assume that some approximation to invariant manifold $\widetilde{\Psi}(a, \mathcal{K})$ is found (here $a = \{a_1, \ldots, a_6\}$ are some coordinates on this manifold). The next step is the constructing of the macroscopic dynamic equations.

In order to comply with dissipativity and slowness by means of the recipe from the previous section, we need to find six lowest eigenvectors of the operator J_{d}. We shall always assume in the sequel that the hydrodynamic interaction parameter κ is small enough so that the dissipativity of J_{d} (13.40) is not violated.

Let us consider two classes of functions: $\mathcal{C}_1 = \{w_0(Q^2)\}$ and $\mathcal{C}_2 = \{w_1(Q^2)\,\overset{\circ}{\mathbf{QQ}}\}$, where $w_{0,1}$ are functions of Q^2 and the notation $\circ$ indicates the traceless part, e.g. for the dyad $\mathbf{QQ}$:

$$(\overset{\circ}{\mathbf{QQ}})_{ij} = Q_i Q_j - \frac{1}{3}\delta_{ij}Q^2 \ .$$

Since the sets $\mathcal{C}_1$ and $\mathcal{C}_2$ are invariant with respect to operator J_{d}, i.e. $J_{\mathrm{d}}\mathcal{C}_1 \subset \mathcal{C}_1$ and $J_{\mathrm{d}}\mathcal{C}_2 \subset \mathcal{C}_2$, and densities $\mathbf{FQ} = f\,\overset{\circ}{\mathbf{QQ}} + (1/3)\mathbf{1}fQ^2$ of the moments comprising the stress tensor $\boldsymbol{\tau}_p$ (13.32) belong to the space $\mathcal{C}_1 \oplus \mathcal{C}_2$, we shall seek the desired eigenvectors in the classes $\mathcal{C}_1$ and $\mathcal{C}_2$. Namely, we intend to find one lowest isotropic eigenvector $\Psi_{\mathrm{eq}}m_0(Q^2)$ of the eigenvalue $-\lambda_0$ ($\lambda_0 > 0$) and five nonisotropic eigenvectors $m_{ij} = \Psi_{\mathrm{eq}}m_1(Q^2)(\overset{\circ}{\mathbf{QQ}})_{ij}$ corresponding to another eigenvalue $-\lambda_1$ ($\lambda_1 > 0$). The method of derivation and analytic evaluation of these eigenvalues were discussed in the subsection "Auxiliary formulas, 1". For now we assume that these eigenvectors are known.

In the next step we parameterize the given manifold $\widetilde{\Psi}$ by the values of the functionals:

$$M_0 = \langle \Psi_{\mathrm{eq}}m_0, \widetilde{\Psi}\rangle_{\mathrm{s}} = \int m_0 \widetilde{\Psi}\,\mathrm{d}\mathbf{Q}\ ,$$

$$\overset{\circ}{\mathbf{M}} = \langle \Psi_{\mathrm{eq}}m_1\,\overset{\circ}{\mathbf{QQ}}, \widetilde{\Psi}\rangle_{\mathrm{s}} = \int m_1\,\overset{\circ}{\mathbf{QQ}}\,\widetilde{\Psi}\,\mathrm{d}\mathbf{Q}\ . \tag{13.76}$$

Once the desired parameterization $\widetilde{\Psi}(M_0, \overset{\circ}{\mathbf{M}}, \mathcal{K})$ is obtained, the dynamic equations are found as:

$$\frac{DM_0}{D\widehat{t}} + \lambda_0 M_0 = \left\langle (\overset{\circ}{\hat{\boldsymbol{\gamma}}} : \overset{\circ}{\mathbf{QQ}})m_0' \right\rangle \tag{13.77}$$

$$\overset{\circ}{\mathbf{M}}_{[1]} + \lambda_1\,\overset{\circ}{\mathbf{M}} = -\frac{1}{3}\mathbf{1}\hat{\boldsymbol{\gamma}} : \overset{\circ}{\mathbf{M}} - \frac{1}{3}\hat{\boldsymbol{\gamma}}\langle m_1 Q^2\rangle + \left\langle \overset{\circ}{\mathbf{QQ}}\,(\overset{\circ}{\hat{\boldsymbol{\gamma}}} : \overset{\circ}{\mathbf{QQ}})m_1' \right\rangle\ ,$$

where all averages are calculated with the distribution function $\widetilde{\Psi}$, i.e. $\langle \bullet \rangle = \int \bullet\,\widetilde{\Psi}\,\mathrm{d}\mathbf{Q}$, $m_{0,1}' = \mathrm{d}m_{0,1}(Q^2)/\mathrm{d}(Q^2)$ and the subscript [1] represents the upper convective derivative of a tensor:

$$\boldsymbol{\Lambda}_{[1]} = \frac{D\boldsymbol{\Lambda}}{D\widehat{t}} - \left\{\widehat{\mathbf{k}}\cdot\boldsymbol{\Lambda} + \boldsymbol{\Lambda}\cdot\widehat{\mathbf{k}}^{\dagger}\right\}\ .$$

The parameters $\lambda_{0,1}$, which are the absolute values of eigenvalues of the operator J_{d}, are calculated from the formulas (for the definition of operators G_1 and G_2 see subsection "Auxiliary formulas, 1"):

$$\lambda_0 = -\frac{\langle m_0 G_0 m_0 \rangle_e}{\langle m_0 m_0 \rangle_e} > 0 \,, \tag{13.78}$$

$$\lambda_1 = -\frac{\langle Q^4 m_1 G_1 m_1 \rangle_e}{\langle m_1 m_1 Q^4 \rangle_e} > 0 \,, \tag{13.79}$$

where we have introduced the notation for the equilibrium average:

$$\langle y \rangle_e = \int \Psi_{eq} y \, d\mathbf{Q} \,. \tag{13.80}$$

Equations for the components of the polymeric stress tensor $\boldsymbol{\tau}_p$ (13.32) are constructed as a change of variables $\{M_0, \overset{\circ}{\mathbf{M}}\} \rightarrow \boldsymbol{\tau}_p$. The use of the projector $\widetilde{P}$ makes this operation straightforward:

$$\frac{D\boldsymbol{\tau}_p}{D\widehat{t}} = -nk_B T \int \mathbf{F}\mathbf{Q}\widetilde{P}J\widetilde{\Psi}(M_0(\boldsymbol{\tau}_p, \mathcal{K}), \overset{\circ}{\mathbf{M}}(\boldsymbol{\tau}_p, \mathcal{K}), \mathcal{K}) \, d\mathbf{Q} \,. \tag{13.81}$$

Here, the projector $\widetilde{P}$ is associated with the parameterization by the variables M_0 and $\overset{\circ}{\mathbf{M}}$:

$$\widetilde{P} = \frac{\partial \widetilde{\Psi}}{\partial M_0} \langle \Psi_{eq} m_0, \bullet \rangle_s + \sum_{kl} \frac{\partial \widetilde{\Psi}}{\partial \overset{\circ}{\mathbf{M}}_{kl}} \langle \Psi_{eq} m_1 (\mathbf{QQ})_{kl}, \bullet \rangle_s \,. \tag{13.82}$$

We note that sometimes it is easier to make a transition to the variables $\boldsymbol{\tau}_p$ after solving the equations (13.77) rather than to construct explicitly and solve equations in terms of $\boldsymbol{\tau}_p$. This allows to avoid inverting the functions $\boldsymbol{\tau}_p(M_0, \overset{\circ}{\mathbf{M}})$ and to deal with simpler equations.

Zero-Order Constitutive Equation

In this subsection we derive the closed constitutive equations based on the zero-order manifold $\Psi_{(0)}$ found as the appropriate solution to (13.75). Following the approach described in subsection, "Linear zero-order equations", we construct such a solution as the linear expansion near the equilibrium state Ψ_{eq} (13.60). After parameterization by the values of the variables M_0 and $\overset{\circ}{\mathbf{M}}$ associated with the eigenvectors $\Psi_{eq} m_0$ and $\Psi_{eq} m_1 \overset{\circ}{\mathbf{QQ}}$ we find:

$$\Psi_{(0)} = \Psi_{eq} \left(1 + M_0 \frac{m_0}{\langle m_0 m_0 \rangle_e} + \frac{15}{2} \overset{\circ}{\mathbf{M}} : \overset{\circ}{\mathbf{QQ}} \frac{m_1}{\langle m_1 m_1 Q^4 \rangle_e} \right) \,. \tag{13.83}$$

With the help of the projector (13.82):

$$P_M^{(0)} = \Psi_{eq} \left\{ \frac{m_0}{\langle m_0 m_0 \rangle_e} \langle m_0, \bullet \rangle_e + \frac{15}{2} \frac{m_1}{\langle m_1 m_1 Q^4 \rangle_e} \overset{\circ}{\mathbf{QQ}} : \langle m_1 \overset{\circ}{\mathbf{QQ}}, \bullet \rangle_e \right\}$$
$$\tag{13.84}$$

and using the formula (13.81) we obtain:

$$\frac{D\mathrm{tr}\boldsymbol{\tau}_\mathrm{p}}{D\hat{t}} + \lambda_0 \mathrm{tr}\boldsymbol{\tau}_\mathrm{p} = a_0 \left(\overset{\circ}{\boldsymbol{\tau}}_\mathrm{p} : \hat{\boldsymbol{\gamma}}\right), \tag{13.85}$$

$$\overset{\circ}{\boldsymbol{\tau}}_{\mathrm{p}[1]} + \lambda_0 \overset{\circ}{\boldsymbol{\tau}}_\mathrm{p} = b_0 \left[\overset{\circ}{\boldsymbol{\tau}}_\mathrm{p} \cdot \hat{\boldsymbol{\gamma}} + \hat{\boldsymbol{\gamma}} \cdot \overset{\circ}{\boldsymbol{\tau}}_\mathrm{p}\right] - \frac{1}{3}\mathbf{1}(\overset{\circ}{\boldsymbol{\tau}}_\mathrm{p} : \hat{\boldsymbol{\gamma}}) + (b_1 \mathrm{tr}\boldsymbol{\tau}_\mathrm{p} - b_2 nk_\mathrm{B}T)\hat{\boldsymbol{\gamma}},$$

where the constants b_i, a_0 are defined by the following equilibrium averages:

$$a_0 = \frac{\langle fm_0 Q^2 \rangle_\mathrm{e}\langle m_0 m_1 Q^4 m_1' \rangle_\mathrm{e}}{\langle fm_0 Q^4 \rangle_\mathrm{e} \langle m_0^2 \rangle_\mathrm{e}},$$

$$b_0 = \frac{2}{7}\frac{\langle m_1 m_2' Q^6 \rangle_\mathrm{e}}{\langle m_1^2 Q^4 \rangle_\mathrm{e}},$$

$$b_1 = \frac{1}{15}\frac{\langle fm_1 Q^4 \rangle_\mathrm{e}}{\langle fm_0 Q^2 \rangle_\mathrm{e}}\left\{2\frac{\langle m_0 m_2' Q^4 \rangle_\mathrm{e}}{\langle m_1^2 Q^4 \rangle_\mathrm{e}} + 5\frac{\langle m_0 m_1 Q^2 \rangle_\mathrm{e}}{\langle m_1 m_1 Q^4 \rangle_\mathrm{e}}\right\},$$

$$b_2 = \frac{1}{15}\frac{\langle fm_1 Q^4 \rangle_\mathrm{e}}{\langle m_1 m_1 Q^4 \rangle_\mathrm{e}}\left\{2\langle m_2' Q^4 \rangle_\mathrm{e} + 5\langle m_1 Q^2 \rangle_\mathrm{e}\right\}. \tag{13.86}$$

We remind that $m_{0,1}' = \partial m_{0,1}/\partial(Q^2)$. These formulas were obtained using the auxiliary results from subsection "Auxiliary formulas, 2".

Revised Oldroyd 8 Constant Constitutive Equation for the Stress

It is remarkable that when rewritten in terms of the full stresses, $\boldsymbol{\tau} = -\nu_\mathrm{s}\dot{\boldsymbol{\gamma}} + \boldsymbol{\tau}_\mathrm{p}$, the dynamic system (13.85) takes the form:

$$\boldsymbol{\tau} + c_1 \boldsymbol{\tau}_{[1]} + c_3 \left\{\dot{\boldsymbol{\gamma}} \cdot \boldsymbol{\tau} + \boldsymbol{\tau} \cdot \dot{\boldsymbol{\gamma}}\right\} + c_5 (\mathrm{tr}\boldsymbol{\tau})\dot{\boldsymbol{\gamma}} + \mathbf{1}\left(c_6 \boldsymbol{\tau} : \dot{\boldsymbol{\gamma}} + c_8 \mathrm{tr}\boldsymbol{\tau}\right)$$

$$= -\nu\left\{\dot{\boldsymbol{\gamma}} + c_2 \dot{\boldsymbol{\gamma}}_{[1]} + c_4 \dot{\boldsymbol{\gamma}} \cdot \dot{\boldsymbol{\gamma}} + c_7 (\dot{\boldsymbol{\gamma}} : \dot{\boldsymbol{\gamma}})\mathbf{1}\right\}, \tag{13.87}$$

where the parameters ν, c_i are given by the following relationships:

$$\nu = \lambda_\mathrm{r}\nu_\mathrm{s}\mu, \qquad \mu = 1 + nk_\mathrm{B}T\lambda_1 b_2/\nu_\mathrm{s},$$
$$c_1 = \lambda_\mathrm{r}/\lambda_1, \qquad c_2 = \lambda_\mathrm{r}/(\mu\lambda_1),$$
$$c_3 = -b_0\lambda_\mathrm{r}/\lambda_0, \qquad c_4 = -2b_0\lambda_\mathrm{r}/(\mu\lambda_1),$$
$$c_5 = \frac{\lambda_\mathrm{r}}{3\lambda_1}(2b_0 - 3b_1 - 1), \qquad c_6 = \frac{\lambda_\mathrm{r}}{\lambda_1}(2b_0 + 1 - a_0),$$
$$c_7 = \frac{\lambda_\mathrm{r}}{\lambda_1\mu}(2b_0 + 1 - a_0), \qquad c_8 = \frac{1}{3}(\lambda_0/\lambda_1 - 1). \tag{13.88}$$

In the last two formulas we returned to the original dimensional quantities: time t and gradient of velocity tensor $\boldsymbol{k} = \nabla\boldsymbol{v}$, and at the same time we kept the notations for the dimensional convective derivative, $\boldsymbol{\Lambda}_{[1]} = D\boldsymbol{\Lambda}/Dt - \boldsymbol{k} \cdot \boldsymbol{\Lambda} - \boldsymbol{\Lambda} \cdot \boldsymbol{k}^\dagger$.

Note that all the parameters (13.88) are related to the entropic spring law f due to (13.86). Thus, the constitutive relation for the stress τ (13.87) is fully derived from the microscopic kinetic model.

If the constant c_8 were equal to zero, then (13.87) would be recognized as *the Oldroyd 8 constant* model [362], proposed by Oldroyd about 40 years ago on a phenomenological basis. Nonzero value of c_8 indicates a difference between λ_r/λ_0 and λ_r/λ_1 which are the relaxation times of trace $\operatorname{tr}\tau$ and of the traceless components $\overset{\circ}{\tau}$ of the stress tensor τ.

Higher-Order Constitutive Equations

In this subsection we discuss some properties of corrections to the revised Oldroyd 8 constant constitutive equation (that is, the zero-order model) (13.87). Let $P_M^{(0)}$ (13.84) be the projector onto the zero-order manifold $\Psi_{(0)}$ (13.83). The invariance equation (13.73) for the first-order correction $\Psi_{(1)} = \Psi_{(0)} + \Psi_1$ takes the form:

$$L\Psi_1 = -\left(1 - P_M^{(0)}\right)(J_{\mathrm{d}} + J_{\mathrm{h}})\Psi_{(0)} \qquad (13.89)$$

$$P_M^{(0)}\Psi_1 = 0$$

where $L = (1 - P_M^{(0)})J_{\mathrm{d}}(1 - P_M^{(0)})$ is the symmetric operator. If the manifold $\Psi_{(0)}$ is parameterized by the functionals $M_0 = \int g_0 \Psi_{(0)} \, d\mathbf{Q}$ and $\overset{\circ}{\mathbf{M}} = \int m_1 \overset{\circ}{\mathbf{QQ}} \Psi_{(0)} \, d\mathbf{Q}$, where $\Psi_{\mathrm{eq}} m_0$ and $\Psi_{\mathrm{eq}} \overset{\circ}{\mathbf{QQ}} m_1$ are lowest eigenvectors of J_{d}, then the general form of the solution is given by:

$$\Psi_1 = \Psi_{\mathrm{eq}}\left\{ z_0 M_0(\overset{\circ}{\dot{\gamma}} : \overset{\circ}{\mathbf{QQ}}) + z_1(\overset{\circ}{\mathbf{M}}:\overset{\circ}{\mathbf{QQ}})(\overset{\circ}{\dot{\gamma}} : \overset{\circ}{\mathbf{QQ}}) \right.$$

$$\left. + z_2\{\overset{\circ}{\dot{\gamma}}\cdot \overset{\circ}{\mathbf{M}} + \overset{\circ}{\mathbf{M}}\cdot\overset{\circ}{\dot{\gamma}}\} : \overset{\circ}{\mathbf{QQ}} + z_3 \overset{\circ}{\dot{\gamma}} : \overset{\circ}{\mathbf{M}} + \frac{1}{2}\overset{\circ}{\dot{\gamma}} : \overset{\circ}{\mathbf{QQ}} \right\}. \qquad (13.90)$$

The terms z_0 through z_3 are the functions of Q^2 found as the solutions to some linear differential equations.

We observe two features of the new manifold:

- first, it remains *linear* in variables M_0 and $\overset{\circ}{\mathbf{M}}$;
- second, it contains the dependence on the rate of strain tensor $\overset{\circ}{\dot{\gamma}}$.

As the consequence, the transition to variables τ is given by the linear relations:

$$-\frac{\overset{\circ}{\tau}_{\mathrm{p}}}{nk_{\mathrm{B}}T} = r_0 \overset{\circ}{\mathbf{M}} + r_1 M_0 \overset{\circ}{\dot{\gamma}} + r_2\{\overset{\circ}{\dot{\gamma}}\cdot \overset{\circ}{\mathbf{M}} + \overset{\circ}{\mathbf{M}}\cdot\overset{\circ}{\dot{\gamma}}\} + r_3 \overset{\circ}{\dot{\gamma}}\cdot\overset{\circ}{\dot{\gamma}}, \qquad (13.91)$$

$$-\frac{\operatorname{tr}\tau_{\mathrm{p}}}{nk_{\mathrm{B}}T} = p_0 M_0 + p_1 \overset{\circ}{\dot{\gamma}} : \overset{\circ}{\mathbf{M}},$$

where r_i and p_i are some constants. Finally, the equations in terms of τ should be also linear. It can be shown that the first-order correction to the modified Oldroyd 8 constants model (13.87) will be transformed into the equations of the following general structure:

$$\tau + c_1\tau_{[1]} + \left\{\boldsymbol{\Gamma}_1 \cdot \tau \cdot \boldsymbol{\Gamma}_2 + \boldsymbol{\Gamma}_2^\dagger \cdot \tau \cdot \boldsymbol{\Gamma}_1^\dagger\right\}$$
$$+\boldsymbol{\Gamma}_3(\mathrm{tr}\tau) + \boldsymbol{\Gamma}_4(\boldsymbol{\Gamma}_5 : \tau) = -\nu_0\boldsymbol{\Gamma}_6 , \qquad (13.92)$$

where $\boldsymbol{\Gamma}_1$ through $\boldsymbol{\Gamma}_6$ are tensors dependent on the rate-of-strain tensor $\dot{\boldsymbol{\gamma}}$ and its first convective derivative $\dot{\boldsymbol{\gamma}}_{[1]}$, constant c_1 is the same as in (13.88) and ν_0 is a positive constant.

Because the explicit form of the tensors $\boldsymbol{\Gamma}_i$ is quite extensive we do not present them here. Instead we offer several general remarks about the structure of the first- and higher-order corrections:

1. Since the manifold (13.90) does not depend on the vorticity tensor $\omega = \boldsymbol{k} - \boldsymbol{k}^\dagger$, the latter enters the equations (13.92) only via convective derivatives of τ and $\dot{\boldsymbol{\gamma}}$. This is sufficient to acquire the frame indifference, since all the tensorial quantities in dynamic equations are indifferent in any time dependent reference frame [361].

2. When $\boldsymbol{k} = 0$, the first order equations (13.92) as well as equations for any order reduce to linear relaxation dynamics of slow modes:

$$\frac{D\overset{\circ}{\tau}}{Dt} + \frac{\lambda_1}{\lambda_r}\overset{\circ}{\tau} = 0 ,$$
$$\frac{D\mathrm{tr}\tau}{Dt} + \frac{\lambda_0}{\lambda_r}\mathrm{tr}\tau = 0,$$

which is obviously concordant with the dissipativity and the slowness requirements.

3. In all higher-order corrections one will be always left with linear manifolds if the projector associated with functionals $M_0[\Psi]$ and $\overset{\circ}{\mathbf{M}}[\Psi]$ is used in every step. It follows that the resulting constitutive equations will always take a linear form (13.92), where all tensors $\boldsymbol{\Gamma}_i$ depend on higher order convective derivatives of $\dot{\boldsymbol{\gamma}}$ (the highest possible order is limited by the order of the correction). Similarly to the first and zero orders the frame indifference is guaranteed if the manifold does not depend on the vorticity tensor unless the latter is incorporated in any frame invariant time derivatives. It is reasonable to eliminate the dependence on vorticity (if any) at the stage of constructing the solution to iteration equations (13.73).

4. When the force $\boldsymbol{F}$ is linear, $\boldsymbol{F} = \boldsymbol{Q}$, we are led to Oldroyd-B model ((13.87) with $c_i = 0$ for $i = 3, \ldots, 8$). This follows from the fact that the spectrum of the corresponding operator J_d is more degenerated, in particular $\lambda_0 = \lambda_1 = 1$ and the corresponding lowest eigenvectors correspond to the simple dyad $\Psi_\mathrm{eq}\boldsymbol{Q}\boldsymbol{Q}$.

13.5.7 Tests on the FENE Dumbbell Model

In this section we specify the choice of the force law as the FENE spring
(13.35), and present results of test calculations for the revised Oldroyd 8
constants (13.85) equations on the examples of two simple viscometric flows.
We introduce the extensibility parameter of FENE dumbbell model b:

$$b = \widehat{\mathbf{Q}}_0^2 = \frac{H\mathbf{Q}_0^2}{k_{\mathrm{B}}T} \; . \tag{13.93}$$

It was estimated [151] that b is proportional to the length of polymeric molecule and has a meaningful variation interval 50–1000. The limit $b \to \infty$
corresponds to the Hookean case and therefore to the Oldroyd-B constitutive
equation.

In our test calculations we compare our results with the Brownian dynamic (BD) simulation data made on FENE dumbbell equations [363], and
also with one popular approximation to the FENE model known as *FENE-P*
(FENE-Peterelin) model [151, 364, 365]. The latter is obtained by selfconsis-
tent approximation to the FENE force:

$$\mathbf{F} = \frac{1}{1 - \langle \mathbf{Q}^2 \rangle / b} \mathbf{Q} \; . \tag{13.94}$$

This force law, like the Hookean case, allows the exact moment closure leading
to nonlinear constitutive equations [151,365]. Specifically, we use the modified
variant of the FENE-P model, which matches the dynamics of the original
FENE near equilibrium better than the classical variant. This is achieved by
a slight modification of Kramers definition of the stress tensor:

$$\boldsymbol{\tau}_{\mathrm{p}} = n k_{\mathrm{B}} T (1 - \theta b)\mathbf{1} - \langle \mathbf{FQ} \rangle \; . \tag{13.95}$$

The case $\theta = 0$ gives the classical definition of FENE-P, while a more thorough estimation [354, 365] is $\theta = (b(b+2))^{-1}$.

Constants

The specific feature of the FENE model is that the length of dumbbells $\mathbf{Q}$ can
vary only in a bounded domain of $\mathbf{R}^3$, namely inside a sphere $S_b = \{Q^2 \le b\}$.
The sphere S_b defines the domain of integration for averages $\langle \bullet \rangle_{\mathrm{e}} = \int_{S_b} \Psi_{\mathrm{eq}} \bullet$
$d\mathbf{Q}$, where the equilibrium distribution reads $\Psi_{\mathrm{eq}} = c^{-1}\left(1 - Q^2/b\right)^{b/2}$, $c = $
$\int_{S_b} \left(1 - Q^2/b\right)^{b/2} d\mathbf{Q}$.

In order to find constants for the zero-order model (13.85) we do the following: First, we analytically compute the lowest eigenfunctions of operator
J_{d}: $g_1(Q^2)\, \mathbf{QQ}$ and $g_0(Q^2)$ without hydrodynamic interaction ($\kappa = 0$). The
functions g_0 and g_1 were computed by a procedure presented in Subsect.
"Auxiliary formulas, 1" with the help of the *Maple V.3* [366]. Second, we

calculate the perturbations terms $h_{0,1}$ by formulas (13.68) introducing the account of hydrodynamic interaction. Table 13.1 presents the constants $\lambda_{0,1}$, a_i, b_i (13.79) (13.86) of the zero-order model (13.85) without hydrodynamic interaction ($\kappa = 0$) for several values of extensibility parameter b. The relative error $\delta_{0,1}$ (see Subsect. "Auxiliary formulas, 1") of approximation for these calculations did not exceed the value 0.02. Table 13.2 shows the linear correction terms for constants from Table 13.1 which take into account the hydrodynamic interaction: $\lambda_{0,1}^h = \lambda_{0,1}(1 + \kappa(\delta\lambda_{0,1}))$, $a_i^h = a_i(1 + \kappa(\delta a_i))$, $b_i^h = b_i(1 + \kappa(\delta b_i))$. The latter are calculated by substituting the perturbed functions $m_{0,1} = g_{0,1} + \kappa h_{0,1}$ into (13.79) and (13.86), and expanding them up to first-order in κ. One can observe, since $\kappa > 0$, the effect of hydrodynamic interaction results in the reduction of the relaxation times.

Table 13.1. Values of constants to the revised Oldroyd 8 constants model computed on the base of the FENE dumbbells model

b	λ_0	λ_1	b_0	b_1	b_2	a_0
20	1.498	1.329	−0.0742	0.221	1.019	0.927
50	1.198	1.135	−0.0326	0.279	1.024	0.982
100	1.099	1.068	−0.0179	0.303	1.015	0.990
200	1.050	1.035	0.000053	0.328	1.0097	1.014
∞	1	1	0	1/3	1	1

Table 13.2. Corrections due to hydrodynamic interaction to the constants of the revised Oldroyd 8 constants model based on FENE force

b	$\delta\lambda_0$	$\delta\lambda_1$	δb_0	δb_1	δb_2	δa_0
20	−0.076	−0.101	0.257	−0.080	−0.0487	−0.0664
50	−0.0618	−0.109	−0.365	0.0885	−0.0205	−0.0691
100	−0.0574	−0.111	−1.020	0.109	−0.020	−0.0603

Dynamic Problems

The rest of this section concerns the computations for two particular flows. The shear flow is defined by

$$\mathbf{k}(t) = \dot{\gamma}(t) \begin{bmatrix} 0 & 1 & 0 \\ 0 & 0 & 0 \\ 0 & 0 & 0 \end{bmatrix}, \tag{13.96}$$

where $\dot{\gamma}(t)$ is the shear rate, and the elongation flow corresponds to the choice:

$$\boldsymbol{k}(t) = \dot{\varepsilon}(t) \begin{bmatrix} 1 & 0 & 0 \\ 0 & -1/2 & 0 \\ 0 & 0 & -1/2 \end{bmatrix} , \qquad (13.97)$$

where $\dot{\varepsilon}(t)$ is the elongation rate.

In test computations we look at viscometric material functions defined through the components of the polymeric part of the stress tensor $\boldsymbol{\tau}_\mathrm{p}$. Namely, for shear flow they are the shear viscosity ν, the first and the second normal stress coefficients ψ_1, ψ_2, and for the elongation flow the only function is the elongation viscosity $\bar{\nu}$. In dimensionless form they are written as:

$$\widehat{\nu} = \frac{\nu - \nu_\mathrm{s}}{n k_\mathrm{B} T \lambda_\mathrm{r}} = -\frac{\boldsymbol{\tau}_{\mathrm{p},12}}{\overline{\gamma} n k_\mathrm{B} T}, \qquad (13.98)$$

$$\widehat{\psi}_1 = \frac{\psi_1}{n k_\mathrm{B} T \lambda_\mathrm{r}^2} = \frac{\boldsymbol{\tau}_{\mathrm{p},22} - \boldsymbol{\tau}_{\mathrm{p},11}}{\overline{\gamma}^2 n k_\mathrm{B} T}, \qquad (13.99)$$

$$\widehat{\psi}_2 = \frac{\psi_2}{n k_\mathrm{B} T \lambda_\mathrm{r}^2} = \frac{\boldsymbol{\tau}_{\mathrm{p},33} - \boldsymbol{\tau}_{\mathrm{p},22}}{\overline{\gamma}^2 n k_\mathrm{B} T}, \qquad (13.100)$$

$$\vartheta = \frac{\bar{\nu} - 3\nu_\mathrm{s}}{n k_\mathrm{B} T \lambda_\mathrm{r}} = \frac{\boldsymbol{\tau}_{\mathrm{p},22} - \boldsymbol{\tau}_{\mathrm{p},11}}{\overline{\varepsilon} n k_\mathrm{B} T}, \qquad (13.101)$$

where $\overline{\gamma} = \dot{\gamma} \lambda_\mathrm{r}$ and $\overline{\varepsilon} = \dot{\varepsilon} \lambda_\mathrm{r}$ are dimensionless shear and elongation rates. Characteristic values of the latter parameters $\overline{\gamma}$ and $\overline{\varepsilon}$ allow to estimate the parameter ε_1 (13.43). For all flows considered below the second flow parameter (Deborah number) ε_2 is equal to zero.

Let us consider the steady state values of viscometric functions in steady shear and elongation flows: $\dot{\gamma} = const$, $\dot{\varepsilon} = const$. For the shear flow the steady values of these functions are found from (13.85) as follows:

$$\widehat{\nu} = b_2/(\lambda_1 - c\overline{\gamma}^2) , \qquad \widehat{\psi}_1 = 2\widehat{\nu}/\lambda_1 , \qquad \widehat{\psi}_2 = 2b_0\widehat{\nu}/\lambda_1 ,$$

where $c = 2/3(2b_0^2 + 2b_0 - 1)/\lambda_1 + 2b_1 a_0/\lambda_0$. Estimations for the constants (see Table I) show that $c \leq 0$ for all values of b (case $c = 0$ corresponds to $b = \infty$), thus all three functions are monotonically decreasing in absolute value with the increase of $\overline{\gamma}$, besides the case $b = \infty$. Although they qualitatively correctly predict the shear thinning for large shear rates due to a power law, but the exponent -2 in the limit of large $\overline{\gamma}$ deviations from the values -0.66 for $\widehat{\nu}$ and -1.33 for $\widehat{\psi}_1$ observed in Brownian dynamic simulations [363]. It is explained by the fact that slopes of shear thining lie out of the applicability domain of our model. A comparison with BD simulations and modified FENE-P model is shown in Fig. 13.1.

The predictions for the second normal stress coefficient indicate one more difference between the revised Oldroyd 8 constant equation and FENE-P model. FENE-P model shows zero values for $\widehat{\psi}_2$ in any shear flow, either steady or time dependent, while the model (13.85), as well as BD simulations (see Fig. 9 in [363]) predict small, but nonvanishing values for this quantity. Namely, due to the model (13.85) in shear flows the following relation $\widehat{\psi}_2 =$

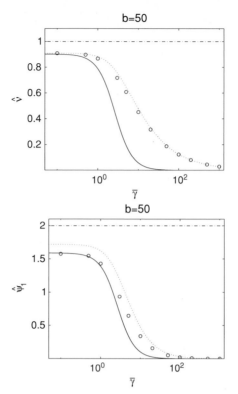

Fig. 13.1. Dimensionless shear viscosity $\widehat{\nu}$ and first normal stress coefficient $\widehat{\psi}_1$ vs. shear rate: (———) revised Oldroyd 8 constant model; ($\cdots\cdots$) FENE-P model; ($\circ\,\circ\,\circ$) BD simulations on the FENE model; ($-\cdot-\cdot-$) Hookean dumbbell model

$b_0\widehat{\psi}_1$ is always valid, with proportionality coefficient b_0 small and mostly negative, which leads to small and mostly negative values of $\widehat{\psi}_2$.

In the elongation flow the steady state value to ϑ is found as:

$$\vartheta = \frac{3b_2}{\lambda_1 - \frac{5}{6}(2b_0 + 1)\bar{\varepsilon} - 7b_1 a_0 \bar{\varepsilon}^2/\lambda_0} \, . \tag{13.102}$$

The denominator has one root on positive semi-axis

$$\bar{\varepsilon}_* = -\frac{5\lambda_0(2b_0 + 1)}{84b_1 a_0} + \left(\left(\frac{5\lambda_0(2b_0 + 1)}{84b_1 a_0}\right)^2 + \frac{\lambda_1\lambda_0}{7b_1 a_0}\right)^{1/2}, \tag{13.103}$$

which defines a singularity point for the dependence $\vartheta(\bar{\varepsilon})$. The BD simulations [363] on the FENE dumbbell models shows that there is no divergence of elongation viscosity for all values of elongation rate (see Fig. 13.2). For the Hookean spring, $\bar{\varepsilon}_* = 1/2$ while in our model (13.85) the singularity

Table 13.3. Singular values of elongation rate

b	20	50	100	120	200	∞
$\bar{\varepsilon}^*$	0.864	0.632	0.566	0.555	0.520	0.5

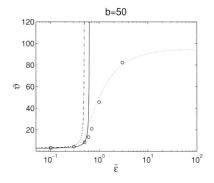

Fig. 13.2. Dimensionless elongation viscosity vs. elongation rate: (———) revised Oldroyd 8 constant model, ($\cdots\cdots$) FENE-P model, ($\circ\,\circ\,\circ$) BD simulations on the FENE model; ($-\cdot-\cdot-$) Hookean dumbbell model

point shifts to higher values with respect to decreasing values of b as it is demonstrated in Table 13.3.

The Fig. 13.3 gives an example of dynamic behavior for elongation viscosity in the instant start-up of the elongational flow. Namely, it shows the evolution of initially vanishing polymeric stresses after instant jump of elongation rate at the time moment $t = 0$ from the value $\bar{\varepsilon} = 0$ to the value $\bar{\varepsilon} = 0.3$.

It is possible to conclude that the revised Oldroyd 8 constants model (13.85) with estimations given by (13.86) for small and moderate rates of strain up to $\varepsilon_1 = \lambda_r |\dot{\gamma}|/(2\lambda_1) \sim 0.5$ yields a good approximation to the original FENE dynamics. The quality of the approximation in this interval is the same or better than the one of the nonlinear FENE-P model.

13.5.8 The Main Results of this Example are as Follows:

(i) We have developed a systematic method of constructing constitutive equations from the kinetic models of polymeric solutions. The method is free from a'priori assumptions about the form of the spring force and is consistent with the basic physical requirements: frame invariance and dissipativity of the internal motions of the fluid. The method extends the method of invariant manifold onto equations coupled with external fields. Two characteristic parameters of fluid flows were distinguished in order to account for the effect of the presence of external fields. The

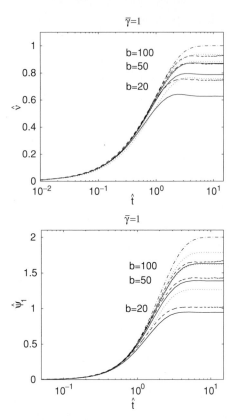

Fig. 13.3. Time evolution of elongation viscosity after inception of the elongation flow with elongation rate $\bar{\varepsilon} = 0.3$: (———) revised Oldroyd 8 constant model, (······) FENE-P model, (− − −) BD simulations on FENE model; (− · − · −) Hookean dumbbell model

iterative Newton scheme for obtaining a slow invariant manifold of the system driven by the flow with relatively low values of both characteristic parameters was developed.

(ii)   We demonstrated that the revised phenomenological Oldroyd 8 constants constitutive equations represent the slow dynamics of microscopic elastic models with any nonlinear spring force in the limit when the rate of strain and frequency of time variation of the flow are sufficiently small and microscopic states are taken not far from the equilibrium.

(iii) The corrections to the zero-order manifold lead generally to linear in stresses equations but with highly nonlinear dependence on the rate of strain tensor and its convective derivatives.

(iv) The zero-order constitutive equation is compared to the direct Brownian dynamics simulation for FENE dumbbell model as well as to predictions of FENE-P model. This comparison shows that the zero-order constitutive equation gives the correct predictions in the domain of its validity, but does not exclude qualitative discrepancy occurring out of this domain, particularly in elongation flows.

This discrepancy calls for a further development, in particular, the use of nonlinear manifolds for derivation of zero-order model. The reason is in the necessity to provide concordance with the requirement of the positivity of distribution function. It may lead to nonlinear constitutive equation on any order of correction.

13.6 Example: Explosion of Invariant Manifold, Limits of Macroscopic Description for Polymer Molecules, Molecular Individualism, and Multimodal Distributions

Derivation of macroscopic equations from the simplest dumbbell models is revisited [109]. It is demonstrated that the onset of the macroscopic description is sensitive to the flows. For the FENE-P model, small deviations from the Gaussian solution undergo a slow relaxation before the macroscopic description sets on. Some consequences of these observations are discussed. A new class of closures is discussed, the *kinetic multipeak polyhedra*. Distributions of this type are expected in kinetic models with a multidimensional instability as universally, as the Gaussian distribution appears for stable systems. The number of possible relatively stable states of a nonequilibrium system grows as 2^m, and the number of macroscopic parameters is in order mn, where n is the dimension of configuration space, and m is the number of independent unstable directions in this space. The elaborated class of closures and equations describes effects of "molecular individualism".

13.6.1 Dumbbell Models and the Problem of the Classical Gaussian Solution Stability

We shall again consider the simplest case of dilute polymer solutions represented by dumbbell models. The dumbbell model reflects the two features of real-world macromolecules to be orientable and stretchable by a flowing solvent [151].

Let us consider the simplest one-dimensional kinetic equation for the configuration distribution function $\Psi(q, t)$, where q is the reduced vector connecting the beads of the dumbbell. This equation is slightly different from the usual Fokker–Planck equation. It is nonlinear, because of the dependence of potential energy U on the moment $M_2[\Psi] = \int q^2 \Psi(q) \, dq$. This dependence allows us to get the exact quasiequilibrium equations on M_2, but these equations are not always solving the problem: this quasiequilibrium manifold may become unstable when the flow is present [109]. Here is this model:

$$\partial_t \Psi = -\partial_q \{\alpha(t) q \Psi\} + \frac{1}{2} \partial_q^2 \Psi \ . \tag{13.104}$$

Here

$$\alpha(t) = \kappa(t) - \frac{1}{2} f(M_2(t)) \ , \tag{13.105}$$

$\kappa(t)$ is the given time-dependent velocity gradient, t is the reduced time, and the function $-fq$ is the reduced spring force. Function f may depend on the second moment of the distribution function $M_2 = \int q^2 \Psi(q, t) \, dq$. In particular, the case $f \equiv 1$ corresponds to the linear Hookean spring, while $f = [1 - M_2(t)/b]^{-1}$ corresponds to the self-consistent finite extension nonlinear elastic spring (the FENE-P model [365]). The second moment M_2 occurs in the FENE-P force f as the result of the pre-averaging approximation to the original FENE model (with nonlinear spring force $f = [1 - q^2/b]^{-1}$). The parameter b changes the characteristics of the force law from Hookean at small extensions to a confining force for $q^2 \to b$. Parameter b is roughly equal to the number of monomer units represented by the dumbell and should therefore be a large number. In the limit $b \to \infty$, the Hookean spring is recovered. Recently, it has been demonstrated that FENE-P model appears as first approximation within a systematic self-confident expansion of nonlinear forces [29].

Equation (13.104) describes an ensemble of non-interacting dumbells subject to a pseudo-elongational flow with fixed kinematics. As is well known, the Gaussian distribution function,

$$\Psi^G(M_2) = \frac{1}{\sqrt{2\pi M_2}} \exp \left[-\frac{q^2}{2M_2} \right] \ , \tag{13.106}$$

solves equation (13.104) provided the second moment M_2 satisfies

$$\frac{dM_2}{dt} = 1 + 2\alpha(t) M_2 \ . \tag{13.107}$$

Solution (13.106) and (13.107) is the valid macroscopic description if all other solutions of the equation (13.104) are rapidly attracted to the family of Gaussian distributions (13.106). In other words [11], the special solution (13.106) and (13.107) is the macroscopic description if equation (13.106) is the stable invariant manifold of the kinetic equation (13.104). If not, then the Gaussian solution is just a member of the family of solutions, and equation (13.107) has no meaning of the macroscopic equation. Thus, the complete answer to the question of validity of the equation (13.107) as the macroscopic equation requires a study of dynamics in the neighborhood of the manifold (13.106). Because of the simplicity of the model (13.104), this is possible to a satisfactory level even for M_2-dependent spring forces.

13.6.2 Dynamics of the Moments and Explosion of the Gaussian Manifold

In [109] it was shown, that there is a possibility of "explosion" of the Gaussian manifold: with the small initial deviation from it, the solutions of the equation (13.104) are fast going far from the manifold, and then slowly come back to the stationary point which is located on the Gaussian manifold. The distribution function Ψ is stretched fast, but looses the Gaussian form, and after that the Gaussian form recovers slowly with the new value of M_2. Let us describe briefly the results of [109].

Let $M_{2n} = \int q^{2n} \Psi \, dq$ denote the even moments (odd moments vanish by symmetry). We consider deviations $\mu_{2n} = M_{2n} - M_{2n}^{G}$, where $M_{2n}^{G} = \int q^{2n} \Psi^{G} \, dq$ are moments of the Gaussian distribution function (13.106). Let $\Psi(q, t_0)$ be the initial condition to (13.104) at time $t = t_0$. Introducing functions,

$$p_{2n}(t, t_0) = \exp\left[4n \int_{t_0}^{t} \alpha(t') \, dt'\right] , \tag{13.108}$$

where $t \geq t_0$, and $2n \geq 4$, the exact time evolution of the deviations μ_{2n} for $2n \geq 4$ reads

$$\mu_4(t) = p_4(t, t_0)\mu_4(t_0) , \tag{13.109}$$

and

$$\mu_{2n}(t) = \left[\mu_{2n}(t_0) + 2n(4n - 1) \int_{t_0}^{t} \mu_{2n-2}(t')p_{2n}^{-1}(t', t_0) \, dt'\right] p_{2n}(t, t_0) , \tag{13.110}$$

for $2n \geq 6$. Equations (13.108), (13.109) and (13.110) describe evolution near the Gaussian solution for arbitrary initial condition $\Psi(q, t_0)$. Notice that explicit evaluation of the integral in (13.108) requires solution to the moment equation (13.107) which is not available in the analytical form for the FENE-P model.

It is straightforward to conclude that any solution with a non-Gaussian initial condition converges to the Gaussian solution asymptotically as $t \to \infty$ if

$$\lim_{t \to \infty} \int_{t_0}^{t} \alpha(t') \, dt' < 0 . \tag{13.111}$$

However, even if this asymptotic condition is met, deviations from the Gaussian solution may survive for considerable *finite* times. For example, if for some finite time T, the integral in (13.108) is estimated as $\int_{t_0}^{t} \alpha(t') \, dt' > \alpha(t - t_0)$, $\alpha > 0$, $t \leq T$, then the Gaussian solution becomes exponentially unstable during this time interval. If this is the case, the moment equation (13.107) cannot be regarded as the macroscopic equation. Let us consider specific examples.

For the Hookean spring ($f \equiv 1$) under constant elongation ($\kappa = $ const), the Gaussian solution is exponentially stable for $\kappa < 0.5$, and it becomes

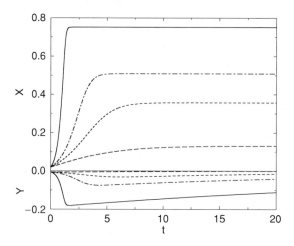

Fig. 13.4. Deviations of reduced moments from the Gaussian solution as a function of reduced time t in pseudo-elongation flow for the FENE-P model. *Upper* part: Reduced second moment $X = M_2/b$. *Lower* part: Reduced deviation of fourth moment from Gaussian solution $Y = -\mu_4^{1/2}/b$. *Solid*: $\kappa = 2$, *dash-dot*: $\kappa = 1$, *dash*: $\kappa = 0.75$, *long dash*: $\kappa = 0.5$. (The figure from the paper [109], computed by P. Ilg.)

exponentially unstable for $\kappa > 0.5$. The exponential instability in this case is accompanied by the well known breakdown of the solution to (13.107) due to the infinite stretching of the dumbbell. The situation is much more interesting for the FENE-P model because this nonlinear spring force does not allow the infinite stretching of the dumbbell [412, 413].

Euations (13.107) and (13.109) were integrated by the 5-th order Runge-Kutta method with adaptive time step. The FENE-P parameter b was set equal to 50. The initial condition was $\Psi(q, 0) = C(1 - q^2/b)^{b/2}$, where C is the normalization (the equilibrium of the FENE model, notoriously close to the FENE-P equilibrium [363]). For this initial condition, in particular, $\mu_4(0) = -6b^2/[(b+3)^2(b+5)]$ which is about 4% off the value of M_4 in the Gaussian equilibrium for $b = 50$. In Fig. 13.4 we demonstrate deviation $\mu_4(t)$ as a function of time for several values of the flow. Function $M_2(t)$ is also given for comparison. For small enough κ we find an adiabatic regime, that is μ_4 relaxes exponentially to zero. For stronger flows, we observe an initial *fast runaway* from the invariant manifold with $|\mu_4|$ growing over three orders of magnitude as compared to its initial value. After the maximum deviation is reached, μ_4 relaxes to zero. This relaxation is exponential as soon as the solution to (13.107) approaches the steady state. However, the time constant of this exponential relaxation $|\alpha_\infty|$ is very small. Specifically, for large κ,

$$\alpha_\infty = \lim_{t\to\infty} \alpha(t) = -\frac{1}{2b} + O(\kappa^{-1}) . \qquad (13.112)$$

Thus, the steady state solution is unique and Gaussian but the stronger is the flow, the larger is the initial runaway from the Gaussian solution, while the return to it thereafter becomes flow-independent. Our observation demonstrates that, though the stability condition (13.111) is met, *significant deviations from the Gaussian solution persist over the times when the solution of (13.107) is already reasonably close to the stationary state.* If we accept the usually quoted physically reasonable minimal value of parameter b of the order 20 then the minimal relaxation time is of order 40 in the reduced time units of Fig. 13.4. We should also stress that the two limits, $\kappa \to \infty$ and $b \to \infty$, are not commutative, thus it is not surprising that the estimation (13.112) does not reduce to the above mentioned Hookean result as $b \to \infty$. Finally, peculiarities of convergence to the Gaussian solution are even furthered if we consider more complicated (in particular, oscillating) flows $\kappa(t)$. Further numerical experiments are presented in [110]. The statistics of FENE-P solutions with random strains was studied recently [368]

13.6.3 Two-Peak Approximation for Polymer Stretching in Flow and Explosion of the Gaussian Manifold

In accordance with [369], the ansatz for Ψ can be suggested in the following form:

$$\Psi^{An}(\{\sigma,\varsigma\},q) = \frac{1}{2\sigma\sqrt{2\pi}}\left(e^{-\frac{(q+\varsigma)^2}{2\sigma^2}} + e^{-\frac{(q-\varsigma)^2}{2\sigma^2}}\right) . \qquad (13.113)$$

Natural inner coordinates on this manifold are σ and ς. Note, that now $\sigma^2 \neq M_2$. The value σ^2 is a dispersion of one of the Gaussian summands in (13.113),

$$M_2(\Psi^{An}(\{\sigma,\varsigma\},q)) = \sigma^2 + \varsigma^2 .$$

To build the thermodynamic projector on the manifold (13.113), the thermodynamic Lyapunov function is necessary. It is necessary to emphasize that equations (13.104) are nonlinear. For such equations, the arbitrariness in the choice of the thermodynamic Lyapunov function is much smaller than for the linear Fokker–Planck equation. Nevertheless, such a thermodynamic Lyapunov function exists. It is the free energy

$$F = U(M_2[\Psi]) - TS[\Psi] , \qquad (13.114)$$

where

$$S[\Psi] = -\int \Psi(\ln\Psi - 1)\,dq ,$$

$U(M_2[\Psi])$ is the potential energy in the mean field approximation, T is the temperature (below we assume $T = 1$).

Note that the Kullback–form entropy [156] $S_k = -\int \Psi \ln\left(\frac{\Psi}{\Psi^*}\right) dq$ also has the form $S_k = -F/T$:

$$\Psi^* = \exp(-U) \,,$$

$$S_k[\Psi] = -\langle U \rangle - \int \Psi \ln \Psi \, dq \,.$$

If $U(M_2[\Psi])$ in the mean field approximation is the convex function of M_2, then the free energy (13.114) is the convex functional too.

For the FENE-P model $U = -\ln[1 - M_2/b]$.

In accordance with thermodynamics the vector I of the flow of Ψ must be proportional to the gradient of the corresponding chemical potential μ:

$$I = -B(\Psi)\nabla_q \mu \,, \qquad (13.115)$$

where $\mu = \frac{\delta F}{\delta \Psi}$, $B \geq 0$. From (13.114) it follows that

$$\mu = \frac{dU(M_2)}{dM_2} \cdot q^2 + \ln \Psi \,;$$

$$I = -B(\Psi)\left[2\frac{dU}{dM_2} \cdot q + \Psi^{-1}\nabla_q \Psi\right] \,. \qquad (13.116)$$

If we assume here $B = \frac{D}{2}\Psi$, then we get

$$I = -D\left[\frac{dU}{dM_2} \cdot q\Psi + \frac{1}{2}\nabla_q \Psi\right] \,;$$

$$\frac{\partial \Psi}{\partial t} = \mathrm{div}_q I = D\frac{dU(M_2)}{dM_2}\partial_q(q\Psi) + \frac{D}{2}\partial^2 q\Psi \,, \qquad (13.117)$$

When $D = 1$ this equation coincides with (13.104) in the absence of the flow, and $dF/dt \leq 0$ due to (13.117).

Let us construct the thermodynamic projector with the help of the thermodynamic Lyapunov function F (13.114). Corresponding entropic scalar product at the point Ψ has the form

$$\langle f|g\rangle_\Psi = \frac{d^2U}{dM_2^2}\Bigg|_{M_2=M_2[\Psi]} \cdot \int q^2 f(q)dq \cdot \int q^2 g(q)\,dq + \int \frac{f(q)g(q)}{\Psi(q)}\,dq \,. \qquad (13.118)$$

When stuying the ansatz (13.113), the scalar product (13.118) constructed for the corresponding point of the Gaussian manifold with $M_2 = \sigma^2$ will be used. This will allow us to investigate the neighborhood of the Gaussian manifold (and to get all the results analytically):

$$\langle f|g\rangle_{\sigma^2} = \frac{d^2U}{dM_2^2}\Bigg|_{M_2=\sigma^2} \cdot \int q^2 f(q)\,dq \cdot \int q^2 g(q)\,dq$$
$$+\sigma\sqrt{2\pi}\int e^{\frac{q^2}{2\sigma^2}} f(q)g(q)\,dq \,. \qquad (13.119)$$

Also we need to know the functional $D_f F$ at the point of Gaussian manifold:

$$D_f F_{\sigma^2}(f) = \left(\frac{dU(M_2)}{dM_2} \bigg|_{M_2=\sigma^2} - \frac{1}{2\sigma^2} \right) \int q^2 f(q)\, dq \,, \qquad (13.120)$$

(subject to the condition $\int f(q)\, dq = 0$). The point

$$\frac{dU(M_2)}{dM_2} \bigg|_{M_2=\sigma^2} = \frac{1}{2\sigma^2} \,,$$

corresponds to the equilibrium.

The tangent space to the manifold (13.113) is spanned by the vectors

$$f_\sigma = \frac{\partial \Psi^{An}}{\partial(\sigma^2)}; \; f_\varsigma = \frac{\partial \Psi^{An}}{\partial(\varsigma^2)} \,; \qquad (13.121)$$

$$f_\sigma = \frac{1}{4\sigma^3\sqrt{2\pi}} \left[e^{-\frac{(q+\varsigma)^2}{2\sigma^2}} \frac{(q+\varsigma)^2 - \sigma^2}{\sigma^2} + e^{-\frac{(q-\varsigma)^2}{2\sigma^2}} \frac{(q-\varsigma)^2 - \sigma^2}{\sigma^2} \right] \,;$$

$$f_\varsigma = \frac{1}{4\sigma^2\varsigma\sqrt{2\pi}} \left[-e^{-\frac{(q+\varsigma)^2}{2\sigma^2}} \frac{q+\varsigma}{\sigma} + e^{-\frac{(q-\varsigma)^2}{2\sigma^2}} \frac{(q-\varsigma)}{\sigma} \right] \,;$$

The Gaussian entropy (free energy) production in the directions f_σ and f_ς (13.120) has a very simple form:

$$DF_{\sigma^2}(f_\varsigma) = DF_{\sigma^2}(f_\sigma) = \frac{dU(M_2)}{dM_2} \bigg|_{M_2=\sigma^2} - \frac{1}{2\sigma^2} \,. \qquad (13.122)$$

The linear subspace $\ker DF_{\sigma^2}$ in $\lin\{f_\sigma, f_\varsigma\}$ is spanned by the vector $f_\varsigma - f_\sigma$.

Let us consider the given vector field $d\Psi/dt = J(\Psi)$ at the point $\Psi(\{\sigma,\varsigma\})$. We need to build the projection of J onto the tangent space $T_{\sigma,\varsigma}$ at the point $\Psi(\{\sigma,\varsigma\})$:

$$P^{th}_{\sigma,\varsigma}(J) = \varphi_\sigma f_\sigma + \varphi_\varsigma f_\varsigma \,. \qquad (13.123)$$

This equation means that the equations for σ^2 and ς^2 will have the form

$$\frac{d\sigma^2}{dt} = \varphi_\sigma \,; \quad \frac{d\varsigma^2}{dt} = \varphi_\varsigma \,. \qquad (13.124)$$

Projection $(\varphi_\sigma, \varphi_\varsigma)$ can be found from the following two equations:

$$\varphi_\sigma + \varphi_\varsigma = \int q^2 J(\Psi)(q)\, dq \,;$$

$$\langle \varphi_\sigma f_\sigma + \varphi_\varsigma f_\varsigma | f_\sigma - f_\varsigma \rangle_{\sigma^2} = \langle J(\Psi) | f_\sigma - f_\varsigma \rangle_{\sigma^2} \,, \qquad (13.125)$$

where $\langle f|g \rangle_{\sigma^2} = \langle J(\Psi)|f_\sigma - f_\varsigma \rangle_{\sigma^2}$, (13.118). First equation of (13.125) means, that the time derivative dM_2/dt is the same for the initial and the reduced equations. Due to the formula for the dissipation of the free energy

(13.120), this equality is equivalent to the persistence of the dissipation in the neighborhood of the Gaussian manifold. Indeed, in according to (13.120) $dF/dt = A(\sigma^2) \int q^2 J(\Psi)(q)\,dq = A(\sigma^2)dM_2/dt$, where $A(\sigma^2)$ does not depend of J. On the other hand, the time derivative of M_2 due to projected equation (13.124) is $\varphi_\sigma + \varphi_\varsigma$, because $M_2 = \sigma^2 + \varsigma^2$.

The second equation in (13.125) means, that J is projected orthogonally on $\ker D_f S \bigcap T_{\sigma,\varsigma}$. Let us use the orthogonality with respect to the entropic scalar product (13.119). The solution of equations (13.125) has the form

$$\frac{d\sigma^2}{dt} = \varphi_\sigma = \frac{\langle J|f_\sigma - f_\varsigma\rangle_{\sigma^2} + M_2(J)(\langle f_\varsigma|f_\varsigma\rangle_{\sigma^2} - \langle f_\sigma|f_\varsigma\rangle_{\sigma^2})}{\langle f_\sigma - f_\varsigma|f_\sigma - f_\varsigma\rangle_{\sigma^2}},$$

(13.126)

$$\frac{d\varsigma^2}{dt} = \varphi_\varsigma = \frac{-\langle J|f_\sigma - f_\varsigma\rangle_{\sigma^2} + M_2(J)(\langle f_\sigma|f_\sigma\rangle_{\sigma^2} - \langle f_\sigma|f_\varsigma\rangle_{\sigma^2})}{\langle f_\sigma - f_\varsigma|f_\sigma - f_\varsigma\rangle_{\sigma^2}},$$

where $J = J(\Psi)$, $M_2(J) = \int q^2 J(\Psi)\,dq$.

It is easy to check, that the formulas (13.126) are indeed defining the projector: if f_σ (or f_ς) is substituted instead of the function J, then we get $\varphi_\sigma = 1, \varphi_\varsigma = 0$ (or $\varphi_\sigma = 0, \varphi_\varsigma = 1$, respectively). Let us substitute the right part of the initial kinetic equations (13.104), calculated at the point $\Psi(q) = \Psi(\{\sigma,\varsigma\},q)$ (see (13.113)) in (13.126) instead of J. We shall get the closed system of equations on σ^2, ς^2 in the neighborhood of the Gaussian manifold.

This system describes the dynamics of the distribution function Ψ. The distribution function is represented as the half-sum of two Gaussian distributions with the averages of distribution $\pm\varsigma$ and mean-square deviations σ. All integrals in the right-hand part of (13.126) are possible to calculate analytically.

The basis (f_σ, f_ς) is convenient to use everywhere except for the points on the Gaussian manifold, $\varsigma = 0$, because if $\varsigma \to 0$, then

$$f_\sigma - f_\varsigma = O\left(\frac{\varsigma^2}{\sigma^2}\right) \to 0.$$

Let us analyze the stability of the Gaussian manifold with respect to the "dissociation" of the Gaussian peak in two peaks (13.113). In order to do this, it is necessary to find the first nonvanishing term in the Taylor series expansion in ς^2 of the right-hand side of the second equation in the system (13.126). The denominator has the order of ς^4, the numerator has, as it is easy to see, the order not less, than ς^6 (because the Gaussian manifold is invariant with respect to the initial system).

With the accuracy up to ς^4:

$$\frac{1}{\sigma^2}\frac{d\varsigma^2}{dt} = 2\alpha\frac{\varsigma^2}{\sigma^2} + o\left(\frac{\varsigma^4}{\sigma^4}\right),$$

(13.127)

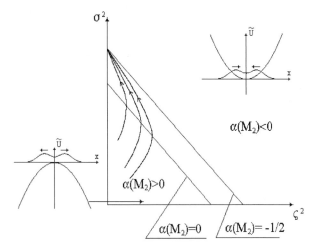

Fig. 13.5. Phase trajectories for the two-peak approximation, FENE-P model. The vertical axis ($\varsigma = 0$) corresponds to the Gaussian manifold. The triangle with $\alpha(M_2) > 0$ is the domain of exponential instability

where

$$\alpha = \kappa - \left.\frac{dU(M_2)}{dM_2}\right|_{M_2=\sigma^2}.$$

Thus, if $\alpha > 0$, then ς^2 grows exponentially ($\varsigma \sim e^{\alpha t}$) and the Gaussian manifold is unstable; if $\alpha < 0$, then ς^2 decreases exponentially and the Gaussian manifold is stable.

Near the vertical axis $d\sigma^2/dt = 1 + 2\alpha\sigma^2$. The form of the phase trajectories is shown qualitative on Fig. 13.5. Note that this result completely agrees with equation (13.109).[2]

For the linear Fokker–Planck equation with a non-linear force law (for example, with the FENE force) the motion in the presence of the flow can be represented as the motion in the effective potential well $\tilde{U}(q) = U(q) - \frac{1}{2}\kappa q^2$. Different variants of the phase portrait for the FENE potential are present on Fig. 13.6. Instability and dissociation of the unimodal distribution functions ("peaks") for the FPE is the general effect when the flow is present. The instability occurs when the matrix $\partial^2\tilde{U}/\partial q_i \partial q_j$ starts to have negative eigenvalues ($\tilde{U}$ is the effective potential energy, $\tilde{U}(q) = U(q) - \frac{1}{2}\sum_{i,j}\kappa_{i,j}q_iq_j$).

13.6.4 Polymodal Polyhedron and Molecular Individualism

What are the possible physical consequences of the instability of the Gaussian manifolds? The discovery of the molecular individualism for dilute polymers

---

[2] Pavel Gorban calculated the projector (13.126) analytically without Taylor expansion and with the same, but exact result: $d\varsigma^2/dt = 2\alpha\varsigma^2$, $d\sigma^2/dt = 1 + 2\alpha\sigma^2$.

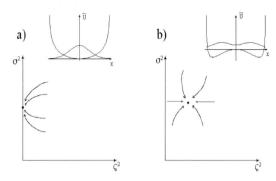

Fig. 13.6. Phase trajectories for the two-peak approximation, FENE model: (**a**) A stable equilibrium on the vertical axis, one stable peak; (**b**) A stable equilibrium with $\varsigma > 0$, stable two-peak configuration

in the elongational flow [370] was the challenge to theory from the very beginning. "Our data should serve as a guide in developing improved microscopic theories for polymer dynamics"... was the concluding sentence of the paper [370]. P.G. de Gennes invented the notion "molecular individualism" [371]. He stressed that in this case the usual averaging procedures are not applicable. At the highest strain rates distinct conformation shapes with different dynamics were observed [370]. Further works for the shear flow demonstrated not only shape differences, but large temporal fluctuations [372].

Equation for the molecules in a flow are known. These are the Fokker–Planck equations with external force. The theory of the molecular individualism is hidden inside these equations. Following the logic of model reduction we should solve two problems: to construct the slow manifold, and to project the equation on this manifold. The second problem is solved: the thermodynamic projector is necessary for this projection.

How to solve the first problem? We can find a hint in previous subsections. The Gaussian distributions form the invariant manifold for the FENE-P model of polymer dynamics, but this manifold can become unstable in the presence of a flow. We propose to model this instability as dissociation of the Gaussian peak into two peaks. This dissociation describes appearance of an unstable direction in the configuration space.

In the one-dimensional FENE-P model of the preceding section the polymer molecule is represented by one coordinate: the stretching of the molecule (the connector vector between the beads). There is a simple mean field generalized models for multidimensional configuration spaces of molecules. In these models, dynamics of distribution functions is described by the Fokker–Planck equation in a quadratic potential well. The matrix of coefficients of this quadratic potential depends on the matrix of the second order moments of the distribution function. The Gaussian distributions form the invariant manifold for these models, and the first dissociation of the Gaussian peak after the emergence of the unstable direction in the configuration space has the same nature and the same description, as for the one-dimensional models of molecules considered below.

At a higher strain, new unstable directions can appear, and corresponding dissociations of Gaussian peaks form a *cascade* of dissociation. For m unstable directions we get the Gaussian parallelepiped: The distribution function is represented as a sum of 2^m Gaussian peaks located in the vertices of parallelepiped:

$$\Psi(q) = \frac{1}{2^m (2\pi)^{n/2} \sqrt{\det \Sigma}} \tag{13.128}$$

$$\times \sum_{\varepsilon_i = \pm 1,\, (i=1,\dots,m)} \exp\left(-\frac{1}{2} \left(\Sigma^{-1} \left(q + \sum_{i=1}^{m} \varepsilon_i \varsigma_i \right),\, q + \sum_{i=1}^{m} \varepsilon_i \varsigma_i \right) \right),$$

where n is the dimension of the configurational space, $2\varsigma_i$ is the vector of the ith edge of the parallelepiped, Σ is the one-peak covariance matrix (in this model, Σ is the same for all peaks). The macroscopic variables for this model are:

1. The covariance matrix Σ for a peak;
2. The set of vectors ς_i (or the parallelepiped edges).

The stationary polymodal distribution for the Fokker–Planck equation corresponds to the persistence of several local minima of the function $\tilde{U}(q)$. The multidimensional case is different from one-dimensional because it has the huge amount of possible configurations. An attempt to describe this picture quantitative meet the following obstacle: we do not know the details of the potential U, on the other hand, the effect of molecular individualism [370–372] seems to be universal in its essence, that is, independent of details of interactions.

We should find a mechanism that is as general, as the effect. The simplest dumbbell model which we have discussed in the previous subsection does not explain the effect, but it gives us a hint: the flow can violate the stability of unimodal distributions. If we assume that the whole picture is hidden inside a multidimensional Fokker–Planck equation for a large molecule in a flow, then we can use this hint in such a way: when the flow strain grows, there appears a sequence of bifurcations, and for each of them a new unstable direction arises. For the qualitative description of such a picture we can apply a language of normal forms [373], subject to a certain modification.

The bifurcation in dimension one with appearance of two point of minima from one point has the simplest polynomial representation: $U(q, \alpha) = q^4 + \alpha q^2$. If $\alpha \geq 0$, then this potential has one minimum, if $\alpha < 0$, then there are two points of minima. The normal form of degenerated singularity is $U(q) = q^4$. Such polynomial forms as $q^4 + \alpha q^2$ are very simple, but they have inconvenient asymptotic at $q \to \infty$. For our goals it is more appropriate to use logarithms of convex combinations of Gaussian distributions instead of polynomials. It is the same class of jets near the bifurcation, but with given quadratic asymptotic $q \to \infty$. If one needs another class of asymptotic, it is possible just to change the choice of the basic peak. All normal forms of the critical form of functions, and families of versal deformations are well investigated and known [373].

Let us represent the deformation of the probability distribution under the strain in multidimensional case as a cascade of peak dissociation. The number of peaks will duplicate on the each step. The possible cascade of peaks dissociation is presented qualitatively on Fig. 13.7. The important property of this qualitative picture is the linear complexity of dynamical description with exponential complexity of geometrical picture. Let m be the number of bifurcation steps in the cascade. Then

– For description of parallelepiped it is sufficient to describe m edges;
– There are 2^{m-1} geometrically different conformations associated with 2^m vertex of parallelepiped (central symmetry halved this number).

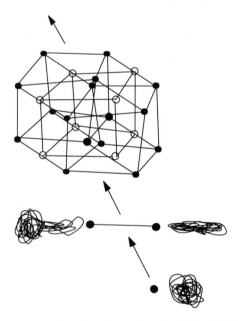

Fig. 13.7. Cartoon representing the steps of molecular individualism. Black dots are vertices of Gaussian parallelepiped. Zero, one, and four-dimensional polyhedrons are drawn. Presented is also the three-dimensional polyhedron used to draw the four-dimensional object. Each new dimension of the polyhedron adds as soon as the corresponding bifurcation occurs. Quasi-stable polymeric conformations are associated with each vertex. First bifurcation pertinent to the instability of a dumbbell model in elongational flow is described in the text

Another important property is the *threshold* nature of each dissociation: It appears in points of stability loss for new directions, in these points the dimension of unstable direction increases.

Positions of peaks correspond to parallelepiped vertices. Different vertices in configuration space present different geometric forms. So, it seems *plausible*[3] that observed different forms ("dumbbels", "half-dumbbels", "kinked", "folded" and other, not classified forms) correspond to these vertices of parallelepiped. Each vertex is a metastable state of a molecule and has its own basin of attraction. A molecule goes to the vertex which depends strongly on details of initial conditions.

The simplest multidimensional dynamic model is the Fokker–Planck equation with quadratic mean field potential. This is direct generalization of the FENE-P model: the quadratic potential $U(q)$ depends on the tensor of second moments $\boldsymbol{M}_2 = \langle q_i q_j \rangle$ (here the angle brackets denote the averaging). This dependence should provide the finite extensibility. This may be, for example, a simple matrix generalization of the FENE-P energy:

$$U(q) = \sum_{ij} K_{ij} q_i q_j, \ \mathbf{K} = \mathbf{K}^0 + \phi(\boldsymbol{M}_2/b), \ \langle U(q) \rangle = \mathrm{tr}(\mathbf{K}\boldsymbol{M}_2/b)$$

where b is a constant (the limit of extensibility), $\mathbf{K}^0$ is a constant matrix, $\boldsymbol{M}_2$ is the matrix of second moments, and ϕ is a positive analytical monotone increasing function of one variable on the interval $(0, 1)$, $\phi(x) \to \infty$ for $x \to 1$ (for example, $\phi(x) = -\ln(1-x)/x$, or $\phi(x) = (1-x)^{-1}$).

For quadratic multidimensional mean field models persists the qualitative picture of Fig. 13.5: there is *non-stationary moleqular individualism for stationary "molecular collectivism"*. The stationary distribution is the Gaussian distribution, and on the way to this stationary point there exists an unstable region, where the distribution dissociates onto 2^m peaks (m is the number of unstable degrees of freedom).

Dispersion of individual peak in unstable region increases too. This effect can deform the observed situation: If some of the peaks have significant intersection, then these peaks join into new extended classes of observed molecules. The stochastic walk of molecules between connected peaks can be observed as "large non-periodical fluctuations". This walk can be unexpected fast, because it can be effectively a *motion in a low-dimensional space*, for example, in one-dimensional space (in a neighborhood of a part of one-dimensional skeleton of the polyhedron).

---

[3] We can not *prove* it now, and it is necessary to determine the status of proposed qualitative picture: it is much more general than a specific model, it is the mechanism which acts in a wide class of models. The cascade of instabilities can appear and, no doubt, it appears for the Fokker–Planck equation for a large molecule in a flow. But it is not proven yet that the effects observed in well-known experiments have exactly this mechanism. This proof requires quantitative verification of a specific model. And now we talk not about a proven, but about the plausible mechanism which typically appears for systems with instabilities.

We discussed the important example of ansatz: the multipeak models. Two examples of these type of models demonstrated high efficiency during decades: the Tamm–Mott-Smith bimodal ansatz for shock waves, and the the Langer–Bar-on–Miller [374–376] approximation for spinodal decomposition.

The multimodal polyhedron appears every time as an appropriate approximation for distribution functions for systems with instabilities. We create such an approximation for the Fokker–Planck equation for polymer molecules in a flow. Distributions of this type are expected to appear in each kinetic model with multidimensional instability as universally, as Gaussian distribution appears for stable systems. This statement needs a clarification: everybody knows that the Gaussian distribution is stable with respect to convolutions, and the appearance of this distribution is supported by central limit theorem. Gaussian polyhedra form a stable class: convolution of two Gaussian polyhedra is a Gaussian polyhedron, convolution of a Gaussian polyhedron with a Gaussian distribution is a Gaussian polyhedron with the same number of vertices. On the other hand, a Gaussian distribution in a potential well appears as an exponent of a quadratic form which represents the simplest stable potential (a normal form of a nondegenerated critical point). Families of Gaussian parallelepipeds appear as versal deformations with given asymptotic for systems with cascade of simplest bifurcations.

The usual point of view is: The shape of the polymers in a flow is either a coiled ball, or a stretched ellipsoid, and the Fokker–Planck equation describes the stretching from the ball to the ellipsoid. It is not the whole truth, even for the FENE-P equation, as it was shown in [109, 369]. The Fokker–Planck equation describes the shape of a probability cloud in the space of conformations. In the flow with increasing strain this shape changes from the ball to the ellipsoid, but, after some thresholds, this ellipsoid transforms into a multimodal distribution which can be modeled as the peak parallelepiped. The peaks describe the finite number of possible molecule conformations. The number of this distinct conformations grows for a parallelepiped as 2^m with the number m of independent unstable direction. Each vertex has its own basin of attraction. A molecule goes to the vertex which depends strongly on details of initial conditions.

These models pretend to be the kinetic basis for the theory of molecular individualism. The detailed computations will be presented in following works, but some of the qualitative features of the models are in agreement with some of qualitative features of the picture observed in experiment [370–372]: effect has the threshold character, different observed conformations depend significantly on the initial conformation and orientation.

Some general questions remain open:

– Of course, appearance of 2^m peaks in the Gaussian parallelepiped is possible, but some of these peaks can join in following dynamics, hence the first question is: what is the typical number of significantly different peaks for a m–dimensional instability?

– How can we decide what scenario is more realistic from the experimental
  point of view: the proposed universal kinetic mechanism, or the scenario
  with long living metastable states (for example, the relaxation of knoted
  molecules in the flow can give an other picture than the relaxation of
  unknoted molecules)?

– The analysis of random walk of molecules from peak to peak should be
  done, and results of this analysis should be compared with observed large
  fluctuations.

The systematic discussion of the difference between the Gaussian elipsoid
(and its generalizations) and the Gaussian multipeak polyhedron (and its
generalizations) seems to be necessary. This polyhedron appears generically
as the effective ansatz for kinetic systems with instabilities.

14 Dimension of Attractors Estimation

How can we prove that all the attractors of an infinite-dimensional system belong to a finite-dimensional manifold? How can we estimate the dimensions of attractor? There exist two methods to obtain such estimations.

First, if we find that *k-dimensional volumes are contracted* due to dynamics, then (after some additional technical steps) we can claim that the *Hausdorff dimension* of the maximal attractor is less than k.

Second, if we find a representation of our system as a nonlinear kinetic system with *conservation of supports* of distributions, then (again, after some additional technical steps) we can state that the asymptotics is finite-dimensional. This conservation of support has a *quasi-biological interpretation, inheritance* (if a gene was not presented initially in an isolated population without mutations, then it cannot appear at later time). The finite-dimensional asymptotic demonstrates effects of *"natural" selection*.

In this chapter we describe these approaches.

14.1 Lyapunov Norms, Finite-Dimensional Asymptotics and Volume Contraction

In a general case, it is impossible to prove the existence of a global Lyapunov function on the basis of local data. We can only verify or falsify the hypothesis about a given function, is it a global Lyapunov function, or is it not. On the other hand, there exists a more strict stability property which can be verified or falsified (in principle) with local data analysis. This is a Lyapunov norm existence.

A norm $\|\bullet\|$ is the *Lyapunov norm* for the system (13.1), if for any two solutions $x^{(1)}(t)$, $x^{(2)}(t)$, $t \geq 0$, the function $\|x^{(1)}(t) - x^{(2)}(t)\|$ is non-increasing in time.

A linear operator A is *dissipative* with respect to a norm $\|\bullet\|$, if $\exp(At)$ $(t \geq 0)$ is a semigroup of contractions: $\|\exp(At)x\| \leq \|x\|$ for any x and $t \geq 0$. A family of linear operators $\{A_\alpha\}_{\alpha \in K}$ is *simultaneously dissipative*, if all operators A_α are dissipative with respect to some norm $\|\bullet\|$ (it should be stressed that in this definition one requires *one* norm for *all* A_α, $\alpha \in K$). The mathematical theory of simultaneously dissipative operators for finite-dimensional spaces was developed in [299–303].

Alexander N. Gorban and Iliya V. Karlin: *Invariant Manifolds for Physical and Chemical Kinetics*, Lect. Notes Phys. **660**, 419–456 (2005)
www.springerlink.com   © Springer-Verlag Berlin Heidelberg 2005

Let the system (13.1)

$$\frac{dx}{dt} = J(x) + J_{ex}(x,t), x \subset U$$

be defined in a convex set $U \subset E$, and A_x be Jacobi operator at the point x: $A_x = D_x(J(x) + J_{ex}(x))$. This system has a Lyapunov norm, if the family of operators $\{A_x\}_{x \in U}$ is simultaneously dissipative. If one can choose such $\varepsilon > 0$ that for all A_x, $t > 0$, any vector z, and this Lyapunov norm

$$\| \exp(A_x t)z \| \leq \exp(-\varepsilon t)\|z\| ,$$

then for any two solutions $x^{(1)}(t)$, $x^{(2)}(t)$, $t \geq 0$ of equations (13.1)

$$\|x^{(1)}(t) - x^{(2)}(t)\| \leq \exp(-\varepsilon t)\|x^{(1)}(0) - x^{(2)}(0)\| .$$

The existence of the Lyapunov norm is a very strong restriction on nonlinear systems, and such systems are not widespread in applications. But if we go from the distance contraction to the contraction of the k-dimensional volumes ($k = 2, 3, \ldots$) [311], the situation changes dramatically. There exist many kinetic systems with a monotonous contraction of the k-dimensional volumes for sufficiently large k (see, for example, [173, 317, 318, 334]). Let $x(t)$, $t \geq 0$ be a solution of equation (13.1). Let us write the first approximation equation for small variation of $x(t)$:

$$\frac{d\Delta x}{dt} = A_{x(t)}\Delta x . \tag{14.1}$$

This is a linear system with coefficients dependent on t. Let us study how the system (14.1) changes the k-dimensional volumes. For k-dimensional parallelepiped with the edges $x^{(1)}, x^{(2)}, \ldots x^{(k)}$ we can define an element of the kth exterior power, the oriented volume:

$$x^{(1)} \wedge x^{(2)} \wedge \ldots \wedge x^{(k)} \in \underbrace{E \wedge E \wedge \ldots \wedge E}_{k} = \wedge^k E ,$$

(this is an antisymmetric tensor or k-vector). A norm in the kth exterior power of the space E is a measure of the k-dimensional volumes (one of the possible measures). Dynamics of volumes induced by the system (14.1) is given by equations

$$\frac{d}{dt}(\Delta x^{(1)} \wedge \Delta x^{(2)} \wedge \ldots \wedge \Delta x^{(k)}) = (A_{x(t)}\Delta x^{(1)}) \wedge \Delta x^{(2)} \wedge \ldots \wedge \Delta x^{(k)}$$
$$+\Delta x^{(1)} \wedge (A_{x(t)}\Delta x^{(2)}) \wedge \ldots \wedge \Delta x^{(k)} + \ldots$$
$$+\Delta x^{(1)} \wedge \Delta x^{(2)} \wedge \ldots \wedge (A_{x(t)}\Delta x^{(k)})$$
$$= (\wedge_D^k A_{x(t)})(\Delta x^{(1)} \wedge \Delta x^{(2)} \wedge \ldots \wedge \Delta x^{(k)}) . \tag{14.2}$$

Here $\wedge_D^k A_{x(t)}$ are operators of the *induced differential action* of $A_{x(t)}$ on the kth exterior power of E. Again, a decrease of $\|\Delta x^{(1)} \wedge \Delta x^{(2)} \wedge \ldots \wedge \Delta x^{(k)}\|$ in time is equivalent to dissipativity of all the operators $\wedge_D^k A_{x(t)}$, $t \geq 0$ in the norm $\|\bullet\|$. Existence of such a norm for all $\wedge_D^k A_{x(t)}(x \in U)$ is equivalent to the contraction of the volumes of all parallelepipeds due to first approximation system (14.1) for any solution $x(t)$ of equations (13.1). We shall call such a system the *k-contraction*. If one can choose such $\varepsilon > 0$ that for all $A_x(x \in U)$, any vector $z \in E \wedge E \wedge \ldots \wedge E$, and this norm,

$$\|\exp\left(\wedge_D^k A_{x(t)} t\right) z\| \leq \exp(-\varepsilon t)\|z\|\,,$$

then the volumes of the parallelepipeds contract exponentially as $\exp(-\varepsilon t)$.

For such systems we can estimate the Hausdorff dimension of the attractor (under some additional technical conditions about solutions boundedness): *it cannot exceed k*. It is necessary to stress here that this estimation of the Hausdorff dimension does not solve the problem of construction of the invariant manifold containing this attractor, and one needs special technique and additional restriction on the system in order to obtain this manifold (see [318, 342, 346, 347]).

Let us remind here the definition of the *Hausdorff dimension* of metric space (subset of a normed space, for example). Let X be a metric space, and d be a number. The d-dimensional *Hausdorff measure* of X, $H^d(X)$, is the infimum of positive numbers y such that for every $r > 0$, X can be covered by a countable family of closed balls, each of diameter less than r, such that the sum of the dth powers of their diameters is less than y. Note that $H^d(X)$ may be infinite, and d need not be an integer. The Hausdorff dimension $D_H(X)$ of X is the infimum of such $d \geq 0$ that the $H^d(X) = 0$:

$$D_H(X) = \inf\{d \geq 0 | H^d(X) = 0\}\,.$$

The simplest way for construction of the slow invariant manifold becomes available for systems with a dominance of the linear part in higher dimensions. Let an infinite-dimensional system have a form: $\dot u + Au = R(u)$, where the operator A is self-adjoint, and has discrete spectrum λ_i ($\lambda_1 < \lambda_2 < \ldots$, $\lambda_i \to \infty$) with sufficiently big gaps between the eigenvalues λ_i, and $R(u)$ is continuous. Let E_i be the eigenspace of A for the eigenvalue λ_i, and all the E_i are finite-dimensional spaces. One can build the slow manifold as the graph over the space $\oplus_{i=1}^k E_i$ for some k. Indeed, let the basis consists of eigenvectors of A. In this basis $\dot u_i = -\lambda_i u_i + R_i(u)$, and it seems plausible that, for some k and sufficiently big i, the functions $u_i(t)$ exponentially fast tend to $u_i(u_1(t), \ldots u_k(t))$, if $R_i(u)$ are bounded and continuous in a suitable sense. Here $u_i(u_1, \ldots u_k)$ are some smooth functions that describe the manifold with internal coordinates $u_1, \ldots u_k$.

Variants of rigorous theorems about systems with such a dominance of the linear part in higher dimensions may be found in literature (see, for example, the textbook [100]). Even if all the sufficient conditions hold, efficient computation of these manifold remains the problem, and various strategies of calculations are proposed: from the Euler method for the manifold correction [25] to various algorithms of discretization [13, 105, 343].

The simplest condition of simultaneous dissipativity for the family of operators $\{A_x\}$ can be obtained in a following way: let us take a norm $\|\bullet\|$. If all operators A_x are dissipative with respect to this norm, then the family A_x is (obviously) simultaneously dissipative in this norm. So, we can verify or falsify a hypothesis about the simultaneous dissipativity for a given norm. Simplest examples are provided by the quadratic and l^1 norms.

For the quadratic norm associated with a scalar product $\langle | \rangle$ the dissipativity of the operator A is equivalent to nonpositivity of the spectrum of the operator $A + A^+$, where A^+ is the adjoint to A operator with respect to scalar product $\langle | \rangle$.

For the l^1 norm with weights $\|x\| = \sum_i w_i |x_i|$, $w_i > 0$, the condition of the operator's A dissipativity in this norm is the weighted diagonal dominance for columns of the matrix $A = (a_{ij})$:

$$a_{ii} < 0, \; w_i |a_{ii}| \geq \sum_{j,\, j \neq i} w_j |a_{ji}| \, .$$

For the exponential contraction, it is necessary and sufficient that some gap exists in the dissipativity inequalities:

–   For the quadratic norm,

$$\sigma(A + A^+) < \varepsilon < 0 \, ,$$

where $\sigma(A + A^+)$ is the spectrum of $A + A^+$;
–   For the l^1 norm with weights,

$$a_{ii} < 0, \; w_i |a_{ii}| \geq \sum_{j,\, j \neq i} w_j |a_{ji}| + \varepsilon, \; \varepsilon > 0 \, .$$

Sufficient conditions of simultaneous dissipativity can have a different form (not only the form of dissipativity checking with respect to a given norm) [300–303], but the problem of necessary and sufficient conditions in a general case remains open.

The dissipativity conditions for operators $\wedge_D^k A_x$ of the differential induced action of A_x on the kth exterior power of E have a similar form. If we know the spectrum of $A_x + A_x^+$, then it is easy to find the spectrum of $\wedge_D^k A_x + (\wedge_D^k A_x)^+$. For example, for simple discrete spectrum each eigenvalue of this operator is a sum of k distinct eigenvalues of $A_x + A_x^+$.

A basis of the kth exterior power of E can be constructed from the basis $\{e_i\}$ of E: it is

$$\{e_{i_1 i_2 \ldots i_k}\} = \{e_{i_1} \wedge e_{i_2} \wedge \ldots \wedge e_{i_k}\}, \; i_1 < i_2 < \ldots < i_k \, .$$

For the l^1 norm with weights in the kth exterior power of E the set of weights is $\{w_{i_1 i_2 \ldots i_k} > 0, \; i_1 < i_2 < \ldots < i_k\}$. The norm of a vector z is

$$\|z\| = \sum_{i_1 < i_2 < \ldots < i_k} w_{i_1 i_2 \ldots i_k} |z_{i_1 i_2 \ldots i_k}| \, .$$

The dissipativity conditions for operators $\wedge_D^k A_{x(t)}$ of the induced differential action of A in the l^1 norm with weights have the form:

$$a_{i_1 i_1} + a_{i_2 i_2} + \ldots + a_{i_k i_k} < 0 \, ,$$

$$w_{i_1 i_2 \ldots i_k} |a_{i_1 i_1} + a_{i_2 i_2} + \ldots + a_{i_k i_k}| \geq \sum_{l=1}^{k} \sum_{j, \, j \neq i_1, i_2, \ldots i_k} w_{i_1 i_2 \ldots i_k}^{l.j} |a_{j i_l}|$$

for any $i_1 < i_2 < \ldots < i_k$,   (14.3)

where $w_{i_1 i_2 \ldots i_k}^{l.j} = w_I$, the multiindex I consists of indexes i_p $(p \neq l)$, and j.

For infinite-dimensional systems the problem of volume contraction and Lyapunov norms for exterior powers of E consists of three parts: the geometrical part that concerns the choice of norm for simultaneous dissipativity of operator families, the topological part that concerns the topological non-equivalence of the constructed norms, and estimation of the bounded set (the so-called *absorbing set*) containing a compact attractor. This appropriate apriori estimations of the bounded convex positively invariant set $V \subset U$ where the compact attractor is situated may become a difficult problem.

The estimation of the attractor dimension based on Lyapunov norms in the exterior powers is rather rough. This is a local estimation. Refined estimations are based on global Lyapunov exponents (Lyapunov or Kaplan-Yorke dimension [308, 309]). There are many different measures of the dimension [307, 310], and much effort is invested towards better estimates of various dimensions [345].

Estimations of the dimension of attractors was given for various systems: from the Navier-Stokes hydrodynamics [316] to the climate dynamics [312]. An introduction and a review of many results is presented in the book [318]. The local estimations remain the main tool for the estimation of the attractors dimension, because global estimations for complex systems are much more complicated and often just unattainable because of the computational complexity.

14.2 Examples: Lyapunov Norms for Reaction Kinetics

In this section we consider the reaction kinetics systems which obey the mass action law. The direct and reverse reactions will be represented in the stoichiometric equations separately, and the reaction mechanisms under consideration can include an elementary reaction without its reverse reaction.

Neither detailed balance, nor other dissipation requirements are presumed. Such schemes appear, for example, in modeling of catalytic reactions, where only part of the components are included into equations, and concentrations of other components are used as parameters [81].

The simplest class of nonlinear kinetic (open) systems with Lyapunov norms was described in the paper [304]. These are reaction systems without interactions of various substances. The stoichiometric equation of each elementary reaction has a form

$$\alpha_r A_{i_r} \to \sum_j \beta_{rj} A_j \ , \tag{14.4}$$

where r enumerates reactions, i_r is defined for each reaction α_r, β_{rj} are non-negative stoichiometric coefficients (usually they are integer), A_i are symbols of substances.

In the left hand part of equation (14.4) there is one initial reagent, though $\alpha_r > 1$ is possible (there may be several copies of A_{i_r}, for example $3A \to 2B + C$). This explains the notion of the reaction without interaction of different substances.

For the mass action law (2.67) kinetic equations for the kinetic scheme (14.4) have a form

$$\frac{dc_i}{dt} = - \sum_{r:\ i=i_r} \alpha_r k_r c_i^{\alpha_r} + \sum_r \beta_{ri} k_r c_{i_r}^{\alpha_r} \ , \quad k_r > 0 \ . \tag{14.5}$$

Let there exist a positive balance for the reaction scheme (14.4):

$$b_{i_r} \alpha_r = \sum_j b_j \beta_{rj}$$

for some $b_i > 0$ and all r. In this case kinetic equations for the reaction system (14.4) have a Lyapunov norm [304]. This is the weighted l^1 norm: $\|x\| = \sum_i b_i |x_i|$, where $b_i > 0$ are the coefficients from the linear conservation law $\sum_i b_i c_i = $ const. There exists no quadratic Lyapunov norm for general reaction systems without interaction of different substances.

Let us call a reaction mechanism *dissipative*, if it allows a universal Lyapunov norm[1], that is, the Lyapunov norm which is independent of reaction rate constants.

The reaction mechanism with only one elementary reaction,

$$\sum_i \alpha_i A_i \to \sum_i \beta_i A_i \ , \tag{14.6}$$

is dissipative if and only if for any i if $\alpha_i > 0$, then $\alpha_i > \beta_i$.

---

[1] For the motion on the planes with given values of the linear conservation laws.

The reaction mechanism with two elementary reaction without autocatalysis ($\alpha_{ri} \times \beta_{ri} \equiv 0$),

$$\sum_i \alpha_{ri} A_i \to \sum_i \beta_{ri} A_i, (r = 1, 2) \qquad (14.7)$$

is *not* dissipative if and only if for two of components A_i, A_j at least one of the following conditions holds:

1. Branching

$$\alpha_{1i}, \beta_{1j}, \alpha_{2j}, \beta_{2i} > 0, \ \frac{\beta_{1j}}{\alpha_{1i}} \frac{\beta_{2i}}{\alpha_{2j}} > 1 , \qquad (14.8)$$

2. Loss disproportion

$$\alpha_{1i}, \alpha_{1j}, \alpha_{2i}, \alpha_{2j} > 0, \ \frac{\alpha_{1i}}{\alpha_{1j}} \neq \frac{\alpha_{2i}}{\alpha_{2j}} , \qquad (14.9)$$

3. Gain–loss connection

$$\alpha_{1i}, \alpha_{1j}, \alpha_{2i}, \beta_{2j} > 0, \ \frac{\alpha_{1i}}{\alpha_{1j}} \frac{\beta_{2j}}{\alpha_{2i}} > 1 . \qquad (14.10)$$

These elementary *obstacles to dissipativity* alow to prove the non-dissipativity of many reaction mechanisms.

It is possible to describe all the dissipative reaction mechanisms. For example, for three components A_1, A_2, A_3 with the linear conservation law $c_1 + c_2 + c_3 = $ const, for bounded stoichiometric numbers ($\alpha_{ri}, \beta_{ri} \leq 3$), and without autocatalysis ($\alpha_{ri} \times \beta_{ri} \equiv 0$) there are three such maximal mechanisms, and any other dissipative mechanism is a subset of one of these maximal schemes (may be after permutation of the components) [305]. The first maximal scheme consists of all the reactions without interaction of different substances and with given conservation law:

$$\alpha_r A_{i_r} \to \beta_{r1} A_1 + \beta_{r2} A_2 + \beta_{r3} A_3, \ \alpha_r \leq 3, \ \beta_{r1} + \beta_{r2} + \beta_{r3} = \alpha_r . \quad (14.11)$$

The corresponding Lyapunov norm is $\|c\| = |c_1| + |c_2| + |c_3|$.

The second maximal scheme includes a reaction with interaction of different substances, $A_1 + A_2 \to 2A_3$:

$$A_1 \to A_2, \ A_1 \to A_3, \ A_2 \to A_1, \ A_2 \to A_3 ,$$
$$2A_1 \to A_2 + A_3, \ 2A_2 \to A_1 + A_3, \ 2A_3 \to A_1 + A_2 ,$$
$$3A_1 \to A_2 + 2A_3, \ 3A_1 \to 2A_2 + A_3, \ 3A_2 \to A_1 + 2A_3, \ 3A_2 \to 2A_1 + A_3 ,$$
$$A_1 + A_2 \to 2A_3 . \qquad (14.12)$$

The corresponding norm is $\|c\| = |c_1| + |c_2| + |c_1 + c_2 + c_3|$.

The third maximal scheme includes another reaction with interaction of different substances, $A_1 + 2A_2 \to 3A_3$:

$A_1 \to A_2$, $A_1 \to A_3$, $A_2 \to A_3$,

$2A_1 \to A_2 + A_3$, $2A_2 \to A_1 + A_3$,

$3A_1 \to A_2 + 2A_3$, $3A_1 \to 2A_2 + A_3$, $3A_2 \to A_1 + 2A_3$, $3A_3 \to A_1 + 2A_2$,

$A_1 + 2A_2 \to 3A_3$.   (14.13)

The corresponding norm is $\|c\| = 2|c_1| + |c_2| + |c_1 + c_2 + c_3|$.

14.3 Examples: Infinite-Dimensional Systems With Finite-Dimensional Attractors

In this section we list some of known examples of infinite-dimensional systems that are k-contractions (that is, they contract k-dimensional volume for some k) and have finite-dimensional attractors.

The most celebrated example are the *Navier–Stokes equations* for incompressible fluid in a bounded domain Ω of R^3 or R^2:

$$\mathrm{div}\, \boldsymbol{u} = 0 \ ,$$

$$\frac{\partial \boldsymbol{u}}{\partial t} + (\boldsymbol{u}, \boldsymbol{\nabla})\boldsymbol{u} - \nu \triangle \boldsymbol{u} + \boldsymbol{\nabla} p = \boldsymbol{f} \ , \qquad (14.14)$$

where $\boldsymbol{u}(\boldsymbol{x}, t)$ is the velocity field, ν is the kinematic viscosity, $\triangle = \boldsymbol{\nabla}^2$ is the Laplace operator, $(\, , \,)$ is the standard inner product, p is the pressure, $\boldsymbol{f}$ is the volume force (normalized to unit density).

The two main cases are studied, the flow with the Dirichlet boundary conditions $\boldsymbol{u} = 0$ on $\partial \Omega$, or the flow with periodic boundary conditions.

For three-dimensional flows it is possible to prove conditional theorems of such a kind [313]: If X is a bounded attractor, then

$$D_H(X) \leq \mathrm{const}\, \mathrm{Re}^3 \ , \qquad (14.15)$$

where const is an universal constant and Re is the Reynolds number. The usual definition of Re is

$$\mathrm{Re} = \frac{LU}{\nu} \ , \qquad (14.16)$$

where L and U are the reference length and reference velocity.

There is no natural "typical velocity" in a turbulent flow and therefore there is no obvious unique choice of the Reynolds number. The different versions of the Reynolds numbers are discussed, for instance, in [313, 314].

The classical physical estimate of the number of degrees of freedom for flows with large Reynolds numbers is $\sim \mathrm{Re}^{9/4}$ [315]. This estimate was obtained mathematically rigorously for the homogeneous decaying turbulence [316]. That is, for the Navier–Stokes equations without a forse $\boldsymbol{f}$ and after a special time-dependent transformation of space and time scales.

For two-dimensional flows it is possible to prove the existence of attractors, because in this case the situation with existence and uniqueness of solutions is much more clear. There is a family of estimates of attractor dimension [318]. The simplest and at the same time most important form of such an estimate is

$$D_H(X) \le \text{const}\,(G+1) , \qquad (14.17)$$

where G is the generalized Grashof number:

$$G = \frac{L^2\|f\|}{\nu^2} , \qquad (14.18)$$

$\|f\|$ is the L^2-norm of f in Ω, L is the reference length. Of course, this definition allows some useful variations, as well, as the Reynolds number.

For special classes of solutions these estimates could be made more precise. For example, for the boundary and pressure-gradient driven incompressible fluid flows in elongated two-dimensional channels the following estimates of the dimension of the attractor for the solutions of the Navier–Stokes equations are obtained [320]. For boundary driven shear flows and flux driven channel flows the upper bounds for the degrees of freedom was found in the form $c\alpha\,\text{Re}^{3/2}$ where c is a universal constant, α denotes the aspect ratio of the channel (length/width), and Re is the Reynolds number based on the channel width and the imposed "outer" velocity scale. For fixed pressure gradient driven channel flows an upper bound of form $c'\,\text{Re}^2$ was obtained, where c' is another universal positive constant and the Reynolds number is based on a velocity defined by the infimum, over all possible trajectories, of the time averaged mass flux per unit channel width.

The estimates of the global attractor dimension were obtained for so-called *generalized Navier–Stokes* equations characterized by polynomial dependence between the stress tensor and the symmetric velocity gradient [319].

The maximal attractor X for the *complex Ginzburg–Landau equation* periodic on the interval $[0,1]$

$$\frac{\partial A}{\partial t} = RA + (1+i\nu)\frac{\partial^2 A}{\partial x^2} - (1+i\mu)A|A|^2 \qquad (14.19)$$

has a finite dimension [321, 322]:

– For $|\mu| \le \sqrt{3}$
$$D_H(X) < 2\sqrt{3}(R^{1/2}/2\pi) + 1 ;$$

– For $|\mu| > \sqrt{3}$

$$D_H(X) < (\sqrt{3}/\pi)|\mu|R + 3|\mu|^{1/2}R^{1/2}/2\pi + 2 .$$

The *Kuramoto–Sivashinsky equation* was invented in order to describe waves in chemically reacting media:

$$\frac{\partial c}{\partial t} = -\nu \triangle^2 c - \triangle c - \frac{1}{2}(\nabla c)^2 \ . \tag{14.20}$$

where $\nu > 0$ and $\triangle$ is the Laplace operator. The term $-\triangle c$ in the right hand side of this equation describes the negative diffusion, and the solutions remain regular due to the term $-\nu \triangle^2 c$.

The global dynamical properties of the Kuramoto–Sivashinsky equation were studied in [323]. In particular, in the one-dimensional case for *even* solutions with the L-periodic boundary conditions, the following estimate for the dimension of the attractor X which includes all bounded attractors was obtained:

$$D_H(X) \le 5.4 \tilde{L}^{13/8} \ , \tag{14.21}$$

where $\tilde{L} = L/(2\pi\sqrt{\nu})$ is the dimensionless size of the box (or of the "pattern cell").

For the two- and three-dimensional cases the estimations of attractor dimension include an a priori estimate of the L^2 norm of the c gradient $R = \overline{\lim}_{t\to\infty} \|\nabla c(t)\|$ [323].

Dynamics of the *nonlocal Kuramoto–Sivashinsky equation* (in the one-dimensional case) was studied in [324]

$$\frac{\partial c}{\partial t} = -\frac{\partial^4 c}{\partial x^4} - \frac{\partial^2 c}{\partial x^2} - c\frac{\partial c}{\partial x} - \alpha H\left(\frac{\partial^3 c}{\partial x^3}\right) \ , \tag{14.22}$$

where $\alpha > 0$, and

$$H(f) = \frac{1}{\pi} \int_{-\infty}^{\infty} \frac{f(\xi)}{x - \xi} \, d\xi \ , \tag{14.23}$$

(the integral is understood in the sense of the Cauchy principle value). This equation arises in the modeling of the flow of a thin film of viscous liquid down an inclined plane in electric field.

For large α and l the obtained attractor dimension estimate is

$$D_H(X) \le O(\alpha^{3/2}\tilde{L}^{3/2} + \tilde{L}^2) \ . \tag{14.24}$$

The *Cahn–Hilliard equation* describes the evolution of a conserved concentration field during phase separation:

$$\frac{\partial c}{\partial t} = \triangle(-\nu \triangle c + f(c)) \ , \tag{14.25}$$

where $f(c)$ is a polynomial of an odd order. Usually, $f(c) = -\alpha c + \beta c^3$, $\alpha, \beta > 0$. Like the Kuramoto–Sivashinsky equation, the Cahn–Hilliard equation contains the regularizing term $-\nu \triangle^2 c$ and the negative diffusion term $-\alpha \triangle c$ in the right hand side. The operator $-\nu \triangle^2$ effectively suppresses the short waves (with the logarithmic decrement $\sim k^4$ for the wave vector k); it allows to prove the existence of the finite-dimensional attractors and to estimate their dimensions for various hypothesis about f and boundary conditions [325–329]. The Cahn–Hilliard equation can be presented as a gradient system that minimizes the free energy (possibly, to a local minimum). Hence, the limiting behavior of this system is rather simple; it always tends to a stationary point (to the stable or metastable equilibrium).

It should be stressed that the maximal attractor (4.1)

$$X = \bigcap_{t \geq 0} T_t(Y) \,,$$

where Y is the positively invariant absorbing set, and T_t is the map of the phase flow (the shift operator for the time t), estimates the limit behavior from above and often does not satisfy the intuitive expectation (that is, it is too large). For example, if the absorbing set Y includes a saddle point then the corresponding maximal attractor X includes the whole unstable manifold of this point, and the dimension of X is not less than the dimension of this unstable manifold. For instance, when all the solutions tend to fixed points from a finite set, the dimension of maximal attractor might be nonzero and rather large. Therefore, the study of the *structure* of the attractor becomes important. For the *viscous Cahn–Hilliard equation*

$$(1 - \alpha)\frac{\partial c}{\partial t} = \triangle \left(-\nu \triangle c + f(c) + \alpha \frac{\partial c}{\partial t} \right) \,, \tag{14.26}$$

with usual no-flow boundary conditions the structure of the global attractor was studied in [330] (for the one-dimensional case). The dimension of the unstable manifolds was calculated for all stationary states. In the unstable case, the flow on the global attractor is shown to be semi-conjugate to the flow of the global attractor of the *Chaffee–Infante equation*

$$\frac{\partial c}{\partial t} = \frac{\partial^2 c}{\partial x^2} + \lambda^2 (c - c^3) \tag{14.27}$$

with zero boundary conditions. The connection between these equations is in accordance with their physical sense [331, 332].

In the *reaction–diffusion equations* the only source of instabilities can be the reaction part. The diffusion suppresses short waves (with the logarithmic decrement $\sim k^2$ for wave vector k). Let us consider the following system in a bounded domain $\Omega \subset R^n$ with the zero or non-flow boundary conditions [333, 335]:

$$\frac{\partial c}{\partial t} = \nu d \triangle c - f(c) + g(x) \,, \tag{14.28}$$

where $c(x,t) = (c^1, \ldots, c^N)$, $\nu > 0$, $d = d_{ij}$ is a real $N \times N$ matrix with positive symmetric part $(d + d^+)/2 \geq 1$, $f(c) = (f^1(c), \ldots, f^N(c))$, and $g(x) = (g^1(x), \ldots, g^N(x))$. Here, $f(c)$ is the reaction part, and $g(x)$ describes the external sources and sinks.

Let $f(c)$ be a smooth map, $D_c f(c) = (\partial f^i / \partial c^j)$, and the following conditions hold

$$\sum_i \gamma_i |c^i|^{p_i} - C_1 \leq (c, f(c)) ; \qquad (14.29)$$

$$\sum_i |f^i(c)|^{q_i} \leq C_2 \left(\sum_i |c^i|^{p_i} + 1 \right) ; \qquad (14.30)$$

$$(D_c f(c)w, w) \geq -C_3 |w|^2 , \qquad (14.31)$$

for some $C_1, C_2, C_3 > 0$, $\gamma_i > 0$ $p_i \geq 2$, $1/p_i + 1/q_i = 1$, any c and all $w \in R^N$. Then the dimension of the attractor X for (14.28) admits the following estimate:

$$D_H(X) \leq N|\Omega| \left(\frac{C_3}{\nu c_0(\Omega)} \right)^{n/2} , \qquad (14.32)$$

where n is the space dimension, $|\Omega|$ is the measure of $\Omega \subset R^n$, and $c_0(\Omega)$ depends on the shape of Ω only, that is, $c_0(\lambda\Omega) = c_0(\Omega)$ for all $\lambda > 0$. The constant C_3 estimates the possible increment of the kinetic trajectories instability from above.

The reaction–diffusion systems remain the basic example of the finite-dimension volumes contractions, because of transparent structure: the diffusion generates contraction (with the logarithmic decrement $\sim k^2$ for wave vector k), and the reaction can cause finite-dimensional instabilities. It is well-studied, see, for instance, [318, 333–336].

For the *dissipative wave equations* the estimates of attractor dimensions were found in [333, 337–339] and discussed in [318, 335]. Let us consider the following equation in a bounded domain $\Omega \subset R^n$

$$\frac{\partial^2 u}{\partial t^2} + \gamma \frac{\partial u}{\partial t} = \Delta u - f(u) + g(x) \qquad (14.33)$$

with zero boundary conditions. This equation contains the damping term $\gamma \frac{\partial u}{\partial t}$ with $\gamma > 0$. The essential difference between (14.33) and the reaction–diffusion equations (14.28) is in the character of dissipation. According to the dissipative wave equations short waves decay as $\exp(-\gamma t)$, the logarithmic decrement has the constant limit for $k^2 \to \infty$. It is the same situation, as for the Grad equations (see Chap. 8), or for the model kinetic equations (see Chap. 2).

Let $0 < \lambda_1 \leq \lambda_3 \leq \lambda_4 \leq \ldots$ be the eigenvalues of the operator $-\Delta$ in Ω with zero boundary conditions (in the order of their values). Assume that $g \in L^2(\Omega)$ and the smooth function $f(u)$ satisfies the following conditions for any u

$$F(u) \geq -mu^2 - C_m ; \qquad (14.34)$$
$$f(u)u - \varsigma F(u) + mu^2 \geq -C_m ; \qquad (14.35)$$
$$|f'_u(u)| \leq C_0(1 + |u|^p) , \qquad (14.36)$$

where $F(u) = \int_0^u f(v)\, dv$, $\lambda_1 > m > 0$, $\varsigma > 0$, C_m and C_0 are some constants, $2/(n-2) > \rho > 0$ for $n \geq 3$, and $\rho > 0$ is an arbitrary positive number for $n = 1, 2$.

The important examples of (14.33) give the nonlinearities

$$f(u) = \beta \sin u$$

and

$$f(u) = \beta |u|^\rho u .$$

In order to present more explicit expressions for dimension estimates let us restrict ourselves by the *sin-Gordon equation* with $f(u) = \beta \sin u$. More general consideration can be found elsewhere (see, for example, [318, 335]).

Let us introduce the following dimensionless numbers:

$$J = \frac{\beta^2}{\lambda_1 \gamma^2}, \quad D = \frac{\gamma^2}{\lambda_1} \tag{14.37}$$

For the sin-Gordon equation the dimension estimate of the attractor is $D_H(X) \leq m$, where the number m (see [318], p. 364) is the first integer such that

$$m \geq 2^7 \lambda_1 \sum_{i=1}^m \lambda_i^{-1} J(1 + D^2) . \tag{14.38}$$

For large i, $\lambda_i \sim \text{const} \lambda_1 i^{2/n}$, hence, for all i, $\lambda_1/\lambda_i < \text{const}\, i^{-2/n}$,

$$\lambda_1 \sum_{i=1}^m \lambda_i^{-1} < C m^{1 - \frac{2}{n}} \tag{14.39}$$

for $n > 2$,

$$\lambda_1 \sum_{i=1}^m \lambda_i^{-1} < C \ln m \tag{14.40}$$

for $n = 2$, and

$$\lambda_1 \sum_{i=1}^m \lambda_i^{-1} < C \tag{14.41}$$

for $n = 1$. Here, in (14.39)–(14.40) constants depend only on the shape of Ω. Following these inequalities we can take as the upper dimension estimate the first integer m such that

$$m \geq (2^7 C J(1 + D^2))^{n/2} \tag{14.42}$$

for $n > 2$,

$$m \geq 2^7 C J(1 + D^2) \ln m \tag{14.43}$$

for $n = 2$, and

$$m \geq 2^7 C J(1 + D^2) \tag{14.44}$$

for $n = 1$.

The estimates for attractor dimension were also obtained for some other systems: thermohydraulics (including the Boussinesq equation), magnetohydrodynamics (see [318]), the Smoluchowski equation arising in the modeling of nematic liquid crystalline polymers [340], and others. Various dimensions (not only the Hausdorff dimension) were estimated.

14.4 Systems with Inheritance: Dynamics of Distributions with Conservation of Support, Natural Selection and Finite-Dimensional Asymptotics

14.4.1 Introduction: Unusual Conservation Law

In the 1970th-1980th years, theoretical studies developed one more "common" field belonging simultaneously to physics, biology and mathematics. For physics it is (so far) a part of the theory of approximations of a special kind, demonstrating, in particular, interesting mechanisms of discreteness in the course of evolution of distributions with initially smooth densities. But what for physics is merely a convenient approximation, is a fundamental law in biology (inheritance), whose consequences comprehended informally (selection theory [115, 377–380, 383, 384])[2] permeate most of the sections of this science.

Consider a community of animals. Let it be biologically isolated. Mutations can be neglected in the first approximation. In this case new genes do not emerge.

And here is an example is from physics. Let waves with wave vectors k be excited in some system. Denote K a set of wave vectors k of excited waves. Let the waves interaction do not lead to generation of waves with new $k \notin K$. Such an approximation is applicable to a variety of situations. For the wave turbulence it was described in detail in [385, 386].

What is common in these examples is the evolution of a distribution with a support not increasing in time.

What does not increase must, as a rule, decrease, if the decrease is not prohibited. This naive thesis can be converted into rigorous theorems for the case under consideration [115]. The support is proved to decrease in the limit $t \to \infty$, if it was sufficiently large initially. Considered usually are such system, for which at finite times the distributions support conserves and decrease only in the limit $t \to \infty$. Conservation of the support usually results in the following effect: dynamics of initially infinite-dimensional system at $t \to \infty$ can be described by finite-dimensional systems.

---

[2] We do not try to review the scientific literature about the evolution, and mention here only the references that are especially important for our understanding of the selection theory and applications.

The simplest and most common in applications class of equations for which the distributions support does not grow in time, is constructed as follows. To each distribution μ is assigned a function k_μ by which the distribution can be multiplied. Written down is equation

$$\frac{\mathrm{d}\mu}{\mathrm{d}t} = k_\mu \times \mu \,. \tag{14.45}$$

The multiplier k_μ is called a *reproduction coefficient*. The right-hand side is the product of the function k_μ with the distribution μ, hence $\mathrm{d}\mu/\mathrm{d}t$ should be zero where μ is equal to zero, therefore the support of μ is conserved in time (over the finite times).

Let us remind the definition of the support. Each distribution on a compact space X is a continuous linear functional on the space of continuous functions $C(X)$[3]. The space $C(X)$ is the Banach space endowed with the norm

$$\|f\| = \max_{x \in X} |f(x)| \,. \tag{14.46}$$

Usually, when X is a bounded closed subset of a finite-dimensional space, we represent this functional as the integral

$$\mu[f] = \int \mu(x)f(x)\,\mathrm{d}x \,,$$

where $\mu(x)$ is the (generalized) density function of the distribution μ. The support of μ, suppμ, is the smallest closed subset of X with the following property: if $f(x) = 0$ on suppμ, then $\mu[f] = 0$, i.e. $\mu(x) = 0$ outside suppμ.

Strictly speaking, the space on which μ is defined and the distribution class it belongs to, should be specified. One should also specify are properties of the mapping $\mu \mapsto k_\mu$ and answer the question of existence and uniqueness of solutions of (14.45) under given initial conditions. In specific situations the answers to these questions are not difficult.

Let us start with the simplest example

$$\frac{\partial \mu(x,t)}{\partial t} = \left[f_0(x) - \int_a^b f_1(x)\mu(x,t)\,\mathrm{d}x \right] \mu(x,t) \,, \tag{14.47}$$

where the functions $f_0(x)$ and $f_1(x)$ are positive and continuous on the closed segment $[a, b]$. Let the function $f_0(x)$ reaches the global maximum on the segment $[a, b]$ at a single point x_0. If $x_0 \in$ supp$\mu(x, 0)$, then

$$\mu(x,t) \to \frac{f_0(x_0)}{f_1(x_0)}\delta(x - x_0), \text{ when } t \to \infty \,, \tag{14.48}$$

---

[3] We follow the Bourbaki approach [393]: a measure is a continuous functional, an integral. The book [393] contains all the necessary notions and theorems (and much more material than we need here).

where $\delta(x - x_0)$ is the δ-function.

We use in the space of measures the *weak convergence*, i.e. the convergence of averages:

$$\mu_i \to \mu^* \text{ if and only if } \int \mu_i \varphi(x) \, \mathrm{d}x \to \int \mu^* \varphi(x) \, \mathrm{d}x \qquad (14.49)$$

for all continuous functions $\varphi(x)$. This weak convergence of measures generates the *weak topology* on the space of measures (the weak topology of conjugated space).

If $f_0(x)$ has several global maxima, then the right-hand side of (14.48) can be a sum of a finite number of δ-functions. Here a natural question arises: is it worth to pay attention to such a possibility? Should not we deem improbable for $f_0(x)$ to have more than one global maximum? Indeed, such a case seems to be very unlikely to occur. More details about this are given below.

Equations in the form (14.45) allow the following biological interpretation: μ is the distribution of the number (or of a biomass, or of another extensive variable) over inherited units: species, varieties, supergenes, genes. Whatever is considered as the *inherited unit* depends on the context, on a specific problem. The value of $k_\mu(x)$ is the reproduction coefficient of the inherited unit x under given conditions. The notion of "given conditions" includes the distribution μ, the reproduction coefficient depends on μ. Equation (14.47) can be interpreted as follows: $f_0(x)$ is the specific birth-rate of the inherited unit x (below, for the sake of definiteness, x is a variety, following the spirit of the famous Darwin's book [377]), the death rate for the representatives of all inherited units (varieties) is determined by one common factor depending on the density $\int_a^b f_1(x)\mu(x,t)\,\mathrm{d}x$; $f_1(x)$ is the individual contribution of the variety x into this death-rate.

On the other hand, for systems of waves with a parametric interaction, $k_\mu(x)$ can be the amplification (decay) rate of the wave with the wave vector x.

The first step in the routine of a dynamical system investigation is the question about fixed points and their stability. And the first observation concerning the system (14.45) is that asymptotically stable can be only the steady-state distributions, whose support is discrete (i.e. the sums of δ-functions). This can be proved for all the consistent formalizations, and can be understood as follows.

Let the "total amount" (integral of $|\mu|$ over U) be less than $\varepsilon > 0$ but not equal to zero in some domain U. Substitute distribution μ by zero on U, the rest remains as it is. It is natural to consider this disturbance of μ as ε-small. However, if the dynamics is described by (14.45), there is no way back to the undisturbed distribution, because the support cannot increase. If the steady state distribution μ^* is asymptotically stable, then for some $\varepsilon > 0$ any ε-small perturbation of μ^* relaxes back to μ^*. This is possible only in the case if for any domain U the integral of $|\mu^*|$ over U is either 0 or greater than ε. Hence, this asymptotically stable distribution μ^* is the sum of the finite number of δ-functions:

$$\mu^*(x) = \sum_{i=1}^{q} N_i \delta(x - x_i) \tag{14.50}$$

with $|N_i| > \varepsilon$ for all i.

So we have: the support of asymptotically stable distributions for the system (14.45) is always discrete. This simple observation has many strong generalizations to general ω-limit points, to equations for vector measures, etc.

Dynamic systems where the phase variable is a distribution μ, and the distribution support is the integral of motion, frequently occur both in physics and in biology. Because of their attractive properties they are frequently used as approximations: we try to find the "main part" of the system in the form (14.45), and represent the rest as a small perturbation of the main part.

In biology such an approximation is essentially all the classical genetics, and also the formal contents of the theory of natural selection. The initial diversity is "thinned out" in time, and the limit distribution supports are described by some extremal principles (principles of optimality).

Conservation of the support in equation (14.45) can be considered as inheritance, and, consequently, we call the system (14.45) and its nearest generalizations "systems with inheritance". Traditional division of the process of transferring biological information into inheritance and mutations, small in any admissible sense, can be compared to the description according to the following pattern: system (14.45) (or its nearest generalizations) plus small disturbances. Beyond the limits of such a description, talking about inheritance loses the conventional sense.

The first study of the dynamics systems with inheritance was due to J.B.S. Haldane. He used the simplest examples, studied steady-state distributions, and obtained the extremal principle for them. His pioneering book "The Causes of Evolution" (1932) [378] gives the clear explanation of the connections between the inheritance (the conservation of distributions support) and the optimality of selected varieties.

Haldane's work was followed by entirely independent series of works on the S-approximation in the spin wave theory and on the wave turbulence [385–387], which studied wave configurations in the approximation of "inherited" wave vector, and by "Synergetics" [392], where the "natural selections" of modes is one of the basic concepts.

At the same time, a series of works on biological kinetics was done (see, for example, [115, 381–383]). The studies addressed not only steady-states, but also common limit distributions [115, 382] and waves in the space of inherited units [381]. For the steady-states a new type of stability was described – the *stable realizability* (see below).

The purpose of this paper is to present general results of the theory of systems with inheritance: optimality principles the for limit distributions, theorems about selection, and estimations of the limit diversity (estimates of number of points in the supports of the limit distributions), drift effect and drift equations. Some of these result were published in preprints in Russian [382] (and, partially, in the Russian book [115]).

14.4.2 Optimality Principle for Limit Diversity

Description of the limit behavior of a dynamical system does not necessarily reduce to enumerating stable fixed points and limit cycles. The possibility of stochastic oscillations is common knowledge, while the domains of structurally unstable (non coarse) systems discovered by S. Smale [394][4] have so far not been mastered in applied and natural sciences).

The leading rival to adequately formalize the limit behavior is the concept of the "ω-limit set". It was discussed in detail in the classical monograph [395]. The fundamental textbook on dynamical systems [396] and the introductory review [397] are also available.

If $f(t)$ is the dependence of the position of point in the phase space on time t (i.e. the *motion* of the dynamical system), then the ω-limit points are such points y, for which there exist such sequences of times $t_i \to \infty$, that $f(t_i) \to y$.

The set of all ω-limit points for the given motion $f(t)$ is called the ω-limit set. If, for example, $f(t)$ tends to the equilibrium point y^* then the corresponding ω-limit set consists of this equilibrium point. If $f(t)$ is winding onto a closed trajectory (the limit cycle), then the corresponding ω-limit set consists of the points of the, cycle and so on.

General ω-limit sets are not encountered oft in specific situations. This is because of the lack of efficient methods to find them in a general situation. Systems with inheritance is a case, where there are efficient methods to estimate the limit sets from above. This is done by the optimality principle.

Let $\mu(t)$ be a solution of (14.45). Note that

$$\mu(t) = \mu(0) \exp \int_0^t k_{\mu(\tau)} \, d\tau \ . \tag{14.51}$$

---

[4] "Structurally stable systems are not dense". Without exaggeration we can say that so entitled work [394] opened a new era in the understanding of dynamics. Structurally stable (rough) systems are those whose phase portraits do not change qualitatively under small perturbations. Smale constructed such structurally unstable system that any other system close enough to it is also structurally unstable. This result defeated hopes for a classification if not all, but at least "almost all" dynamical systems. Such hopes were associated with the success of the classification of two-dimensional dynamical systems, among which structurally stable systems are dense.

Here and below we do not display the dependence of distributions μ and of the reproduction coefficients k on x when it is not necessary. Fix the notation for the average value of $k_{\mu(\tau)}$ on the segment $[0, t]$

$$\langle k_{\mu(t)} \rangle_t = \frac{1}{t} \int_0^t k_{\mu(\tau)} \, d\tau \, . \tag{14.52}$$

Then the expression (14.51) can be rewritten as

$$\mu(t) = \mu(0) \exp(t \langle k_{\mu(t)} \rangle_t) \, . \tag{14.53}$$

If μ^* is the ω-limit point of the solution $\mu(t)$, then there exists such a sequence of times $t_i \to \infty$, that $\mu(t_i) \to \mu^*$. Let it be possible to chose a convergent subsequence of the sequence of the average reproduction coefficients $\langle k_{\mu(t)} \rangle_t$, which corresponds to times t_i. We denote as k^* the limit of this subsequence. Then, the following statement is valid: on the support of μ^* the function k^* vanishes and on the support of $\mu(0)$ it is non-positive:

$$k^*(x) = 0 \text{ if } x \in \text{supp}\mu^* \, ,$$
$$k^*(x) \le 0 \text{ if } x \in \text{supp}\mu(0) \, . \tag{14.54}$$

Taking into account the fact that $\text{supp}\mu^* \subseteq \text{supp}\mu(0)$, we come to the formulation of **the optimality principle** (14.54): *The support of limit distribution consists of points of the global maximum of the average reproduction coefficient on the initial distribution support. The corresponding maximum value is zero.*

We should also note that not necessarily all points of maximum of k^* on $\text{supp}\mu(0)$ belong to $\text{supp}\mu^*$, but all points of $\text{supp}\mu^*$ are the points of maximum of k^* on $\text{supp}\mu(0)$.

If $\mu(t)$ tends to the fixed point μ^*, then $\langle k_{\mu(t)} \rangle_t \to k_{\mu^*}$ as $t \to \infty$, and $\text{supp}\mu^*$ consists of the points of the global maximum of the corresponding reproduction coefficient k_{μ^*} on the support of μ^*. The corresponding maximum value is zero.

If $\mu(t)$ tends to the limit cycle $\mu^*(t)$ $(\mu^*(t + T) = \mu^*(t))$, then all the distributions $\mu^*(t)$ have the same support. The points of this support are the points of maximum (global, zero) of the averaged over the cycle reproduction coefficient

$$k^* = \langle k_{\mu^*(t)} \rangle_T = \frac{1}{T} \int_0^T k_{\mu^*(\tau)} \, d\tau \, , \tag{14.55}$$

on the support of $\mu(0)$.

The supports of the ω-limit distributions are specified by the functions k^*. It is obvious where to get these functions from for the cases of fixed points and limit cycles. There are at least two questions: what ensures the existence of average reproduction coefficients at $t \to \infty$, and how to use the described extremal principle (and how efficient is it). The latter question is the subject to be considered in the following sections. In the situation to follow the answers to these questions have the validity of theorems. Let X be a space on which the distributions are defined. Assume it to be a compact metric space (for example, a closed bounded subset of Euclidian space). The distribution μ is identified with the Radon measure, that is the continuous linear functional on the space of continuous functions on $X, C(X)$. We use the conventional notation for this linear functional as the integral of the function φ as $\int \varphi(x)\mu(x)\,dx$. Here $\mu(x)$ is acting as the distribution density, although, of course, the arbitrary X has no initial (Lebesgue, for example) dx.

The sequence of continuous functions $k_i(x)$ is considered to be convergent if it converges uniformly. The sequence of measures μ_i is called convergent if for any continuous function $\varphi(x)$ the integrals $\int \varphi(x)\mu_i(x)\,dx$ converge (weak convergence (14.49)). The mapping $\mu \mapsto k_\mu$ assigning the reproduction coefficient k_μ to the measure μ is assumed to be continuous. And, finally, the space of measures is assumed to have a bounded[5] set M which is positively invariant relative to system (14.45): if $\mu(0) \in M$, then $\mu(t) \in M$ (and is non-trivial). This M will serve as the phase space of system (14.45).

Most of the results about systems with inheritance use **the theorem about weak compactness**: *The bounded set of measures is precompact with respect to the the weak convergence (i.e., its closure is compact).* Therefore, the set of corresponding reproduction coefficients $k_M = \{k_\mu | \mu \in M\}$ is precompact, the set of averages (14.52) is precompact, because it is the subset of the closed convex hull $\overline{\text{conv}}(k_M)$ of the compact set. This compactness allows us to claim the existence of the *average reproduction coefficient* k^* for the description of the ω-limit distribution μ^* with the optimality principle (14.54).

14.4.3 How Many Points Does the Limit Distribution Support Hold?

The limit distribution is concentrated in the points of (zero) global maximum of the average reproduction coefficient. The average is taken along the solution, but the solution is not known beforehand. With the convergence towards a fixed point or to a limit cycle this difficulty can be circumvented. In the general case the extremal principle can be used without knowing the solution, in the following way [115]. Considered is a set of all dependencies $\mu(t)$ where μ belongs to the phase space, the bounded set M. The set of all averages over t is $\{\langle k_{\mu(t)}\rangle_t\}$. Further, taken are all limits of sequences formed by these averages – the set of averages is closed. The result is the closed convex hull $\overline{\text{conv}}(k_M)$ of the compact set k_M. This set involves all possible averages (14.52) and all their limits. In order to construct it, the true solution $\mu(t)$ is not needed.

---

[5] The set of measures M is bounded, if the sets of integrals $\{\mu[f]|\mu \in M, \|f\| \leq 1\}$ is bounded, where $\|f\|$ is the norm (14.46).

The weak optimality principle is expressed as follows. Let $\mu(t)$ be a solution of (14.45) in M, μ^* is any of its ω-limit distributions. Then in the set $\overline{\text{conv}}(k_M)$ there is such a function k^* that its maximum value on the support $\text{supp}\mu_0$ of the initial distribution μ_0 equals to zero, and $\text{supp}\mu^*$ consists of the points of the global maximum of k^* on $\text{supp}\mu_0$ only (14.54).

Of course, in the set $\overline{\text{conv}}(k_M)$ usually there are many functions that are irrelevant to the time average reproduction coefficients for the given motion $\mu(t)$. Therefore, the weak extremal principle is really weak – it gives too many possible supports of μ^*. However, even such a principle can help to obtain useful estimates of the number of points in the supports of ω-limit distributions.

It is not difficult to suggest systems of the form (14.45), in which any set can be the limit distribution support. The simplest example: $k_\mu \equiv 0$. Here ω-limit (fixed) is any distribution. However, almost any arbitrary small perturbation of the system destroys this pathological property.

In the realistic systems, especially in biology, the coefficients fluctuate and are never known exactly. Moreover, the models are in advance known to have a finite error which cannot be exterminated by the choice of the parameters values. This gives rise to an idea to consider not individual systems (14.45), but ensembles of similar systems [115].

Having posed the questions of how many points can the support of ω-limit distributions have, estimate the maximum for each individual system from the ensemble (in its ω-limit distributions), and then, estimate the minimum of these maxima over the whole ensemble – (*the minimax estimation*). The latter is motivated by the fact, that if the inherited unit has gone extinct under some conditions, it will not appear even under the change of conditions.

Let us consider an ensemble that is simply the ε-neighborhood of the given system (14.45). The minimax estimates of the number of points in the support of ω-limit distribution are constructed by approximating the dependencies k_μ by finite sums

$$k_\mu = \varphi_0(x) + \sum_{i=1}^{n} \varphi_i(x)\psi_i(\mu) . \qquad (14.56)$$

Here φ_i depend on x only, and ψ_i depend on μ only. Let $\varepsilon_n > 0$ be the distance from k_μ to the nearest sum (14.56) (the "distance" is understood in the suitable rigorous sense, which depends on the specific problem). So, we reduced the problem to the estimation of the diameters $\varepsilon_n > 0$ of the set $\overline{\text{conv}}(k_M)$.

The minimax estimation of the number of points in the limit distribution support gives the answer to the question, "How many points does the limit distribution support hold": *If $\varepsilon > \varepsilon_n$ then, in the ε-vicinity of k_μ, the minimum of the maxima of the number of points in the ω-limit distribution support does not exceed n.*

In order to understand this estimate it is sufficient to consider system (14.45) with k_μ of the form (14.56). The averages (14.52) for any dependence

$\mu(t)$ in this case have the form

$$\langle k_{\mu(t)}\rangle_t = \frac{1}{t}\int_0^t k_{\mu(\tau)}\,\mathrm{d}\tau = \varphi_0(x) + \sum_{i=1}^n \varphi_i(x)a_i \ . \qquad (14.57)$$

where a_i are some numbers. The ensemble of the functions (14.57) for various a_i forms a n-dimensional linear manifold. How many of points of the global maximum (equal to zero) could a function of this family have?

Generally speaking, it can have any number of maxima. However, it seems obvious, that "usually" one function has only one point of global maximum, while it is "improbable" the maximum value is zero. At least, with an arbitrary small perturbation of the given function, we can achieve for the point of the global maximum to be unique and the maximum value be non-zero.

In a one-parametric family of functions there may occur zero value of the global maximum, which cannot be eliminated by a small perturbation, and individual functions of the family may have two global maxima.

In the general case we can state, that "usually" each function of the n-parametric family (14.57) can have not more than $n - 1$ points of the zero global maximum (of course, there may be less, and for the majority of functions of the family the global maximum, as a rule, is not equal to zero at all). What "usually" means here requires a special explanation given in the next section.

In application k_μ is often represented by an integral operator, linear or nonlinear. In this case the form (14.56) corresponds to the kernels of integral operators, represented in a form of the sums of functions' products. For example, the reproduction coefficient of the following form

$$k_\mu = \varphi_0(x) + \int K(x,y)\mu(y)\,\mathrm{d}y \ ,$$

$$\text{where } K(x,y) = \sum_{i=1}^n \varphi_i(x)g_i(y) \ , \qquad (14.58)$$

has also the form (14.56) with $\psi_i(\mu) = \int g_i(y)\mu(y)\,\mathrm{d}y$.

The linear reproduction coefficients occur in applications rather frequently. For them the problem of the minimax estimation of the number of points in the ω-limit distribution support is reduced to the question of the accuracy of approximation of the linear integral operator by the sums of kernels-products (14.58).

14.4.4 Selection Efficiency

The first application of the extremal principle for the ω-limit sets is the theorem of the selection efficiency. The dynamics of a system with inheritance indeed leads in the limit $t \to \infty$ to a selection. In the typical situation, a diversity in the limit $t \to \infty$ becomes less than the initial diversity. There is an efficient selection for the "best". The basic effects of selection are formulated below.

Theorem of Selection Efficiency

1. *Almost always the support of any ω-limit distribution is nowhere dense in X (and it has the Lebesgue measure zero for Euclidean space).*
2. *Let $\varepsilon_n > 0$, $\varepsilon_n \to 0$ be an arbitrary chosen sequence. The following statement is almost always true for system (14.45). Let the support of the initial distribution be the whole X. Then the support of any ω-limit distribution μ^* is almost finite. This means that it is approximated by finite sets faster than $\varepsilon_n \to 0$: for any $\delta > 0$ there is such a number N that for any ω-limit distribution μ^* there exists a finite set S_N of N elements such that $\mathrm{dist}(S_N, \mathrm{supp}\,\mu^*) < \delta\varepsilon_N$, where dist is the Hausdorff distance:*

$$\mathrm{dist}(A, B) = \max\{\sup_{x \in A} \inf_{x \in B} \rho(x, y), \sup_{x \in B} \inf_{x \in A} \rho(x, y)\}, \qquad (14.59)$$

where $\rho(x, y)$ is the distance between points.
3. *In the previous statement for any chosen sequence $\varepsilon_n > 0$, $\varepsilon_n \to 0$, almost all systems (14.45) have ω-limit distributions with supports that can be approximated by finite sets faster than $\varepsilon_n \to 0$. The order is important: "for any sequence almost all systems..." But if we use only the recursive (algorithmic) analogue of sequences, then we can easily prove the statement with the reverse order: "almost all systems for any sequence..." This is possible because the set of all recursive enumerable countable sets is also countable and not continuum. This observation is very important for algorithmic foundations of probability theory [398]. Let L be a set of all sequences of real numbers $\varepsilon_n > 0$, $\varepsilon_n \to 0$ with the property: for each $\{\varepsilon_n\} \in L$ the rational subgraph $\{(n, r) : \varepsilon_n > r \in Q\}$ is recursively enumerable. For almost all systems (14.45) and any $\{\varepsilon_n\} \in L$ the support of any ω-limit distribution μ^* is approximated by finite sets faster than $\varepsilon_n \to 0$.*

These properties hold for the continuous reproduction coefficients. It is well-known, that it is dangerous to rely on the genericity among continuous functions. For example, almost all continuous functions are nowhere differentiable. But the properties 1 and 2 hold also for the smooth reproduction coefficients on the manifolds and sometimes allow to replace the "almost finiteness" by simply finiteness. In order to appreciate this theorem, note that:

1. Support of an arbitrary ω-limit distribution μ^* consist of points of global maximum of the average reproduction coefficient on a support of the initial distribution. The corresponding maximum value is zero.
2. Almost always a function has only one point of global maximum, and corresponding maximum value is not 0.

3. In a one-parametric family of functions almost always there may occur zero values of the global maximum (at one point), which cannot be eliminated by a small perturbation, and individual functions of the family may stably have two global maximum points.
4. For a generic n-parameter family of functions, there may exist stably a function with $n - 1$ points of global maximum and with zero value of this maximum.
5. Our phase space M is compact. The set of corresponding reproduction coefficients k_M in $C(X)$ for the given map $\mu \to k_\mu$ is compact too. The average reproduction coefficients belong to the closed convex hull of this set $\overline{\mathrm{conv}}(k_M)$. And it is compact too.
6. A compact set in a Banach space can be approximated by its projection on an appropriate finite-dimensional linear manifold with an arbitrary accuracy. Almost always the function on such a manifold may have only $n - 1$ points of global maximum with zero value, where n is the dimension of the manifold.

The rest of the proof is purely technical. The easiest demonstration of the "natural" character of these properties is the demonstration of instability of exclusions: If, for example, a function has several points of global maxima then with an arbitrary small perturbation (for all usually used norms) it can be transformed into a function with the unique point of global maximum. However "stable" does not always mean "dense". In what sense the discussed properties of the system (14.45) are usually valid? "Almost always", "typically", "generically" a function has only one point of global maximum. This sentence should be given an rigorous meaning. Formally it is not difficult, but haste is dangerous when defining "genericity".

Here are some examples of correct but useless statements about "generic" properties of function: Almost every continuous function is not differentiable; Almost every C^1-function is not convex. Their meaning for applications is most probably this: the genericity used above for continuous functions or for C^1-function is irrelevant to the subject.

Most frequently the motivation for definitions of genericity is found in such a situation: given n equations with m unknowns, what can we say about the solutions? The answer is: in a typical situation, if there are more equations, than the unknowns ($n > m$), there are no solutions at all, but if $n \leq m$ (n is less or equal to m), then, either there is a ($m - n$)-parametric family of solutions, or there are no solutions.

The best known example of using this reasoning is the *Gibbs phase rule* in classical chemical thermodynamics. It limits the number of co-existing phases. There exists a well-known example of such reasoning in mathematical biophysics too. Let us consider a medium where n species coexist. The medium is assumed to be described by m parameters. In the simplest case, the medium is a well-mixed solution of m substances. Let the organisms interact through the medium, changing its parameters – concentrations of m

substances. Then, in a steady state, for each of the coexisting species we have an equation with respect to the state of the medium. So, the number of such species cannot exceed the number of parameters of the medium. In a typical situation, in the m-parametric medium in a steady state there can exist not more than m species. This is the *Gause concurrent exclusion principle* [399]. This fact allows numerous generalizations. Theorem of the natural selection efficiency may be considered as its generalization too.

Analogous assertion for a non-steady state coexistence of species in the case of equations (11) is not true. It is not difficult to give an example of stable coexistence under oscillating conditions of n species in the m-parametric medium at $n > m$. But, if k_μ are linear functions of μ, then for non-stable conditions we have the concurrent exclusion principle, too. In that case, the average in time of reproduction coefficient $k_{\mu(t)}$ is the reproduction coefficient for the average $\mu(t)$ because of linearity. Therefore, the equation for the average reproduction coefficient,

$$k^*(x) = 0 \text{ for } x \in \mathrm{supp}\mu^* , \tag{14.60}$$

transforms into the following equation for the reproduction coefficient of the average distribution

$$k^*(\langle\mu\rangle) = 0 \text{ for } x \in \mathrm{supp}\mu^* \tag{14.61}$$

(the *Volterra averaging principle* [400]). This system has as many linear equations as it has coexisting species. The averages can be non-unique. Then all of them satisfy this system, and we obtain the non-stationary Gause principle. And again, it is valid "almost always".

Formally, various definitions of genericity are constructed as follows. All systems (or cases, or situations and so on) under consideration are somehow parameterized – by sets of vectors, functions, matrices etc. Thus, the "space of systems" Q can be described. Then the "thin sets" are introduced into Q, i.e. the sets, which we shall later neglect. The union of a finite or countable number of thin sets, as well as the intersection of any number of them should be thin again, while the whole Q is not thin. There are two traditional ways to determine thinness.

1. A set is considered thin when it has measure zero. This is good for a finite-dimensional case, when there is the standard Lebesgue measure – the length, the area, the volume.
2. But most frequently we deal with the functional parameters. In that case it is common to restore to the second definition, according to which the sets of first category are negligible. The construction begins with nowhere dense sets. The set Y is nowhere dense in Q, if in any nonempty open set $V \subset Q$ (for example, in a ball) there exists a nonempty open subset $W \subset V$ (for example, a ball), which does not intersect with Y. Roughly speaking, Y is "full of holes" – in any neighborhood of any point of the set Y there is an open hole. Countable union of nowhere dense sets is called

the set of the first category. The second usual way is to define thin sets as the *sets of the first category.*

But even the real line R can be divided into two sets, one of which has zero measure, the other is of the first category. The genericity in the sense of measure and the genericity in the sense of category considerably differ in the applications where both of these concepts can be used. The conflict between the two main views on genericity stimulated efforts to invent new and stronger approaches.

In Theorem of selection efficiency a very strong genericity was used. Systems (14.45) were parameterized by continuous maps $\mu \mapsto k_\mu$. Denote by Q the space of these maps $M \to C(X)$ with the topology of uniform convergence on M. So, it is a Banach space. We shall call the set Y in the Banach space Q *completely thin,* if for any compact set K in Q and arbitrary positive $\varepsilon > 0$ there exists a vector $q \in Q$, such that $\|q\| < \varepsilon$ and $K + q$ does not intersect Y. So, a set, which can be moved out of intersection with any compact by an arbitrary small translation, is *completely negligible.* In a finite-dimensional space there is only one such set – the empty one. In an infinite-dimensional Banach space compacts and closed subspaces with infinite codimension provide us examples of completely negligible sets. *In Theorem of selection efficiency "usually" means "the set of exceptions is completely thin".*

14.4.5 Gromov's Interpretation of Selection Theorems

In his talk [401], M. Gromov offered a geometric interpretation of the selection theorems. Let us consider dynamical systems in the standard m-simplex σ_m in $m + 1$-dimensional space R^{m+1}:

$$\sigma_m = \{x \in R^{m+1} | x_i \geq 0, \sum_{i=1}^{m+1} x_i = 1\} .$$

We assume that simplex σ_m is positively invariant with respect to these dynamical systems: if the motion starts in σ_m at some time t_0 then it remains in σ_m for $t > t_0$. Let us consider the motions that start in the simplex σ_m at $t = 0$ and are defined for $t > 0$.

For large m, almost all volume of the simplex σ_m is concentrated in a small neighborhood of the center of σ_m, near the point $c = \left(\frac{1}{m}, \frac{1}{m}, \ldots, \frac{1}{m}\right)$. Hence, one can expect that a typical motion of a general dynamical system in σ_m for sufficiently large m spends almost all the time in a small neighborhood of c.

Let us consider dynamical systems with an additional property ("inheritance"): all the faces of the simplex σ_m are also positively invariant with respect to the systems with inheritance. It means that if some $x_i = 0$ initially at the time $t = 0$ then $x_i = 0$ for $t > 0$ for all motions in σ_m. The essence of selection theorems is as follows: a typical motion of a typical dynamical system with inheritance spends almost all the time in a small neighborhood of low-dimensional faces, even if it starts near the center of the simplex.

Let us denote by $\partial_r \sigma_m$ the union of all r-dimensional faces of σ_m. Due to the selection theorems, a typical motion of a typical dynamical system with inheritance spends almost all time in a small neighborhood of $\partial_r \sigma_m$ with $r \ll m$. It should not obligatory reside near just one face from $\partial_r \sigma_m$, but can travel in neighborhood of different faces from $\partial_r \sigma_m$ (the drift effect). The minimax estimation of the number of points in ω-limit distributions through the diameters $\varepsilon_n > 0$ of the set $\overline{\mathrm{conv}}(k_M)$ is the estimation of r.

14.4.6 Drift Equations

To this end, we talked about the support of an individual ω-limit distribution. Almost always it is small. But this does not mean, that the union of these supports is small even for one solution $\mu(t)$. It is possible that a solution is a finite set of narrow peaks getting in time more and more narrow, moving slower and slower, but not tending to fixed positions, rather continuing to move along its trajectory, and the path covered tends to infinity as $t \to \infty$.

This effect was not discovered for a long time because the slowing down of the peaks was thought as their tendency to fixed positions. There are other difficulties related to the typical properties of continuous functions, which are not typical for the smooth ones. Let us illustrate them for the distributions over a straight line segment. Add to the reproduction coefficients k_μ the sum of small and narrow peaks located on a straight line distant from each other much more than the peak width (although it is ε-small). However small is chosen the peak's height, one can choose their width and frequency on the straight line in such a way that from any initial distribution μ_0 whose support is the whole segment, at $t \to \infty$ we obtain ω-limit distributions, concentrated at the points of maximum of the added peaks.

Such a model perturbation is small in the space of continuous functions. Therefore, it can be put as follows: *by small continuous perturbation the limit behavior of system (14.45) can be reduced onto a ε-net for sufficiently small ε.* But this can not be done with the small smooth perturbations (with small values of the first and the second derivatives) in the general case. The discreteness of the net, onto which the limit behavior is reduced by small continuous perturbations, differs from the discreteness of the support of the individual ω-limit distribution. For an individual distribution the number of points is estimated, roughly speaking, by the number of essential parameters (14.56), while for the conjunction of limit supports – by the number of stages in approximation of k_μ by piece-wise constant functions.

Thus, in a typical case the dynamics of systems (14.45) with smooth reproduction coefficients transforms a smooth initial distributions into the ensemble of narrow peaks. The peaks become more narrow, their motion slows down, but not always they tend to fixed positions.

The equations of motion for these peaks can be obtained in the following way [115]. Let X be a domain in the n-dimensional real space, and the initial distributions μ_0 be assumed to have smooth density. Then, after sufficiently large time t, the position of distribution peaks are the points of the average reproduction coefficient maximum $\langle k_\mu \rangle_t$ (14.52) to any accuracy set in advance. Let these points of maximum be x^α, and

$$q_{ij}^\alpha = -t \frac{\partial^2 \langle k_\mu \rangle_t}{\partial x_i \partial x_j}\bigg|_{x=x^\alpha} .$$

It is easy to derive the following differential relations

$$\sum_j q_{ij}^\alpha \frac{\mathrm{d}x_j^\alpha}{\mathrm{d}t} = \frac{\partial k_{\mu(t)}}{\partial x_i}\bigg|_{x=x^\alpha} ;$$

$$\frac{\mathrm{d}q_{ij}^\alpha}{\mathrm{d}t} = -\frac{\partial^2 k_{\mu(t)}}{\partial x_i \partial x_j}\bigg|_{x=x^\alpha} . \tag{14.62}$$

These relations do not form a closed system of equations, because the right-hand parts are not functions of x_i^α and q_{ij}^α. For sufficiently narrow peaks there should be separation of the relaxation times between the dynamics *on* the support and the dynamics *of* the support: the relaxation of peak amplitudes (it can be approximated by the relaxation of the distribution with the finite support, $\{x^\alpha\}$) should be significantly faster than the motion of the locations of the peaks, the dynamics of $\{x^\alpha\}$. Let us write the first term of the corresponding asymptotics [115].

For the finite support $\{x^\alpha\}$ the distribution is $\mu = \sum_\alpha N_\alpha \delta(x - x^\alpha)$. Dynamics of the finite number of variables, N_α obeys the system of ordinary differential equations

$$\frac{\mathrm{d}N_\alpha}{\mathrm{d}t} = k_\alpha(\boldsymbol{N})N_\alpha , \tag{14.63}$$

where $\boldsymbol{N}$ is vector with components N_α, $k_\alpha(\boldsymbol{N})$ is the value of the reproduction coefficient k_μ at the point x^α:

$$k_\alpha(\boldsymbol{N}) = k_\mu(x^\alpha) \text{ for } \mu = \sum_\alpha N_\alpha \delta(x - x^\alpha) .$$

Let the dynamics of the system (14.63) for a given set of initial conditions be simple: the motion $\boldsymbol{N}(t)$ goes to the stable fixed point $\boldsymbol{N} = \boldsymbol{N}^*(\{x^\alpha\})$. Then we can take in the right hand side of (14.62)

$$\mu(t) = \mu^*(\{x^\alpha(t)\}) = \sum_\alpha N_\alpha^* \delta(x - x^\alpha(t)) . \tag{14.64}$$

Because of the time separation we can assume that (i) relaxation of the amplitudes of peaks is completed and (ii) peaks are sufficiently narrow, hence, the difference between true $k_{\mu(t)}$ and the reproduction coefficient for the measure (14.64) with the finite support $\{x^\alpha\}$ is negligible. Let us use the notation $k^*(\{x^\alpha\})(x)$ for this reproduction coefficient. The relations (14.62) transform into the ordinary differential equations

$$\sum_j q_{ij}^\alpha \frac{\mathrm{d}x_j^\alpha}{\mathrm{d}t} = \left. \frac{\partial k^*(\{x^\beta\})(x)}{\partial x_i} \right|_{x=x^\alpha} ;$$

$$\frac{\mathrm{d}q_{ij}^\alpha}{\mathrm{d}t} = - \left. \frac{\partial^2 k^*(\{x^\beta\})(x)}{\partial x_i \partial x_j} \right|_{x=x^\alpha} . \qquad (14.65)$$

For many purposes it may be useful to switch to the logarithmic time $\tau = \ln t$ and to new variables

$$b_{ij}^\alpha = \frac{1}{t} q_{ij}^\alpha = - \left. \frac{\partial^2 \langle k(\mu) \rangle t}{\partial x_i \partial x_j} \right|_{x=x^\alpha} .$$

For large t we obtain from (14.65)

$$\sum_j b_{ij}^\alpha \frac{\mathrm{d}x_j^\alpha}{\mathrm{d}\tau} = \left. \frac{\partial k^*(\{x^\beta\})(x)}{\partial x_i} \right|_{x=x^\alpha} ;$$

$$\frac{\mathrm{d}b_{ij}^\alpha}{\mathrm{d}\tau} = - \left. \frac{\partial^2 k^*(\{x^\alpha\})(x)}{\partial x_i \partial x_j} \right|_{x=x^\beta} - b_{ij}^\alpha . \qquad (14.66)$$

The way of constructing the drift equations (14.65,14.66) for a specific system (14.45) is as follows:

1. For finite sets $\{x^\alpha\}$ one studies systems (14.63) and finds the equilibrium solutions $\boldsymbol{N}^*(\{x^\alpha\})$;
2. For given measures $\mu^*(\{x^\alpha(t)\})$ (14.64) one calculates the reproduction coefficients $k_\mu(x) = k^*(\{x^\alpha\})(x)$ and first derivatives of these functions in x at points x^α. That is all, the drift equations (14.65,14.66) are set up.

The drift equations (14.65,14.66) describe the dynamics of the peaks positions x^α and of the coefficients q_{ij}^α. For given x^α, q_{ij}^α and N_α^* the distribution density μ can be approximated as the sum of narrow Gaussian peaks:

$$\mu = \sum_\alpha N_\alpha^* \sqrt{\frac{\det Q^\alpha}{(2\pi)^n}} \exp\left(-\frac{1}{2} \sum_{ij} q_{ij}^\alpha (x_i - x_i^\alpha)(x_j - x_j^\alpha) \right) , \qquad (14.67)$$

where Q^α is the inverse covariance matrix (q_{ij}^α).

If the limit dynamics of the system (14.63) for finite supports at $t \to \infty$ can be described by a more complicated attractor, then instead of reproduction coefficient $k^*(\{x^\alpha\})(x) = k_{\mu^*}$ for the stationary measures μ^* (14.64) one can use the average reproduction coefficient with respect to the corresponding *Sinai–Ruelle–Bowen measure* [396,397]. If finite systems (14.63) have several attractors for given $\{x^\alpha\}$, then the dependence $k^*(\{x^\alpha\})$ is multi-valued, and there may be bifurcations and hysteresis with the function $k^*(\{x^\alpha\})$ transition from one sheet to another. There are many interesting effects concerning peaks' birth, desintegration, divergence, and death, and the drift equations (14.65,14.66) describe the motion in a non-critical domain, between these critical effects.

Inheritance (conservation of support) is never absolutely exact. Small variations, mutations, immigration in biological systems are very important. Excitation of new degrees of freedom, modes diffusion, noise are present in physical systems. How does small perturbation in the inheritance affect the effects of selection? The answer is usually as follows: there is such a value of perturbation of the right-hand side of (14.45), at which they would change nearly nothing, just the limit δ-shaped peaks transform into sufficiently narrow peaks, and zero limit of the velocity of their drift at $t \to \infty$ substitutes by a small finite one.

The simplest model for "inheritance + small variability" is given by a perturbation of (14.45) with diffusion term

$$\frac{\partial \mu(x,t)}{\partial t} = k_{\mu(x,t)} \times \mu(x,t) + \varepsilon \sum_{ij} d_{ij}(x) \frac{\partial^2 \mu(x,t)}{\partial x_i \partial x_j} . \tag{14.68}$$

where $\varepsilon > 0$ and the matrix of diffusion coefficients d_{ij} is symmetric and positively definite.

There are almost always no qualitative changes in the asymptotic behaviour, if ε is sufficiently small. With this the asymptotics is again described by the drift equations (14.65, 14.66), modified by taking into account the diffusion as follows:

$$\sum_j q_{ij}^\alpha \frac{\mathrm{d}x_j^\alpha}{\mathrm{d}t} = \left. \frac{\partial k^*(\{x^\beta\})(x)}{\partial x_i} \right|_{x=x^\alpha} ;$$

$$\frac{\mathrm{d}q_{ij}^\alpha}{\mathrm{d}t} = - \left. \frac{\partial^2 k^*(\{x^\beta\})(x)}{\partial x_i \partial x_j} \right|_{x=x^\alpha} - 2\varepsilon \sum_{kl} q_{ik}^\alpha d_{kl}(x^\alpha) q_{lj}^\alpha . \tag{14.69}$$

Now, as distinct from (14.65), the eigenvalues of the matrices $Q^\alpha = (q_{ij}^\alpha)$ cannot grow infinitely. This is prevented by the quadratic terms in the right-hand side of the second equation (14.69).

Dynamics of (14.69) does not depend on the value $\varepsilon > 0$ qualitatively, because of the obvious scaling property. If ε is multiplied by a positive number ν, then, upon rescaling $t' = \nu^{-1/2}t$ and $q_{ij}^{\alpha'} = \nu^{-1/2}q_{ij}^\alpha$, we have the same system again. Multiplying $\varepsilon > 0$ by $\nu > 0$ changes only peak's velocity values by a factor $\nu^{1/2}$, and their width by a factor $\nu^{1/4}$. The paths of peaks' motion do not change at this for the drift approximation (14.69) (but the applicability of this approximation may, of course, change).

14.4.7 Three Main Types of Stability

Stable steady-state solutions of equations of the form (14.45) may be only the sums of δ-functions – this was already mentioned. There is a set of specific conditions of stability, determined by the form of equations.

Consider a stationary distribution for (14.45) with a finite support

$$\mu^*(x) = \sum_\alpha N_\alpha^* \delta(x - x^{*\alpha}) .$$

Steady state of μ^* means, that

$$k_{\mu^*}(x^{*\alpha}) = 0 \text{ for all } \alpha . \tag{14.70}$$

The *internal stability* means, that this distribution is stable with respect to perturbations not increasing the support of μ^*. That is, the vector N_α^* is the stable fixed point for the dynamical system (14.63). Here, as usual, it is possible to distinguish between the Lyapunov stability, the asymptotic stability and the first approximation stability (negativeness of real parts for the eigenvalues of the matrix $\partial \dot{N}_\alpha^*/\partial N_\alpha^*$ at the stationary points).

The *external stability* means stability to an expansion of the support, i.e. to adding to μ^* of a small distribution whose support contains points not belonging to $\mathrm{supp}\mu^*$. It makes sense to speak about the external stability only if there is internal stability. In this case it is sufficient to restrict ourselves with δ-functional perturbations. The external stability has a very transparent physical and biological sense. It is stability with respect to *introduction* into the systems of a new inherited unit (gene, variety, specie...) in a small amount.

The *necessary condition for the external stability* is: the points $\{x^{*\alpha}\}$ are points of the global maximum of the reproduction coefficient $k_{\mu^*}(x)$. It can be formulated as the optimality principle

$$k_{\mu^*}(x) \leq 0 \text{ for all } x; \ k_{\mu^*}(x^{*\alpha}) = 0 . \tag{14.71}$$

The *sufficient condition for the external stability* is: the points $\{x^{*\alpha}\}$ and only these points are points of the global maximum of the reproduction coefficient $k_{\mu^*}(x^{*\alpha})$. At the same time it is the condition of the external stability in the first approximation and the optimality principle

$$k_{\mu^*}(x) < 0 \text{ for } x \notin \{x^{*\alpha}\}; \ k_{\mu^*}(x^{*\alpha}) = 0 . \tag{14.72}$$

The only difference from (14.71) is the change of the inequality sign from $k_{\mu^*}(x) \leq 0$ to $k_{\mu^*}(x) < 0$ for $x \notin \{x^{*\alpha}\}$. The necessary condition (14.71) means, that the small δ-functional addition will not grow in the first approximation. According to the sufficient condition (14.72) such a small addition will exponentially decrease.

If X is a finite set then the combination of the external and the internal stability is equivalent to the standard stability for a system of ordinary differential equations.

For the continuous X there is one more kind of stability important from the applications viewpoint. Substitute δ-shaped peaks at the points $\{x^{*\alpha}\}$ by narrow Gaussians and shift slightly the positions of their maxima away from the points $x^{*\alpha}$. How will the distribution from such initial conditions evolve? If it tends to μ without getting too distant from this steady state distribution, then we can say that the third type of stability – *stable realizability* – takes place. It is worth mentioning that the perturbation of this type is only weakly small, in contrast to perturbations considered in the theory of internal and external stability. Those perturbations are small by their norms[6].

In order to formalize the condition of stable realizability it is convenient to use the drift equations in the form (14.66). Let the distribution μ^* be internally and externally stable in the first approximations. Let the points $x^{*\alpha}$ of global maxima of $k_{\mu^*}(x)$ be non-degenerate in the second approximation. This means that the matrices

$$b_{ij}^{*\alpha} = - \left(\frac{\partial^2 k_{\mu^*}(x)}{\partial x_i \partial x_j} \right)_{x=x^{*\alpha}} \tag{14.73}$$

are strictly positively definite for all α.

Under these conditions of stability and non-degeneracy the coefficients of (14.66) can be easily calculated using Taylor series expansion in powers of $(x^\alpha - x^{*\alpha})$. The stable realizability of μ^* in the first approximation means that the fixed point of the drift equations (14.66) with the coordinates

$$x^\alpha = x^{*\alpha} , \quad b_{ij}^\alpha = b_{ij}^{*\alpha} \tag{14.74}$$

is stable in the first approximation. It is the usual stability for the system (14.66) of ordinary differential equations.

14.4.8 Main Results About Systems with Inheritance

1. If a kinetic equation has the quasi-biological form (14.45) then it has a rich system of invariant manifolds: for any closed subset $A \subset X$ the set of distributions $M_A = \{\mu \mid \operatorname{supp}\mu \subseteq A\}$ is invariant with respect to the system (14.45). These invariant manifolds form important algebraic structure, the summation of manifolds is possible:

$$M_A \oplus M_B = M_{A\cup B} .$$

(Of course, $M_{A\cap B} = M_A \cap M_B$).

---

[6] Let us remind that the norm of the measure μ is $\|\mu\| = \sup_{|f|\leq 1} \mu[f]$. If one shifts the δ-measure of unite mass by any nonzero distance ε, then the norm of the perturbation is 2. Nevertheless, this perturbation weakly tends to 0 with $\varepsilon \to 0$.

2. Typically, all the ω-limit points belong to invariant manifolds $\boldsymbol{M}_A$ with finite A. The finite-dimensional approximations of the reproduction coefficient (14.56) provides the minimax estimation of the number of points in A.

3. For systems with inheritance (14.45) a solution typically tends to be a finite set of narrow peaks getting in time more and more narrow, moving slower and slower. It is possible that these peaks do not tend to fixed positions, rather they continue moving, and the path covered tends to infinity at $t \to \infty$. This is the *drift effect*.

4. The equations for peak dynamics, the drift equations, (14.65,14.66,14.69) describe dynamics of the shapes of the peaks and their positions. For systems with small variability ("mutations") the drift equations (14.69) has the scaling property: the change of the intensity of mutations is equivalent to the change of the time scale.

5. Three specific types of stability are important for the systems with inheritance: internal stability (stability with respect to perturbations without extension of distribution support), external stability (stability with respect to small one-point extension of distribution support), and stable realizability (stability with respect to weakly small[7] perturbations: small extensions and small shifts of the peaks).

Some exact results of the mathematical selection theory can be found in [402,403]. There exist many physical examples of systems with inheritance [385–391]. A wide field of ecological applications was described in the book [383]. An introduction into adaptive dynamics was given in notes [404] that illustrate largely by way of examples, how standard ecological models can be put into an evolutionary perspective in order to gain insight in the role of natural selection in shaping life history characteristics. The cell division self-synchronization below demonstrates effects of unusual inherited unit, it is the example of a "phase selection".

14.5 Example: Cell Division Self-Synchronization

The results described above admit for a whole family of generalizations. In particular, it seems to be important to extend the theorems of selection to the case of vector distributions, when $k_\mu(x)$ is a linear operator at each μ, x. It is possible also to make generalizations for some classes of non-autonomous equations with explicit dependencies of $k_\mu(x)$ on t.

Availability of such a network of generalizations allows to construct the reasoning as follows: *what* is inherited (i.e. for what the law of conservation of support holds) is the subject of selection (i.e. with respect to these variables at $t \to \infty$ the distribution becomes discrete and the limit support can be described by the optimality principles).

---

[7] That is, small in the weak topology.

This section gives a somewhat unconventional example of inheritance and selection, when the reproduction coefficients are subject additional conditions of symmetry.

Consider a culture of microorganisms in a certain medium (for example, pathogenous microbes in the organism of a host). Assume, for simplicity, the following: *let the time period spent by these microorganisms for the whole life cycle be identical.*

At the end of the life cycle the microorganism disappears and new several microorganisms appear in the initial phase. Let T be the time of the life cycle. Each microorganism holds the value of the inherited variable, it is "the moment of its appearance (mod T)". Indeed, if the given microorganism emerges at time τ $(0 < \tau \leq T)$, then its first descendants appear at time $T + \tau$, the next generation – at the moment $2T + \tau$, then $3T + \tau$ and so on.

It is natural to assume that the phase τ (mod T) is the inherited variable. This implies selection of phases and, therefore, survival of their discrete number $\tau_1, \ldots \tau_m$, only. But results of the preceding sections cannot be applied directly to this problem. The reason is the additional symmetry of the system with respect to the phase shift. But the typicalness of selection and the instability of the uniform distribution over the phases τ (mod T) can be shown for this case, too. Let us illustrate it with the simplest model.

Let the difference between the microorganisms at each time moment be related to the difference in the development phases only. Let us also assume that the state of the medium can be considered as a function of the distribution $\mu(\tau)$ of microorganisms over the phases $\tau \in]0, T]$ (the quasi-steady state approximation for the medium). Consider the system at discrete times nT and assume the coefficient connecting μ at moments nT and $nT + T$ to be the exponent of the linear integral operator value:

$$\mu_{n+1}(\tau) = \mu_n(\tau)\exp\left[k_0 - \int_0^T k_1(\tau - \tau')\mu_n(\tau')\,d\tau'\right]. \qquad (14.75)$$

Here, $\mu_n(\tau)$ is the distribution at the moment nT, $k_0 = $ const, $k_1(\tau)$ is a periodic function of period T.

The uniform steady-state $\mu^* \equiv n^* = $ const is:

$$n^* = \frac{k_0}{\int_0^T k_1(\theta)\,d\theta}. \qquad (14.76)$$

In order to examine stability of the uniform steady state μ^* (14.76), the system (14.75) is linearized. For small deviations $\Delta\mu(\tau)$ in linear approximation

$$\Delta\mu_{n+1}(\tau) = \Delta\mu_n(\tau) - n^* \int_0^T k_1(\tau - \tau')\Delta\mu_n(\tau')\,d\tau'. \qquad (14.77)$$

Expand $k_1(\theta)$ into the Fourier series:

$$k_1(\theta) = b_0 + \sum_{n=1}^{\infty}\left(a_n \sin\left(2\pi n\frac{\theta}{T}\right) + b_n \cos\left(2\pi n\frac{\theta}{T}\right)\right). \qquad (14.78)$$

Denote by A operator of the right-hand side of (14.77). In the basis of functions

$$e_{s\,n} = \sin\left(2\pi n\frac{\theta}{T}\right), \quad e_{c\,n} = \cos\left(2\pi n\frac{\theta}{T}\right)$$

on the segment $]0,T]$ the operator A is block-diagonal. The vector e_0 is eigenvector, $Ae_0 = \lambda_0 e_0$, $\lambda_0 = 1 - n^* b_0 T$. On the two-dimensional space, generated by vectors $e_{s\,n}$, $e_{c\,n}$ the operator A is acting as a matrix

$$A_n = \begin{pmatrix} 1 - \frac{Tn^*}{2}b_n & -\frac{Tn^*}{2}a_n \\ \frac{Tn^*}{2}a_n & 1 - \frac{Tn^*}{2}b_n \end{pmatrix}. \qquad (14.79)$$

The corresponding eigenvalues are

$$\lambda_{n\,1,2} = 1 - \frac{Tn^*}{2}(b_n \pm ia_n). \qquad (14.80)$$

For the uniform steady state μ^* (14.76) to be unstable it is sufficient that the absolute value of at least one eigenvalue $\lambda_{n\,1,2}$ be larger than 1: $|\lambda_{n\,1,2}| > 1$. If there is at least one negative Fourier cosine-coefficient $b_n < 0$, then $\mathrm{Re}\lambda_n > 1$, and thus $|\lambda_n| > 1$.

Note now, that almost all periodic functions (continuous, smooth, analytical – this does not matter) have negative Fourier cosine-coefficient. This can be understood as follows. The sequence b_n tends to zero at $n \to \infty$. Therefore, if all $b_n \geq 0$, then, by changing b_n at sufficiently large n, we can make b_n negative, and the perturbation value can be chosen less than any previously set positive number. On the other hand, if some $b_n < 0$, then this coefficient cannot be made non-negative by sufficiently small perturbations. Moreover, the set of functions that have all Fourier cosine-coefficient non-negative is completely thin, because for any compact of functions K (for most of norms in use) the sequence $B_n = \max_{f \in K}|b_n(f)|$ tends to zero, where $b_n(f)$ is the nth Fourier cosine-coefficient of function f.

The model (14.75) is revealing, because for it we can trace the dynamics over large times, if we restrict ourselves with a finite segment of the Fourier series for $k_1(\theta)$. Describe it for

$$k_1(\theta) = b_0 + a\sin\left(2\pi\frac{\theta}{T}\right) + b\cos\left(2\pi\frac{\theta}{T}\right). \qquad (14.81)$$

Assume further that $b < 0$ (then the homogeneous distribution $\mu^* \equiv \frac{k_0}{b_0 T}$ is unstable) and $b_0 > \sqrt{a^2 + b^2}$ (then the $\int \mu(\tau)\,d\tau$ cannot grow unbounded in time). Introduce notations

$$M_0(\mu) = \int_0^T \mu(\tau)\,d\tau, \ M_c(\mu) = \int_0^T \cos\left(2\pi\frac{\tau}{T}\right)\mu(\tau)\,d\tau\ ,$$

$$M_s(\mu) = \int_0^T \sin\left(2\pi\frac{\tau}{T}\right)\mu(\tau)\,d\tau, \ \langle\mu\rangle_n = \frac{1}{n}\sum_{m=0}^{n-1}\mu_m\ , \tag{14.82}$$

where μ_m is the distribution μ at the discrete time m.

In these notations,

$$\mu_{n+1}(\tau) = \mu_n(\tau)\exp\left[k_0 - b_0 M_0(\mu_n) - (aM_c(\mu_n) + bM_s(\mu_n))\sin\left(2\pi\frac{\tau}{T}\right)\right.$$
$$\left. + (aM_s(\mu_n) - bM_c(\mu_n))\cos\left(2\pi\frac{\tau}{T}\right)\right]\ . \tag{14.83}$$

Represent the distribution $\mu_n(\tau)$ through the initial distribution $\mu_0(\tau)$ and the functionals M_0, M_c, M_s values for the average distribution $\langle\mu\rangle_n$):

$$\mu_n(\tau) = \mu_0(\tau)$$
$$\times\exp\left\{n\left[k_0 - b_0 M_0(\langle\mu\rangle_n) - (aM_c(\langle\mu\rangle_n) + bM_s(\langle\mu\rangle_n))\sin\left(2\pi\frac{\tau}{T}\right)\right.\right.$$
$$\left.\left. + (aM_s(\langle\mu\rangle_n) - bM_c(\langle\mu\rangle_n))\cos\left(2\pi\frac{\tau}{T}\right)\right]\right\}\ . \tag{14.84}$$

The exponent in (14.84) is either independent of τ, or there is a function with the single maximum on $]0, T]$. The coordinate $\tau_n^{\#}$ of this maximum is easily calculated

$$\tau_n^{\#} = -\frac{T}{2\pi}\arctan\frac{aM_c(\langle\mu\rangle_n) + bM_s(\langle\mu\rangle_n)}{aM_s(\langle\mu\rangle_n) - bM_c(\langle\mu\rangle_n)} \tag{14.85}$$

Let the non-uniform smooth initial distribution μ_0 has the whole segment $[0, T]$ as its support. At the time progress the distributions $\mu_n(\tau)$ takes the shape of ever narrowing peak. With high accuracy at large a we can approximate $\mu_n(\tau)$ by the Gaussian distribution (approximation accuracy is understood in the weak sense, as closeness of mean values):

$$\mu_n(\tau) \approx M_0\sqrt{\frac{q_n}{\pi}}\exp[-q_n(\tau - \tau_n^{\#})^2], \ M_0 = \frac{k_0}{k_1(0)} = \frac{k_0}{b_0 + b}\ , \tag{14.86}$$
$$q_n^2 = n^2\left(\frac{2\pi}{T}\right)^4\left[(aM_c(\langle\mu\rangle_n) + bM_s(\langle\mu\rangle_n))^2 + (aM_s(\langle\mu\rangle_n) - bM_c(\langle\mu\rangle_n))^2\right]\ .$$

Expression (14.86) involves the average measure $\langle\mu\rangle_n$ which is difficult to compute. However, we can operate without direct computation of $\langle\mu\rangle_n$. At $q_n \gg \frac{1}{T^2}$ we can compute q_{n+1} and $\tau_{n+1}^{\#}$:

$$\mu_{n+1} \approx M_0\sqrt{\frac{q_n + \Delta q}{\pi}}\exp\left[-(q_n + \Delta q)((\tau - \tau_n^{\#} - \Delta\tau^{\#})^2\right]\ ,$$
$$\Delta q \approx -\frac{1}{2}bM_0\left(\frac{2\pi}{T}\right)^2\ , \ \Delta\tau^{\#} \approx \frac{1}{q}M_0\frac{2\pi}{T}\ . \tag{14.87}$$

The accuracy of these expression grows with time n. The value q_n grows at large n almost linearly, and $\tau_n^{\#}$, respectively, as the sum of the harmonic series (mod T), i.e. as $\ln n$ (mod T). The drift effect takes place: location of the peak $\tau_n^{\#}$, passes at $n \to \infty$ the distance diverging as $\ln n$.

Of interest is the case, when $b > 0$ but

$$|\lambda_1|^2 = \left(1 + n^* b \frac{T}{2}\right)^2 + \left(n^* a \frac{T}{2}\right)^2 > 1 .$$

With this, homogeneous distribution $\mu^* \equiv n^*$ is not stable but μ does not tend to δ-functions. There are smooth stable "self-synchronization waves" of the form

$$\mu_n = \gamma \exp\left[q \cos\left((\tau - n\Delta\tau^{\#})\frac{2\pi}{T}\right)\right] .$$

At small $b > 0$ ($b \ll |a|$, $bM_0 \ll a^2$) we can find explicit form of approximated expressions for q and $\Delta\tau^{\#}$:

$$q \approx \frac{a^2 M_0}{2b} , \quad \Delta\tau^{\#} \approx \frac{bT}{\pi a} . \tag{14.88}$$

At $b > 0, b \to 0$, smooth self-synchronization waves become ever narrowing peaks, and their steady velocity approaches zero. If $b = 0$, $|\lambda_1|^2 > 1$ then the effect of selection takes place again, and for almost all initial conditions μ_0 with the support being the whole segment $[0, T]$ the distribution μ_n takes

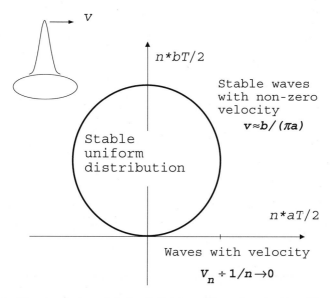

Fig. 14.1. The simplest model of cell division self-synchronization: The parametric portrait

at large n the form of a slowly drifting almost Gaussian peak. It becomes narrower with the time, and the motion slows down. Instead of the linear growth of q_n which takes place at $b < 0$ (14.87), for $b = 0$, $q_{n+1} - q_n \approx \mathrm{const} q_n^{-1}$ and q_n grows as $\mathrm{const}\sqrt{n}$.

The parametric portrait of the system for the simple reproduction coefficient (14.81) is presented in Fig. 14.1.

As usual, a small desynchronization transforms δ-functional limit peaks to narrow Gaussian peaks, and the velocity of peaks tends to small but nonzero velocity instead of zero. The systems with small desynchronization can be described by equations of the form (14.69).

There are many specific mechanisms of synchronization and desynchronysation in physics and biology (see, for example [405–409]). We described here very simple mechanism: it requires only that the time of the life cycle is fixed, in this case in a generic situation we should observe the self-synchronization. Of course, the real-world situation can be much more complicated, with a plenty of additional factors, but the basic mechanism of the "phase selection" works always if the life cycle has more or less fixed duration.

15 Accuracy Estimation and Post-Processing in Invariant Manifolds Construction

The *post-processing* algorithms are developed for the accuracy control and enhancement of approximate invariant manifold.

15.1 Formulas for Dynamic and Static Post-Processing

Assume that for the dynamical system (3.1)

$$\frac{\mathrm{d}x}{\mathrm{d}t} = J(x)$$

an approximate invariant manifold is constructed and the slow motion equations are derived:

$$\frac{\mathrm{d}x_{\mathrm{sl}}}{\mathrm{d}t} = P_{x_{\mathrm{sl}}}(J(x_{\mathrm{sl}})), x_{\mathrm{sl}} \in \Omega_{\mathrm{sl}} . \tag{15.1}$$

Here, $P_{x_{\mathrm{sl}}}$ is the projector onto the tangent space $T_{x_{\mathrm{sl}}}$ of Ω_{sl} parallel to the plain of fast motions. Suppose that we have solved the system (15.1) and have obtained $x_{\mathrm{sl}}(t)$. Let us consider the following two questions:

- How well this solution approximates the true solution $x(t)$ with the same initial condition?
- Is it possible to use the solution $x_{\mathrm{sl}}(t)$ for its refinement?

It should be stressed that these questions can be asked only if the slow system (15.1) is obtained as a result of reduction, that is, with the help of the projector $P_{x_{\mathrm{sl}}}$ that identifies fast fibers ($\ker P_{x_{\mathrm{sl}}}$). These question are meaningless if "some" closure approximation is used without a specification what means "fast" and "slow" in this approximation. In the latter case one can only hope that the closure is a good guess, that it is thermodynamically consistent, etc, but nothing can be done on its refinement.

These two questions are interconnected. The first question states the problem of the *accuracy estimation*. The second states the problem of *post-processing* [348–351].

The simplest ("naive") estimation is given by the "invariance defect":

$$\Delta_{x_{\mathrm{sl}}} = (1 - P_{x_{\mathrm{sl}}})J(x_{\mathrm{sl}}) , \tag{15.2}$$

Alexander N. Gorban and Iliya V. Karlin: *Invariant Manifolds for Physical and Chemical Kinetics*, Lect. Notes Phys. **660**, 457–466 (2005)
www.springerlink.com

which can be compared with $J(x_{\text{sl}})$. For example, this estimate is given by $\epsilon = \|\Delta_{x_{\text{sl}}}\|/\|J(x_{\text{sl}})\|$ using some appropriate norm.

Probably, the most comprehensive answer to the above questions can be given by solving the following equation:

$$\frac{\mathrm{d}(\delta x)}{\mathrm{d}t} = \Delta_{x_{\text{sl}}(t)} + D_x J(x)|_{x_{\text{sl}}(t)}\delta x \ . \tag{15.3}$$

This linear equation describes the dynamics of the variation $\delta x(t) = x(t) - x_{\text{sl}}(t)$ in the linear approximation. The solution with zero initial condition $\delta x(0) = 0$ allows to estimate the robustness of x_{sl}, as well as the error. Using $x_{\text{sl}}(t) + \delta x(t)$ instead of $x_{\text{sl}}(t)$ gives the required solution refinement. This *dynamical post-processing* [350] allows to refine the solution substantially. However, the price for this is solving equation (15.3) with variable coefficients. The dynamical post-processing can be addressed by a whole hierarchy of simplifications, both dynamic and static. Let us mention some of them, starting from the dynamic ones.

(1) **Freezing coefficients**. In the equation (15.3) the linear operator $D_x J(x)|_{x_{\text{sl}}(t)}$ is replaced by its value in some distinguished point x^* (for example, in the equilibrium) or it is frozen somehow else. As a result, one gets the equation with constant coefficients and the explicit integration formula:

$$\delta x(t) = \int_0^t exp(D^*(t-\tau))\Delta_{x_{\text{sl}}(\tau)}\,\mathrm{d}\tau \ , \tag{15.4}$$

where D^* is the "frozen" operator and $\delta x(0) = 0$.

Another important way of freezing is to replace (15.3) by some *model equation*, i.e. substituting $-\frac{1}{\tau^*}$ instead of $D_x J(x)$, where τ^* is the relaxation time. In this case the formula for $\delta x(t)$ has a very simple form:

$$\delta x(t) = \int_0^t e^{\frac{\tau - t}{\tau^*}}\Delta_{x_{\text{sl}}(\tau)}\,\mathrm{d}\tau \ . \tag{15.5}$$

(2) **One-dimensional Galerkin-type approximation**. Another "scalar" approximation is given by projecting (15.3) on $\Delta(t) = \Delta_{x_{\text{sl}}(t)}$. Using the ansatz

$$\delta x(t) = \delta(t)\Delta(t) \ , \tag{15.6}$$

substituting it into (15.3), and projecting the result orthogonally on $\Delta(t)$ we obtain

$$\frac{\mathrm{d}\delta}{\mathrm{d}t} = 1 + \delta\frac{\langle\Delta|D\Delta\rangle - \langle\Delta|\dot\Delta\rangle}{\langle\Delta|\Delta\rangle} \ , \tag{15.7}$$

where $\langle|\rangle$ is an appropriate scalar product, which can depend on the point x_{sl} (for example, the entropic scalar product), $D = D_x J(x)|_{x_{\text{sl}}(t)}$ or the self-adjoint linearizarion of this operator, or some approximation of it, and $\dot\Delta = \mathrm{d}\Delta(t)/\mathrm{d}t$.

A "hybrid" between equations (15.7) and (15.3) has a rather simple form (but it is more difficult for computations than (15.7)):

$$\frac{\mathrm{d}(\delta x)}{\mathrm{d}t} = \Delta(t) + \frac{\langle\Delta|D\Delta\rangle}{\langle\Delta|\Delta\rangle}\delta x \ . \tag{15.8}$$

Here one uses the normalized matrix element $\frac{\langle\Delta|D\Delta\rangle}{\langle\Delta|\Delta\rangle}$ instead of the linear operator $D = D_x J(x)|_{x_{\mathrm{sl}}(t)}$.

Both equations (15.7) and (15.8) can be solved explicitly:

$$\delta(t) = \int_0^t \mathrm{d}\tau \exp\left(\int_\tau^t k(\theta)\,\mathrm{d}\theta\right) , \tag{15.9}$$

$$\delta x(t) = \int_0^t \Delta(\tau)\mathrm{d}\tau \exp\left(\int_\tau^t k_1(\theta)\,\mathrm{d}\theta\right) , \tag{15.10}$$

where $k(t) = \frac{\langle\Delta|D\Delta\rangle - \langle\Delta|\dot\Delta\rangle}{\langle\Delta|\Delta\rangle}$, $k_1(t) = \frac{\langle\Delta|D\Delta\rangle}{\langle\Delta|\Delta\rangle}$.

The projection of $\Delta_{x_{\mathrm{sl}}}(t)$ on the slow motion is equal to zero, hence, for the post-processing of the slow motion, the one-dimensional model (15.7) should be supplemented by one more iteration in order to find the first non-vanishing term in $\delta x_{\mathrm{sl}}(t)$:

$$\frac{\mathrm{d}(\delta x_{\mathrm{sl}}(t))}{\mathrm{d}t} = \delta(t) P_{x_{\mathrm{sl}}(t)}(D_x J(x)|_{x_{\mathrm{sl}}(\tau)})(\Delta(t)) \ ;$$

$$\delta x_{\mathrm{sl}}(t) = \int_0^t \delta(\tau) P_{x_{\mathrm{sl}}(\tau)}(D_x J(x)|_{x_{\mathrm{sl}}(\tau)})(\Delta(\tau))\,\mathrm{d}\tau \ . \tag{15.11}$$

where $\delta(t)$ is the solution of (15.7).

(3) For a **static post-processing**, one uses stationary points of dynamic equations (15.3), or of their simplified versions (15.4),(15.7). Instead of (15.3) one gets:

$$D_x J(x)|_{x_{\mathrm{sl}}(t)}\delta x = -\Delta_{x_{\mathrm{sl}}(t)} \tag{15.12}$$

with the additional condition $P_{x_{\mathrm{sl}}}\delta x = 0$. This is exactly the iteration equation of the Newton method for solving the invariance equation. A clarification is in order here. Static post-processing (15.12) as well as other post-processing formulas should not be confused with the Newton method and others for correcting the approximately invariant manifold. Here, only the single trajectory $x_{\mathrm{sl}}(t)$ on the manifold is corrected, not the whole manifold.

The corresponding stationary problems for the model equations and for the projections of (15.3) on Δ are obvious. We only mention that in the projection on Δ one gets a step of the relaxation method for the invariant manifold construction.

In the following Example it will be demonstrated how one can use function $\Delta(x_{\mathrm{sl}}(t))$ in the accuracy estimation of macroscopic equations in the dynamics of polymer solution.

15.2 Example: Defect of Invariance Estimation and Switching from the Microscopic Simulations to Macroscopic Equations

A method which recognizes the onset and breakdown of the macroscopic description in microscopic simulations was developed in [29, 268, 414]. The method is based on the invariance of the macroscopic dynamics relative to the microscopic dynamics, and it is demonstrated for a model of dilute polymeric solutions where it decides switching between Direct Brownian Dynamics simulations and integration of constitutive equations.

15.2.1 Invariance Principle and Micro-Macro Computations

Derivation of reduced (macroscopic) dynamics from the microscopic dynamics is the dominant theme of non-equilibrium statistical mechanics. At the present time, this very old theme demonstrates new facets in view of a massive use of simulation techniques on various levels of description. A two-side benefit of this use is expected: On the one hand, simulations provide data on molecular systems which can be used to test various theoretical constructions about the transition from micro to macro description. On the other hand, while the microscopic simulations in many cases are based on limit theorems [such as, for example, the central limit theorem underlying the Direct Brownian Dynamics simulations (BD)] they are extremely time-consuming in any real situation, and a timely recognition of the onset of a macroscopic description may considerably reduce computational efforts.

In this subsection, we aim at developing a 'device' which is able to recognize the onset and the breakdown of a macroscopic description in the course of microscopic computations.

Let us first present the main ideas of the construction in an abstract setting. We assume that the microscopic description is set up in terms of microscopic variables ξ. In the examples considered below, microscopic variables are distribution functions over the configuration space of polymers. The microscopic dynamics of variables ξ is given by the microscopic time derivative $\mathrm{d}\xi/\mathrm{d}t = \dot{\xi}(\xi)$. We also assume that the set of macroscopic variables $\boldsymbol{M}$ is chosen. Typically, the macroscopic variables are some lower-order moments if the microscopic variables are distribution functions. The reduced (macroscopic) description assumes (a) The dependence $\xi(\boldsymbol{M})$, and (b) The macroscopic dynamics $\mathrm{d}\boldsymbol{M}/\mathrm{d}t = \dot{\boldsymbol{M}}(\boldsymbol{M})$. We do not discuss here in any detail the way one gets the dependence $\xi(\boldsymbol{M})$, however, we should remark that, typically, it is based on some (explicit or implicit) idea about decomposition of motions into slow and fast, with $\boldsymbol{M}$ as slow variables. With this, such tools as maximum entropy principle, quasi-stationarity, cumulant expansion etc. become available for constructing the dependence $\xi(\boldsymbol{M})$.

Let us compare the microscopic time derivative of the function $\xi(\boldsymbol{M})$ with its macroscopic time derivative due to the macroscopic dynamics:

$$\Delta(\boldsymbol{M}) = \frac{\partial \xi(\boldsymbol{M})}{\partial \boldsymbol{M}} \cdot \dot{\boldsymbol{M}}(\boldsymbol{M}) - \dot{\xi}(\xi(\boldsymbol{M})) \,. \tag{15.13}$$

If the *defect of invariance* $\Delta(\boldsymbol{M})$ (15.13) is equal to zero on the set of admissible values of the macroscopic variables M, it is said that the reduced description $\xi(\boldsymbol{M})$ is invariant. Then the function $\xi(\boldsymbol{M})$ represents the invariant manifold in the space of microscopic variables. The invariant manifold is relevant if it is stable. Exact invariant manifolds are known in a very few cases (for example, the exact hydrodynamic description in the kinetic Lorentz gas model [202], in Grad's systems [40, 42], and one more example will be mentioned below). Corrections to the approximate reduced description through minimization of the defect of invariance is a part of the so-called method of invariant manifolds [11]. We here consider a different application of the invariance principle for the purpose mentioned above.

The time dependence of the macroscopic variables can be obtained in two different ways: First, if the solution of the microscopic dynamics at time t with initial data at t_0 is ξ_{t,t_0}, then evaluation of the macroscopic variables on this solution gives $\boldsymbol{M}_{t,t_0}^{\text{micro}}$. On the other hand, solving dynamic equations of the reduced description with initial data at t_0 gives $\boldsymbol{M}_{t,t_0}^{\text{macro}}$. Let $\|\Delta\|$ be a value of defect of invariance with respect to some norm, and $\epsilon > 0$ is a fixed tolerance level. Then, if at the time t the following inequality is valid,

$$\|\Delta(\boldsymbol{M}_{t,t_0}^{\text{micro}})\| < \epsilon \,, \tag{15.14}$$

this indicates that the accuracy provided by the reduced description is not worse than the true microscopic dynamics (the macroscopic description *sets on*). On the other hand, if

$$\|\Delta(\boldsymbol{M}_{t,t_0}^{\text{macro}})\| > \epsilon \,, \tag{15.15}$$

then the accuracy of the reduced description is insufficient (the reduced description *breaks down*), and we must use the microscopic dynamics.

Thus, evaluating the defect of invariance (15.13) on the current solution to macroscopic equations, and checking the inequality (15.15), we are able to answer the question whether we can trust the solution without looking at the microscopic solution. If the tolerance level is not exceeded then we can safely integrate the macroscopic equation. We now proceed to a specific example of this approach. We consider a well-known class of microscopic models of dilute polymeric solutions

15.2.2 Application to Dynamics of Dilute Polymer Solution

A well-known problem of the non-Newtonian fluids is the problem of establishing constitutive equations on the basis of microscopic kinetic equations. We here consider a model introduced by Lielens et al. [410]:

$$\dot{f}(q,t) = -\partial_q \left\{ \kappa(t)qf - \frac{1}{2}f\partial_q U(q^2) \right\} + \frac{1}{2}\partial_q^2 f \ . \qquad (15.16)$$

With the potential $U(x) = -(b/2)\ln(1 - x/b)$ equation (15.16) becomes the one-dimensional version of the FENE dumbbell model which is used to describe the elongational behavior of dilute polymer solutions.

The reduced description seeks a closed time evolution equation for the stress $\tau = \langle q\partial_q U(q^2)\rangle - 1$. Due to its non-polynomial character, the stress τ for the FENE potential depends on all moments of f. We have shown in [411] how such potentials can be approximated systematically by a set of polynomial potentials $U_n(x) = \sum_{j=1}^{n} \frac{1}{2j}c_j x^j$ of degree n with coefficients c_j depending on the even moments $M_j = \langle q^{2j}\rangle$ of f up to order n, with $n = 1, 2, \ldots$, formally converging to the original potential as n tends to infinity. In this approximation, the stress τ becomes a function of the first n even moments of f, $\tau(\boldsymbol{M}) = \sum_{j=1}^{n} c_j M_j - 1$, where the set of macroscopic variables is denoted by $\boldsymbol{M} = \{M_1, \ldots, M_n\}$.

The first two potentials approximating the FENE potential are:

$$U_1(q^2) = U'(M_1)q^2 \qquad (15.17)$$

$$U_2(q^2) = \frac{1}{2}(q^4 - 2M_1 q^2)U''(M_1) + \frac{1}{2}(M_2 - M_1^2)q^2 U'''(M_1) \ , \quad (15.18)$$

where U', U'' and U''' denote the first, second and third derivative of the potential U, respectively. The potential U_1 corresponds to the well-known FENE–P model. The kinetic equation (15.16) with the potential U_2 (15.18) will be termed the FENE–P+1 model below. Direct Brownian Dynamics simulation (BD) of the kinetic equation (15.16) with the potential U_2 for the flow situations studied in [410] demonstrates that it is a reasonable approximation to the true FENE dynamics whereas the corresponding moment chain is of a simpler structure. In [29] this was shown for a periodic flow, while Fig. 15.1 shows results for the flow

$$\kappa(t) = \begin{cases} 100t(1-t)e^{-4t} & 0 \le t \le 1 \ ; \\ 0 & \text{else} \ . \end{cases} \qquad (15.19)$$

The quality of the approximation indeed increases with the order of the polynomial.

For any potential U_n, the invariance equation can be studied directly in terms of the full set of the moments, which is equivalent to studying the distribution functions. The kinetic equation (15.16) can be rewritten equivalently in terms of moment equations,

$$\dot{M}_k = F_k(M_1, \ldots, M_{k+n-1}) \ ; \qquad (15.20)$$

$$F_k = 2k\kappa(t)M_k + k(2k-1)M_{k-1} - k\sum_{j=1}^{n} c_j M_{k+j-1} \ .$$

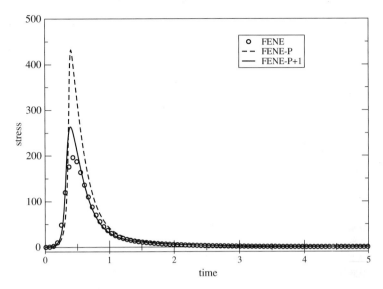

Fig. 15.1. Stress τ versus time from direct Brownian dynamics simulation: symbols – FENE, *dashed* line – FENE–P, *solid* line – FENE–P+1

We seek functions $M_k^{\mathrm{macro}}(\boldsymbol{M})$, $k = n+1, \ldots$ which are form-invariant under the dynamics:

$$\sum_{j=1}^{n} \frac{\partial M_k^{\mathrm{macro}}(\boldsymbol{M})}{\partial M_j} F_j(\boldsymbol{M}) = F_k(M_1, \ldots, M_n, M_{n+1}(\boldsymbol{M}), \ldots, M_{n+k}(\boldsymbol{M})) .$$

(15.21)

This set of invariance equations states the following: The time derivative of the form $M_k^{\mathrm{macro}}(\boldsymbol{M})$ when computed due to the closed equation for $\boldsymbol{M}$ (the first contribution on the left hand side of (15.21), or the 'macroscopic' time derivative) equals the time derivative of M_k as computed by true moment equation with the same form $M_k(\boldsymbol{M})$ (the second contribution, or the 'microscopic' time derivative), and this equality should hold whatsoever values of the moments $\boldsymbol{M}$ are.

Equations (15.21) in case $n = 1$ (FENE–P) are solvable exactly with the result

$$M_k^{\mathrm{macro}} = a_k M_1^k , \qquad \text{with} \quad a_k = (2k-1)a_{k-1}, \ a_0 = 1 .$$

This dependence corresponds to the Gaussian solution in terms of the distribution functions. As expected, the invariance principle give just the same result as the usual method of solving the FENE–P model.

Let us briefly discuss the potential U_2, considering a simple closure approximation

$$M_k^{\mathrm{macro}}(M_1, M_2) = a_k M_1^k + b_k M_2 M_1^{k-2} ,$$

(15.22)

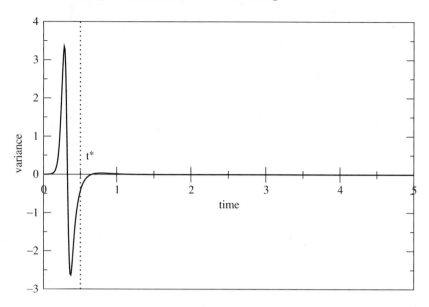

Fig. 15.2. Defect of invariance Δ_3/b^3, (15.23), versus time extracted from BD simulation (the FENE–P+1 model) for the flow situation of (15.19)

where $a_k = 1-k(k-1)/2$ and $b_k = k(k-1)/2$. The function M_3^{macro} closes the moment equations for the two independent moments M_1 and M_2. Note, that M_3^{macro} differs from the corresponding moment M_3 of the actual distribution function by the neglect of the 6-th cumulant. The defect of invariance of this approximation is a set of functions Δ_k where

$$\Delta_3(M_1, M_2) = \frac{\partial M_3^{\text{macro}}}{\partial M_1}F_1 + \frac{\partial M_3^{\text{macro}}}{\partial M_2}F_2 - F_3 \, , \qquad (15.23)$$

and analogously for $k \geq 3$. In the sequel, we make all conclusions based on the defect of invariance Δ_3 (15.23).

It is instructive to plot the defect of invariance Δ_3 versus time, assuming the functions M_1 and M_2 are extracted from the BD simulation (see Fig. 15.2). We observe that the defect of invariance is a nonmonotonic function of the time, and that there are three pronounced domains: From $t_0 = 0$ to t_1 the defect of invariance is almost zero which means that the ansatz is reasonable. In the intermediate domain, the defect of invariance jumps to high values (so the quality of approximation is poor). However, after some time $t = t^*$, the defect of invariance again becomes negligible, and remains so for later times. Such behavior is typical of so-called "kinetic layer".

Instead of attempting to improve the closure, the invariance principle can be used directly to switch from the BD simulation to the solution of the macroscopic equation without loosing the accuracy to a given tolerance. Indeed, the defect of invariance is a function of M_1 and M_2, and it can be easily

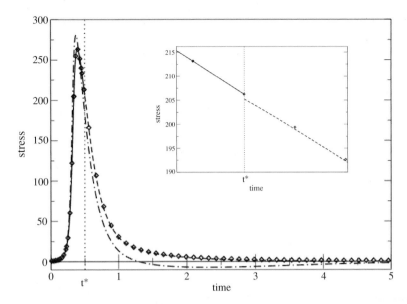

Fig. 15.3. Switching from the BD simulations to macroscopic equations after the defect of invariance has reached the given tolerance level (the FENE–P+1 model): symbols – the BD simulation, solid line – the BD simulation from time $t = 0$ up to time $t = t^*$, *dashed* line – integration of the macroscopic dynamics with initial data from BD simulation at time $t = t^*$. For comparison, the *dot-dashed* line gives the result for the integration of the macroscopic dynamics with equilibrium conditions from $t = 0$. Inset: Transient dynamics at the switching from BD to macroscopic dynamics on a finer time scale

evaluated both on the data from the solution to the macroscopic equation, and the BD data. If the defect of invariance *exceeds* some given tolerance on the macroscopic solution this signals to switch to the BD integration. On the other hand, if the defect of invariance becomes *less* than the tolerance level on the BD data signals that the BD simulation is not necessary anymore, and one can continue with the integration of the macroscopic equations. This reduces the necessity of using BD simulations only to get through the kinetic layers. A realization of this hybrid approach is demonstrated in Fig. 15.3: For the same flow we have used the BD dynamics only for the first period of the flow while integrated the macroscopic equations in all the later times. The quality of the result is comparable to the BD simulation whereas the total integration time is much shorter. The *transient dynamics* at the point of switching from the BD scheme to the integration of the macroscopic equations (shown in the inset in Fig. 15.3) deserves a special comment: The initial conditions at t^* are taken from the BD data. Therefore, we cannot expect that at the time t^* the solution is already on the invariant manifold, rather, at best, close to it. Transient dynamics therefore signals the *stability* of the invariant manifold we expect: Even though the macroscopic solution starts not on this manifold, it nevertheless attracts to it. The transient dynamics becomes progressively less pronounced if the switching is done at later times. The stability of the invariant manifold in case of the FENE–P model is studied in detail in [109].

The present approach of combined microscopic and macroscopic simulations can be realized on the level of moment closures (which then needs reconstruction of the distribution function from the moments at the switching from macroscopic integration to BD procedures), or for parametric sets of distribution functions if they are available [410]. It can be used for a rigorous construction of domain decomposition methods in various kinetic problems.

16 Conclusion

It is useful to construct slow invariant manifolds. Effective model reduction becomes unfeasible without them for complex kinetic systems.

Why should we attempt to reduce the description in the times of super-computers?

– First, in order to gain insight. In the process of reducing the description we is often able to extract the essential, and the mechanisms of the processes under study become more transparent.
– Second, once we obtain the detailed description of the system, then we can try to solve the initial-value problem for this system. But what should one do in the case where the system represents just a small part of the huge number of interacting systems? For example, a complex chemical reaction system may represent just a point in a three-dimensional flow.
– Third, without reducing the kinetic model, it is impossible to construct this model. This statement seems paradoxical only at the first glance: How can it come, the model is first simplified, and is constructed only after the simplification is done? However, in practice, the statement of the problem typical for a mathematician (Let the system of differential equations be *given*, then ...) is rather rarely applicable for detailed kinetics. On the contrary, the thermodynamic data (energies, enthalpies, entropies, chemical potentials etc) for sufficiently rarefied systems are quite reliable. Final identification of the model is always done on the basis of comparison with the experiment and with the help of fitting. For this purpose, it is extremely important to reduce the dimension of the system, and to reduce the number of tunable parameters.
– And, finally, for every supercomputer there exist problems that are too complicated. Model reduction makes these problems less complicated and sometimes gives us the possibility to solve them.

It is useful to apply thermodynamics and the quasiequilibrium concept while seeking slow invariant manifolds. Although open systems are important for many applications, it is useful to begin their study and model reduction with the analysis of closed (sub)systems. The thermodynamics equips then these systems with the Lyapunov functions (entropy, free energy, free enthalpy, depending on the context). These Lyapunov functions are usually known much better than the right hand sides of kinetic equations

Alexander N. Gorban and Iliya V. Karlin: *Invariant Manifolds for Physical and Chemical Kinetics*, Lect. Notes Phys. **660**, 467–468 (2005)
www.springerlink.com

(in particular, this is the case in reaction kinetics). Using a Lyapunov function, one constructs the initial approximation to the slow manifold, that is, the quasiequilibrium manifold, and also one constructs the thermodynamic projector.

The thermodynamic projector is the unique operator which transforms the arbitrary vector field equipped with the given Lyapunov function into a vector field with the same Lyapunov function (and also this happens on any manifold which is not tangent to the level of the Lyapunov function).

The quasi-chemical approximation is an extremely rich toolbox for assembling equations. It enables one to construct and study wide classes of evolution equations equipped with prescribed Lyapunov functions and Onsager reciprocity relations.

The **method of natural projector** is an attractive method of model reduction for dissipative systems, and at the same time it gives a clue to the problem of irreversibility. The formula for *entropy production*,

$$\sigma \sim \frac{\text{defect of invariance}}{\text{curvature}}$$

clarifies the geometrical sense of the dissipation. Here, "defect of invariance" is the defect of invariance of the quasiequilibrium manifold, and "curvature" is the curvature of the **film of nonequilibrium states** in the direction of the defect of invariance of the quasiequilibrium manifold

Slow invariant manifolds of thermodynamically closed systems are useful for constructing slow invariant manifolds of the corresponding open systems. The necessary technique is developed.

The k-**contractions** and the **quasi-biological representation** allow to find finite-dimensional asymptotics for some of infinite-dimensional systems.

The **postprocessing** of the invariant manifold construction is important both for the estimation of the accuracy and for the accuracy improvement.

The main result of this book can be formulated as follows: **It is possible indeed to construct invariant manifolds.** The problem of constructing invariant manifolds can be formulated as the invariance equation, subject to additional conditions of slowness (stability). The Newton method with incomplete linearization, relaxation methods, the method of natural projector, and the method of invariant grids enables educated approximations to the slow invariant manifolds. These methods were tested on a recently discovered class of **exactly solvable reduction problems**.

It becomes more and more evident at the present time that the constructive methods of invariant manifold are useful for a wide range of subjects, spanning from applied hydrodynamics to physical and chemical kinetics.

References

1. Van Kampen, N.G., Elimination of fast variables, Physics Reports, **124** (1985), 69–160.
2. Bogolyubov, N.N., Dynamic theory problems in statistical physics, Gostekhizdat, Moscow, Leningrad, 1946.
3. Lyapunov A.M., The general problem of the stability of motion, Taylor & Francis, London, 1992.
4. Kolmogorov, A.N., On conservation of conditionally periodic motions under small perturbations of the Hamiltonian. Dokl. Akad. Nauk SSSR, **98** (1954), 527–530.
5. Arnold, V.I., Proof of a theorem of A.N. Kolmogorov on the invariance of quasi–periodic motions under small perturbations of the Hamiltonian. (English translation) Russian Math Surveys, **18** (1963), 9–36.
6. Moser, J., Convergent series expansions for quasi–periodic motions, Math. Ann., **169** (1967), 136–176.
7. Moser, J., On invariant manifolds of vector fields and symmetric partial differential equations, Differential Anal., Bombay Colloq. (1964), 227–236.
8. Sacker, R.J., A new approach to the perturbation theory of invariant surfaces, Comm. Pure. Appl. Math., **18** (1965), 717–732.
9. Gorban, A.N., Karlin, I.V., Thermodynamic parameterization, Physica A, **190** (1992), 393–404.
10. Gorban, A.N., Karlin, I.V., Uniqueness of thermodynamic projector and kinetic basis of molecular individualism, Physica A, **336**, 3–4 (2004), 391–432. Preprint online: http://arxiv.org/abs/cond-mat/0309638.
11. Gorban, A.N., Karlin, I.V., Method of invariant manifolds and regularization of acoustic spectra, Transport Theory and Stat. Phys., **23** (1994), 559–632.
12. Gorban, A.N., Karlin, I.V., Zinovyev, A.Yu., Constructive methods of invariant manifolds for kinetic problems, Phys. Reports, **396**, 4–6 (2004), 197–403. Preprint online: http://arxiv.org/abs/cond-mat/0311017.
13. Roberts, A.J., Low–dimensional modelling of dynamical systems applied to some dissipative fluid mechanics, in: Nonlinear dynamics from lasers to butterflies, World Scientific, Lecture Notes in Complex Systems, **1**, (2003), Rowena Ball and Nail Akhmediev, eds, 257–313.
14. Gorban, A.N., Karlin, I.V., The constructing of invariant manifolds for the Boltzmann equation, Adv.Model. and Analysis C, **33(3)** (1992), 39–54.
15. Ehrenfest, P., Ehrenfest-Afanasyeva, T., in: Mechanics Enziklopädie der Mathematischen Wissenschaften, Vol.4., Leipzig, 1911. (Reprinted in: Ehrenfest, P., Collected Scientific Papers, North–Holland, Amsterdam, 1959, pp. 213–300.)
16. Hilbert, D. Begründung der kinetischen Gastheorie, Mathematische Annalen, **72** (1912), 562–577.

17. Karlin, I.V., Dukek, G., Nonnenmacher, T.F., Invariance principle for extension of hydrodynamics: Nonlinear viscosity, Phys. Rev. E, **55(2)** (1997), 1573–1576.
18. Santos, A., Nonlinear viscosity and velocity distribution function in a simple longitudinal flow, Phys. Rev. E **62** (2000), 6597–6607.
19. Santos, A., Comments on nonlinear viscosity and Grad's moment method, Phys. Rev. E **67** (2003), 053201.
20. Garzó, V., Santos, A., Kinetic theory of gases in shear flows. nonlinear transport, Book series: Fundamental Theories of Physics, Vol. 131, Kluwer, Dordrecht, 2003.
21. Karlin, I.V., Gorban, A.N., Dukek, G., Nonnenmacher, T.F. Dynamic correction to moment approximations, Phys. Rev. E, **57** (1998), 1668–1672.
22. Gorban, A.N., Karlin, I.V., Method of invariant manifold for chemical kinetics, Chem. Eng. Sci., **58**, 21 (2003), 4751–4768. Preprint online: http://arxiv.org/abs/cond-mat/0207231.
23. Gorban, A.N., Karlin, I.V., Zmievskii, V.B., Dymova S.V., Reduced description in reaction kinetics, Physica A, **275(3–4)** (2000), 361–379.
24. Karlin, I.V., Zmievskii, V.B., Invariant closure for the Fokker–Planck equation, 1998. Preprint online: http://arxiv.org/abs/adap-org/9801004.
25. Foias, C., Jolly, M.S., Kevrekidis, I.G., Sell, G.R., Titi, E.S. On the computation of inertial manifolds, Physics Letters A, **131**, 7–8 (1988), 433–436.
26. Gorban, A.N., Karlin, I.V., Zmievskii, V.B., Nonnenmacher, T.F., Relaxational trajectories: global approximations, Physica A, **231** (1996), 648–672.
27. Gorban, A.N., Karlin, I.V., Zmievskii, V.B., Two–step approximation of space–independent relaxation, Transp.Theory Stat. Phys., **28(3)** (1999), 271–296.
28. Guckenheimer, J., Vladimirsky, A., A fast method for approximating invariant manifolds, SIAM Journal on Applied Dynamical Systems, **3**, 3 (2004), 232–260.
29. Gorban, A.N., Karlin, I.V., Ilg, P., and Öttinger, H.C., Corrections and enhancements of quasi–equilibrium states, J.Non–Newtonian Fluid Mech. **96** (2001), 203–219.
30. Gorban, A.N., Karlin, I.V., Öttinger, H.C., and Tatarinova, L.L., Ehrenfest's argument extended to a formalism of nonequilibrium thermodynamics, Phys. Rev. E **63** (2001), 066124.
31. Gorban, A.N., Karlin, I.V., Reconstruction lemma and fluctuation–dissipation theorem, Revista Mexicana de Fisica **48**, Supl. 1 (2002), 238–242.
32. Gorban, A.N., Karlin, I.V., Macroscopic dynamics through coarse–graining: A solvable example, Phys. Rev. E, **56** (2002), 026116.
33. Gorban, A.N., Karlin, I.V., Geometry of irreversibility, in: Recent Developments in Mathematical and Experimental Physics, Volume C: Hydrodynamics and Dynamical Systems, Ed. F. Uribe, Kluwer, Dordrecht, 2002, 19–43.
34. Karlin, I.V., Tatarinova, L.L., Gorban, A.N., Öttinger, H.C., Irreversibility in the short memory approximation, Physica A, **327**, 3–4 (2003), 399–424. Preprint online: http://arXiv.org/abs/cond-mat/0305419 v1 18 May 2003.
35. Karlin, I.V., Ricksen, A., Succi, S., Dissipative quantum dynamics from Wigner distributions, in: Quantum Limits to the Second Law: First International Conference on Quantum Limits to the Second Law, San Diego, California (USA), 29–31 July 2002, AIP Conference Proceedings, **643**, 19–24.

36. Wigner, E., On the quantum correction for thermodynamic equilibrium, Phys. Rev., **40** (1932), 749–759.

37. Caldeira, A.O., Leggett, A.J. Influence of damping on quantum interference: An exactly soluble model, Phys. Rev. A, **31** (1985), 1059–1066.

38. Filinov, V.S., Wigner approach to quantum statistical mechanics and quantum generalization molecular dynamics method. 1, Mol. Phys., **88** (1996), 1517–1528; 2, ibidem 1529-1539.

39. Calzetta, E.A., Hu, B.L., Correlation entropy of an interacting quantum field and H-theorem for the $O(N)$ model, Phys. Rev. D, **68** (2003), 065027.

40. Gorban, A.N., Karlin, I.V., Short–wave limit of hydrodynamics: a soluble example, Phys. Rev. Lett. **77** (1996), 282–285.

41. Karlin, I.V., Exact summation of the Chapman-Enskog expansion from moment equations, J. Physics A: Math. Gen., **33** (2000), 8037–8046.

42. Karlin, I.V., Gorban, A.N., Hydrodynamics from Grad's equations: What can we learn from exact solutions?, Ann. Phys. (Leipzig) **11** (2002), 783–833. Preprint online: http://arXiv.org/abs/cond-mat/0209560.

43. Gorban, A.N., Karlin, I.V., Structure and approximations of the Chapman–Enskog expansion, Sov. Phys. JETP **73** (1991), 637–641.

44. Gorban, A.N., Karlin, I.V., Structure and approximations of the Chapman–Enskog expansion for linearized Grad equations, Transport Theory and Stat. Phys. **21** (1992), 101–117.

45. Karlin, I.V., Simplest nonlinear regularization, Transport Theory and Stat. Phys., **21** (1992), 291–293.

46. Fenichel, N., Persistence and smooothness of invariant manifolds for flows, Indiana Univ. Math. J., **21** (1971), 193–226.

47. Hirsch, M.W., Pugh, C., Shub, M., Invariant manifolds, Lecture Notes in Mathematics, V. 583, Springer, NY, 1977.

48. Jones, D.A., Stuart, A.M., Titi, E.S., Persistence of invariant sets for dissipative evolution equations, Journal of Mathematical Analysis and Applications, **219**, 2 (1998), 479–502.

49. De la Llave, R., Invariant manifolds associated to invariant subspaces without invariant complements: a graph transform approach, Mathematical Physics Electronic Journal, **9** (2003). http://www.ma.utexas.edu/mpej/MPEJ.html

50. Poincaré, H.: Les méthodes nouvelles de la mécanique céleste. Vols. 1–3. Gauthier–Villars, Paris, 1892/1893/1899.

51. Beyn, W.-J., Kless, W. Numerical Taylor expansions of invariant manifolds in large dynamical systems, Numerische Mathematik **80** (1998), 1–38.

52. Kazantzis, N., Singular PDEs and the problem of finding invariant manifolds for nonlinear dynamical systems, Physics Letters, A **272** (4) (2000), 257–263.

53. Shirkov, D.V., Kovalev, V.F., Bogoliubov renormalization group and symmetry of solution in mathematical physics, Physics Reports, **352** (2001), 219–249. Preprint online: http://arxiv.org/abs/hep-th/0001210.

54. Zinn-Justin, J., Quantum field theory and critical phenomena, Clarendon Press, Oxford, 1989.

55. Pashko O., Oono, Y., The Boltzmann equation is a renormalization group equation, Int. J. Mod. Phys. B, **14** (2000), 555–561.

56. Kunihiro T., A geometrical formulation of the renormalization group method for global analysis, Prog. Theor. Phys. **94** (1995), 503–514; Erratum: ibid. **95** (1996), 835. Preprint online: http://arxiv.org/abs/hep-th/9505166.

57. Ei, S.-I., Fujii, K., Kunihiro, T., Renormalization–group method for reduction of evolution equations; invariant manifolds and envelopes, Annals Phys. **280** (2000), 236–298. Preprint online: http://arxiv.org/abs/hep-th/9905088.

58. Hatta Y., Kunihiro T. Renormalization group method applied to kinetic equations: roles of initial values and time, Annals Phys. **298** (2002), 24–57. Preprint online: http://arxiv.org/abs/hep-th/0108159.

59. Degenhard A., Rodrigues-Laguna J. Towards the evaluation of the relevant degrees of freedom in nonlinear partial differential equations, J. Stat. Phys., **106**, No. 516 (2002), 1093–1119.

60. Forster, D., Nelson D.R., Stephen, M.J., Long–time tails and the large–eddy behavior of a randomly stirred fluid, Phys. Rev. Lett. **36** (1976), 867–870.

61. Forster, D., Nelson D.R., Stephen, M.J., Large–distance and long–time properties of a randomly stirred fluid, Phys. Rev. A **16** (1977), 732–749.

62. Adzhemyan, L.Ts., Antonov, N.V., Kompaniets, M.V., Vasil'ev, A.N., Renormalization–group approach to the stochastic Navier Stokes equation: Two–loop approximation, International Journal of Modern Physics B, **17**, 10 (2003), 2137–2170.

63. Chen, H., Succi, S., Orszag, S., Analysis of subgrid scale turbulence using Boltzmann Bhatnagar-Gross-Krook kinetic equation, Phys. Rev. E, **59** (1999), R2527–R2530.

64. Chen, H., Kandasamy, S., Orszag, S., Shock, R., Succi, S., Yakhot, V., Extended Boltzmann Kinetic Equation for Turbulent Flows, Science, **301** (2003), 633–636.

65. Degond, P., Lemou, M., Turbulence Models for Incompressible Fluids Derived from Kinetic Theory, Journal of Mathematical Fluid Mechanics, **4**, 3 (2002), 257–284.

66. Ansumali, S., Karlin, I.V., Succi, S., Kinetic theory of turbulence modeling: Smallness parameter, scaling and microscopic derivation of Smagorinsky model, Physica A, (2004), to appear. Preprint online: http://arxiv.org/abs/cond-mat/0310618.

67. Smagorinsky, J., General Circulation Experiments with the Primitive Equations: I. The Basic Equations, Mon. Weather Rev., **91** (1963), 99–164.

68. Bricmont, J., Gawedzki, K., Kupiainen, A., KAM theorem and quantum field theory. Commun. Math. Phys. **201** (1999), 699–727. E-print mp_arc 98–526, online: http://mpej.unige.ch/mp_arc/c/98/98-517.ps.gz.

69. Gorban, A.N., Karlin, I.V., Methods of nonlinear kinetics, in: Encyclopedia of Life Support Systems, Encyclopedia of Mathematical Sciences, EOLSS Publishers, Oxford, 2004, http://www.eolss.net/. Preprint online: http://arXiv.org/abs/cond-mat/0306062.

70. Chapman, S., Cowling, T., Mathematical theory of non-uniform gases, Third edition, Cambridge University Press, Cambridge, 1970.

71. Galkin V.S., Kogan M.N., Makashev N.K., Chapman-Enskog generalized method, Dokl. Akademii Nauk SSSR, **220** (1975), 304–307.

72. Bobylev, A.V., The Chapman–Enskog and Grad methods for solving the Boltzmann equation, Sov. Phys. Dokl., **27** (1982), No. 1, 29–31.

73. Bobylev, A.V., Exact-solutions of the nonlinear Boltzmann-equation and the theory of relaxation of a Maxwellian gas, Theor. Math. Phys., **60** (1984), 820–841.

74. Bobylev, A.V., Quasi-stationary hydrodynamics for the Boltzmann equation, J. Stat. Phys. **80**, (1995), 1063–1083.

75. Ernst, M.H., Nonlinear Model-Boltzmann equations and exact solutions, Physics Reports, **78** (1981), 1–171.
76. García-Colín, L.S., Green, M.S., Chaos, F., The Chapman-Enskog solution of the generalized Boltzmann equation Physica, **32**, 2 (1966), 450–478.
77. Bowen, J.R., Acrivos, A., Oppenheim, A.K., Singular perturbation refinement to quasi–steady state approximation in chemical Kinetics. Chemical Engineering Science, **18** (1963), 177–188.
78. Segel, L.A., Slemrod, M., The quasi–steady–state assumption: A case study in perturbation. SIAM Rev., **31** (1989), 446–477.
79. Fraser, S.J., The steady state and equilibrium approximations: A geometrical picture. J. Chem. Phys., **88**(8) (1988), 4732–4738.
80. Roussel, M.R., Fraser, S.J., Geometry of the steady–state approximation: Perturbation and accelerated convergence methods, J. Chem. Phys., **93** (1990), 1072–1081.
81. Yablonskii, G.S., Bykov, V.I., Gorban, A.N., Elokhin, V.I., Kinetic models of catalytic reactions. Comprehensive Chemical Kinetics, Vol. 32, Compton R. G. ed., Elsevier, Amsterdam (1991).
82. Vasil'eva A.B., Butuzov V.F., Kalachev L.V., The boundary function method for singular perturbation problems, SIAM (1995).
83. Strygin V.V., Sobolev V.A., Spliting of motion by means of integral manifolds. Nauka, Moscow (1988).
84. Roos, H.G., Stynes, M., Tobiska, L., numerical methods for singularly perturbed differential equations: Convection–diffusion and flow problems, Springer Verlag, 1996.
85. Mishchenko, E.F., Kolesov, Y.S., Kolesov, A.U., Rozov, N.Kh., Asymptotic methods in singularly perturbed systems, Consultants Bureau, 1994.
86. Novozhilov, I.V., Fractional analysis: Methods of motion decomposition, Birkhäuser, Boston, 1997.
87. Milik, A., Singular perturbation on the Web, 1997. http://www.ima.umn.edu/~milik/singdir.html#geo:sing.
88. Gear, C.W., Numerical initial value problems in ordinary differential equations, Prentice–Hall, Englewood Cliffs, NJ (1971).
89. Rabitz, H., Kramer, M., Dacol, D., Sensitivity analysis in chemical kinetics, Ann. Rev. Phys. Chem., 34, 419–461 (1983).
90. Lam, S.H., Goussis, D.A., The CSP Method for Simplifying Kinetics, International Journal of Chemical Kinetics, **26** (1994), 461–486.
91. Valorani, M., Goussis, D.A., Explicit time-scale splitting algorithm for stiff problems: Auto-ignition of gaseous mixtures behind a steady shock, Journal of Computational Physics, **169** (2001), 44–79.
92. Valorani, M., Najm, H.N., Goussis, D.A., CSP analysis of a transient flame-vortex interaction: time scales and manifolds, Combustion and Flame **134** (2003), 35–53.
93. Maas, U., Pope, S.B., Simplifying chemical kinetics: intrinsic low– dimensional manifolds in composition space, Combustion and Flame, **88** (1992), 239–264.
94. Kaper, H.G., Kaper, T.J., Asymptotic analysis of two reduction methods for systems of chemical reactions, Physica D, **165** (2002), 66–93.
95. Zagaris, A., Kaper, H.G., Kaper, T.J. Analysis of the computational singular perturbation reduction method for chemical kinetics, Journal of Nonlinear Science, **14**, 1 (2004), 59–91. Preprint on-line: http://arxiv.org/abs/math.DS/0305355.

474   References

96. Debussche A., Temam, R., Inertial manifolds and slow manifolds. Appl. Math. Lett., **4**, 4 (1991), 73–76.
97. Foias, C., Prodi, G., Sur le comportement global des solutions non stationnaires des equations de Navier-Stokes en dimension deux, Rend. Sem. Mat. Univ. Padova. **39** (1967), 1–34.
98. Ladyzhenskaya, O.A., A dynamical system generated by Navier-Stokes equations, J. of Soviet Mathematics, **3** (1975), 458–479.
99. Chueshov, I.D., Theory of functionals that uniquely determine the asymptotic dynamics of infinite-dimentional dissipative systems, Russian Math. Surveys., **53**, 4 (1998), 731–776.
100. Chueshov, I.D., Introduction to the theory of infinite-dimensional dissipative systems, The Electronic Library of Mathematics, 2002, http://rattler. cameron.edu/EMIS/monographs/Chueshov/. [Translated from Russian edition, ACTA Scientific Publishing House, Kharkov, Ukraine, 1999].
101. Dellnitz, M., Junge, O., Set oriented numerical methods for dynamical systems, in: B. Fiedler, G. Iooss and N. Kopell (eds.): Handbook of Dynamical Systems II: Towards Applications, World Scientific, 2002, 221–264. http://math-www.upb.de/~agdellnitz/papers/handbook.pdf.
102. Dellnitz, M., Hohmann, A. The computation of unstable manifolds using subdivision and continuation, in H.W. Broer et al. (eds.), Progress in Nonlinear Differential Equations and Their Applications 19:449–459, Birkhäuser, Basel / Switzerland, 1996.
103. Broer, H.W., Osinga, H.M., and Vegter, G. Algorithms for computing normally hyperbolic invariant manifolds, Z. angew. Math. Phys. **48** (1997), 480–524.
104. Garay, B.M., Estimates in discretizing normally hyperbolic compact invariant manifolds of Ordinary Differential Equations, Computers and Mathematics with Applications, **42** (2001), 1103–1122.
105. Gorban, A.N., Karlin, I.V., Zinovyev, A.Yu., Invariant grids for reaction kinetics, Physica A, **333** (2004), 106–154. Preprint online: http://www.ihes.fr/ PREPRINTS/P03/Resu/resu-P03–42.html.
106. Zmievskii, V.B., Karlin, I.V., Deville, M., The universal limit in dynamics of dilute polymeric solutions, Physica A, **275(1−2)** (2000), 152–177.
107. Theodoropoulos, C., Qian, Y.H., Kevrekidis, I.G., Coarse stability and bifurcation analysis using time-steppers: a reaction-diffusion example, Proc. Nat. Acad. Sci., **97** (2000), 9840–9843.
108. Kevrekidis, I.G., Gear, C.W.., Hyman, J.M., Kevrekidis, P.G., Runborg, O., Theodoropoulos, C., Equation-free, coarse-grained multiscale computation: enabling microscopic simulators to perform system-level analysis, Comm. Math. Sci., **14** (2003), 715–762.
109. Ilg P., Karlin, I.V., Validity of macroscopic description in dilute polymeric solutions, Phys. Rev. E **62** (2000), 1441–1443.
110. Ilg, P., De Angelis, E., Karlin, I.V., Casciola, C.M., Succi, S., Polymer dynamics in wall turbulent flow, Europhys. Lett., **58** (2002), 616–622.
111. Boltzmann, L., Lectures on gas theory, University of California Press, 1964.
112. Cercignani, C., The Boltzmann equation and its applications, Springer, New York, 1988.
113. Cercignani, C., Illner, R., Pulvirent, M., The mathematical theory of dilute gases, Springer, New York, 1994.

114. Stueckelberg E.C.G., Theoreme H et unitarite de S, Helv. Phys. Acta **25**, 5 (1952), 577–580.

115. Gorban, A.N., Equilibrium encircling. Equations of chemical kinetics and their thermodynamic analysis, Nauka, Novosibirsk, 1984.

116. Bhatnagar, P.L., Gross, E.P., Krook, M., A model for collision processes in gases. I. Small amplitude processes in charged and neutral one-component systems, Phys. Rev., **94**, 3 (1954), 511–525.

117. Gorban, A.N., Karlin, I.V., General approach to constructing models of the Boltzmann equation, Physica A, **206** (1994), 401–420.

118. Lebowitz, J., Frisch, H., Helfand, E., Non–equilibrium distribution functions in a fluid, Physics of Fluids, **3** (1960), 325.

119. DiPerna, R.J., Lions, P.L., On the Cauchy problem for Boltzmann equation: Global existence and weak stability, Ann. Math, **130** (1989), 321–366.

120. Enskog, D., Kinetische theorie der Vorange in massig verdunnten Gasen. I Allgemeiner Teil, Almqvist and Wiksell, Uppsala, 1917.

121. Pöschel, Th., Brilliantov, N. V. Kinetic integrals in the kinetic theory of dissipative gases, In: T. Pöschel, N. Brilliantov (eds.) "Granular Gas Dynamics", Lecture Notes in Physics, Vol. 624, Springer, Berlin, 2003, 131–162.

122. Broadwell, J.E., Study of shear flow by the discrete velocity method, J. Fluid Mech. **19** (1964), 401–414.

123. Broadwell, J.E., Shock structure in a simple discerte velocity gas, Phys.Fluids, **7** (1964), 1243–1247.

124. Palczewski, A., Schneider, J., Bobylev, A.V., A consistency result for a discrete-velocity model of the Boltzmann equation, SIAM Journal on Numerical Analysis, **34**, 5 (1997), 1865–1883.

125. Zwanzig, R., Ensemble method in the theory of irreversibility. J. Chem. Phys., **33**, 5 (1960), 1338–1341.

126. Robertson, B., Equations of motion in nonequilibrium statistical mechanics, Phys. Rev., **144** (1966), 151–161.

127. Bird, G.A., Molecular gas dynamics and the direct simulation of gas flows, Clarendon Press, Oxford, 1994.

128. Oran, E.S., Oh, C.K., Cybyk, B.Z., Direct simulation Monte Carlo: recent advances and applications, Annu Rev. Fluid Mech., **30** (1998), 403–441.

129. Gatignol, R. Theorie cinetique des gaz a repartition discrete de vitesses. Lecture notes in physics, V. 36, Springer, Berlin, etc, 1975.

130. Frisch, U., Hasslacher, B., Pomeau, Y., Lattice–gas automata for the Navier–Stokes equation, Phys. Rev. Lett., **56** (1986), 1505–1509.

131. Mcnamara, Gr., Zanetti, G., Use of the Boltzmann-equation to simulate lattice-gas automata, Phys. Rev. Lett., **61** (1988), 2332–2335.

132. Higuera, F., Succi, S., Benzi, R., Lattice gas – dynamics with enhanced collisions, Europhys. Lett., **9** (1989), 345–349.

133. Benzi, R., Succi, S., Vergassola, M., The lattice Boltzmann-equation - theory and applications Physics Reports, **222**, 3 (1992), 145–197.

134. Chen, S., Doolen, G.D., Lattice Boltzmann method for fluid flows, Annu. Rev. Fluid. Mech. **30** (1998), 329–364.

135. Succi, S., The lattice Boltzmann equation for fluid dynamics and beyond, Clarendon Press, Oxford, 2001.

136. Succi, S., Karlin, I.V., Chen H., Role of the H theorem in lattice Boltzmann hydrodynamic simulations, Rev. Mod. Phys., **74** (2002), 1203–1220.

137. Karlin, I.V., Gorban, A.N., Succi, S., Boffi, V., Maximum entropy principle for lattice kinetic equations, Phys. Rev. Lett., **81** (1998), 6–9.

138. Karlin, I.V., Ferrante, A., Öttinger, H.C., Perfect entropy functions of the Lattice Boltzmann method, Europhys. Lett., **47** (1999), 182–188.

139. Ansumali, S., Karlin, I.V., Stabilization of the Lattice Boltzmann method by the H theorem: A numerical test, Phys. Rev. E, **62** (6), (2000), 7999–8003.

140. Ansumali, S., Karlin, I.V., Entropy function approach to the lattice Boltzmann method, J. Stat. Phys., **107** (1/2) (2002), 291–308.

141. Ansumali, S., Karlin, I.V., Öttinger, H.C., Minimal entropic kinetic models for hydrodynamics, Europhys. Lett., **63** (2003), 798–804.

142. Ansumali, S., Karlin, I.V., Kinetic Boundary condition for the lattice Boltzmann method, Phys. Rev. E, **66** (2002), 026311.

143. Ansumali, S., Chikatamarla, S.S., Frouzakis, C.E., Boulouchos, K., Entropic lattice Boltzmann simulation of the flow past square cylinder, Int. J. Mod. Phys. C, **15** (2004), 435–445.

144. Shan, X., He, X., Discretization of the velocity space in the solution of the Boltzmann equation, Phys. Rev. Lett., **80** (1998), 65–67.

145. Van Beijeren, H, Ernst, M.H., Modified Enskog equation, Physica A, **68**, 3 (1973), 437–456.

146. Marsden, J.E., Weinstein, A., The Hamiltonian structure of the Maxwell-Vlasov equations, Physica D, **4** (1982), 394–406.

147. Braun W, Hepp K, Vlasov dynamics and its fluctuations in 1-N limit of interacting classical particles, Comm. Math. Phys., **56**, 2 (1977), 101–113.

148. Van Kampen, N.G., Stochastic processes in physics and chemistry, North–Holland, Amsterdam 1981.

149. Risken, H., The Fokker–Planck equation, Springer, Berlin, 1984.

150. Hänggi P., Thomas H., Stochastic Processes: Time Evolution, Symmetries and Linear Response, Physics Reports, **88** (1982), 207–319.

151. Bird, R.B., Curtiss, C.F., Armstrong, R.C., Hassager, O., Dynamics of Polymer Liquids, 2nd edn., Wiley, New York, 1987.

152. Doi, M., Edwards, S.F., The theory of polymer dynamics, Clarendon Press, Oxford, 1986.

153. Öttinger, H.C., Stochastic processes in polymeric fluids, Springer, Berlin, 1996.

154. Grmela, M., Öttinger, H.C., Dynamics and thermodynamics of complex fluids. I. Development of a general formalism, Phys. Rev. E **56** (1997), 6620–6632.

155. Öttinger, H.C., Grmela, M., Dynamics and thermodynamics of complex fluids. II. Illustrations of a general formalism, Phys. Rev. E, **56** (1997), 6633–6655.

156. Kullback, S., Information theory and statistics, Wiley, New York, 1959.

157. Plastino, A.R., Miller, H.G., Plastino, A., Minimum Kullback entropy approach to the Fokker-Planck equation, Physical Review E **56** (1997). 3927–3934.

158. Gorban, A.N., Karlin, I.V., Family of additive entropy functions out of thermodynamic limit, Phys. Rev. E, **67** (2003), 016104. Preprint online: http://arxiv.org/abs/cond-mat/0205511.

159. Gorban, A.N., Karlin, I.V., Öttinger H.C., The additive generalization of the Boltzmann entropy, Phys. Rev. E, **67**, 067104 (2003). Preprint online: http://arxiv.org/abs/cond-mat/0209319.

160. Gorban, P., Monotonically equivalent entropies and solution of additivity equation, Physica A, **328** (2003), 380-390. Preprint online: http://arxiv.org/pdf/cond-mat/0304131.

161. Tsallis, C., Possible generalization of Boltzmann-Gibbs statistics. J. Stat. Phys., **52** (1988), 479–487.
162. Abe, S., Okamoto, Y. (Eds.), Nonextensive statistical mechanics and its applications, Springer, Heidelberg, 2001.
163. Dukek, G., Karlin, I.V., Nonnenmacher, T.F., Dissipative brackets as a tool for kinetic modeling, Physica A, **239(4)** (1997), 493–508.
164. Orlov, N.N., Rozonoer, L.I., The macrodynamics of open systems and the variational principle of the local potential, J. Franklin Inst., **318** (1984), 283–314 and 315–347.
165. Volpert, A.I., Hudjaev, S.I., Analysis in classes of discontinuous functions and the equations of mathematical physics. Dordrecht: Nijhoff, 1985.
166. Ansumali S., Karlin, I.V., Single relaxation time model for entropic Lattice Boltzmann methods, Phys. Rev. E, **65** (2002), 056312.
167. Bykov, V.I., Yablonskii, G.S., Akramov, T.A., The rate of the free energy decrease in the course of the complex chemical reaction. Dokl. Akad. Nauk USSR, **234**, 3 (1977) 621–634.
168. Struchtrup, H., Weiss, W., Maximum of the local entropy production becomes minimal in stationary processes, Phys. Rev. Lett., **80** (1998), 5048–5051.
169. Grmela, M., Karlin, I.V., Zmievski, V.B., Boundary layer minimum entropy principles: A case study, Phys. Rev. E, **66** (2002), 011201.
170. Dimitrov, V.I., Simple kinetics, Nauka, Novosibirsk, 1982.
171. Prigogine, I., Thermodynamics of irreversible processes, Interscience, New York, 1961.
172. Lifshitz, E.M., Pitaevskii L.P., Physical kinetics (Landau L.D. and Lifshitz E.M. Course of Theoretical Physics, V. 10), Pergamon Press, Oxford, 1968.
173. Constantin, P., Foias, C., Nicolaenko, B., Temam, R., Integral manifolds and inertial manifolds for dissipative partial differential equations, Applied Math. Sci., 1988, Vol. 70 (Springer Verlag, New York).
174. Robinson, J.C., A concise proof of the "geometric" construction of inertial manifolds, Phy. Lett. A, **200** (1995), 415–417.
175. Ryashko, L.B., Shnol, E.E., On exponentially attracting invariant manifolds of ODEs, Nonlinearity, **16** (2003), 147–160.
176. Walter, W., An elementary proof of the Cauchy–Kovalevsky Theorem, Amer. Math. Month- ly **92** (1985), 115–126.
177. Evans, L.C., Partial differential equations, AMS, Providence, RI, USA, 1998.
178. Dubinskii, Ju.A., Analytic pseudo–differential operators and their applications. Kluwer Academic Publishers, Book Series: Mathematics And its Applications Soviet Series: Volume 68, 1991.
179. Levermore, C.D., Oliver, M., Analyticity of solutions for a generalized Euler equation, J. Differential Equations **133** (1997), 321–339.
180. Oliver, M., Titi, E.S., On the domain of analyticity for solutions of second order analytic nonlinear differential equations, J. Differential Equations **174** (2001), 55–74.
181. Arnold, V.I., Geometrical methods in the theory of differential equations, Springer– Verlag, New York–Berlin, 1983.
182. Arnold, V.I., Vogtmann, K., Weinstein, A., *Mathematical methods of classical mechanics*, Springer Verlag, 1989.
183. Bogoliubov, N.N., Mitropolskii, Yu.A., Asymptotic Methods in the Theory of Nonlinear Oscillations, Fizmatgiz, Moscow, 1958 (in Russian).

184. Kazantzis, N., Kravaris, C., Nonlinear observer design using Lyapunov's auxiliary theorem, Systems Control Lett., **34** (1998), 241–247.

185. Krener, A.J., Xiao, M., Nonlinear observer design in the Siegel domain, SIAM J. Control Optim. Vol. **41**, 3 (2002), 932–953.

186. Kazantzis, N., Good, Th., Invariant manifolds and the calculation of the long–term asymptotic response of nonlinear processes using singular PDEs, Computers and Chemical Engineering **26** (2002), 999–1012.

187. Onsager, L., Reciprocal relations in irreversible processes. I. Phys. Rev. **37** (1931), 405–426; II. Phys. Rev. **38** (1931), 2265–2279.

188. Nettleton, R.E., Freidkin, E.S., Nonlinear reciprocity and the maximum entropy formalism, Physica A, **158**, 2 (1989), 672–690.

189. Grmela, M., Reciprocity relations in thermodynamics, Physica A, **309**, 3–4 (2002), 304–328.

190. Berdichevsky, V.L., Structure of equations of macrophysics, Phys. Rev. E, **68**, 6 (2003), 066126.

191. Wehrl, A., General properties of entropy, Rev. Mod. Phys. **50**, 2 (1978), 221–260.

192. Schlögl, F., Stochastic measures in nonequilibrium thermodynamics, Phys. Rep. **62**, 4 (July 1980), 267–380.

193. Jaynes E.T., Information theory and statistical mechanics, in: Statistical Physics. Brandeis Lectures, V.3, K. W. Ford, ed., New York: Benjamin, 1963, pp. 160–185.

194. Grabert, H. Projection operator techniques in nonequilibrium statistical mechanics, Springer Verlag, Berlin, 1982.

195. Zubarev, D., Morozov, V., Röpke, G. Statistical mechanics of nonequilibrium processes, V.1, Basic concepts, kinetic theory, Akademie Verlag, Berlin, 1996, V.2, Relaxation and hydrodynamic processes, Akademie Verlag, Berlin, 1997.

196. Evans, M.W., Grigolini, P., Pastori Parravicini, G. (Eds.), Memory function approaches to stochastic problems in condensed matter, Advances in Chemical Physics, V. 62, J. Wiley & Sons, New York etc., 1985.

197. Uhlenbeck, G.E., in: Fundamental problems in statistical mechanics II, edited by E.G.D. Cohen, North Holland, Amsterdam, 1968.

198. Glimm, J., Jaffe, A., Quantum Physics: A Functional Integral Point of View, Springer, NY, 1981.

199. Parisi, G., Statistical Field Theory, Addison-Wesley, Reading, Massachusetts, 1988.

200. Grad, H.Principles of the kinetic theory of gases, in: S. Flügge, ed., Handbuch der Physics, Band 12, Springer, Berlin, 205–294.

201. Grad, H., On the kinetic theory of rarefied gases, Comm. Pure and Appl. Math. **24**, (1949), 331–407.

202. Hauge, E.H., Exact and Chapman-Enskog Solutions of the Boltzmann Equation for the Lorentz Model Phys. Fluids **13** (1970), 1201–1208.

203. Titulaer, U.M., A systematic solution procedure for the Fokker-Planck equation of a Brownian particle in the high–friction case, Physica A, **91**, 3–4 (1978), 321–344.

204. Widder, M.E., Titulaer, U.M., Two kinetic models for the growth of small droplets from gas mixtures, Physica A, **167**, 3 (1990), 663–675.

205. Karlin, I.V., Dukek, G., Nonnenmacher, T.F., Gradient expansions in kinetic theory of phonons, Phys. Rev. B **55** (1997), 6324–6329.

206. Narayanamurti, V., Dynes, R.C., Phys. Rev. B **12** (1975), 1731–1738.
207. Narayanamurti, V., Dynes, R.C., Andres, K., Propagation of sound and second sound using heat pulses Phys. Rev. B **11** (1975), 2500–2524.
208. Guyer, R.A., Krumhansl, J.A., Dispersion relation for 2nd sound in solids, Phys. Rev. **133** (1964), A1411–A1417.
209. Guyer, R.A., Krumhansl, J.A., Solution of linearized phonon Boltzmann equation, Phys. Rev. **148** (1966), 766–778.
210. Guyer, R.A., Krumhansl, J.A., Thermal conductivity 2nd sound and phonon hydrodynamic phenomena in nonmetallic crystals, Phys. Rev. **148** (1966), 778–788.
211. H.Beck, in: Dynamical Properties of Solids, Vol. 2, G.K.Horton and A.A.Maradudin, eds., North-Holland, Amsterdam, 1975, p. 207.
212. Dreyer, W., Struchtrup, H., Heat pulse experiments revisited, Continuum Mech. Thermodyn. **5** (1993), 3-50.
213. Ranninger, J., Heat-Pulse Propagation in Ionic Lattices, Phys. Rev. B **5** (1972), 3315–3321.
214. Paszkiewicz, T., Exact and approximate generalized diffusion equation for the Lorentz gas, Physica A, **123** (1984), 161–174.
215. Jasiukiewicz Cz., Paszkiewicz, T., The explicit time-dependence of moments of the distribution function for the Lorentz gas with planar symmetry in k-space, Physica A, **145** (1987), 239–254.
216. Jasiukiewicz Cz., Paszkiewicz, T., Woźny, J., Crossover from kinetic to diffusive behavior for a class of generalized models of the Lorentz gas, Physica A, **158** (1989), 864–893.
217. Jasiukiewicz Cz., Paszkiewicz, T., Relaxation of initial spatially unhomogeneous states of phonon gases scattered by point mass defects embedded in isotropic media, Z. Phys. B, **77** (1989), 209–218.
218. F.Uribe and E.Piña, Comment on "Invariance principle for extension of hydrodynamics: Nonlinear viscosity", Phys. Rev. E **57** (1998), 3672–3673.
219. Karlin, I.V., Exact summation of the Chapman–Enskog expansion from moment equations, J. Phys. A: Math.Gen. **33** (2000), 8037–8046.
220. Slemrod M., Constitutive relations for monatomic gases based on a generalized rational approximation to the sum of the Chapman–Enskog expansion, Arch. Rat. Mech. Anal, **150** (1) (1999), 1–22.
221. Slemrod M., Renormalization of the Chapman–Enskog expansion: Isothermal fluid flow and Rosenau saturation J. Stat. Phys, **91**, 1–2 (1998), 285–305.
222. Gibbs, G.W., Elementary Principles of Statistical Mechanics, Dover, 1960.
223. Kogan, A.M., Rozonoer, L.I., On the macroscopic description of kinetic processes, Dokl. AN SSSR **158** (3) (1964), 566–569.
224. Kogan, A.M., Derivation of Grad–type equations and study of their properties by the method of entropy maximization, Prikl. Math. Mech. **29** (1) (1965), 122–133.
225. Rozonoer, L.I., Thermodynamics of nonequilibrium processes far from equilibrium, in: Thermodynamics and Kinetics of Biological Processes (Nauka, Moscow, 1980), 169–186.
226. Karkheck, J., Stell, G., Maximization of entropy, kinetic equations, and irreversible thermodynamics Phys. Rev. A **25**, 6 (1984), 3302–3327.
227. Alvarez-Romero, J.T., García-Colín, L.S., The foundations of informational statistical thermodynamics revisited, Physica A, **232**, 1–2 (1996), 207–228.

228. Eu, B.C., Kinetic theory and irreversible thermodynamics, Wiley, New York, 1992.

229. Bugaenko, N.N., Gorban, A.N., Karlin, I.V., Universal Expansion of the Triplet Distribution Function, Teoreticheskaya i Matematicheskaya Fisika, **88**, 3 (1991), 430–441 (Transl.: Theoret. Math. Phys. (1992) 977–985).

230. Levermore C.D., Moment Closure Hierarchies for Kinetic Theories, J. Stat. Phys. **83** (1996), 1021–1065.

231. Balian, R., Alhassid, Y., Reinhardt, H., Dissipation in many–body systems: A geometric approach based on information theory, Physics Reports **131**, 1 (1986), 1–146.

232. Degond, P., Ringhofer, C., Quantum moment hydrodynamics and the entropy principle, J. Stat. Phys., **112** (2003), 587–627.

233. Gorban, A.N., Karlin, I.V., Quasi–equilibrium approximation and non-standard expansions in the theory of the Boltzmann kinetic equation, in: "Mathematical Modelling in Biology and Chemistry. New Approaches", ed. R. G. Khlebopros, Nauka, Novosibirsk, P. 69–117 (1991).[in Russian]

234. Gorban, A.N., Karlin, I.V., Quasi–equilibrium closure hierarchies for the Boltzmann equation [Translation of the first part of the paper [233]], Preprint, 2003, Preprint online: http://arXiv.org/abs/cond-mat/0305599.

235. Jou, D., Casas-Vázquez, J., Lebon, G., Extended irreversible thermodynamics, Springer, Berlin, 1993.

236. Müller, I., Ruggeri, T., Extended Thermodynamics, Springer, NY, 1993.

237. Gorban, A., Karlin, I., New methods for solving the Boltzmann equations, AMSE Press, Tassin, France, 1994.

238. Hirschfelder, J.O., Curtiss C.F., Bird, R.B., Molecular theory of gases and liquids, J. Wiley, NY, 1954.

239. Dorfman, J., van Beijeren, H., in: Statistical Mechanics B, B. Berne, ed., Plenum, NY, 1977.

240. Résibois, P., De Leener, M., Classical kinetic theory of fluids, Wiley, NY, 1977.

241. Ford, G., Foch, J., in: Studies in Statistical Mechanics, G. Uhlenbeck and J. de Boer, eds., V. 5, North Holland, Amsterdam, 1970.

242. Van Rysselberge, P., Reaction rates and affinities, J. Chem. Phys., **29**, 3 (1958), 640–642.

243. Feinberg, M., Chemical kinetics of a sertain class, Arch. Rat. Mech. Anal., **46**, 1 (1972), 1–41.

244. Bykov, V.I., Gorban, A.N., Yablonskii, G.S., Description of nonisothermal reactions in terms of Marcelin – de Donder kinetics and its generalizations, React. Kinet. Catal. Lett., **20**, 3–4 (1982), 261–265.

245. De Donder, T., Van Rysselberghe, P., Thermodynamic theory of affinity. A book of principles. Stanford: University Press, 1936.

246. Karlin, I.V., On the relaxation of the chemical reaction rate, in: Mathematical Problems of Chemical Kinetics, eds. K.I. Zamaraev and G.S. Yablonskii, Nauka, Novosibirsk, 1989, 7–42. [In Russian].

247. Karlin, I.V., The problem of reduced description in kinetic theory of chemically reacting gas, Modeling, Measurement and Control C, **34(4)** (1993), 1–34.

248. Gorban, A.N., Karlin, I.V., Scattering rates versus moments: Alternative Grad equations, Phys. Rev. E **54** (1996), R3109.

249. Treves, F., Introduction to pseudodifferential and Fourier integral operators, Plenum, NY, (1982).

250. Shubin, M.A., Pseudodifferential operators and spectral theory, Nauka, Moscow, (1978).

251. Dedeurwaerdere, T., Casas-Vázquez, J., Jou, D., Lebon, G., Foundations and applications of a mesoscopic thermodynamic theory of fast phenomena Phys. Rev. E, **53**, 1 (1996), 498–506.

252. Rodríguez, R.F., García-Colín, L.S., López de Haro, M., Jou, D., Pérez-García, C., The underlying thermodynamic aspects of generalized hydrodynamics, Phys. Lett. A, **107**, 1 (1985), 17–20.

253. Struchtrup, H., Torrilhon M., Regularization of Grad's 13 Moment Equations: Derivation and Linear Analysis, Phys. Fluids, **15** (2003), 2668–2680.

254. Ilg, P., Karlin, I.V., Öttinger H.C., Canonical distribution functions in polymer dynamics: I. Dilute solutions of flexible polymers, Physica A, **315** (2002), 367–385.

255. Krook, M, Wu, T.T., Formation of Maxwellian tails, Phys. Rev. Lett, **36** (1976), 1107–1109.

256. Krook, M, Wu, T.T., Exact solutions of Boltzmann-equation, Phys Fluids, **20** (1977), 1589–1595.

257. Ernst, M.H., Hendriks, E.M., Exactly solvable nonlinear Boltzmann-equation Phys. Lett. A **70** (1979), 183–185.

258. Hendriks, E.M., Ernst, M.H., The Boltzmann-equation for very hard particles Physica A, **120** (1983), 545–565.

259. Carleman, T., Sur la thèorie de l'èquation intègro-diffèrentielle de Boltzmann, Acta. Math. **60** (1933), 91–146.

260. Arkeryd, L., Boltzmann-equation. 1. Existence, Arch. Rat. Mech. Anal. **45** (1972), 1; Boltzmann-equation. 2. Full initial value-problem, ibid. **45** (1972), 17.

261. Truesdell, C., Muncaster, R., Fundamentals of Maxwell's Kinetic Theory of a Simple Monatomic Gas, Academic Press, NY, 1980.

262. Bobylev, A.V., Exact solutions to Boltzmann equations, Dokl. Akad. Nauk SSSR **225** (1975), 1296–1299; One class of invariant solutions to Boltzmann-equation, ibid. **231** (1976), 571–574.

263. Bobylev, A.V., Cercignani, C. Self-similar solutions of the Boltzmann equation and their applications, J. Stat. Phys., **106** (2002), 1039–1071.

264. Tjon, J.A., Approach to Maxwellian distribution, Phys. Lett. A, **70** (1979), 369–371.

265. Cornille, H., Nonlinear Kac model - spatially homogeneous solutions and the Tjon effect, J. Stat. Phys., **39** (1985), 181–213.

266. Ilg, P., Karlin, I.V., Kröger, M., Öttinger H.C., Canonical distribution functions in polymer dynamics: II Liquid–crystalline polymers, Physica A, **319** (2003), 134–150.

267. Ilg, P, Kröger, M., Magnetization dynamics, rheology, and an effective description of ferromagnetic units in dilute suspension, Phys. Rev. E **66** (2002) 021501. Erratum, Phys. Rev. E **67** (2003), 049901(E).

268. Ilg, P., Karlin, I.V., Combined micro–macro integration scheme from an invariance principle: application to ferrofluid dynamics, J. Non–Newtonian Fluid Mech, 2004, to appear. Ppeprint online: http://arxiv.org/abs/cond-mat/0401383.

269. Courant, R., Friedrichs, K.O., Lewy, H., On the partial difference equations of mathematical physics., IBM Journal (March 1967), 215–234.

270. Ames, W.F., Numerical Methods for Partial Differential Equations, 2nd ed. (New York: Academic Press), 1977.

271. Richtmyer, R.D., and Morton, K.W., Difference methods for initial value problems, 2nd ed., Wiley–Interscience, New York, 1967.

272. Gorban, A.N., Zinovyev, A.Yu., Visualization of data by method of elastic maps and its applications in genomics, economics and sociology. Institut des Hautes Etudes Scientifiques, Preprint. IHES M/01/36. (2001) . Online: http://www.ihes.fr/PREPRINTS/M01/Resu/resu-M01-36.html.

273. Jolliffe, I.T., Principal component analysis, Springer–Verlag, 1986.

274. Callen, H.B., Thermodynamics and an introduction to thermostatistics, Wiley, New York, 1985.

275. Use of Legendre transforms in chemical thermodynamics (IUPAC Technical Report), Prepared for publication by R.A. Alberty. Pure Appl.Chem., **73**, 8 (2001), pp. 1349–1380. Online: http://www.iupac.org/publications/pac/2001/pdf/7308x1349.pdf.

276. Aizenberg, L., Carleman's formulas in complex analysis: Theory and applications, (Mathematics and its applications; V. 244), Kluwer, 1993.

277. Gorban, A.N., Rossiev, A.A., Wunsch, D.C.II, Neural network modeling of data with gaps: method of principal curves, Carleman's formula, and other, The talk was given at the USA–NIS Neurocomputing opportunities workshop, Washington DC, July 1999 (Associated with IJCNN'99). Preprint online: http://arXiv.org/abs/cond-mat/0305508.

278. Gorban, A.N., Rossiev, A.A., Neural network iterative method of principal curves for data with gaps, Journal of Computer and System Sciences International, **38**, 5 (1999), 825–831.

279. Dergachev, V.A., Gorban, A.N., Rossiev, A.A., Karimova, L.M., Kuandykov, E.B., Makarenko, N.G., Steier, P., The filling of gaps in geophysical time series by artificial neural networks, Radiocarbon, **43**, 2A (2001), 365–371.

280. Gorban A., Rossiev A., Makarenko N., Kuandykov Y., Dergachev V., Recovering data gaps through neural network methods, International Journal of Geomagnetism and Aeronomy, **3**, 2 (2002), 191–197.

281. Lewis, R.M., A unifying principle in statistical mechanics, J. Math. Phys., **8** (1967), 1448–1460.

282. Chorin, A.J., Hald O.H., Kupferman, R., Optimal prediction with memory, Physica D 166 (2002) 239–257.

283. Hoover, W.G., Time reversibility, computer simulation, and chaos, Advansed series in nonlinear dynamics, V. 13, World Scientific, Singapore, 1999.

284. Sone, Y., Kinetic theory and fluid dynamics, Birkhäuser, Boston, 2002.

285. McKean, H.P. Jr., J. Math. Phys. **8**, 547 (1967).

286. Montroll, E.W., Lebowitz, J.L. (Eds.), Studies in Statistical Mechanics, V.IX, North-Holland, 1981.

287. Del Río-Correa, J.L., García-Colín, L.S., Increase-in-entropy law, Phys. Rev. E **48** (1993), 819–828.

288. Leontovich, M.A., An Introduction to thermodynamics, GITTL Publ., Moscow, 1950 (in Russian).

289. Lebowitz, J.L., Bergmann, P.G., New approach to nonequilibrium processes, Phys. Rev., **99** (1955), 578–587.

290. Lebowitz, J.L., Bergmann, P.G., Irreversible Gibbsian Ensembles, Annals of Physics, 1:1, 1957.

291. Lebowitz, J.L., Stationary Nonequilibrium Gibbsian Ensembles, Phys. Rev., **114** (1959), 1192–1202.

292. Lebowitz, J.L., Botzmann's entropy and time's arrow, Physics Today, **46** 9 (1993), 32–38.

293. Leff, H.S., Rex, A.F. (Eds.), Maxwell's Demon 2: Entropy, Classical and Quantum Information, Computing, 2nd edition, IOP, Philadelphia, 2003.

294. Von Baeyer, H.C., Maxwell's Demon: Why Warmth Disperses and Time Passes, Random House, 1998.

295. Pour-El, M.B., Richards, J.I., Computability in Analysis and Physics, Springer Verlag, NY, 1989.

296. Copeland, B.J., The Church-Turing Thesis, In: The Stanford Encyclopedia of Philosophy, E.N. Zalta (Ed.) (Fall 2002 Edition), On-line: http://plato.stanford.edu/archives/fall2002/entries/church-turing/.

297. Feynman, R., The Character of Physical Law, Cox and Wyman, London, 1965. Lecture No. 5.

298. Gorban, A.N., Karlin, I.V., *Geometry of irreversibility: Film of nonequilibrium states*, The lecture given on the V Russian National Seminar "Modeling of Nonequilibrium systems", Krasnoyarsk, Oct. 18–20, 2002, Printed by Krasnoyarsk State Technical University Press, 2002. [In Russian].

299. Gorban, A.N., Bykov, V.I., Yablonskii, G.S., Essays on chemical relaxation, Novosibirsk: Nauka, 1986.

300. Verbitskii, V.I., Gorban, A.N., Utjubaev, G.Sh., Shokin, Yu.I., Moore effect in interval spaces, Dokl. AN SSSR. **304**, 1 (1989), 17–21.

301. Bykov, V.I., Verbitskii, V.I., Gorban, A.N., On one estimation of solution of Cauchy problem with uncertainty in initial data and rigt part, Izv. vuzov, Ser. mat., N. 12 (1991), 5–8.

302. Verbitskii, V.I., Gorban, A.N., Simultaneously dissipative operators and their applications, Sib. Mat. Jurnal, **33**, 1 (1992), 26–31.

303. Gorban, A.N., Shokin, Yu.I., Verbitskii, V.I., Simultaneously dissipative operators and the infinitesimal Moore effect in interval spaces, Preprint (1997). Preprint online: http://arXiv.org/abs/physics/9702021.

304. Gorban, A.N., Bykov, V.I., Yablonskii, G.S., Thermodynamic function analogue for reactions proceeding without interaction of various substances, Chemical Engineering Science, **41**, 11 (1986), 2739–2745.

305. Gorban, A.N., Verbitskii, V.I., Thermodynamic restriction and quasithermodynamic conditions in reaction kinetics, in: Mathematical problems of chamical kinetics, K.I. Zamaraev, G.S. Yablonskii (eds.), Nauka, Novosibirsk, 1989, 43–83.

306. Grassberger, P., On the Hausdorff Dimension of Fractal Attractors, J. Stat. Phys. **26** (1981), 173–179.

307. Grassberger, P. and Procaccia, I., Measuring the Strangeness of Strange Attractors, Physica D **9** (1983), 189–208.

308. Frederickson, P., Kaplan, J.L., Yorke, E.D., Yorke, J.A., The Lyapunov dimension of strange attractors. J. Differ. Equations **49** (1983), 185–207.

309. Ledrappier F., Young, L.-S., The metric entropy of diffeomorphisms: I. Characterization of measures satisfying Pesin's formula; II. Relations between entropy, exponents and dimensions, Annals of Mathematics, **122** (1985), 509–539, 540–574.

310. Hentschel, H.G.E., Procaccia, I., The infinite number of generalized dimensions of fractals and strange attractors, Physica D: Nonlinear Phenomena, **8** 3 (1983), 435–444.

311. Ilyashenko, Yu.S., On dimension of attractors of k–contracting systems in an infinite dimensional space, Vest. Mosk. Univ. Ser. 1 Mat. Mekh., No. 3 (1983), 52–58.

312. Nicolis, C., Nicolis, G., Is there a climate attractor?, Nature, **311** (1984), 529–532.

313. Constantin, P., Foias, C., Temam, R., Attractors representing turbulent flows, Memoirs of the American Mathematical Society, V. 53, No. 314, Providence, 1985.

314. Dubois, T., Jauberteau, F., Temam, R., Dynamic multilevel methods and the numerical simulation of turbulence, Cambridge Univ. Press, Cambridge, 1999.

315. Landau, L.D., Lifshitz, E.M., Fluid Mechanics (Landau L.D. and Lifshitz E.M. Course of Theoretical Physics, V.6), Pergamon Press, Oxford, 1993.

316. Foias, C., Manley, O.P., Temam, R., An estimate of the Hausdorff dimension of the attractor for homogeneous decaying turbulence, Physics Letters A, **122** 3–4 (1987), 140–144.

317. Foias, C., Sell, G.R., Temam R., Inertial manifolds for dissipative nonlinear evolution equations, Journal of Differential Equations, **73** (1988), 309–353.

318. Temam R., Infinite–dimensional dynamical systems in mechanics and physics, Applied Math. Sci., Vol 68, Springer Verlag, New York, 1988 (Second edition, 1997).

319. Málek, J., Prazák, D., Finite fractal dimension of the global attractor for a class of non-Newtonian fluids, Applied Mathematics Letters, **13**, 1 (2000), 105–110.

320. Doering, C.R., Wang, X., Attractor dimension estimates for two-dimensional shear flows, Physica D: Nonlinear Phenomena Volume 123, Issues 1–4, 15 November 1998, Pages 206–222.

321. Doering, C.R., Gibbon, J.D., Holm, D.D., Nicolaenko, B., Finite dimensionality in the complex Ginzburg–Landau equation, Contemporary Mathemathics, **99**, 1989, 117–141.

322. Ghidaglia, J.M., Héron, B., Dimension of the attractors associated to the Ginzburg–Landau partial differential equation, Physica D, **28**, 3 (1987), 282–304.

323. Nicolaenko, B. Scheurer, B.,Temam, R., Some global dynamical properties of the Kuramoto–Sivashinsky equations: Nonlinear stability and attractors, Physica D, **16**, 2 (1985), 155–183.

324. Duan, J., Ervin, V.J., Dynamics of a Nonlocal Kuramoto–Sivashinsky Equation, J., Diff. Equ., **143** (1998), 243–266.

325. Nicolaenko, B. Scheurer, B.,Temam, R., Some global dynamical properties of a class of pattern formation equations, Comm. Partial Diff. Equ., **14** (1989), 245–297.

326. Debussche, A., Dettori, L., On the Cahn–Hilliard equation with a logarithmic free energy, Nonlinear Anal, **24** (1995), 1491–1514.

327. Li, D., Zhong, Ch., Global Attractor for the Cahn–Hilliard System with Fast Growing Nonlinearity, Journal of Differential Equations, **149** (1998), 191–210.

328. Miranville, A., Zelik, S., Robust exponential attractors for singularly perturbed phase–field type equations, Electronic J., of Diff. Eqns., **2002** (2002), No. 63, 1–28

329. Miranville, A., Piétrus, A., Rakotoson, J.M., Dynamical aspect of a generalized Cahn–Hilliard equation based on a microforce balance, Asymptotic Anal., **16** (1998), 315–345.

330. Grinfeld, M., Novick-Cohen, A., The viscous Cahn-Hilliard equation: Morse decomposition and structure of the global attractor, Trans. Amer. Math. Soc., **351** (1999), 2375–2406.

331. Cahn, J.W., Hilliard, J.E., Free energy of a nonuniform systems. I. Interfacial free energy, J. Chem. Phys., **28** (1958), 258–267.

332. Allen, S., Cahn, J.W., A microscopic theory for antiphase boundary motion and its application to antiphase domain coarsening, Acta Metall., **27** (1979), 1084–1095.

333. Babin, A.V., Vishik, M.I., Attractors of evolutionary equations (Studies in mathematics and its application, V. 25), Elsevier, NY, 1992.

334. Vishik, M.I., Asymptotic behaviour of solutions of evolutionary equations, Cambridge University Press, 1993.

335. Chepyzhov, V.V., Vishik M.I., Attractors for equations of mathematical physics, AMS Colloquium Publications, V. 49, American Mathematical Society, Providence, 2002.

336. Efendiev, M., Miranville, A., The dimension of the global attractor for dissipative reaction-diffusion systems, Applied Mathematics Letters, **16**, 3 (2003), 351–355.

337. Haraux, A, Two remarks on dissipative hyperbolic problems, in: Nonlinear partial differential equations and their applications (H. Brezis, J.L. Lions, eds), Research Notes Maths., Vol. 112, Pitman, Boston, 1985, 161–179.

338. Ghidaglia, J.M., Temam, R., Attractors for damped nonlinear hyperbolic equations, J. Math. Pures Appl., **66** (1987), 273–319.

339. Ladyzhenskaya, O.A., On finding the minimal global attractors for the Navier–Stokes equation and other PDEs, Uspekhi Mat. Nauk, **42** (1987), 25–60; Engl. transl. Russian Math Surveys, **42** (1987).

340. Constantin, P., Kevrekidis, I., Titi, E.S., Remarks on a Smoluchowski equation, Discrete and Continuous Dynamical Systems, **11** (2004), 101–112.

341. Foias, C., Sell, G.R., Titi, E.S., Exponential tracking and approximation of inertial manifolds for dissipative nonlinear equations Journal of Dynamics and Differential Equations, **1** (1989), 199–244.

342. Jones, D.A., Titi, E.S., C^1 Approximations of inertial manifolds for dissipative nonlinear equations, Journal of Differential Equations, **127**, 1 (1996), 54–86.

343. Robinson, J.C., Computing inertial manifolds, Discrete and Continuous Dynamical Systems, **8**, 4 (2002), 815-833.

344. Christofides, P.D., Nonlinear and robust control of partial differential equation systems: Methods and applications to transport–reaction processes, Birkhäuser, Boston, 2001.

345. Chepyzhov, V.V., Ilyin, A.A., A note on the fractal dimension of attractors of dissipative dynamical systems, Nonlinear Analysis, **44** (2001), 811–819.

346. Marion, M., Temam, R., Nonlinear Galerkin methods, SIAM J. Numer. Anal., **26** (1989), 1139–1157.

347. Jones, C., Kaper, T., Kopell, N., Tracking invariant manifolds up to exponentially small errors, SIAM J. Math., Anal. **27** (1996), 558–577.

348. Yinnian He, Mattheij, R.M.M., Stability and convergence for the reform post-processing Galerkin method, Nonlinear Anal. Real World Appl., **4** (2000), 517–533.

349. Garsia-Archilla, B., Novo, J., Titi E.S., Postprocessing the Galerkin method: a novel approach to approximate inertial manifolds, SIAM J. Numer. Anal., **35** (1998), 941–972.

350. Margolin, L.G., Titi, E.S., Wynne, S., The postprocessing Galerkin and nonlinear Galerkin methods – a truncation analysis point of view, SIAM, Journal of Numerical Analysis, **41**, 2 (2003), 695–714.

351. Novo, J., Titi, E.S., Wynne, S., Efficient methods using high accuracy approximate inertial manifolds, Numerische Mathematik, **87** (2001), 523–554.

352. Fenichel, N., Geometric singular perturbation theory for ordinary differential equations, J. Diff Eq., **31** (1979), 59–93

353. Jones, C.K.R.T., Geometric singular perturbation theory, in: Dynamical Systems (Montecatini Terme, 1904), L. Arnold (ed.), Lecture Notes in Mathematics, **1609**, Springer-Verlag. Berlin, 1994, 44–118.

354. Bird, R.B., Wiest, J.M., Constitutive equations for polymeric liquids, Annu. Rev. Fluid Mech. **27** (1995), 169.

355. Warner, H.R., Kinetic theory and rheology of dilute suspensions of finitely extendible dumbbells, Ind. Eng. Chem. Fundamentals **11** (1972), 379.

356. Oseen, C.W., Ark. f. Mat. Astr. og Fys. **6** No. 29 (1910) 1.

357. Burgers, J.M., Verhandelingen Koninkl. Ned. Akad. Wetenschap. **16** (Sect. 1, Chap. 3) (1938), 113.

358. Rotne, J., Prager, S., Variational treatment of hydrodynamic interaction, J. Chem. Phys., **50** (1969) 4831.

359. Yamakawa, H., Transport properties of polymer chain in dilute solution: Hydrodynamic interaction, J. Chem. Phys. **53** (1970), 436.

360. Noll, W., A mathematical theory of the mechanical behavior of continuous media, Arch. Ratl. Mech. Anal., **2** (1958), 197.

361. Astarita, G., Marrucci, G., Principles of non-Newtonian fluid mechanics, McGraw–Hill, London, 1974.

362. Oldroyd, J.G., Non–Newtonian effects in steady motion of some idealized elastico–viscous liquids, Proc. Roy. Soc. A245 (1958), 278.

363. Herrchen M., Öttinger, H.C., A detailed comparison of various FENE dumbbell models, J. Non–Newtonian Fluid Mech. **68** (1997), 17.

364. Kröger, M., Simple models for complex nonequilibrium fluids, Physics Reports, **390**, 6 (2004), 453–551.

365. Bird, R.B., Dotson, R.B., Jonson, N.J., Polymer solution rheology based on a finitely extensible bead–spring chain model, J. Non–Newtonian Fluid Mech. **7** (1980), 213–235; Corrigendum **8** (1981), 193,

366. Char B.W., et al., Maple V Language Reference Manual, Springer–Verlag, New York, 1991.

367. Kato, T., Perturbation theory for linear operators, Springer–Verlag, Berlin, 1976.

368. Thiffeault, J.-L., Finite extension of polymers in turbulent flow, Physics Letters A **308**, 5-6 (2003), 445–450.

369. Gorban, A.N., Gorban, P.A., Karlin I.V., Legendre integrators, postprocessing and quasiequilibrium, J. Non–Newtonian Fluid Mech., **120** (2004), 149–167. Preprint on-line: http://arxiv.org/pdf/cond-mat/0308488.

370. Perkins, Th.T., Smith, D.E., Chu, S., Single polymer dynamics in an elongational flow, Science, **276**, 5321 (1997), 2016-2021.

371. De Gennes, P.G., Molecular individualism, Science, **276**, 5321 (1997), 1999-2000.

372. Smith, D.E., Babcock, H.P., Chu, S., Single-polymer dynamics in steady shear flow. Science **283** (1999), 1724–1727.

373. Arnold, V.I., Varchenko, A.N., Gussein-Zade, S.M., Singularities of differentiable maps, Brickhäuser, Boston, 1985-1988. 2 vol.

374. Langer, J.S., Bar-on, M., Miller, H.D., New computational method in the theory of spinodal decomposition, Phys. Rev. A, **11**, 4 (1975), 1417–1429.

375. Grant, M., San Miguel, M., Vinals, J., Gunton, J.D., Theory for the early stages of phase separation: The long-range-force limit, Phys. Rev. B, **31**, 5 (1985), 3027–3039.

376. Kumaran, V., Fredrickson, G.H., Early stage spinodal decomposition in viscoelastic fluids, J. Chem. Phys., **105**, 18 (1996), 8304–8313.

377. Darwin, Ch., On the origin of species by means of natural selection, or preservation of favoured races in the struggle for life: A Facsimile of the First Edition, Harvard, 1964. http://www.literature.org/authors/darwin-charles/the-origin-of-species/

378. Haldane, J.B.S., The Causes of Evolution, Princeton Science Library, Princeton University Press, 1990.

379. Mayr, E., Animal Species and Evolution. Cambridge, MA: Harvard University Press, 1963.

380. Ewens, W.J., Mathematical Population Genetics. Springer-Verlag, Berlin, 1979.

381. Rozonoer, L.I., Sedyh, E.I., On the mechanisms of of evolution of self-reproduction systems, 1, Automation and Remote Control, **40**, 2 (1979), 243–251; 2, ibid., **40**, 3 (1979), 419–429; 3, ibid, **40**, 5 (1979), 741–749.

382. Gorban, A.N., Dynamical systems with inheritance, in: Some problems of community dynamics, R.G. Khlebopros (ed.), Institute of Physics RAS, Siberian Branch, Krasnoyarsk, 1980 [in Russian].

383. Semevsky, F.N., Semenov S.M., Mathematical modeling of ecological processes, Gidrometeoizdat, Leningrad, 1982 [in Russian].

384. Gorban, A.N., Khlebopros, R.G., Demon of Darwin: Idea of optimality and natural selection, Nauka (FizMatGiz), Moscow, 1988 [in Russian].

385. Zakharov, V.E., L'vov, V.S., Starobinets, S.S., Turbulence of spin-waves beyond threshold of their parametric-excitation, Uspekhi Fizicheskikh Nauk **114**, 4 (1974), 609–654; English translation Sov. Phys. - Usp. **17**, 6 (1975), 896–919.

386. Zakharov, V.E., L'vov, V.S., Falkovich, G.E., Kolmogorov spectra of turbulence, vol. 1 Wave Turbulence. Springer, Berlin, 1992.

387. L'vov, V.S., Wave turbulence under parametric excitation applications to magnets, Springer, Berlin, Heidelberg, 1994.

388. Ezersky A.B., Rabinovich M.I., Nonlinear-wave competition and anisotropic spectra of spatiotemporal chaos of Faraday ripples, Europhysics Letters **13**, 3 (1990), 243–249.

389. Krawiecki, A, Sukiennicki, A., Marginal synchronization of spin-wave amplitudes in a model for chaos in parallel pumping, Physica Status Solidi B–Basic Research **236**, 2 (2003), 511–514.

390. Vorobev, V.M., Selection of normal variables for unstable conservative media, Zhurnal Tekhnicheskoi Fiziki, **62**, 8 (1992), 172–175.

391. Seminozhenko, V.P., Kinetics of interacting quasiparticles in strong external fields. Phys. Reports, **91**, 3 (1982), 103–182.

392. Haken, H., Synergetics, an introduction. Nonequilibrium phase transitions and self–organization in physics, chemistry and biology, Springer, Berlin, Heidelberg, New York, 1978.

393. Bourbaki, N., Elements of mathematics - Integration I, Springer, Berlin, Heidelberg, New York, 2003.

394. Smale, S., Structurally stable systems are not dense, Amer. J. Math., **88** (1966), 491–496.

395. Birkhoff, G.D., Dynamical systems, AMS Colloquium Publications, Providence, 1927. Online: http://www.ams.org/online _bks/coll9/

396. Hasselblatt, B., Katok, A. (Eds.), Handbook of Dynamical Systems, Vol. 1A, Elsevier, 2002.

397. Katok, A., Hasselblat, B., Introduction to the Modern Theory of Dynamical Systems, Encyclopedia of Math. and its Applications, Vol. 54, Cambridge University Press, 1995.

398. Levin, L.A., Randomness conservation inequalities; Information and independence in mathematical theories, Information and Control, **61**, 1 (1984), 15–37.

399. Gause, G.F., The struggle for existence, Williams and Wilkins, Baltimore, 1934. Online: http://www.ggause.com/Contgau.htm.

400. Volterra, V., Lecons sur la théorie mathematique de la lutte pour la vie, Gauthier-Villars, Paris, 1931.

401. Gromov, M., A dynamical model for synchronisation and for inheritance in micro-evolution: a survey of papers of A.Gorban, The talk given in the IHES seminar, "Initiation to functional genomics: biological, mathematical and algorithmical aspects", Institut Henri Poincaré, November 16, 2000.

402. Kuzenkov, O.A., Weak solutions of the Cauchy problem in the set of Radon probability measures, Differential Equations, **36**, 11 (2000), 1676–1684.

403. Kuzenkov, O.A., A dynamical system on the set of Radon probability measures, Differential Equations, **31**, 4 (1995), 549–554.

404. Diekmann, O. A beginner's guide to adaptive dynamics, in: Mathematical modelling of population dynamics, Banach Center Publications, V. 63, Institute of Mathematics Polish Academy of Sciences, Warszawa, 2004, 47–86.

405. Blekhman, I.I., Synchronization in science and technology. ASME Press, N.Y., 1988.

406. Pikovsky, A., Rosenblum, M., Kurths, J., Synchronization: A Universal Concept in Nonlinear Science, Cambridge University Press, 2002.

407. Josić, K., Synchronization of chaotic systems and invariant manifolds, Nonlinearity 13 (2000) 1321–1336.

408. Mosekilde, E., Maistrenko, Yu., Postnov, D., Chaotic synchronization: Applications to living systems, World Scientific, Singapore, 2002.

409. Cooper, S., Minimally disturbed, multi-cycle, and reproducible synchrony using a eukaryotic "baby machine", Bioessays **24** (2002), 499–501.

410. Lielens, G., Halin, P., Jaumin, I., Keunings,R., Legat, V., New closure approximations for the kinetic theory of finitely extensible dumbbells, J.Non–Newtonian Fluid Mech. **76** (1998), 249–279.

411. Ilg, P., Karlin, I.V., Öttinger, H.C., Generating moment equations in the Doi model of liquid–crystalline polymers, Phys. Rev. E, **60** (1999), 5783–5787.

412. Phan–Thien, N., Goh, C.G., Atkinson, J.D., The motion of a dumbbell molecule in a torsional flow is unstable at high Weissenberg number, J. Non–Newtonian Fluid Mech. **18**, 1 (1985), 1–17.

413. Goh, C.G., Phan–Thien, N., Atkinson, J.D., On the stability of a dumbbell molecule in a continuous squeezing flow, Journal of Non–Newtonian Fluid Mechanics, **18**, 1 (1985), 19–23.
414. Karlin, I.V., Ilg, P., Öttinger, H.C., Invariance principle to decide between micro and macro computation, in: Recent Developments in Mathematical and Experimental Physics, Volume C: Hydrodynamics and Dynamical Systems, Ed. F. Uribe, Kluwer, Dordrecht, 2002, 45–52.

Mathematical Notation and Some Terminology

- The *operator* L from space W to space E: $L : W \to E$.
- The *kernel* of a linear operator $L : W \to E$ is a subspace $\ker L \subset W$ that transforms by L into 0: $\ker L = \{x \in W \mid Lx = 0\}$.
- The *image* of a linear operator $L : W \to E$ is a subspace $\operatorname{im} L = L(W) \subset E$.
- *Projector* is a linear operator $P : E \to E$ with the property $P^2 = P$. Projector P is *orthogonal* one, if $\ker P \perp \operatorname{im} P$ (the kernel of P is orthogonal to the image of P).
- If $F : U \to V$ is a map of domains in normed spaces ($U \subset W, V \subset E$) then the *differential* of F at a point x is a linear operator $D_x F : W \to E$ with the property: $\|F(x + \delta x) - F(x) - (D_x F)(\delta x)\| = o(\|\delta x\|)$. This operator (if it exists) is the best linear approximation of the map $F(x + \delta x) - F(x)$.
- The differential of the function $f(x)$ is the linear functional $D_x f$. The *gradient* of the function $f(x)$ can be defined, if there is a given scalar product $\langle \mid \rangle$, and if there exists a Riesz representation for functional $D_x f$: $(D_x f)(a) = \langle \operatorname{grad}_x f \mid a \rangle$. The gradient $\operatorname{grad}_x f$ is a vector.
- The *second differential* of a map $F : U \to V$ is a bilinear operator $D_x^2 F : W \times W \to E$ which can be defined by Taylor formula: $F(x + \delta x) = F(x) + (D_x F)(\delta x) + \frac{1}{2}(D_x^2 F)(\delta x, \delta x) + o(\|\delta x\|^2)$.
- The differentiable map of domains in normed spaces $F : U \to V$ is an *immersion*, if for any $x \in U$ the operator $D_x F$ is injective: $\ker D_x F = \{0\}$. In this case the image of F (i.e. $F(U)$) is called the *immersed manifold*, and the image of $D_x F$ is called the *tangent space* to the immersed manifold $F(U)$. We use the notation T_x for this tangent space: $\operatorname{im} D_x F = T_x$.
- The subset U of the vector space E is *convex*, if for every two points $x_1, x_2 \in U$ it contains the segment between x_1 and x_2: $\lambda x_1 + (1 - \lambda)x_2 \in U$ for every $\lambda \in [0, 1]$.
- The function f, defined on the convex set $U \subset E$, is *convex*, if its *epigraph*, i.e. the set of pairs $\operatorname{Epi} f = \{(x, g) \mid x \in U, g \geq f(x)\}$, is the convex set in $E \times R$. The twice differentiable function f is convex if and only if the quadratic form $(D_x^2 f)(\delta x, \delta x)$ is nonnegative.
- The convex function f is called *strictly convex* if in the domain of definition there is no line segment on which it is constant and finite ($f(x) = const \neq \infty$). The sufficient condition for the twice differentiable function f to be

strictly convex is that the quadratic form $(D_x^2 f)(\delta x, \delta x)$ is positive defined (i.e. it is positive for all $\delta x \neq 0$).

- We use summation convention for vectors and tensors, $c_i g_i = \sum_i c_i g_i$, when it cannot cause a confusion, in more complicated cases we use the sign $\sum$.

Index

Lecture Notes in Physics

For information about Vols. 1–613
please contact your bookseller or Springer
LNP Online archive: springerlink.com

Printing: Strauss GmbH, Mörlenbach
Binding: Schäffer, Grünstadt